NEPA and Environmental Planning

Tools, Techniques, and
Approaches for Practitioners

NEPA and Environmental Planning

Tools, Techniques, and Approaches for Practitioners

Charles H. Eccleston

CRC Press
Taylor & Francis Group
Boca Raton   London   New York

CRC Press is an imprint of the
Taylor & Francis Group, an **informa** business

CRC Press
Taylor & Francis Group
6000 Broken Sound Parkway NW, Suite 300
Boca Raton, FL 33487-2742

First issued in paperback 2020

ISBN-13: 978-0-367-57752-0 (pbk)
ISBN-13: 978-0-8493-7559-0 (hbk)

Visit the Taylor & Francis Web site at
http://www.taylorandfrancis.com

and the CRC Press Web site at
http://www.crcpress.com

To Our Son, Justin

We trained hard, but every time we began to form up teams we would be reorganized. I was to learn later in life that we tend to meet any new situation by reorganizing; and a wonderful method it can be for creating the illusion of progress while producing confusion, inefficiency, and demoralization.

Petronius Arbiter, 210 B.C.

Contents

Preface

From one perspective NEPA may be seen as the capstone of national environmental policy; more importantly, it should be viewed as the foundation for the future.

Lynton Caldwell Congressional Testimony, 1998

National Environmental Policy Act (NEPA) analyses are frequently prepared that examine every conceivable impact, significant or otherwise, in excruciating detail, yet blatantly ignore the most basic regulatory requirements regarding planning, decision-making, and efficiency. Such practice can lead to increased costs and delays and sometimes even to poorly planned projects. This book is not about preparing bigger analyses; it is about producing better ones.

This book integrates three separate but interrelated themes into a single comprehensive framework: (1) environmental planning procedures and practices, (2) NEPA, and (3) emergency response planning. As a special feature, this book provides the reader with the first comprehensive approach using a NEPA (or a NEPA-like) process for identifying potential natural or human-induced threats and evaluating potential consequences and alternatives; its intention is to help planners and decision-makers in preparing for and mitigating the worst effects of potential terrorist attacks, natural disasters, and other emergencies. The processes described in these chapters have widespread implications for the ways civilian agencies, communities, and the corporate world plan and prepare to mitigate the impacts of such threats.

OBJECTIVES OF THIS BOOK

Emphasis is placed on addressing problems and dilemmas that have traditionally hindered NEPA and other environmental planning. This book is unique in that it focuses on providing practitioners and decision-makers with state-of-the-art tools, techniques, and approaches for resolving the following 10 elemental problems (Top 10 Problems) encountered in environmental impact assessment and decision-making:

- *Sufficiency Test*—A defensible and systematic tool for determining how much information and description are necessary or sufficient to adequately describe a topic, action, or impact in a NEPA document (Section 2.9).
- *Smithsonian Solution*—A defensible and streamlined tool for determining to what extent a project can be changed (new circumstances or information) before that change triggers the need for new or supplemental NEPA documentation (Section 8.7).
- *Action-Impact Model*—There is a great diversity in both the quality and methodologies used to assess the environmental consequences of a proposal. In professional practice, the spectrum ranges from use of no systematic methodology to employing highly complex approaches that are frequently too difficult for practical use in evaluating many common problems. The author's action-impact model provides novices and practitioners alike with a systematic yet easy-to-use approach for assessing environmental impacts (Section 4.4).
- *Integrated NEPA/ISO 14001 EMS*—A systematic approach for managing implementation of NEPA decisions. An integrated NEPA/ISO 14001 environmental management system (EMS) and adaptive management system (AMS) can significantly improve project implementation while reducing cost and delays (Sections 2.7 and 2.8).
- *Nonfederal Assessment Tool*—A systematic tool for determining what constitutes a "federal action." This tool provides a defensible and streamlined method for consistently

determining when a nonfederal action becomes "federalized" triggering the requirements of NEPA (Section 6.5).

- *Decision-Based Scoping*—A decision-identification tree (DIT) in conjunction with decision-based scoping (DBS) can result in a superior scoping process that more closely supports the agency's final decision (Section 8.2).
- *Integrated Federal Planning*—NEPA provides a flexible and general-purpose process for integrating federal project or program plans, such as a pollution prevention program or assessing potential terrorist events into a single comprehensive planning process (Chapters 4, 10, and 11).
- *Cumulative Impact Assessment*—The significant-departure principle (SDP) provides a systematic and defensible method for addressing the assessment of cumulative impacts and resolving the cumulative impact paradox (Section 9.3).
- *P-EIS Scoping (P-EIS) Tool*—Programmatic EISs (P-EISs) are often bloated, lengthy, and costly. This defensible and streamlining tool provides a systematic technique for determining the appropriate scope of a P-EIS (Section 8.8).
- *Risk–Uncertainty Significance Test*—One of the most difficult challenges in assessing the significance of an impact is determining when a very low probability event with potentially high or catastrophic consequences is significant and must be evaluated in an EIS. This test provides decision-makers and practitioners with a systematic and defensible procedure for determining significance based on the severity and frequency of an event (Section 10.6).

The subject matter has been carefully organized to assist the reader in locating issues, problems, and topics of interest in a quick and effective manner. Some of the important features that distinguish this book include, but are not limited to, the following:

- Describing NEPA and environmental planning from the perspective of a true planning process as opposed to simply conducting a document preparation effort. A step-by-step approach is used to provide a detailed description of the entire environmental and NEPA planning process.
- Providing the reader with a strategic approach using a NEPA (or a NEPA-like) process for identifying threats and evaluating potential consequences and alternatives, in an effort to prepare for and mitigate the effects of potential terrorist attacks or natural disasters.
- Describing the latest concepts, direction, and guidance from the Council on Environmental Quality (CEQ), professional societies, and practitioners for improving and streamlining NEPA implementation and the field of environmental planning practice.

AUDIENCE

This book is designed for use by beginners and experts alike. It begins with fundamental elements and advances progressively into more complex issues in later chapters. Professionals interested in streamlining or improving the effectiveness of their environmental planning processes will find this text particularly useful.

The book is also aimed at professionals in government and consulting, and those in the private sectors who are involved in some way with NEPA and project planning and who seek to master it. NEPA practitioners may use the book as a resource to review quickly any issues of special interest. The book provides a comprehensive examination of all key NEPA planning and compliance requirements.

The organization of the book equally lends itself to individuals who desire a step-by-step introduction to the disciplines of NEPA and environmental planning. Each chapter includes a list of questions or problems that make the book an ideal text for use in undergraduate and graduate environmental and planning classes.

Chapters 10 and 11 provide government and corporate planners and managers with a detailed process for identifying threats and evaluating potential consequences and alternatives for implementing an emergency response and recovery program. Taken together, these chapters describe the process for preparing emergency response plans for mitigating the impacts of terrorist attacks, accidents, and natural disasters.

Professionals, individuals, and groups who will find this book interesting include environmental planners, project engineers, analysts, scientists, regulators, decision-makers, students and educators, and lawyers. Citizen and advocacy groups involved in NEPA compliance issues will also find this book particularly useful in their role of reviewing and perhaps challenging an agency's NEPA compliance process.

If you have technical questions or issues, or need assistance, you may contact the author at env_planning@msn.com.

Charles H. Eccleston

Acknowledgments

The author is indebted to the many people who reviewed and provided comments on this book. Although space constraints make it difficult to mention all individuals by name, I would like to call attention to the following professionals.

Peyton Doub, a biologist, seasoned NEPA practitioner, and close associate, performed a review of Chapter 7.

I am indebted to Victoria Leighton, a recognized expert on Business Continuity Planning, who wrote Chapter 11 of this book.

Many thanks go to Professor Daniel Mandelker, author of the textbook *NEPA Law and Litigation*. Dr. Mandelker conducted a legal review on much of this book and provided detailed comments on Chapters 5, 6, and 9.

Keith Bradley (president of KBA Enviroscience, Inc.), who teaches environmental law and regulatory classes around the country, provided a thorough critique of Chapter 4.

Matthew McMillen, a particularly well-qualified expert in NEPA practice, performed an in-depth review of the Preface, Introduction, and Chapters 1, 2, and 8. As in the past, Mr. McMillen made many improvements to the quality and accuracy of this text.

Many thanks go to Donald Sayre, author of the celebrated text *Inside ISO 14000* (St. Lucie Press) and an internationally respected expert on ISO 14001, who reviewed Chapter 2, with emphasis on the integrated NEPA/ISO 14001 methodology. He also reviewed Chapter 13 and contributed to the sections on root-cause analysis (Chapter 2) and sustainability (Chapter 13).

I am indebted to Dr. Robert Smythe, owner and principal of Potomac Resource Consultants and a former senior staff member of the President's Council on Environmental Quality (CEQ). Dr. Smythe reviewed and contributed important comments on Chapters 12 and 13.

Special thanks go to Fred March of Sandia National Laboratories, who is the author of two widely read books on NEPA. A friend and close associate, he reviewed and provided important comments on Chapters 3 and 10.

Particular mention goes to Wendy Read, a close associate of the late Dr. Lynton Caldwell, who performed a painstaking review of the entire manuscript, edited the draft document and corrected technical errors, and provided numerous suggestions for improving the book.

I am indebted to my late friend and mentor, Professor Lynton Caldwell, the father of NEPA, who wrote the early endorsement to the publisher for the draft manuscript before he passed away.

Finally, special thanks go to my wife, Alicia, who provided inspiration, support, and technical assistance in preparing this book.

Author

Charles H. Eccleston, Ph.D., is a NEPA consultant for Enercon Services who specializes in NEPA and nuclear consulting. He is recognized in *Who's Who in America* and *Who's Who in the World* as a leading authority in NEPA and for his innovative environmental achievements. He is an elected member of the Board of Directors of the National Association of Environmental Professionals (NAEP) and chairs the NAEP's Policy Committee. His previous books include *The NEPA Planning Process, Environmental Impact Statements, Effective Environmental Assessments,* and *Megacrises: A Survivors Guide to the Future.* He teaches and lectures on NEPA and can be reached at env_planning@msn.com.

List of Acronyms

AEC	Atomic Energy Commission
AM	adaptive management
APA	Administrative Procedures Act
AR	associate reviewers
BCP	business continuity planning
BIA	business impact analysis
CAA	Clean Air Act
CATX	categorical exclusion
CBA	cost–benefit analysis
CEQ	Council on Environmental Quality
CERCLA	Comprehensive Environmental Response, Compensation, and Liability Act
CIA	cumulative impact assessment
CFR	Code of Federal Regulations
CRR	cultural resources review
CZMA	Coastal Zone Management Act
DBS	decision-based scoping
DIT	decision-identification tree
DOD	U.S. Department of Defense
DOE	U.S. Department of Energy
EA	environmental assessment
ECO	environmental compliance officers
EIA	environmental impact assessment
EIS	environmental impact statement
EJ	environmental justice
EPA	U.S. Environmental Protection Agency
EPCRA	Emergency Planning and Community Right-to-Know Act
ERC	environmental review coordinator
ESA	Endangered Species Preservation Act
FIP	facility implementation plan
FONSI	finding of no significant impact
FS	forest service
IAIA	International Association of Impact Assessment
IP	EIS implementation plan
ISO	International Organization for Standardization
EMS	environmental management system
NAAQS	National Ambient Air Quality Standards
NEPA	National Environmental Policy Act of 1969
NHPA	National Historic Preservation Act
NNIP	nonnative invasive plants
NOA	notice of availability
NOI	notice of intent
NRC	National Regulatory Commission
NRDA	Natural Resource Damage Assessment
NRHP	National Register of Historic Places
OFA	Office of Federal Activities

P-EIS	programmatic EIS
PR	principal reviewer
PSD	prevention of significant deterioration
RCRA	Resource Conservation and Recovery Act
RFI/CMS	RCRA facility investigation/corrective measures study
RI/FS	remedial investigation/feasibility study
RMP	resource management plans
ROD	record of decision
SDP	significant departure principle
S-EIS	supplemental EIS
SEA	strategic environmental assessment
SEPA	State Environmental Policy Act
SHPO	State Historic Preservation Office
USC	United States Code
VE	value engineering

Introduction

NEPA is equal parts philosophy and law, and that's what makes it so beautiful.

Stan Flitner, Diamond Tail Ranch, Greybull, Wyoming

Before beginning an introduction to the National Environmental Policy Act (NEPA), let us ask ourselves what the discipline of environmental planning, and NEPA in particular, have in common with pike fish.

NEPA AND THE PIKE SYNDROME

As many accomplished fishermen will testify, pike are long fish with razor sharp teeth that viciously attack smaller prey. Revered as challenging and tenacious fighters, these fish are prized by sportsmen around the world.

But most of these fishermen would be surprised to learn that pike have also been the subject of some eye-opening experiments. If a bell jar filled with minnows is lowered into an aquarium already containing them, the fish will repeatedly lunge at the minnows, striking their heads hard on the bell jar glass as they do so. Hurt by this experience, the pike will eventually give up and ignore the minnows.

Now let us try an interesting variation to this experiment. If the bell jar is removed from the aquarium, allowing the minnows to swim freely, the pike will continue to ignore them! The large fish have been conditioned to leave the minnows alone; they are unable to adapt to the new situation. Now surrounded by an abundant food supply of minnows, the pike may eventually starve to death rather than attack their natural prey. The pike are unable to comprehend that what they learned earlier is no longer applicable.

This experiment is known as the infamous Pike Syndrome.[1] In scientific circles it has become a metaphor for fixed, unyielding, conditioned thinking. Over eons, this syndrome has probably been responsible for the extinction of countless species.

This syndrome may also help to explain an underlying theme of this book—why resistance has been encountered for accepting new and potentially more effective approaches for implementing planning. Some quarters continue to experience inefficiencies in their planning processes. As projects are increasingly managed with a focus on efficiency, practitioners of NEPA can little afford to continue the way of the pike. To keep up, more effective and streamlined approaches must be integrated into NEPA practice.

It is for reasons such as this that this book has been written. As federal funding has been tightened, emphasis is increasingly shifting toward doing more with less. What is needed are modern approaches, specifically designed to address problems that have traditionally hindered federal planning. In large measure, this book is a response to problems that have traditionally plagued both the NEPA process and environmental planning.

In addition to introducing the reader to the details of NEPA, this text provides an arsenal of peer-reviewed tools, techniques, and approaches for enhancing the practice of environmental planning.

THE NEED TO STREAMLINE NEPA

Some criticism of NEPA efficiency is warranted, but many of the problems and much of the criticism directed its way are either misdirected or, in some cases, completely unjustified. Not only direct costs, but also indirect project savings, for which accounting is often more difficult, must be included in any objective cost–benefit assessment of NEPA. Consider the following case.

Costly or a Cost Savings?

The Department of Energy's (DOE) Hanford Site in Richland, Washington, has been characterized as the most contaminated area in the Western Hemisphere. Radioactive groundwater is migrating toward the Columbia River. Contaminated facilities and soil sites are located across the site. Enormous underground tanks have leaked high-level radioactive waste into the soil column. Some local citizens claim that their health has been jeopardized as a result of past activities.

As of this writing, approximately $2 billion is being allocated annually to the management and cleanup of the Hanford Site; it has been estimated that the total price tag could exceed $100 billion.

NEPA simply requires that federal officials examine the possible consequences of their proposed action (and alternatives) before making a final decision to proceed. Had NEPA and other modern environmental statutes been on the books and enforced back when these Hanford activities were taking place, they could have saved taxpayers untold billions of dollars in later natural resource damages assessments (NRDA)[2] and cleanup costs; many ill-conceived actions and mistakes could have been prevented.

Thus, much of the enormous Hanford cleanup bill and those at hundreds of other sites across the country could have been largely avoided if NEPA had simply existed at that time and had been used to comprehensively plan waste management strategies and projects.

NEPA PROBLEMS

Notwithstanding what it has to offer, it would be incorrect to omit the fact that NEPA is not without its problems. However, these problems are not generally inherent to NEPA's statutory requirements. Instead, problems associated with NEPA generally result from the way in which agencies have chosen to implement their statutory obligations. In other words, most NEPA problems are management related. With respect to those problems that can be tied directly to poor management practices, many of the root causes can be traced simply to a reluctance to adopt and implement modern or innovative tools, techniques, and approaches that are essential in effectively managing planning processes.

Project proponents generally have a strong vested interest in the outcome of a decision. As such, they are generally not shy about exercising a strong, sometimes overbearing, influence in favor of a particular course of action that will benefit their own objectives. The author refers to this as *project momentum* or *project inertia*. Project inertia can slant decision-making in favor of the proposed action, even where more optimum alternatives might exist in terms of both environmental and non-environmental advantages.

All too frequently, NEPA is implemented more as a permitting requirement for documenting decisions already made than as a true decision-making process. Moreover, many planning failures can frequently be traced directly to the fact that NEPA has either not been properly integrated with other federal planning processes or not pursued during the early planning phase as required by the NEPA regulations (Regulations).[3] Optimal plans and alternatives have often been overlooked because NEPA was not performed using an interdisciplinary, open, systematic, or rigorous process.[4] In such circumstances, NEPA's proven ability (when properly implemented) to provide a framework for comprehensive long-term planning often goes unexploited.

THANK GOD FOR NEPA!

Consistent with the goal of Section 101 of NEPA, environmentally questionable projects are often modified during the environmental impact statement (EIS) process. In many cases, the NEPA process has been so instrumental in flushing out potential environmental problems and providing more cost-effective alternatives that some projects have been canceled and many others substantially modified. Consider the following recent example.

The DOE prepared an EIS for expansion of the U.S. Strategic Petroleum Reserve. In the course of NEPA process, geotechnical studies indicated that one of the candidate sites was unreasonable.

As a result, the alternatives were changed, and new combinations for expansion at other sites were identified that could better serve the reserve's mission. DOE also made design changes related to one of the candidate sites that would protect endangered species and critical habitat. Now consider a second and more salient success story.

In the early 1990s, the DOE was seriously considering a proposal to build a multibillion new production reactor (NPR) to produce tritium for the nation's aging nuclear weapons stockpile. At the time, the DOE was under intense political pressure to move forward with the project. Former Naval Admiral and DOE Secretary James Watkins initially expressed resistance to the NEPA process and particularly the need to consider a "no-action alternative." However, the EIS that was completed was instrumental in determining a cheaper, faster, and more effective means for securing a national tritium supply. Based principally on NEPA requirements and other related studies, the Bush administration made a decision in 1992 to cancel the multibillion dollar NPR. Watkins later exclaimed before a House Armed Services Committee hearing:

> *Thank God for NEPA*, because there were so many pressures to make a selection for a technology that might have been forced upon us, and that would have been wrong for the country[5] (emphasis added).

To a large extent, the contribution that NEPA brings to federal decision-making depends on how one chooses to view and exercise this planning process. Five additional, if widely different, case examples that the author has personally witnessed, if not directly participated in, are described below. Because the author has spent much of his career working in the nuclear industry, four of the following examples involve nuclear-related projects at the DOE's Hanford site in Richland, Washington. The first is an example of an ineffective planning process that contributed little to the final decision. The second demonstrates the ineffectiveness of NEPA when it is improperly implemented, even though the proposal was carried out in a relatively cost-efficient manner. The third example vividly details how NEPA can lead to excellent planning, decision-making, and public consensus. The fourth demonstrates how an open pubic participation and scoping process can markedly shape and improve an agency's proposal. The final case demonstrates how NEPA not only can lead to an excellent decision, but also result in sizable cost savings. We will now examine these varied experiences and see what can be learned from them.

CASE 1: A DISMAL FAILURE

In the mid-1990s, the DOE decided to prepare an EIS for a relatively modest proposal to stabilize plutonium at its plutonium finishing plant (PFP), located at the Hanford site near Richland, Washington. A decision was made to prepare an EIS, even though there was substantial reason to believe that in this case a much less rigorous and costly environmental assessment (EA) would suffice. Nonetheless, implemented in a streamlined fashion, the decision to prepare an EIS on such a modest proposal did not necessarily present either a cost or an efficiency problem.

The real problem that emerged was the manner in which the PFP-EIS was prepared. When compared to other DOE actions of a nuclear nature, the activity in question was relatively innocuous, yet this *Documentation Encyclopedium* examined nearly every conceivable impact in virtual microscopic detail, even though they were clearly insignificant. The final EIS was well beyond the Council on Environmental Quality's (CEQ's) recommended page limit of 150 pages for a typical EIS and was barely within the allowed maximum page limit of 300 pages, which is reserved for projects of unusual scope or complexity. Excluded from this page count were appendices, comprising nearly 250 additional pages of largely irrelevant material. Moreover, the font was changed between the draft EIS stage and final EIS to conceal the fact that the EIS had grown so much that if the text size had been left unchanged it would have exceeded CEQ's 300-page limitation. Nonetheless, both the NEPA compliance officer and NEPA document manager expressed delight with the quality and detail of the final document.

As absurd as it may seem, the final multimillion dollar EIS spanned a total of 550 pages (including appendices) only to conclude that every single impact was insignificant. In this case, an EA costing perhaps less than $50,000 could have reached the same conclusion.

Clearly, the amount of detail provided in this EIS was completely out of proportion to the complexity of the action or its potential for significant impacts. An independent consultant later confided, "This EIS was a NEPA miscarriage."

In the end, this EIS contributed little or nothing to the final decision-making process. At least three lessons can be derived from this experience:

1. The EIS contractor was given virtually free rein, and virtually an open checkbook, to prepare an EIS with little direct federal supervision.
2. A NEPA document manager was assigned to manage the project who had little prior experience in preparing and managing NEPA analyses. Added to this, local DOE NEPA oversight neither managed nor provided clear direction that would have prevented such improvidence.
3. Local DOE NEPA oversight viewed this NEPA effort essentially as a document preparation process as opposed to a true planning process for supporting informed decision-making. This largely explains why the EIS ended up as a voluminous compilation or *Documentation Encyclopedium* of detailed and largely irrelevant data that contributed little to the final decision. Perhaps most alarming was that success appears to have been "measured" in how much the EIS "weighed" rather than in how much it contributed to an effective planning and decision-making process.

CASE 2: A BUSING PROBLEM

In another example, a decision was made to discontinue an employee commuter bus system that was used to transport hundreds of workers daily from the surrounding community to government facilities located on the DOE's Hanford site. Some of these workers were transported to work over round-trip distances estimated to be in excess of 100 miles.

As a cost-cutting measure, site management decided to discontinue the mass-transit system before the NEPA process was even initiated. Discontinuing the mass-transit system meant that workers would be forced to use private vehicles to travel a long distance to work and back. Herein lays the NEPA dilemma. DOE NEPA oversight frequently required the preparation of lengthy, rigorous, and expensive EAs even on projects that were clearly nonsignificant. In this case, however, the proposed action involved potentially significant impacts in terms of increased traffic congestion, air emissions, noise, wear and tear to the road infrastructure system, use of petroleum (a vital strategic natural resource), and an increased risk of accidents due to greater vehicular traffic (resulting in potential injuries or even loss of life).

As indicated in the CEQ's list of 10 significant factors, any number of the environmental and safety issues listed above could be viewed as being potentially significant, requiring, at a minimum, the preparation of an EA. Yet, the entire proposal was exempted from a NEPA review by application of a questionable categorical exclusion even though there was substantial reason to believe that it might involve potentially significant impacts, and even "extraordinary circumstances." This case is a perfect example of a failure not only to implement NEPA correctly, but more importantly to protect the human environment against potentially adverse significant impacts which is the ultimate goal of Section 101 of NEPA (i.e., a national policy to protect the environment).

CASE 3: A PUBLIC INVOLVEMENT SUCCESS

An EIS was completed in 1996 for a highly complex and controversial project involving the disposition of "2100 metric tons of radioactive spent nuclear fuel" at the DOE's Hanford site. One of the storage basins in which this fuel was held was so antiquated that water contaminated with radio-active waste had leaked into the underlying soil column. The critical priority given to this project was underscored by the fact that a seismic event could result in an accident having potentially cata-strophic implications. Although the project was controversial, the NEPA process was instrumental in bringing opposing parties together in a unified effort that expeditiously determined the safe disposition of the radioactive fuel.[5]

In the words of Eric Gerber, the Spent Nuclear Fuel project manager:[6]

> A decision was made to involve the stakeholders from the beginning. We discussed pending decisions before they were finalized and actually changed our plans based on stakeholder input. After ... seeing the impact of their recommendations on our decisions, the Project's credibility became established and stakeholder communications shifted from demands to team participation. An illustration of the success of this effort was the completion of the Project's Environmental Impact Statement in eleven months with few stakeholder comments; previously unheard of for major DOE projects.

CASE 4: PUBLIC INPUT MARKEDLY SHAPES AN AGENCY'S PROPOSAL

As an example of how the NEPA process can shape federal projects consider how its public par-ticipation process influenced and improved the following federal proposal. The spread of nonnative invasive plant (NNIP) infestations across public wilderness lands is becoming an issue of significant national concern. Using its public scoping process, a federal agency solicited input, comments, and suggestions on a comprehensive program to control NNIP infestations on its wilderness land. A variety of treatment methods including manual, mechanical, chemical, and biological controls were proposed. Under this program, a programmatic environmental assessment (P-EA) was pro-posed to evaluate the potential environmental effects of these methods in controlling the spread of the NNIP infestations.

The agency and its NEPA contractor were surprised to find that the public provided a number of penetrating, and key concerns and issues that had not previously been considered by the agen-cy's scientific staff. For example, representatives for the local honeybee industry raised concerns that this proposal could adversely impact the population, and thereby their industry, by negatively affecting the bee's nectar sources. As a result, this concern was elevated to a prominent issue for evaluation in the EA. Mitigation measures were also identified for reducing the impacts on the honeybee industry.

Another commenter stated that many NNIP species have been maligned more for their "alien status" than for any real detriment impact upon the native ecosystem. Another commenter added that approximately half of the popularly identified wildflowers are actually "alien" in origin and that the agency should not automatically consider such species for eradication. As a result, the analy-sis was refocused to place increased emphasis on identifying and eradicating NNIPs that actually posed the greatest harm to the natural ecosystem.

A couple of commenters stated that the program was simply too pervasive and overreaching given the amount of damage that the NNIPs actually cause. Based on this suggestion, the agency considered both a small acreage and an alternative with a larger number of treated acreage.

Another commenter replied that if a decision was made to eradicate NNIPs, then emphasis needed to be given to finding suitable replacement plants that could be planted to prevent NNIPs from invading the treated areas that had been disturbed or devegetated. As a result, increased emphasis was placed on finding suitable plants to revegetate the disturbed areas.

This example clearly illustrates how insightful public input through the NEPA process can help shape and improve a federal proposal, allowing decision-makers to consider a number of competing factors such as the actual need for the proposal, cost, private industrial concerns, and of course the environmental impacts.

Case 5: A 435 Million Dollar Success

The companion book *The NEPA Planning Process*[7] details a major project in which NEPA was credited with saving the federal government $435 million. In this case, the DOE issued a draft EIS for the safe interim storage of Hanford tank waste (New Tank EIS) in 1994. The preferred alternative involved the construction of up to six enormous high-level radioactive waste storage tanks with a projected price tag of $435 million. The need for additional storage space was considered urgent, and political momentum was decidedly in favor of initiating tank construction as soon as possible. The project manager was thus determined to implement this project in record time (i.e., an instance of the "project inertia" problem described earlier).

As the NEPA process proceeded, questions soon began to surface regarding the actual need and justification for additional waste tank storage space. Based on the NEPA scoping effort, it slowly became evident that the underlying need for this project was unjustified; the high-level radioactive waste could be processed using existing technology, reducing the volume and hence the need for additional tank storage capacity. Eventually, reconsideration of waste volume projections and management practices forced the project management team to conclude that the construction of additional tanks was unjustified. The project was canceled (e.g., a no-action alternative was chosen).

The cost savings from this single decision alone was $435 million! These savings exceeded the cost of DOE's entire NEPA process for many years to come. Carol Borgstrom, director of DOE's Office of NEPA Policy and Compliance, stated that this was truly a "NEPA success story." A letter to the DOE from the Confederated Tribes of the Umatilla Indian Reservation characterized this EIS as an excellent example for others to follow.

Lessons Learned

How does one account for such differences in success? What can we learn from such experiences? Why were some of these projects outstanding examples of planning achievement, while others were anything but successful? Clearly, differences in philosophy and management oversight account for much of the contrast. Adherence to (or disregard of) CEQ regulatory direction, and good (or bad) methods of professional practice also account for some of the differences in outcomes. Perhaps most important to note were the last two case examples, which described exercises in true, well-conceived planning.

Thus, it can be seen how, to a great extent, NEPA compliance can result in either a success story or a quagmire, depending on how an agency chooses to implement the process. Fortunately, the bad examples of NEPA practice are more than balanced by its innumerable successes—and NEPA practice continues to evolve and improve.

Nuclear Regulatory Commission Regulations

In many cases, NEPA can be improved simply by changing the culture, approaches, or traditional procedures under which it is implemented. For example, the Nuclear Regulatory Commission (NRC) has experienced numerous delays and inefficiencies in its NEPA and nuclear reactor licensing process. Consequently, the NRC has propagated new regulations for licensing nuclear power plants.[8] These changes are based on lessons learned from previous environmental licensing processes. For example, the NRC staff developed a new process for granting early site permits; this new permitting process carries the potential to streamline the NEPA approval process for site selection and ultimate construction of commercial nuclear reactors. Consistent with the NEPA regulations streamlining and integration provisions, NRC has also developed a combined operating license application (COLA) process that is designed to combine the two-stage licensing process for construction and operation of commercial nuclear reactors into a single rigorous but more streamlined process. Such forward thinking is commendable.

HISTORICAL DEVELOPMENT OF NEPA

As far back as 1959, in the U.S. Resources and Conservation Act that he introduced that year in Congress, Senator James E. Murray of Montana had proposed the establishment of a council to oversee environmental policy. But the right moment had not yet arrived. Nonetheless, spurred by a growing number of severe environmental problems caused largely by industrial pollution, support for a national environmental policy continued to gain momentum during the course of the following decade.

THE FIRST ENVIRONMENTAL IMPACT STATEMENT

Interestingly, there was actually a precedent for preparing an EIS some years before the enactment of NEPA. In the early 1960s, as part of its Operation Plowshare, the former Atomic Energy Commission proposed to detonate a string of nuclear explosives to blast out a harbor along the Alaskan coastline. This proposal was promoted by the esteemed physicist, Edward Teller, widely known as the "father" of the H-bomb.

Alarmed by this proposal, Congress required the Atomic Energy Commission to prepare an environmental study to assess the impacts of this proposal. The project was ultimately canceled, in large measure, because of the results of this study, which has been called the "first *de facto* EIS."[9] This project has since been criticized as potentially one of the most environmentally destructive projects ever proposed. Proposals for potentially dangerous projects such as this only increased a growing belief in parts of government and among a growing segment of the public that some type of national environmental policy was needed.

GROWING PROBLEMS: THE GRAND CANYON DEBATE

Prior to NEPA, U.S. federal agencies routinely embarked on airport, highway, timber cutting, and massive dam building projects without performing any rigorous analysis of the potential impacts and with virtually no consultation with the public. By the late 1960s, Congress was hearing disturbing testimony from the scientific community regarding the increasing erosion of environmental quality and the potential for future ecological disaster.

Jackson, Dingell, and Caldwell

Henry ("Scoop") Jackson, the legendary senator from Washington State, was a conservative "hawk" on the Vietnam War, but progressive when it came to environmental issues. Senator Jackson had become deeply concerned about unconstrained timber cutting and the pollution created by oil spills from tankers entering Puget Sound. One of his particularly contentious committee hearings, concerning a proposal by the Bureau of Reclamation to dam the Colorado River above the Grand Canyon, was later described by insiders as a near shouting match. As a result of such events, Senator Jackson concluded it was time to create and institute a new control mechanism in the form of a national policy to protect and preserve environmental quality.

Senator Jackson's leadership in the Senate along with that of Michigan Representative John Dingell in the House eventually led to enactment of The National Environmental Policy Act of 1969. Among other proposals related to environmental policy introduced during the 1960s, Jackson introduced an unsuccessful bill in 1966 that, among other things, provided for the establishment of a council of environmental advisers.

In early 1968, Jackson asked the Conservation Foundation for help in locating someone with the necessary experience in both government and the environment who could provide the assistance essential in formulating a workable policy for the environment. The foundation knew exactly the expert Jackson was looking for. Professor Lynton Keith Caldwell of Indiana University was then a highly regarded political scientist who had for many years traveled the world on assignments for the

United Nations and other important bodies. Since the early 1960s, he had decided to dedicate the rest of his career to the formulation of environment policy.

On July 11, 1968, a few months after accepting an invitation to join Senator Jackson's special committee, Dr. Caldwell, with the assistance of one of Jackson's senior advisers, William Van Ness, prepared the report, *A National Policy for the Environment*. This report was published for the powerful Committee on Interior and Insular Affairs for which Jackson was the chairman. Caldwell was assigned the responsibility for preparing the original draft of the NEPA statute.

Following a Joint House-Senate colloquium held on July 7, 1968, to discuss the report and to increase environmental awareness in Congress, Senator Jackson then reintroduced his environmental policy statute (NEPA) in 1969. At the only Senate hearing on bill S.1075 held on April 16, 1969, Caldwell introduced the idea for the need to include an action-forcing provision as part of the proposed statute: this provision became the EIS. For originating the concept of NEPA including the EIS, and for all his work in crafting the language of the Act, Caldwell, with justification, is often referred to as the "father of NEPA." He has also been referred to as the "principal author of NEPA," "the principal father of NEPA," and the "inventor of the EIS."

In many ways, NEPA was modeled loosely after a landmark statute, the National Employment Act of 1946. The National Employment Act established a Council of Economic Advisers to assist and advise the president on economic matters. The Council of Economic Advisers provided a model that was used to pattern the Council on Environmental Quality (CEQ) under NEPA.

Rather than creating another new and expansive bureaucracy, Congress wisely chose to craft the national policy by supplementing the existing statutory charter of federal agencies. Thus, these agencies would be expected to balance the goal of preserving the environment with other competing factors and policies such as economic growth. In short, agencies would be required to infuse NEPA into their traditional decision-making process. While NEPA created the CEQ to oversee the Act, each agency would assume responsibility for self-enforcement, with the CEQ providing guidance.

Caldwell and the Congressional Hearings on NEPA

Some experts have gone so far as to call Professor Caldwell "the father of the modern environmental movement." This title could be contested, of course, by supporters of Aldo Leopold, Robert Marshall, Rachel Carson, Sigurd Olsen, Senator Gaylor Nelson (the founder of Earth Day), and Stewart Udall, among other well-known names.

Nevertheless, Caldwell (who wrote an endorsement for this book) should certainly be recognized as one of the great visionary environmental leaders of the twentieth century. Although his career, until the early 1960s, had been devoted to public administration, since his youth he had also been an observant lover of nature and an avid bird watcher. Over the years, as he traveled widely, he observed—rather in the same fashion as George Perkins Marsh a century earlier—how other countries were damaging their natural environments and the impact this was having on their human populations. As a result, his long interest in the environment and his belief that it was time to fashion new policy that could help ensure environmental protection continued to grow.

In 1963, in a period when the inclusive or holistic idea of the meaning of environment was little used or understood, his groundbreaking essay, "Environment: A New Focus for Public Policy?," published in the respected journal, *Public Administration Review*, won him the coveted William E. Mosher Award and laid out both a new agenda for public policy and the need for a national policy of environmental protection.[10] The article won him considerable acclaim and launched a new subfield in the field of public administration. He has also been credited as the originator of another scientific subfield, biopolitics.

From 1963, and for the rest of his long career, Caldwell dedicated his work and considerable expertise to writing, teaching, lecturing, and acting as a consultant to a wide variety of government and private institutions in the United States and around the world on environmental policy and related fields. A unique and certainly one of the most important features of his work has been its focus on interdisciplinary synthesis. To him, the environment has always, in his own words, "comprised the relationship of its different parts with each other and their effects on human beings."

Senate Hearings. At the Senate hearing in April 1969, in collaboration with Senator Jackson, Caldwell promoted NEPA's now famous "action forcing," or EIS requirement. Some of his testimony before Senator Jackson's committee follows:[11]

> **Dr. Caldwell**: "I have already suggested, it seems to me, that the Congress indeed has a responsibility and could enunciate [a national environmental] policy. But beyond this, I would urge that in the shaping of such a policy, it has an *action-forcing*, operational aspect. When we speak of policy we ought to think of a statement … that is not merely a statement of things hoped for, [but] a statement which will compel … the Nation … to take the kind of action which will protect and reinforce what I have called the life support system of this country" (emphasis added).

> **Chairman [Senator Jackson]**: "I would like to pursue this policy matter for a moment. I agree with you that realistically what is needed in restructuring the governmental side of this problem is to legislatively create those situations that will bring about an action-forcing procedure the departments must comply with. Otherwise, these lofty declarations are nothing more than that … I believe this is what you were getting at."

> **Dr. Caldwell**: "Yes, exactly so."

> **The Chairman**: "I am wondering if we might not broaden the policy provision in the bill so as to lay down a general requirement that would be applicable to all agencies that have responsibilities that affect the environment rather than trying to go through agency by agency.

> I think the immediate example that comes to my mind and has to yours already by the statement is that the Atomic Energy Commission, in granting permits or licenses in connection with nuclear power-plants, should be required to make an environmental finding …"

As indicated earlier, this action-forcing procedure that Jackson and Caldwell discussed in this hearing eventually manifested itself as the modern EIS.

At a 1995 DOE conference, Professor Caldwell stated:[12]

> The Congress had no explicit constitutional authority to legislate environmental policy *per se*. But the Congress and the President did have authority to define and direct the policies and actions of the Federal agencies. Because agency missions impinged directly or indirectly upon almost every aspect of the American society, a statutory law could be enacted that would be both effective and constitutional. Moreover, a statutory declaration of national policy could be binding upon both the Legislative and Executive branches.

Caldwell went on to add:[12]

> NEPA was thus conceived as a *national policy*, not merely a Congressional or Presidential Policy (emphasis in original text).

Congressional Debate

Early in the congressional debate, some supporters, including Caldwell, realized that an environmental policy statute on its own would not be enforceable. For this reason, they argued for drafting an environmental protection amendment to the constitution. While NEPA eventually gained overwhelming Congressional support, a constitutional amendment did not.

Recognizing that powerful business lobbies would oppose environmental restrictions on the private sector, others urged that a statute could better overcome industrial opposition if it focused on federal government actions. It was widely recognized that the federal government, as the single biggest entity in the United States, and given the vast scope and nature of its actions, accounted for a disproportionably larger share of the nation's environmental problems. They argued that passage of a bill that focused on the actions of the federal government would serve both to demonstrate the seriousness with which Congress viewed environmental protection and establish a precedent for subsequent legislation such as a Clean Water Act and Clean Air Act that would help regulate environmental behavior among the private sectors. An act such as NEPA, even if limited to federal actions, would clearly have profound implications on limiting future environmental degradation.

As the chief sponsor of the bill, Senator Henry Jackson declared that its passage would be

> ... the most important and far-reaching environmental and conservation measure ever enacted ... more than a statement of what we believe as a people and as a nation ... it serves a constitutional function in that people may refer to it for guidance in making decisions where environmental values are found to be in conflict with other values.[13]

For his part, John Dingell, the chief sponsor of the bill, in the House, stated:

> We must consider the natural environment as a whole and assess its quality continuously, if we really wish to make strides in improving it and preserving it.

Individual drafts of NEPA were prepared by both the U.S. House of Representatives and the Senate. The Senate's draft version of the bill was more comprehensive but lacked a clear vision of a national environmental policy. Much of the debate over NEPA centered on the bill's action-forcing mechanisms that appeared in the Senate but not in the House version. Agreeing with Dr. Caldwell's ideas, Senator Jackson was adamant that an action-forcing provision be incorporated. A compromise was eventually struck during a conference committee to include a requirement to prepare a detailed statement that would evaluate the impacts of federal proposals.

In one early version of the bill, the term "finding" was used instead of "detailed statement." Senator Edmund Muskie of Maine, another outspoken supporter of a national environmental policy, negotiated substitution of the phrase "detailed statement," fearing that otherwise the statute would be too weakly worded. This detailed statement would later be known as the EIS. For this reason, perhaps incorrectly, some authorities have given Senator Muskie the credit for being the "father of the EIS."[14]

Some congressmen argued that agencies should not only have to consider alternatives, but in each case should actually be required to select the alternative that was considered to be environmentally benign. But such a provision was considered too restrictive and was successfully resisted. It is for this reason that NEPA contains no requirement forcing agencies to select the alternative that minimizes environmental degradation.[15]

The final bill received the unanimous vote of the Senate Interior Committee and also received widespread support among members of Congress. NEPA was passed by a unanimous vote in the Senate and by an overwhelming majority of the House. In retrospect, it is easy to understand why the Act met with such little resistance. By the end of the 1960s, the public and some elements of the media had widely expressed their desire to curb further environmental degradation, and it was thus in the interests of politicians to heed the wishes of their electorates.

NEPA's Enactment

Although President Richard Nixon originally opposed the Act, he eventually came to embrace it because of its wide popularity. The importance of NEPA was reinforced when the President chose to sign NEPA into law on New Year's Day, 1970. Nixon proclaimed the signing of NEPA as "my first act of the decade." Thus, NEPA holds the noted distinction of being the first law to be enacted in the new decade of the 1970s. NEPA laid the foundation for creation of the U.S. Environmental Protection Agency (EPA), which was established the following year, in part to take on the rapidly growing responsibilities that were already beginning to overwhelm the CEQ.

NEPA's Growing Pains

Beginning in the early 1970s, the CEQ introduced a series of interim guidelines for preparing EISs, pursuant to an executive order. Even so, the years prior to issuance of the CEQ's NEPA regulations were often a difficult time for agencies seeking to comply with the terms of the Act, even though the CEQ's intent was that these guidelines were to be considered binding and nondiscretionary.

In its early years, NEPA was often viewed with total and open disregard by many federal agencies, particularly given that there were no criminal or civil penalties associated with noncompliance. As a result, the federal court system became the default mechanism for ensuring that the Act was enforced. NEPA also provided the environmental movement, including prominent organizations like the Sierra Club and the Natural Resources Defense Counsel, with a mechanism for challenging environmentally questionable projects.

Beginning with the landmark case of *Calvert Cliffs* (1971), the courts concluded that NEPA set "a high standard for the agencies."[16] The Calvert Cliffs case firmly established a precedent that agencies must prepare a detailed statement (EIS) in compliance with NEPA's action-forcing provision. While the CEQ had adopted guidelines under President Nixon, this guidance focused primarily on implementing the EIS process. Lacking definitive direction, the courts often ruled in favor of additional detail. Consequently, many EISs became increasingly "bloated," nonanalytical documents sometimes exceeding thousand pages in length. An early concern was that agencies appeared to view the requirement for an EIS as a perfunctory end unto itself rather than as a vehicle to foster sound, environmentally aware decision-making.

While many agencies openly disregarded NEPA in its formative years, they have generally learned to comply with the Act. A principal driver of NEPA compliance has no doubt been the accumulation of NEPA litigation which contains a clear signal that NEPA is to be taken seriously. Thus, most agencies are now making a good-faith effort to accommodate NEPA's intent into their federal decision-making process. For the most part, NEPA has truly become self-enforcing in the manner intended by Congress. For additional information, the companion book, *The NEPA Planning Process*, provides a more detailed account of the historical forces that led to the enactment of NEPA.[17]

The CEQ Regulations

Eventually, problems with NEPA reached the door of the Oval Office. President Carter responded in 1977 by issuing an executive order directing the CEQ to establish formal NEPA regulations.[18] The earlier guidelines were to be replaced with binding regulations.

General counsel for the CEQ, Nicholas Yost, was the lead author in preparing the CEQ NEPA regulations. In November 1978, 8 years after NEPA was enacted, the CEQ promulgated its formal regulations (40 CFR § 1500–1508). These regulations expanded the scope of the Act to cover the entire NEPA process, rather than simply the EIS process.

These draft regulations provided specific directions for restricting the length of EISs, encouraged agencies to set time limits, and stressed the need both to focus on important issues and to de-emphasize insignificant ones.[19] Additional streamlining directions that are strongly encouraged to help streamline the process include reliance on "scoping" to help agencies' focus on key issues, incorporating material by reference, and combining EISs with other documents.

The draft regulations were issued for public review and comment. Public hearings were held. The CEQ met with every agency in the federal government. Comments were received from nearly 12,000 organizations, private individuals, and state and local agencies. Among those consulted were the Chambers of Commerce, the Building and Construction Trades Department of the AFL-CIO, and the National Resources Defense Counsel. Approximately 500 written comments were received on the draft Regulations, and the majority of these comments were favorable, expressing wide support for the proposed Regulations.[20]

The overwhelming consensus was that the public benefited from NEPA's requirements. However, some comments complained that the process was too lengthy, resulted in large quantities of needless detail, and needed to be further streamlined. As a result, a total of 340 amendments were made to the draft Regulations, before being issued in 1978 as final Regulations.[21]

While there have been many problems over time with the NEPA process, including a host of litigation, the NEPA regulations have proved remarkably resilient. Since their passage, only one provision in the Regulations has been amended. In 1986, a requirement to consider the effects of accidents and other uncertainties was modified to eliminate the requirement to perform a "worst case" analysis. With respect to the CEQ NEPA regulations, the Supreme Court has ruled that "CEQ's interpretation of NEPA is entitled to substantial deference."[22]

It is important to note that under the Regulations each federal agency is required (based on each agency's specific mandates, obligations, and missions) to issue its own NEPA implementing procedures that supplement the minimum requirements set forth in the Regulations (§ 1507.3). These agency-specific NEPA procedures account for the slight differences in each agency's NEPA process. For example, the Federal Highway Administration provides a 30-day comment period (with or without a public meeting) on all Environmental Assessments (EAs) that they develop before a Finding of No Significant Impact (FONSI) is issued while some other agencies have no required comment periods for EAs.[23]

Most agency NEPA implementation procedures are available online at the NEPA Web site (http://ceq.eh.doe.gov/nepa/regs/agency/agencies.cfm). Some are also published in the *Federal Register*. One can also write or call the agency NEPA point of contact(s) and ask for a copy of their procedures. A copy of the NEPA statute and CEQ NEPA Implementing Regulations are provided in Appendices A and B, respectively.

NEPA'S GLOBAL PRECEDENT

Other nations quickly began to realize that environmental pollution and degradation, particularly because of unplanned and uncontrolled development, were not simply an American problem, but a global one. These nations began to recognize that the EIS process inherent in NEPA provided a particularly powerful tool for establishing policies and preparing plans that could avoid environmental pitfalls on their soil as well. Chapter 13 provides a more in-depth discussion on NEPA's effect on the international community.

WORLD'S MOST EMULATED LAW

International agencies and organizations have also followed suit. Today, NEPA's influence permeates virtually every corner of the globe. NEPA has been referred to as one of the most emulated, if not the most emulated, statute in the world.[24,25] More than 25 American states (State Environmental Policy Acts [SEPA]) have emulated NEPA in one form or another.[26] Canter reports that over 100 countries have instituted some form of NEPA's environmental impact assessment (EIA) measures.[27] A book by John Cronin and Robert F. Kennedy, Jr., cites a number that is even greater:

> NEPA, which has now been adopted in some form by over 125 countries, has become one of the great promoters of democracy around the world …[25]

It is indeed remarkable how many foreign EIA processes have come to mirror NEPA's model. This foreign emulation of NEPA is a tribute to the vision that the U.S. Congress established in 1969.

INTERNATIONAL ORGANIZATIONS

In 1974, the Organization of Economic Cooperation and Development (OECD) recommended that its member states adopt EIS-like processes; OECD now uses an EIA process similar to that of NEPA in granting aid to developing nations.[28] Similarly, the United Nations Environmental Program (UNEP) developed EIA guidance and strongly encourages member states to establish EIA processes modeled after NEPA.[29] The World Bank also requires an EIA process on projects that it funds.[30] Other significant examples include the European Economic Community, which requires its member states to comply with an environmental process similar to NEPA.[31] Officials for the North

American Free Trade Agreement (NAFTA) have also considered establishing a requirement to use a NEPA-like process in evaluating transboundary impacts between the borders of the United States, Canada, and Mexico.

DEMOCRACY AND FEDERAL DECISION-MAKING

Prior to NEPA, federal agencies generally based decisions regarding adoption of new programs, plans, or projects principally on political, economic, and technical considerations. Under NEPA, federal agencies are now required to consider and analyze reasonable alternatives to a project, while decision-makers are responsible for seriously considering these alternatives including mitigation measures.

NEPA's sweeping mandate and comprehensive procedural requirements have profoundly affected U.S. policy and decision-making at the highest level.[32] Spensley writes that NEPA has been "a major force in reforming agency decision-making processes."[33]

After the enactment of NEPA, environmental considerations were for the first time required to be considered comprehensively in reaching federal decisions. Decision-makers could no longer legally claim that they were unaware of either the possible deleterious effects or ramifications of their decisions.

Prior to NEPA, federal planning and decision-making were often performed "behind closed doors," far removed from stakeholder input and public scrutiny. NEPA enables citizens to participate in the federal decision-making process through its public scoping, comment, and review process. American citizens as never before are now able to participate in and influence proposals that may affect their lives during the planning process. Arguably, no other statute in modern times has done as much to open up the federal planning and decision-making process to its citizens. The following example vividly illustrates how NEPA allows citizens to participate in and influence federal decisions that affect their lives.

Public Participation

The DOE recently conducted scoping meetings on a supplemental programmatic EIS (PEIS) for managing and reconfiguring its nuclear weapons complex. Nearly 1000 people attended scoping meetings held in 12 diverse locations across the country during 2006. About 350 people provided comments orally at the meetings, and, in addition, DOE received more than 32,000 written comments, mostly via e-mail. Many of the comments asked DOE to add an alternative that assumes continued reduction in the size of the U.S. nuclear stockpile. As a response to public comments, DOE revised the range of alternatives for evaluating a supplemental PEIS on the future configuration of DOE's nuclear weapons complex.[34]

Some commentators also requested that an alternative be added that would implement a 2005 recommendation from the Secretary of Energy Advisory Board Task Force on the Nuclear Weapons Complex Infrastructure. That recommendation was to consolidate most nuclear weapon activities at a single site—a Consolidated Nuclear Production Center. After considering these comments, DOE announced in a recent report to the Congress that it is proposing inclusion of this concept as an alternative to be evaluated in the supplemental PEIS (Report).[35]

Fostering Democracy

NEPA's effect on fostering the development of international democracies is particularly noteworthy. Adoption of international EIA processes similar to NEPA has opened government decision-making processes up to tens of millions of citizens in nations around the world. For example, John Cronin and Robert F. Kennedy, Jr., have pointed out that the elements inherent in NEPA have helped promote democratic principles around the world.[36]

A Scientifically Based Planning Process

Perhaps NEPA's greatest contribution has been to inject a rational, scientifically based planning process throughout all levels of federal decision-making. Robert Bartlett writes:

> NEPA is a great deal more that a mere legal requirement for the preparation of environmental impact statements. NEPA has ... [compelled] bureaucracies to use science-like analysis as a basis for policies and decisions—an attempt to force greater rationality in government decision-making ... implicit in NEPA and underlying its logic as a policy [act] is a distinct form of reasoning—an ecological rationality.

THE NEPA ACT

As described earlier, Congress passed the world's first national environmental policy in 1969 to establish a policy for protecting the environment.[37] NEPA directed all federal agencies to prepare an EIS to rigorously evaluate the environmental impacts and alternatives to any major federal action that could significantly affect the quality of the human environment. In other words, build a highway if needed, but at least consider an alternative route that avoids impacts to wetlands; sponsor a mass transit system, but in doing so, consider methods for minimizing emissions.

The goal was to build into the agency decision-making process an appropriate and careful consideration of all environmental aspects of proposed actions. Professor Caldwell underscored NEPA's precedent setting importance when he wrote:[38]

> ... unless rapid and effective action is taken to prevent further destruction of the planetary biosphere and its living organisms, the early degradation of the human species is a certainty.

While virtually every other environmental law prescribes specific performance standards, NEPA does not contain the detailed technical and substantive requirements of other major environmental legislation. As depicted in Table 1, it differs from virtually every other environmental statute in at least three important ways.

NEPA'S POLICY

The NEPA Act is short, containing just three parts (a statement of "purpose" followed by two titles), and is only about five pages long. Despite its brevity, it has profoundly affected federal decision-making at the highest levels.

TABLE 1
Basic Differences between NEPA and Other Environmental Statutes

1. The principal environmental laws and regulations are concerned with regulating activities whose scope is defined very narrowly. For example, the scope of the Resource Conservation and Recovery Act (RCRA) is primarily oriented toward activities that involve treatment, storage, and disposal of hazardous waste. As indicated above, the scope of NEPA is pervasive, and with few exceptions, virtually every type of action performed by federal agencies is potentially subject to its requirements.
2. The principal environmental statutes and Regulations are concerned with controlling, regulating, or limiting certain specific types of activities. NEPA has little substantive effect in this regard. In lieu of establishing strict controls and Regulations over specific activities, NEPA provides a tool or architectural framework for *planning* future actions. Its intent is not to control and regulate existing activities but to require agencies to evaluate and consider the environmental consequences of their decisions at the proposal stage (prior to beginning an activity).
3. The principal environmental statutes and Regulations clearly define the mechanisms for carrying them out, and delegate, often to the Environmental Protection Agency, enforcement responsibilities including the promulgation of Regulations and (in some cases) policing powers. The language of NEPA leaves the responsibility of enforcement with the respective agencies, subject only to court action.

Title 1 declares a national environmental policy and sets forth procedural requirements that must be followed in pursuing proposed actions while Title II creates the CEQ. While NEPA is briefly summarized below, the reader is encouraged to read the Act in its entirety (see Appendix A).

As stated in the introduction to the Act, the purpose of NEPA is to

> Declare a national policy which will encourage productive and enjoyable harmony between man and his environment; promote efforts which will prevent or eliminate damage to the environment and biosphere, and stimulate the health and welfare of man; enrich the understanding of the ecological systems and natural resources important to the Nation; and to establish a Council on Environmental Quality.

Titles I and II are briefly summarized below.

TITLE I—POLICIES AND GOALS

Title I is the "heart" of NEPA. It announces the world's first national environmental policy. As Section 101 of Title I defines the nation's environmental policy, it is sometimes referred to as the "spirit of the law."

Section 102 of Title I defines a specific procedural or action-forcing mechanism (e.g., the EIS) for carrying out the policy established in Section 101. Because Section 102 specifies the procedural requirements governing preparation of an EIS, it is sometimes referred to as the "letter of the law."

Section 101

Section 101 of NEPA establishes the U.S. national environmental policy, which applies to all Americans. Part A of the section states:

> ... it is the continuing policy of the Federal Government, in cooperation with State and local governments, and other concerned public and private organizations, to use all practicable means and measures, including financial and technical assistance, in a manner calculated to foster and promote the general welfare, to create and maintain conditions under which man and nature can exist in productive harmony, and fulfill the social, economic, and other requirements of present and future generations of Americans.

Part B of Section 101 declares a national policy to use all practical means, including financial and technical means, to promote the general well-being of the environment. Specifically, all practicable means are to be used to

1. fulfill the responsibilities of each generation as trustee of the environment for succeeding generations;
2. assure for all Americans, safe, healthful, productive, esthetically and culturally pleasing surroundings;
3. attain the widest range of beneficial uses of the environment, without degradation, risk to health or safety, or other undesirable or unintended consequences;
4. preserve important historic, cultural, and natural aspects of our national heritage, and maintain, wherever possible, an environment which supports diversity, and variety of individual choice;
5. achieve a balance between population and resource use which will permit high standards of living, and a wide sharing of life's amenities; and
6. enhance the quality of renewable resources and approach the maximum attainable recycling of depletable resources.

While NEPA appears to define a substantive duty to protect environmental quality, court decisions have held that the goals set forth in Title I impose only a procedural duty on federal agencies to consider NEPA's goals in reaching a final decision regarding proposed actions (see Section 1.3). Federal agencies are therefore not bound to comply with the substantive policy requirements of Section 101; that is, they are not obligated to make decisions based simply on preserving environmental quality.

As described in item 5 above, the phrase "life's amenities" requires special mention. In this context, the term "amenities" can be thought of as those aspects and resources that are considered to be beautiful or pleasurable. For example, an environmental amenity might be a natural, unobstructed view of a waterfall.

Productive Harmony. As described earlier, Section 101 laid out the purpose and policy of NEPA while Section 102 laid out the procedural requirements for implementing this policy. NEPA is, and never was, designed to halt general economic or technological progress, or development. On the contrary, as described above, NEPA declares a policy to use all practicable means and measures:

> … to foster and promote the general welfare, to create and maintain conditions under which man and nature can exist in *productive harmony*, and fulfill the social, economic, and other requirements…[35] (emphasis added)

Note the use of the carefully crafted phrase "productive harmony." In fact, the word "productive" is used three times in the statute. Clearly, the architects of NEPA were attempting to strike a balance between environmental goals and those which promote a vibrant, modern, and prosperous society. Unfortunately, this balancing act has all to often been skewed against environmental goals; neglecting the environmental policy intent inherent in the goal of promoting "productive harmony" has been a major factor responsible for the reasons why NEPA has fallen short of achieving its true potential.

Section 102

Under Section 102, all federal agencies shall, "… to the fullest extent possible," interpret and carry out their policies and duties in accordance with NEPA. Among the key provisions which have had far-reaching consequences are the following provisions that direct all federal agencies to

> … utilize a systematic, interdisciplinary approach which will insure the integrated use of the natural and social sciences and the environmental design arts in planning and in decision-making which may have an impact on man's environment … (Section 102[2][A])

> … include in every recommendation or report on proposals for legislation and other major federal actions significantly affecting the quality of the human environment, a detailed statement by the responsible official on (i) the environmental impact of the proposed action … (Section 102[2][C])

The first provision (Section 102[2][A]), shown above, challenges agencies to go beyond their traditional mission-oriented thinking when planning proposed actions.

The second provision forms the core of NEPA's enforcement policy. The Regulations go on to define specific requirements for preparing NEPA's "detailed statement" which is now referred to as an EIS. Through case law, the courts have further clarified the requirements of this provision. Section 102 defines five requirements that the EIS must address:

1. The environmental impacts of the proposed action
2. Any adverse environmental effects which cannot be avoided if the proposed action is implemented
3. Alternatives to the proposed action
4. The relationship between the local short-term uses of man's environment and the maintenance and enhancement of long-term productivity
5. Any irreversible and irretrievable commitments of resources which would be involved in the proposed action should it be implemented

TITLE II—COUNCIL ON ENVIRONMENTAL QUALITY

Title II establishes a CEQ or "Council" as it is often referred to. Under this title, the president, with the Senate's concurrence, is directed to appoint three members to the CEQ. As a part of the Executive Office of the President, the CEQ reports directly to the president, and one member is appointed by the president to act as chairman. Patterned after the Council on Economic Advisers, the CEQ advises the president on environmental matters.

Title II assigns eight specific duties to the CEQ. These duties are the following:

- Advising the president
- Resolving interagency disagreements
- Gathering information about the quality of the environment
- Preparing the annual environmental quality report and making recommendations with respect to policy and legislation

Surprisingly, the Act delegated no enforcement powers to CEQ with respect to the implementation of Title I. As described earlier, an executive order had to be issued by President Carter, granting CEQ authority for issuing implementing regulations.

In reality, CEQ's principal powers are political. As part of the Office of the President, it makes recommendations on matters of policy to "promote the improvement of environmental quality to meet the conservation, social, economic, health, and other requirements and goals of the Nation." Since its inception, the CEQ has been instrumental in shaping many major policies and legislation. Two of its earliest accomplishments, for example, included rejection of the California Minarets Road Proposal and halting construction on the Cross-Florida Barge Canal.[39] In reality, its effectiveness is actually a function of the president and the degree of support he lends to environmental protection. In some administrations, such as Reagan's, the CEQ was virtually ostracized.

THE CEQ REGULATIONS AND OTHER GUIDANCE

The CEQ NEPA regulations (Regulations) have nine parts, which are summarized in Table 2, and are provided in Appendix B. The Regulations focus on how to implement Title I, Section 102(2) of NEPA. The Regulations provide an essential, yet not highly specific, guide to NEPA practice; this can present a challenging working situation to the NEPA practitioner.

In addition to learning the Regulations, the practitioner needs to be aware of key court findings and of available scientific and technical methods for characterizing environmental impacts. The courts have given very strong deference to these regulatory requirements.

As noted earlier, most other environmental regulations are highly prescriptive, defining clear thresholds of compliance, or even prescribing technical means such as specific models or laboratory tests for determining compliance levels. While the NEPA regulations are specific on a number of procedural matters, they leave a large measure of compliance to the agency's discretion, subject only to court authority when challenged. The NEPA regulations have been specifically crafted so as to allow for both innovative implementation methods as well as technical differences in how specific provisions are implemented from one agency to the next.

The implication is that NEPA compliance can be both flexible and variable. Thus, the Regulations allow agencies to respond with different approaches and venues to such factors as the presence or absence of unique project constraints, response to project opposition, detail, and level of analysis in its NEPA documents.

ADDITIONAL REQUIREMENTS AND GUIDANCE

In addition to the Regulations, there is also a body of direction for complying with NEPA which appears in other laws, regulations, guidelines, and executive orders. For example, there is a body of official

TABLE 2
Summary of Parts 1500–1508 of the CEQ NEPA Regulations (40 CFR Parts 1500–1508)

Part	Summary of Regulatory Direction
40 CFR Part 1500: Purpose, Policy, and Mandate	Introduces the purpose and responsibility to comply with NEPA. Specifies means by which agencies shall reduce paperwork and delay: 1. Describes purpose of NEPA and details an agency's duty to comply with NEPA 2. Provides direction for reducing paperwork and delays 3. Specifies agency's authority
40 CFR Part 1501: NEPA and Agency Planning	Prescribes requirements and guidance for integrating NEPA into an agency's decision-making process: 1. Direction to start NEPA early 2. Determining whether to prepare EA or EIS 3. Responsibilities of lead and cooperating agencies 4. Direction on implementing scoping process 5. Placing time limits on the NEPA process
40 CFR Part 1502: Environmental Impact Statement	Provides requirements and instructions for preparing an environmental impact statement: 1. Purpose of an EIS 2. When an EIS is required? 3. When to prepare an EIS? 4. Interdisciplinary preparation requirement 5. Writing in plain English 6. Draft, final, and supplemental statements 7. EIS format and document requirements 8. Circulation of EIS 9. Tiering 10. Incorporation by reference 11. Incomplete or unavailable information 12. Cost–benefit analysis 13. Methodology and scientific accuracy 14. Environmental review and consultation requirements
40 CFR Part 1503: Commenting	Requires proposing agencies to invite comments from other federal agencies, state and local agencies, Indian tribes and the public, and specifies procedures for so doing: 1. Inviting comments 2. Duty to comment and specificity of comments 3. Responding to comments
40 CFR Part 1504: Predecision Referrals to the Council	Defines the procedure for referring interagency disagreements to the CEQ: 1. Criteria for making a referral 2. Procedures for making referrals
40 CFR Part 1505: NEPA and Agency Decision-making	Provides direction for implementing the agency's decision: 1. Agency decision-making procedures 2. Record of decision 3. Implementing the decision, mitigation, and monitoring
40 CFR Part 1506: Other Requirements of NEPA	Provides additional NEPA procedures and requirements: 1. Limitations on actions during NEPA process 2. Cooperating with state agencies

TABLE 2 (Continued)
Summary of Parts 1500–1508 of the CEQ NEPA Regulations (40 CFR Parts 1500–1508)

Part	Summary of Regulatory Direction
	3. Streamlining methods (adoption and combining documents)
	4. Applicants
	5. Public involvement
	6. Legislative EISs
	7. EIS filing requirements
	8. Applying NEPA in emergency situations
40 CFR Part 1507: Agency Compliance	Requires all federal agencies to comply with these Regulations, but states that agencies have flexibility in implementing NEPA:
	1. Compliance requirements
	2. Adopting agency-specific NEPA implementing procedures
40 CFR Part 1508: Terminology and Index	Defines 28 NEPA terms. Many of these definitions are critical in defining how to carry out NEPA procedures.

direction and guidance issued by the President and CEQ, which does not have the force of Regulation but is widely followed and sometimes cited in court. This direction is discussed in the following section.

As outlined in the companion book, *The NEPA Planning Process*, Tables 3 and 4 list supplemental regulations and guidance that affect NEPA practice. Table 3 lists the NEPA Act and CEQ Regulations and official direction. Table 4 lists other laws, regulations, executive orders, and guidance bearing on the implementation of NEPA.

The most widely cited supplemental guidance is the CEQ's *Forty Most Asked Questions*, which clarifies a number of important requirements.[40] Beyond the body of requirements and guidance provided in Tables 3 and 4 are various agency-specific regulations and guidance. For example, the Department of Energy has issued NEPA implementation regulations (10 CFR 1021), promulgated Secretarial Policy, and issued official orders and guidelines.

The reader is encouraged to read all such agency-specific materials prior to initiating NEPA practice for a given agency. Thus, NEPA practitioners are confronted with a large body of direction and guidance of both a general and agency-specific nature. One recently issued E.O. necessitates special mention.

Executive Order 13423

A recently issued executive order (E.O. 13423) builds upon and replaces earlier "Greening the Government" orders and promotes the goal of sustainable practices. The E.O. states that[41]

> … it is the policy of the United States that Federal agencies conduct their environmental, transportation, and energy-related activities … in an environmentally, economically and fiscally sound, integrated, continuously improving, efficient, and sustainable manner.

E.O. 13423 revokes and replaces five earlier E.O.s:

- E.O. 13101, Greening the Government through Waste Prevention, Recycling, and Federal Acquisition
- E.O. 13123, Greening the Government through Energy Efficient Management
- E.O. 13134, Developing and Promoting Biobased Products and Bioenergy
- E.O. 13148, Greening the Government through Leadership in Environmental Management
- E.O. 13149, Greening the Government through Federal Fleet and Transportation Efficiency

Under this E.O., federal agencies are required to implement sustainable practices consistent with the goals set forth in the E.O. These goals include improving energy efficiency and reducing

TABLE 3
NEPA Act and CEQ Regulations and Guidance

The National Environmental Policy Act of 1969, as amended (Public Law 91-190, 42 U.S.C. 4321–4347, January 1, 1970, as amended by Public Law 94-51, July 3, 1975, Public Law 94-83, August 9, 1975, and Public Law 97-258, § 4(b), September 13, 1982)

CEQ, Memorandum for Heads of Federal Agencies, Environmental Review Pursuant to Section 1424(e) of the Safe Drinking Water Act of 1974 and Its Relationship to NEPA, 19 November 1976

CEQ Memorandum for Heads of Agencies: Implementation of Executive Order 11988 on Floodplain Management and Executive Order 11990 on Protection of Wetlands, 21 March 1978

CEQ, Preamble to Proposed CEQ NEPA Regulations, 43 FR 25230, June 9, 1978

CEQ, Preamble to Final CEQ NEPA Regulations, 43 FR 55978, November 29, 1978

CEQ, NEPA Regulations, 40 Code of Federal Regulations, Pts. 1500–1508, 1978; as amended April 25, 1986, to remove the requirement for "worst case analysis" and substitute a process for analyzing "reasonably foreseeable" impacts (40 CFR 1502.22—Incomplete or Unavailable Information), 51 FR 15618

CEQ, Memorandum for NEPA Liaisons: Agency Implementing Procedures under CEQ's NEPA Regulations, 19 January 1979

CEQ, Prime and Unique Agricultural Land, *Federal Register*, Vol. 45, No. 175, September 8, 1980

CEQ, Memorandums for the Heads of Departments and Agencies: Wild and Scenic Rivers and National Trails, 2 August 1979

CEQ, Wild and Scenic Rivers, Memorandum, Interagency Consultations to Avoid or Mitigate Adverse Effects on Rivers in the Nationwide Inventory, 45 FR 59190, October 10, 1980

CEQ, Guidance on Applying Section 404(r) of the Clean Water Act to Federal Projects Which Involve the Discharge of Dredged or Fill Materials into Waters of the U.S., Including Wetlands, 17 November 1980

CEQ, Forty Most Asked Questions Concerning CEQ's National Environmental Policy Act Regulations, 46 FR 18026, March 23, 1981, amended 51 FR 15618, April 25, 1986

CEQ, Scoping Guidance, Memorandum for General Councils, NEPA Liaisons and Participants in Scoping, April 30, 1981, notice of availability published in 46 FR 25461, May 7, 1981

CEQ, Guidance Regarding NEPA Regulations, 48 FR 34263, July 28, 1983. Topics covered: Scoping; Categorical Exclusions; Adoption Process; Contracting Provisions; Selection of Alternatives in Licensing and Permitting Situations; and Tiering

CEQ, National Environmental Policy Act (NEPA) Implementation Procedures; Appendixes I, II and III, 49 FR 49750, December 21, 1984. Appendix I—Federal and Federal State Agency National Environmental Policy Act (NEPA) Contacts. Appendix II—Federal and Federal State Agencies with Jurisdiction by Law and Special Expertise on Environmental Quality Issues. Appendix III—Federal and Federal State Agency Offices for Receiving and Commenting on Other Agencies' Environmental Documents.

CEQ, Memorandum, Guidance on Clean Water Act, November 17, 1990

CEQ, Incorporating Biodiversity Considerations into Environmental Impact Analyses under NEPA, 1993

CEQ, Guidance on Pollution Prevention and the National Environmental Policy Act, published in 58 FR 6478, January 29, 1993

CEQ, The National Environmental Policy Act: A Study of Its Effectiveness after Twenty-Five Years, January 1997

CEQ, Considering Cumulative Impacts under the National Environmental Policy Act, January 1997

CEQ, Guidance on NEPA Analyses for Transboundary Impacts, July 1, 1997

CEQ, Environmental Justice: Guidance under the National Environmental Policy Act, 10 December 1997

CEQ, Memorandum for Heads of Federal Agencies: Designation of Non-Federal Agencies to be Cooperating Agencies in Implementing the Procedural Requirements of NEPA, 28 July 1999

CEQ, Memorandum for Deputy/Assistant Heads of Federal Agencies: Identifying Non-Federal Cooperating Agencies in Implementing the Procedural Requirements of the National Environmental Policy Act, 25 September 2000

CEQ, Memorandum for Heads of Federal Agencies: Cooperating Agencies in Implementing the Procedural Requirements of the National Environmental Policy Act, 30 January 2002

CEQ, Memorandum to Secretary of Agriculture and Secretary of Interior: Guidance for Environmental Assessments of Forest Health Projects, 9 December 2002

CEQ, Exchange of Letters with Secretary of Transportation: Purpose and Need, May 2003 (Parts 1 and 2)

CEQ, Memorandum to Heads of Federal Agencies: Reporting Cooperating Agencies in Implementing the Procedural Requirements of the National Environmental Policy Act, 23 December 2004 (Attachment 1: Cooperating Agency Report Form; Attachment 2: Cooperating Agency Frequently Asked Questions)

CEQ, Guidance on the Consideration of Past Actions in Cumulative Effects Analysis, 24 June 2005

CEQ, Memorandum for Federal NEPA Contacts: Emergency Actions and NEPA, 8 September 2005 (Attachment 1: Emergency Alternative Arrangements Under the National Environmental Policy Act; Attachment 2: Preparing Focused, Concise and Timely Environmental Assessments)

CEQ&OMB, Memorandum on Environmental Conflict Resolution, 28 November 2005

TABLE 4
Additional Statutes, Regulations, Guidance, and Executive Orders Pertaining to the Implementation of NEPA

Directive	Reference
The Environmental Quality Improvement Act of 1970; Title II. Environmental Quality (of Water Quality Improvement Act of 1974)	Public Law 91-224, 42 U.S.C. 4371-4374, April 3, 1970.
Clean Air Act of 1970 § 309	Public Law 91-604 § 12(a), 42 U.S.C. § 7609, 1970
Floodplain management	Executive Order 11988, May 24, 1977 (appears at 42 FR 26951, 3 CFR, 1977)
Coastal zone	NOAA, Federal Consistency with Approved Coastal Management Programs, 15 CFR 930
Protection of wetlands	Executive Order 11990, May 24, 1977 (appears at 42 FR 26961, 3 CFR, 1977)
Protection and enhancement of environmental quality	Executive Order 11514, March 5, 1970 (as amended by Executive Order 11991, Section (2 g) and 3(h), May 24, 1977)
Federal compliance with pollution control standards	Executive Order 12088, October 13, 1978 (appears at 43 FR 47707, 3 CFR, 1978)
Review of EISs	40 CFR 1.37 and 1.39 per § 309 of Clean Air Act
Environmental effects abroad of major federal actions.	Executive Order 21114, January 4, 1979 (appears at 44 FR, 1957, 3 CFR, 1979)
Wild and scenic rivers and national trails	Memorandums from the President Jimmy Carter, Memorandum for the Heads of Departments and Agencies, "Wild and Scenic Rivers and National Trails," August 2, 1979
Incomplete and unavailable information	51 FR, 80, April 25, 1986
Federal actions to address environmental justice in minority populations and low income	Executive Order 12898, February 11, 1994 (appears at 59 FR, 32, February 16, 1994).
Greening the government	Executive Order 13423, Strengthening Federal Environmental, Energy, and Transportation Management, January 24, 2007

greenhouse gas emissions, reducing water consumption intensity, and maintaining cost-effective waste prevention and recycling programs. This order also requires federal agencies to establish environmental management systems (EMSs) to use as a principal mechanism for managing environmental aspects of agency operations.

The E.O. also establishes new responsibilities for CEQ, the Office of Management and Budget, and the Office of the Federal Environmental Executive in overseeing its implementation. Under the E.O., the

- CEQ is to convene a steering committee that includes senior executives designated by the agencies;
- Office of Management and Budget is to issue instructions to the agencies on agency self-evaluation of E.O. implementation;
- Office of the Federal Environmental Executive (within the Environmental Protection Agency) is to monitor agency performance under the E.O., advise CEQ on progress, and submit a biannual report to the president.

PROBLEMS WITH NEPA

Nearly four decades after NEPA was enacted, environmental protection has generally become a widely accepted social goal. Yet, as described earlier, the original model for NEPA has not been a complete success. An unbalanced emphasis on the procedural requirements presented in

Section 102 has sometimes led to an overfocus on compliance at the expense of environmental protection. As Lynton Keith Caldwell has stated:

> The EIS requirement alone is insufficient to achieve the intent declared in NEPA.... The goals and principles declared in Section 101 have been treated as noble rhetoric having little practical significance.[38]

Dr. Caldwell expressed disappointment with NEPA and what it has achieved. Caldwell believed that NEPA had fallen short of its original policy goals. In his view, NEPA has had only a marginal effect on federal agency programs, been hijacked by lawyers, and does not adequately integrate environmental factors into federal decision-making. He believed that NEPA's "substantive" policy requirements are largely ignored by federal agencies and are underenforced by the courts.[42] In concluding that NEPA's promise is yet to be fulfilled, Caldwell wrote:

> The goals declared in NEPA are as valid today as they were in 1969, perhaps more so. NEPA's purpose, "was never the writing of impact statements, but this action-forcing procedure has been a great inducement to ecological rationality in Federal actions, which traditionally have largely ignored environmental consequences."[40]

Yet, among the optimists, the author included, NEPA has had an immeasurable effect in the way it has shaped federal decision-making and bureaucratic behavior. NEPA has opened government decision-making to the "light of day," integrated environmental specialists into federal planning, mandated that an array of alternatives be rigorously considered, provided once disaffected segments of the public with a mechanism for injecting input into federal decision-making, and greatly improved a vast array of decisions which are made on an yearly basis. From the viewpoint of an optimist, this process can only improve, perhaps someday even achieve, the lofty goals Caldwell envisioned.

DELAYS

Some of the commonly cited factors preventing timely completion of NEPA analyses include

- disagreement over the need to prepare an EA or EIS, or failure to start the NEPA process in the early planning stage;
- new circumstances or information which delays the process;
- evolving specifications (i.e., the technical specifications of the project were evolving);
- change in project management;
- insufficient review time;
- categorically excluded actions that were not implemented; and
- funding delays.

Specific methods for reducing some of these delays are addressed in later chapters.

NEPA IMPROVEMENT DIRECTION

Although a university-sponsored study is now a decade old, the continuing focus within the NEPA community (and among critics) on making NEPA documents more comprehensible and useful suggests that there is still need for substantial improvement.[43] A recent report prepared by the Federal Highway Administration in conjunction with the American Association of State Highway and Transportation Officials and the American Council of Engineering Companies provides guidance on preparing better NEPA documents.[44] The report recommends three basic principles to be followed in preparing readable and effective NEPA documents:

> *Principle 1: Tell the story of the project* so that the reader can easily understand the purpose and need ... how each alternative would meet [it], and the strengths and weaknesses associated with each alternative.

Principle 2: Keep the document as brief as possible, using clear, concise writing; an easy-to-use format; effective graphics and visual elements; and discussion of issues and impacts in proportion to their significance.

Principle 3: Ensure that the document meets all legal requirements in a way that is easy to follow for regulators and technical reviewers.

The top issue identified by the group was the unwieldy length and complexity of NEPA documents. To manage document length, the report recommends, applying CEQ guidance:[45]

> ... if only technically trained individuals are likely to understand a particular discussion then it should go in the appendix, and a plain language summary of the analysis and conclusions of that technical discussion should go in the text of the EIS.

While many practitioners may not completely concur, the report endorses combining the discussions of affected environment and environmental consequences in a single chapter to provide an integrated discussion of environmental issues that are important to the proposal and how each alternative affects them.

The report encourages use of headings that use a "question-and-answer" format and which provide context and direct readers to information that is most pertinent to their particular concerns.

To ensure compliance with legal requirements, the report recommends that an EIS demonstrates compliance with key regulatory requirements by listing these requirements, explaining which are applicable, and describing how these have been met (e.g., Section 7 of the Endangered Species Act; Section 106 of the National Historic Preservation Act; Section 404 of the Clean Water Act).

NEPA FROM A CITIZEN'S PERSPECTIVE

Improvement methods have all too frequently centered on "fixing" NEPA at the expense of increasing citizen ownership and making issue resolution a centerpiece for achieving a consensus. In addition to the problems typically described in the literature, Preister describes the following problems from a citizen's perspective that also needs to be considered:[46]

1. *Insufficient scoping*: Practices such as mailings, meetings, or performing scoping too early in the process to solicit meaningful public input are often inadequate to prevent controversy and ensuing conflict. They also frequently fail to identify the full range of interests of the affected population.
2. *Over-reliance on meetings*: Meetings are often unreliable forums when used as the sole or primary means of assessing public interest. Meetings can often establish an atmosphere of polarization because they are used as an opportunity to advocate particular interests rather than as a vehicle for mutual problem solving. Special interests often capture the public process and undermine moderate voices. Moreover, meetings can attract organized groups or renegade voices that do not reflect the broader and frequently more pragmatic views of the community at large. Moderate voices often complain that public meetings do not provide a safe environment and that they are emotionally attacked within public meetings.
3. *Issue stacking*: Issues obtained through scoping are often merely catalogued and saved for later analysis rather than actively addressed and resolved. This can allow controversy to simmer and can provide a route for special interests to influence the process, thereby hardening positions and contributing to further entrenchment. Problem solving opportunities for dialogue, early issue resolution, and building public trust or support are frequently marginalized.
4. *Lack of citizen ownership*: All too often, agencies are driven by project momentum (internal deadlines, constraints posed by other projects, or decisions that have already been made by management). This results in situations where public participation is compromised.

5. *Lack of commitment to community-based solutions*: Agencies need to make an active effort to establish partnerships with special interests and the community at large. This normally requires openness to compromise in a give-and-take atmosphere. Elected officials and federal agency representatives need to reach out in a collaborative effort to work with potentially affected parties. Failure to achieve such relationships can result in conflict, resentment, distrust, and ultimately delays and litigation.

UNIQUE BENEFITS OF NEPA

Some of the benefits that the NEPA planning process can bring the project manager and decision-maker are described in the following sections.

No PERFORMANCE STANDARDS

Some critics have criticized NEPA for consuming resources, yet not mandating specific project performance standards, as do other regulatory and permitting requirements. Yet, the subject agency is still burdened with NEPA's additional procedural requirements.

In reality, what at first appears to be a liability can, in fact, work in the agency's favor. From the standpoint of a project engineer, NEPA is actually one of the most favorable Regulations. As described in the following section, NEPA provides decision-makers and project engineers with a highly flexible planning process while not burdening project proponents with unwieldy requirements as do many other permitting and regulatory requirements.

A FLEXIBLE PROCESS

In contrast to most other environmental laws, NEPA allows an agency to include other factors, such as public sentiment, safety, risk, cost, and schedules, in reaching a final decision.

NEPA imposes no straightjacket on federal decision-makers. Instead, it provides a systematic mechanism for reaching decisions. In questioning the wisdom of a proposal, the best course of action occasionally turns out to be no-action. However, in the majority of cases, a proposal is implemented, although with some modifications that reduce or even eliminate the impacts altogether.

Moreover, NEPA allows agencies to consider alternatives that lie outside their normal jurisdiction. Thus, the analysis can provide the agency or agencies concerned with a rigorous and publicly reviewed basis for seeking a change in an existing law or requirement so that a more sensible or appealing alternative may be pursued.

A TWO-STAGE NEPA/PERMITTING PROCESS

Impending permitting requirements can significantly affect the cost and schedule of a project. An EIS must identify and list all permits, licenses, and other entitlements that must be obtained in implementing an action. Identifying these requirements can be instrumental in allowing the agency to assess and plan for potential permitting requirements during the early planning stage. Thus, NEPA provides planners with the means of reducing surprises and unforeseen risks by ensuring that permitting requirements have been correctly identified and planned for.

NEPA, in combination with regulatory permitting processes, provides a two-pronged approach for protecting the environment (see Figure 1). In the first stage, NEPA can assist an agency in planning future actions so as to avoid impacts that would trigger future lengthy and restrictive permitting requirements. An agency must also consider potential mitigation measures as part of the NEPA process. Under this approach, NEPA can actually act as a tool for streamlining project implementation.

Alternatives and mitigation measures are particularly important aspects of NEPA. Properly planned and implemented, alternatives and mitigation measures can provide a useful tool for

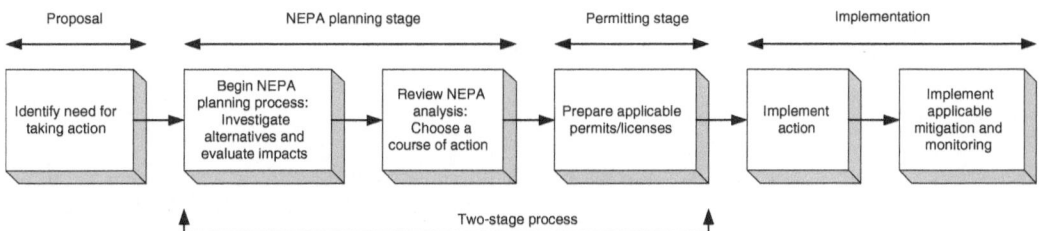

FIGURE 1 Typical two-stage approach: NEPA planning followed by applicable permitting. In actuality some permitting processes may actually run concurrently with NEPA.

reducing costs and project risks such as schedule delays; by incorporating mitigation into the early planning, an agency may avoid subsequent impacts or reduce them to the point where future permits, licenses, and other entitlements can be curtailed and in some cases eliminated altogether.

During the second stage, the agency can control impacts that could not be eliminated through early planning, via permitting and imposing strict regulatory standards (in many instances, some permitting processes may actually run concurrently with NEPA). Thus, a two-stage approach provides a much higher level of protection. Moreover, decisions, mitigation measures, and commitments made during the NEPA planning process can be made part of any permits that are later issued. Under this approach, the subsequent permitting process provides the 'muscle' for ensuring that plans and decisions made during the NEPA process are enforced.

Reducing Project and Noncompliance Risks

The *Aberdeen* court case clearly demonstrated that federal and nonfederal employees can be prosecuted for violating environmental statutes and that the federal government is not obliged to pay for the employee's legal defense.[47] NEPA provides planners and project managers with a means for identifying and minimizing these risks. Applied proactively, NEPA provides an effective tool for identifying and planning environmental compliance over a project's life cycle, reducing long-term project risks and schedule delays, as well as the risk of civil and criminal liability associated with environmental noncompliance.[48]

For example, the NEPA analysis can assist project managers in identifing, planning for, and avoiding violations of the Clean Air Act, Clean Water Act, Resource Conservation and Recovery Act, and National Historic Preservation Act, as well as an array of other environmental statutes that carry civil and criminal penalties.

STRENGTHENING NEPA AND ENVIRONMENTAL PLANNING COMPETENCY

As a final note, the National Association of Environmental Professionals (NAEP) is a multidisciplinary, professional association dedicated to the promotion of ethical practices, technical competency, and professional standards in the environmental fields. It produces a respected peer-reviewed journal. The organization also sponsors a widely acclaimed national environmental conference on an annual basis. In 1977, it initiated an environmental certification program (certified environmental professional [CEP]), which after 14 years was transferred to a subsidiary body—The Academy of Board Certified Environmental Professionals (ABCEP)—which continues to certify environmental professionals.

The NAEP has a membership of over 1000 members and reaches over 5000 environmental practitioners in the United States and around the world. It also has local and regional chapters, and its membership reflects a diversity of employers, including government, industry, utilities, academia, consulting firms, and the private sector in the United States and abroad. Its membership is particularly strong in the disciplines of NEPA and environmental planning; those who are interested in strengthening their professional NEPA and environmental competency are encouraged to join this national organization. The NAEP Web site is www.naep.org.

DISCLAIMER

The author strongly encourages the reader to consult the actual regulatory provisions for details and precise wording of the Regulations cited. Additionally, agency-specific NEPA implementation procedures and agency orders should also be checked for provisions that may provide more specific requirements beyond those presented in the CEQ's NEPA regulations.

The author apologizes for any mistakes or inaccuracies that may appear in this book. The author furthermore stresses the importance of seeking the advice of NEPA and regulatory specialists, independent experts, and legal counsel, particularly in areas involving complex or controversial issues.

PROBLEMS

1. Which senator introduced the NEPA statute in the Senate?
2. With respect to NEPA, what is a "detailed statement"?
3. What is the importance of the *Calvert Cliffs* court case?
4. Who has been called the "principal father of NEPA"?
5. What was the purpose for establishing NEPA?
6. Briefly explain why NEPA is referred to as "world's most emulated law."
7. What section of NEPA is known as the "spirit of the law"?
8. What section of NEPA is known as the "letter of the law"?
9. What are the principal responsibilities of the CEQ?
10. What does 40 CFR Part 1502 cover?
11. What is the importance of the *Aberdeen* court case?
12. Who was Dr. Lynton Caldwell?

REFERENCES

1. Goldberg P., *The Babinski Reflex*, Jeremy P. Tarcher, Inc., Los Angeles, CA, 1990.
2. Section 301(c) of CERCLA, Natural Resource Damages Assessments.
3. 40 CFR 1500.5, 1501.2, and 1506.4.
4. 40 CFR 1501.2, 1501.7, and 1502.14(a).
5. DOE, *Management of Spent Nuclear Fuel from the k-Basins at the Hanford Site, Richland, Washington*, Final, DOE/EIS-0245 (61 FR 3922), 1996.
6. Eric G., Personal communications, 1999.
7. Eccleston C.H., *The NEPA Planning Process: A Comprehensive Guide with Emphasis on Efficiency*, Introduction, John Wiley & Sons, New York, 1999.
8. 10 CFR Part 52 "Early site permits, standard design certifications, and combined licenses for nuclear power plants."
9. O'Neil D., Project Chariot: how Alaska escaped nuclear excavation. *The Bulletin of the Atomic Scientist*, December 1989, 35.
10. Caldwell L.K., *Environment: A New Focus for Public Policy*, Public Administration Review, 23, 1963, 138.
11. *National Environmental Policy Act of 1969: Hearings on S. 1075, S. 237 and S. 1752 before the Senate Comm. on Interior and Insular Affairs*, 91st Congress, 116, 1969.
12. Implementing NEPA: A Non-Technical Political Task DOE Conference: NEPA 25, March 21, 1995.
13. *Hearings on S. 1075, S. 237, and S. 1752 before the Senate Committee on Interior and Insular Affairs*, 91st Congress, 1st Session, 206, 1969; Carl B., A simple formula for crafting better NEPA documents, *Federal Facilities Environmental Journal*, Autumn 1992.
14. 115 Congressional Record 29053 (1969); Yost, N.C. and Rubin J.W., The National Environmental Policy Act, unpublished.
15. Senate Debate on the Conference Report to S. 1075, To Establish a National Policy for the Environment. Congressional Record, December 20, 1969, p. S. 17451.

16. *Calvert Cliffs Coordinating Committee v. AEC*, 449 F.2d 1109, 1117 (D.C. Cir. 1971).
17. Eccleston C.H., *The NEPA Planning Process: A Comprehensive Guide with Emphasis on Efficiency*, Chapter 1, John Wiley & Sons, New York, 1999.
18. Executive Order 11991, *Protection and Enhancement of Environment Quality,* May 24, 1977.
19. Preamble to Final CEQ NEPA Regulations, 43 *Federal Register* 55978, Section 1, November 29, 1978.
20. Preamble to Final CEQ NEPA Regulations, 43 *Federal Register* 55978, Section 3, November 29, 1978.
21. Preamble to Final CEQ NEPA Regulations, 43 *Federal Register* 55978, Section 6, November 29, 1978.
22. *Andrus v. Sierra Club*, 442 U.S. 347 (1979).
23. Federal Highway Administration NEPA Regulations, 23 C.F.R. § 771.119, 2005.
24. Personal communications with Professor Lynton Caldwell. Also, Yost N.C., Don't underline but streamline, *The Environmental Forum*, 41, May/June 2005.
25. Yost N.C., Testimony before the Committee on Resources United States House of Representatives, *Hearing on NEPA: Lessons Learned and Next Steps*, November 17, 2005.
26. Council on Environmental Quality, *The National Environmental Policy Act: A Study of Its Effectiveness after Twenty-Five Years*, 1997, p. 3.
27. Canter L.W. *Environmental Impact Assessment*, Second Edition, McGraw-Hill Inc., New York, 1996.
28. OECD, *Good Practices for Environmental Impact Assessment of Developing Projects*, Development Assistance Committee, Paris, 1992.
29. UNEP, *Environmental Impact Assessment: Basic Procedures for Developing Countries*, Regional Office for Asia and the Pacific, Bangkok, 1988.
30. World Bank, *Environmental Impact Assessment Sourcebook* (Vols. 1–3), World Bank, Washington, D.C., 1991.
31. European Economic Community, Council Directive 85/337/EEC on the assessment of the effects of certain public and private projects on the environment. O.J. European Communities No. L 175, 1985, p. 40.
32. Caldwell L.K. *The National Environmental Policy Act: An Agenda for the Future*, Indiana University Press, Bloomington, IN, 1998, p. 209.
33. Spensley J.W., *National Environmental Policy Act*, Environmental Law Handbook 407, Sullivan T.F.P. (ed.), Fourteenth Edition, Government Institutes, Inc., Rockville, MD; *See also* Caldwell L.K., Beyond NEPA: Future Significance of the National Environmental Policy Act, *The Harvard Environmental Law Review*, 1998, 203–239.
34. DOE, *NEPA Lessons Learned*, March 1, Issue No. 50, 2007.
35. Report on the Plan for Transformation of the National Nuclear Security Administration Nuclear Weapons Complex, January 31, 2007.
36. Cronin J. and Kennedy R.F., Jr., *The Riverkeepers*, Scribner, New York, 1997, pp. 37 and 175.
37. P.L. 91-190, as amended, 42 U.S.C. 4321 et seq.
38. Caldwell L.K., *Environment, A Challenge to Modern Society*, Doubleday & Company, Inc., Garden City, NY, 1971.
39. Davies J.C. and Lettow C.F., The Impacts of Federal Institutional Arrangements, Environmental Law Institute, Federal Environmental Law 126, 133, 1974.
40. CEQ, *Forty Most Asked Questions Concerning CEQ's National Environmental Policy Act Regulations*, 46 FR 18026, March 23, 1981, amended 51 FR 15618, April 25, 1986.
41. E.O. 13423, *Strengthening Federal Environmental, Energy, and Transportation Management*, January 24, 2007.
42. Caldwell L.K., *The National Environmental Policy Act: An Agenda for the Future*, Indiana University Press, Bloomington, IN, 1999.
43. Assessing the impact of environmental impact statements on citizens, *Environmental Impact Assessment Review*, 16(3), May 1996, pp. 171–182.
44. Federal Highway Administration. Improving the Quality of Environmental Documents, May 2006.
45. Council on Environmental Quality, *Forty Most Asked Questions*, Question Number 25.
46. Preister K. and Kent J.A., Using Social Ecology to Meet the Productive Harmony Intent of the National Environmental Policy Act (NEPA), Copyright 2001 by Social Ecology Associates and James Kent Associates. This article is also available on the Internet at http://www.naturalborders.com.
47. *United States v. Dee*, 912 F.2d 741 (4th Cir. 1990).
48. Lillie T.H. and Lindenhofen H.E., NEPA as a tool for reducing risk to programs and program managers, *Federal Facilities Environmental Journal*, Spring 1991.

1 Overview of the NEPA Process and Its Basic Requirements

In his book, *Life on the Mississippi*, Mark Twain humorously described how misunderstandings of certain facts and observations can lead to incorrect, misleading, or even bizarre conclusions:

> In the space of one hundred and seventy-six years the Lower Mississippi has shortened itself two hundred and forty two miles. That is an average of a trifle over one mile and a third per year. Therefore, any calm person, who is not blind or idiotic, can see that in the Old Oolitic Silurian Period, just a million years ago next November, the Lowest Mississippi River was upward of one million three hundred thousand miles long, and stuck out over the Gulf of Mexico like a fishing rod. And by the same token any person can see that seven hundred and forty-two years from now the Lower Mississippi will be only a mile and three quarters long, and Cairo and New Orleans will have joined their streets together and be plodding comfortably along under a single mayor and a mutual board of aldermen. There is something fascinating about science. One gets such wholesale returns of conjecture out of such a trifling investment of facts.

It is partly because of irrational assessments (and decisions) such as this that Senator Henry "Scoop" Jackson and others believed that a scientifically based approach to planning federal actions was sorely needed. This chapter provides an overview of the three fundamental levels of National Environmental Policy Act (NEPA) compliance and describes the basic principles underlying its implementation. Each of the three levels comprising this process is described individually and in more detail in subsequent chapters. For more information on the three levels of NEPA compliance, the reader is directed to the companion books, *The NEPA Planning Process, Effective Environmental Assessments*, and *Environmental Impact Statements*.[1–3]

Before delving into the details of NEPA, it is instructive to present a succinct overview of the entire NEPA planning process. The following section provides a simplified description of the process. It is important to note that minor variations to the description presented herein can and do exist particularly with respect to how individual agencies choose to implement specific details of their respective NEPA processes.

1.1 OVERVIEW OF THE NEPA PLANNING PROCESS

We begin our journey by first defining some fundamental terms. The phrase "NEPA process" means

> The measures necessary for compliance with the requirements of Section 2 and Title I of the NEPA Act (§ 1508.21).

For brevity, the acronym "NEPA" is abbreviated to "Act" throughout this book. Similarly, the term "Council on Environmental Quality (CEQ) NEPA Regulations" is shortened to "Regulations." References to a particular section of these Regulations (40 Code of Federal Regulations [CFR] § 1500–1508) are shortened so that they simply cite the specific section number in the Regulations where the provision is found. Thus, a reference to "40 CFR § 1502.22(b)(2)" is shortened to the more expedient expression: § 1502.22[b][2]. Likewise, the term "proposal" is interpreted to mean the set of "reasonable alternatives," including a no-action alternative (when applicable), and the proposed action.

1.1.1   THREE LEVELS OF NEPA COMPLIANCE

The NEPA planning process can be conceptualized as consisting of three distinct levels of environmental review. Presented from the simplest through the most demanding, these three levels of compliance are as follows:

- *Categorical exclusion (CATX)*: A proposed action is reviewed to determine if it qualifies for a CATX; if so, the action is excluded from further NEPA review and documentation requirements.
- *Environmental assessment (EA)*: An EA is normally prepared to determine if a proposed action qualifies for a finding of no significant impact (FONSI). If an action qualifies for a FONSI, it is exempted from the requirement to prepare an environmental impact statement (EIS).
- *Environmental impact statement (EIS)*: An EIS is most typically prepared for circumstances where a federal action does not qualify for either a CATX or a FONSI.

Until it can be demonstrated that a proposed action is not a major federal action significantly affecting the environment, it should be considered subject to the full requirements of NEPA including, potentially, the requirement for an EIS. However, in practice, many actions can be quickly exempted from the requirement of preparing an EIS by demonstrating that they qualify for either a CATX or a FONSI. A simplified overview of the entire NEPA process is portrayed in Figure 1.1.

1.1.2   COMPLIANCE WITH OTHER LAWS

A common misconception among the uninitiated is that NEPA does not apply to activities that already comply with all other environmental laws, regulations, permits, and standards. This misconception was first considered in the historical case of *Calvert Cliffs* where the court found that compliance with water discharge limitations still did not excuse the agency from NEPA's requirement to evaluate the consequences and alternatives to the proposal.[4] The court's conclusion, reinforced by numerous subsequent decisions, makes sense when one considers the fact that even though a project complies with all other applicable laws and regulations, it may still result in a major significant impact. For example, alternatives or mitigation measures often exist that, if considered, may be found to reduce environmental damage substantially below the level that would occur if an action is merely required to meet minimal regulatory standards. Moreover, unlike most other environmental regulations, NEPA provides a comprehensive process for planning subsequent actions.

1.1.2.1   No "Grandfathering"

In the sense that this term is normally used, NEPA does not recognize the concept of "grandfathering." That is to say that facilities, operations, or other activities that existed or were ongoing prior to NEPA's enactment are not excused from compliance with the Act. However, such activities are not generally considered subject to the requirements of NEPA until there is a proposal involving a new action or a change in an existing facility or operation.

1.1.3   ADOPTING NEPA IMPLEMENTATION PROCEDURES

Agencies are instructed to adopt individualized NEPA implementation procedures that supplement the Regulations, providing additional direction tailored to meet the agencies' individual needs (§ 1505.1, § 1507.1, § 1507.2). Implementation procedures must be adopted within a period of 5 months after the establishment of a new agency.

Agencies are expected to periodically review their policies and procedures in consultation with the CEQ and to revise them as necessary (§ 1507.3). Selected excerpts related to this requirement are presented in Table 1.1.

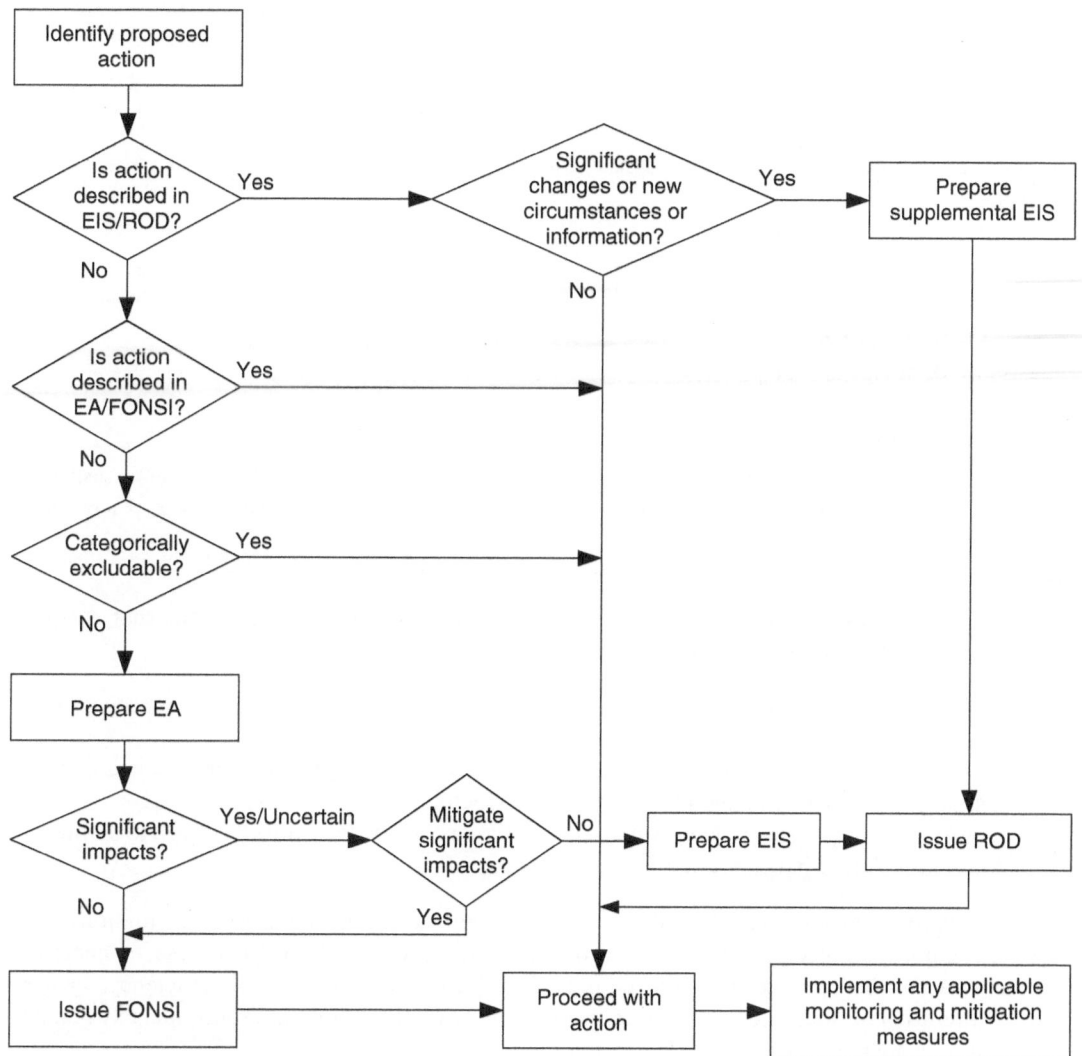

FIGURE 1.1 Typical NEPA process.

The NEPA implementation procedures must identify three categories of NEPA review that are normally performed on typical classes of actions. These categories include classes of actions that normally (§ 1507.3[b][2])

- require an EIS,
- require an EA, but not necessarily an EIS, and
- do not require either an EIS or an EA (i.e., CATX is appropriate).

Some representative examples promulgated by the Department of Energy (DOE) for classes of actions that normally require EAs, but not necessarily EISs,[5] are:

- C4: Upgrading (reconstructing) an existing transmission line
- C8: Protecting fish and wildlife habitat
- C14: Siting, constructing, and operating water treatment facilities

TABLE 1.1

Adopting NEPA Implementation Procedures

Policy (§ 1500.2)

> Federal agencies shall to the fullest extent possible … (b) implement procedures to make the NEPA process more useful to decision-makers and the public …

Agency Decision-Making Procedures (§1505.1)

> Agencies shall adopt procedures … to ensure that decisions are made in accordance with the policies and purposes of the Act …

Public Involvement (§ 1506.6)

> Agencies shall: (a) make diligent efforts to involve the public in preparing and implementing their NEPA procedures …

Procedure Making (§ 1507.3)

> … each agency shall as necessary adopt procedures to supplement these regulations … Such procedures shall not paraphrase these regulations …. Each agency shall consult with the Council while developing its procedures and before publishing them in the *Federal Register* for comment.

Some representative examples promulgated by the DOE for classes of actions that normally require preparation of an EIS are:[5]

- D2: Siting, constructing, operating, and decommissioning of nuclear fuel reprocessing facilities
- D4: Siting, constructing, operating, and decommissioning of power reactors, nuclear material production reactors, and test and research reactors
- D5: Adding main transmission system (i.e., additions of new transmission lines) to a Power Marketing Administration's main transmission grid

One relatively obscure regulatory requirement requires special note. The Regulations grant agencies authority to specify time periods, other than those cited in § 1506.10, if these are necessary to comply with other specific statutory requirements (§ 1507.3[d]). To promote efficiency, agencies should carefully consider this provision in terms of meeting their mission requirements when preparing or revising their NEPA implementation procedures.

An agency's NEPA implementation procedures must be prepared in conformance with both the Act and CEQ's NEPA Regulations. The intent of the direction provided in the implementation procedures is not to paraphrase provisions already laid out in the Regulations, but to supplement the direction provided in them (§ 1507.3). Detailed requirements for preparing and adopting implementation procedures are provided in § 1507.3.

Agencies are expected to consult the CEQ while developing their implementation procedures and before the procedures are published in the *Federal Register* for comment. Both the public and the CEQ must be given the opportunity to review the procedures prior to their adoption. The final procedures must be filed with the CEQ and made available to the public (§ 1507.3).

1.1.4 INITIATING NEPA

The NEPA process must begin early enough so that it can contribute to the decision-making process and does not simply justify decisions already made. This timing requirement has generally been met if the NEPA process is initiated early enough to meet the decision deadline, yet not so early that it cannot meaningfully contribute to the decision-making process (§ 1502.5).

It is important to note that the Act contains no expressed provisions exempting any federal agency from complying with NEPA. Nor does the Act specifically exempt any specific type of federal activity from NEPA's requirements. However, unlike many other environmental laws, the Act provides no explicit penalties for failing to comply. Nevertheless, an agency can be sued for noncompliance under the Administrative Procedures Act (APA, see Section 1.3).

1.1.4.1 Identifying a Proposed Action

The NEPA process is typically initiated when a need to take federal action has been identified (see box labeled "Identify proposed action," Figure 1.1). Note that a limited number of circumstances exist in which a proposal is subject to NEPA's requirements. The reader is referred to the companion book, *The NEPA Planning Process*, for a detailed review of circumstances under which an action may be exempted from these requirements.[6]

Generally, no additional NEPA review is required if a proposal is already adequately described within an existing EIS or EA.

1.1.4.2 Supplemental EISs and the Smithsonian Solution

If an action is described in an existing EIS (see diamond labeled "Is action described in EIS/ROD?," Figure 1.1), a review must be performed to determine if there are any new circumstances or information related to potentially significant impacts (see diamond labeled "Significant changes or new circumstances or information?," Figure 1.1).

Consistent with the rule of reason, Chapter 8 presents a tool referred to by the author as the Smithsonian Solution for determining if new circumstances or information is substantial enough to require preparation of a supplemental EIS. If there are potentially significant new circumstances or information, a supplemental EIS may be required. If there are no new circumstances or information, the agency may proceed with the action (see box labeled "Proceed with action," Figure 1.1).

1.1.4.3 Reviewing Existing EAs

A similar review is prepared for actions that are covered in an existing EA. However, there is generally no mechanism equivalent to the supplemental EIS for actions that are covered in an existing EA. If the action is not adequately covered in an EA, the "No" path is taken to the next diamond labeled "Categorically excludable?."

1.1.5 CATEGORICAL EXCLUSIONS

One of the most effective streamlining methods is the CATX. A CATX is simply a class of actions that have been previously reviewed by an agency, and determined not to result in either an individual or cumulatively significant impact, and are therefore exempt from the requirement to prepare an EA/EIS (§ 1508.4). The CEQ regulations require each federal agency to prepare a list of CATXs as part of its individual NEPA implementation procedures.

When initiating the NEPA process, a review should generally be performed to determine if the proposed action already falls within one of the agency's existing CATXs (see diamond labeled "Categorically excludable?," Figure 1.1). If the proposal is eligible for a CATX, the NEPA review process has been satisfied and no additional review requirements are necessary; the agency may proceed with the action with respect to NEPA's requirements. If the action is not eligible for a CATX, the logic pursues the "No" path to the box labeled "Prepare EA."

1.1.6 ENVIRONMENTAL ASSESSMENTS

To save time and resources, the agency may elect to prepare an EIS without first preparing an EA (this option is not shown in the simplified flowchart). Conversely, an agency may choose to go ahead and prepare an EA even if it plans to eventually prepare an EIS for the proposal.

An agency's NEPA implementing procedures contain categories or types of actions that normally require either an EA or EIS. These procedures should be consulted for guidance in determining if an action normally requires preparation of an EA or EIS.

As just described, an agency may choose to first prepare an EA to determine if the impacts are potentially significant (see box labeled "Prepare EA," Figure 1.1). On completing the EA, the assessment is reviewed by the decision-maker to determine if it could indeed result in a significant impact (see decision diamond labeled "Significant impacts?," Figure 1.1). If it can be demonstrated definitively that no significant impacts would occur, the action qualifies for a FONSI (see box labeled "Issue FONSI," Figure 1.1).

A FONSI can also be issued if the impacts are potentially significant, but can be mitigated to the point of nonsignificance (see decision diamond labeled "Mitigate significant impacts?," Figure 1.1). If the potentially significant impacts cannot be mitigated to the point of nonsignificance, an EIS must be prepared.

If all potential impacts are nonsignificant or can be mitigated to the point of nonsignificance, a FONSI can be issued. Once a FONSI is issued, the agency is free to proceed with the proposed action, subject to any applicable mitigation or monitoring measures. Chapter 7 describes the EA process in detail. For more information on preparing EAs, the reader is referred to the companion text, *Effective Environmental Assessments*.[7]

1.1.6.1   Commonly Required Data

Commonly required data needs for both an EA and EIS include

- project design and engineering data;
- environmental protection standards;
- land use including facilities, infrastructure, and rights-of-way;
- geology and soils;
- surface water and groundwater hydrology (water quality and quantity);
- air quality;
- habitat and species;
- threatened and endangered species;
- wetlands;
- socioeconomics, tribal lands; and
- archaeological resources.

1.1.6.2   Comparison of the Size and Complexity of EAs versus EISs

Most agencies develop EAs that depending on the scope can vary considerably in size and complexity. Small EAs typically[8]

- are developed by one author,
- take from 2 weeks to 2 months to complete,
- vary from 10 to 30 pages in length, and
- cost between $5000 and $20,000.

Large EAs tend to be associated with more controversial, large, or relatively complex projects and are more similar to an EIS in analysis, content, and format. Most mitigated FONSIs are also associated with relatively large EAs. Large EAs typically

- are developed by an interdisciplinary team,
- take from 9 to 18 months to complete,

- vary from 50 to more than 200 pages in length, and
- cost between $50,000 and $200,000.

In contrast, EISs typically

- are developed by an interdisciplinary team,
- take from 1–2 years (sometimes more than 5 years) to complete,
- vary from 200 to more than 2000 pages in length, and
- cost between $250,000 and $2,000,000.

1.1.7 ENVIRONMENTAL IMPACT STATEMENTS

An EIS must be prepared if a proposal could result in a significant impact or if any significant impacts cannot be adequately mitigated to the point of nonsignificance (see box labeled "Prepare EIS?" Figure 1.1). On completing the EIS, the responsible decision-maker reviews the analysis and reaches a decision regarding the appropriate course of action to take. The decision-maker is required to consider the potential impacts in reaching the final decision. A public record of decision (ROD) is prepared, recording the agency's final decision. The agency may then proceed with the action, subject to any applicable mitigation and monitoring measures. Chapter 8 describes the EIS planning process in detail.

1.1.7.1 The EIS Process

With only a few exceptions, an EIS must be prepared for any federal proposal that is not eligible for a FONSI/CATX. The basic purpose of an EIS is to analyze potentially significant effects and investigate alternatives for minimizing, avoiding, or mitigating such impacts.

Preparation of an EIS should always be approached with care as professional experience indicates that a poorly executed process can be characterized by the humorously orchestrated 8-step project process depicted in Figure 1.2.

To promote efficiency, an EIS is required to be concise and to the point:

> Agencies shall focus on significant environmental issues and alternatives and shall reduce paperwork and the accumulation of extraneous background data. Statements shall be concise, clear, and to the point, and shall be supported by evidence that the agency has made the necessary environmental analysis (§ 1502.1).

1.1.7.2 Notice of Intent

The principal steps followed in preparing an EIS are outlined in Figure 1.3. A prescoping effort is frequently performed prior to the formal initiation of an EIS. The formal EIS process is initiated via issuance of a public notice of intent (NOI) in the *Federal Register*. The NOI invites the public to participate in the public scoping process that will determine the scope of analysis to be investigated

```
1. Enthusiasm
2. Apprehension
3. Disillusionment
4. Panic
5. Search for the guilty
6. Persecution of the innocent
7. Praise for the uninvolved
8. Request for additional funding to revise the EIS
```

FIGURE 1.2   Eight steps that characterize a poorly executed EIS.

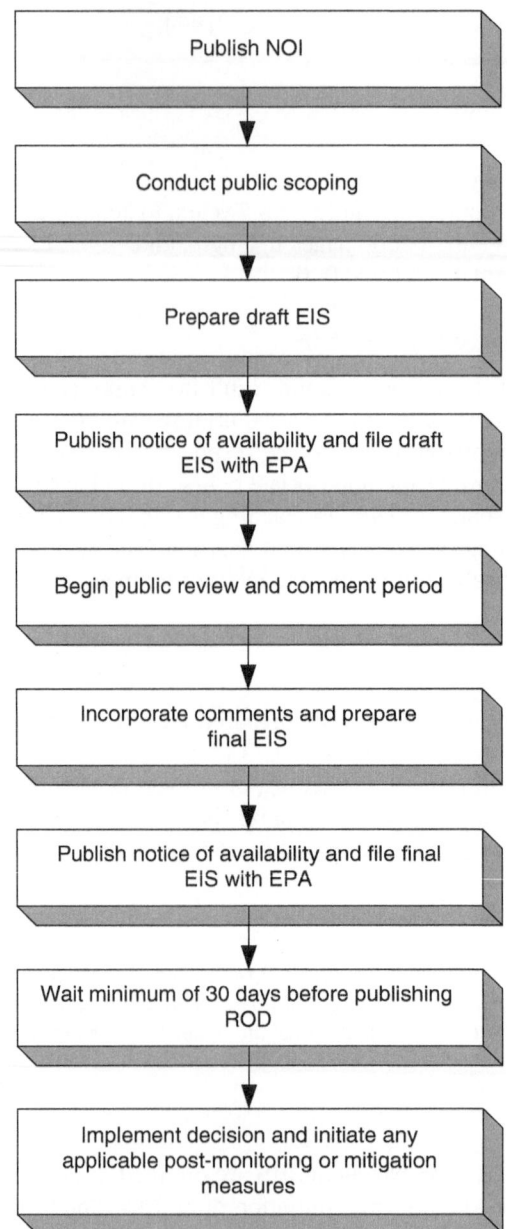

FIGURE 1.3   Typical EIS process.

in the EIS. An interdisciplinary planning effort is performed to define the range of alternatives that will be evaluated in detail. The alternatives are then analyzed.

1.1.7.3   The Draft and Final EIS Stage

A standard EIS process typically involves two phases. The term EIS describes a statement in either its draft or final stage.

A draft EIS is prepared first. Once completed, a notice of availability (NOA) is issued, notifying the public that the draft EIS is being circulated for review and public comment. The draft EIS

TABLE 1.2

NEPA's Essential Features and Requirements

Characteristics	Description
Early	The NEPA planning process must begin during the "early" proposal phase before a final decision has been made.
Systematic	NEPA requires use of a "systematic" planning and decision-making process.
Scientific	A scientifically based process is to be used in planning and analyzing proposals.
Interdisciplinary	NEPA requires use of an "interdisciplinary" approach in the planning and decision-making process.
Impartial	The analysis must be objective and impartial.
Public	NEPA is a "public" process, which involves public input, review, and comment.
Alternatives	Alternatives are to be "rigorously" investigated.
Rigorous	All potentially "significant impacts" and issues are to be examined.
Impacts and issues	All "reasonable alternatives" and "mitigation measures" are to be investigated.

is circulated for public review and comment and filed with the Environmental Protection Agency (EPA), which is responsible for rating the document.

During the second stage, public comments are reviewed and, as appropriate, incorporated into the final EIS. Once comments have been incorporated, a second NOA is published, notifying the public that the final EIS is being issued. The final EIS is filed with the Environmental Protection Agency.

The final EIS is to be used by the responsible decision-maker, together with other relevant decision-making material, in reaching a final decision regarding the course of action to be taken (§ 1502.1). The decision-maker reviews the analysis and selects the alternative or course of action to be pursued.

1.1.7.4 Record of Decision, Monitoring, and Mitigation

A record of decision (ROD) is issued recording the agency's final decision. Unlike an EA, the decision-maker has the latitude to select any adequately analyzed alternative even though it may result in a significant impact. After publishing the ROD, the agency must wait for a minimum period of 30 days before taking any action with respect to the chosen alternative. Once the 30-day waiting period has elapsed the agency (without comment) is free to pursue the selected action described in the ROD.

Any applicable monitoring and mitigation measures mandated in the ROD must be implemented. Chapter 8 describes the entire EIS process in detail. For further information on the regulatory requirements governing the EIS process, the reader is referred to the companion texts, *Environmental Impact Statements* and *NEPA Planning Process*.[1,3]

1.2 BASIC CONCEPTS AND REQUIREMENTS

Table 1.2 lists some of the basic characteristics that NEPA has infused into the federal planning and decision-making process.

Basic concepts and requirements underlying the NEPA planning process are described in the following sections.

1.2.1 COMPLYING TO THE "FULLEST EXTENT POSSIBLE"

Congress authorizes and directs that

> ... to the *fullest extent possible* the policies, regulations, and public laws of the United States shall be interpreted and administered in accordance with the policies set forth in this Act (Sec. 102 [42 USC § 4332]).

Moreover, the Regulations direct that to the "fullest extent possible" Federal agencies are required to

Interpret and administer the policies, regulations, and public laws of the United States in accordance with the policies set forth in the Act and in these regulations (§ 1500.2[a]).

Integrate the requirements of NEPA with other planning and environmental review procedures required by law or by agency practice so that all such procedures run concurrently rather than consecutively (§ 1500.2[c]).

The phrase "fullest extent possible" means that each agency of the federal government must comply with this requirement unless existing law applicable to the agency's operations expressly prohibits or makes compliance impossible (§ 1500.6).

The aforementioned regulatory direction applies to all agencies of the federal government (§ 1500.3 and § 1507.1). Each agency is expected to review its policies and regulations for consistency with NEPA and to revise them as necessary to insure full compliance.

As indicated in Section 105 of NEPA, federal agencies are instructed to interpret the Act as a supplement to their traditional missions:

The policies and goals set forth in this Act are supplementary to those set forth in existing authorizations of Federal agencies (Sec. 105 [42 USC § 4335]).

1.2.1.1 *Calvert Cliffs*

The case of *Calvert Cliffs Coordinating Committee v. Atomic Energy Commission* was one of the first and perhaps most important cases in which the court upheld an agency's responsibility to comply with NEPA.[9] Residents challenged the licensing of a nuclear plant, and during the proceedings the former Atomic Energy Commission (AEC) promulgated rules for complying with NEPA.

One of the remarkable aspects of *Calvert Cliffs* was that plaintiffs were able to raise the issue of NEPA at all. NEPA is a statute without a citizen suit provision. Many judges might have dismissed the suit based on lack of several legal doctrines, including "standing" or the absence of a "private right of action." Instead, Judge Wright concluded that NEPA set standards that were to be "rigorously enforced" by reviewing courts. The Judge wrote that there was a duty "to see that important legislative purposes, heralded in the halls of Congress, are not lost or misdirected in the vast hallways of the federal bureaucracy."[10] This court established several important precedents:

- NEPA makes environmental protection a part of the mandate of every federal agency and department. Agencies are "not only permitted, but compelled to take environmental values into account. Perhaps the greatest importance of NEPA is to require [all] agencies to consider environmental issues just as they consider other matters within their mandates."
- The general substantive policy in Section 101 of NEPA is flexible: "It leaves room for a responsible exercise of discretion and may not require particular substantive results in particular problematic instances."
- The procedural provisions in Section 102 of NEPA are not as flexible. To ensure that an agency balances environmental issues with its other mandates, NEPA Section 102 requires agencies to prepare an EIS.
- As viewed by the court, the procedural duties imposed by NEPA are to be carried out "to the fullest extent possible … Congress did not intend the Act to be a paper tiger."
- Section 102 of NEPA mandates a careful and informed decision-making process and creates judicially enforceable duties.
- NEPA requires that an agency—to the fullest extent possible—consider alternatives to its actions that could reduce environmental damage.
- The AEC's interpretation of its NEPA responsibilities was considered to be "crabbed" and made "a mockery of the Act." As contemplated by the AEC regulations, it is insufficient

for an EIS merely to "accompany" an application through the review process and receive no consideration in the decision-making process.

- Delay in the final operation of the facility may occur, but is not a sufficient reason to reduce or eliminate consideration of environmental factors under NEPA. Some delay is inherent in NEPA compliance; however, it is more consistent with the purposes of the Act to delay a project at a stage where real environmental protection may come about than at a later stage where corrective action may be so costly as to be impossible.

This latter requirement is reinforced in § 1500.6 of the Regulations:

Each agency shall interpret the provisions of the Act as a supplement to its existing authority and as a mandate to view traditional policies and missions in the light of the Act's national environmental objectives. Agencies shall review their policies, procedures, and regulations accordingly and revise them as necessary to insure full compliance with the purposes and provisions of the Act. The phrase "to the fullest extent possible" in section 102 means that each agency of the Federal Government shall comply with that section unless existing law applicable to the agency's operations expressly prohibits or makes compliance impossible (§ 1500.6).

1.2.1.2 Conflict in Statutory Authority

The courts have interpreted the second half of § 1500.6 to mean, in part, that strict observance to NEPA's requirements is not necessarily required where it would prevent an agency from "complying with its duties under existing authority."[11] This provision is reinforced by the landmark case, *Flint Ridge Development Co.*, where plaintiffs challenged the Department of Housing and Urban Development (HUD) for failing to prepare an EIS prior to approving the filing of a disclosure statement under the Interstate Land Sales Full Disclosure Act. Under this act, developers were required to disclose information by filing a statement of record with HUD regarding the title of the land and conditions of the subdivision. The statement of record was to become effective automatically on the 30th day after filing, unless it was found to be materially incomplete or inaccurate.[12]

In the case of *Flint Ridge*, the court held that the Disclosure Act did not give HUD discretion to suspend the effective date of a proposed statement of record for such time as was necessary to prepare an EIS. It was possible to prepare the EIS within a 30-day window. Where a clear and unavoidable conflict in statutory authority existed, NEPA's EIS requirement must yield.

1.2.1.3 Conflicts with Other Statutory Obligations

Section 104 of NEPA goes on to state

Nothing in section 102 ... or 103 ... shall in any way affect the specific statutory obligations of any Federal agency (1) to comply with criteria or standards of environmental quality, (2) to coordinate or consult with any other Federal or State agency, or (3) to act, or refrain from acting contingent upon the recommendations or certification of any other Federal or State agency (Sec. 104 [42 USC § 4334] of NEPA).

For example, under Section 104 of NEPA, an agency is prohibited from preparing an EIS simply as a means for delaying its environmental clean-up responsibilities. Selected excerpts related to the aforementioned requirements are summarized in Table 1.3.

1.2.1.4 Is It Appropriate to Design an Action to Avoid NEPA?

Courts have clearly demonstrated an unwillingness to tolerate attempts to circumvent NEPA. Once NEPA has been triggered, attempts to circumvent compliance with this Act have been viewed by the courts as acts of "bad faith." But is it appropriate to design an action with the intention of avoiding NEPA?

TABLE 1.3
An Agency's Authority and Requirement to Comply with NEPA

Statutory Requirements (Sec. 102(c) of NEPA)

Congress authorizes and directs that, to the fullest extent possible:
(1) the policies, regulations, and public laws of the United States shall be interpreted and administered in accordance with the policies set forth in this Act, and all agencies of the Federal Government shall—include in every recommendation or report on proposals for legislation and other major Federal actions significantly affecting the quality of the human environment, a detailed statement ...

Policy (§ 1500.2)

Federal agencies shall to the fullest extent possible:
(a) Interpret and administer the policies, regulations, and public laws of the United States in accordance with the policies set forth in the Act and ... regulations ...
(f) Use all practical means, consistent with the requirements of the Act ... to avoid or minimize any possible adverse effects.

Mandate (§ 1500.3)

Parts § 1500 through § 1508 of this title provide regulations applicable to and binding on all Federal agencies for implementing the procedural provisions of NEPA ... except where compliance would be inconsistent with other statutory requirements.

Agency Authority (§ 1500.6)

Each agency shall interpret the provision of the Act as a supplement to its existing authority and as a mandate ... Agencies shall review their policies, procedures, and regulations accordingly and revise them as necessary to insure full compliance with the purposes and provisions of the Act. The phrase "to the fullest extent possible" in section 102 means that each agency of the Federal Government shall comply with that section unless existing law applicable to the agency's operations expressly prohibits or makes compliance impossible.

Compliance (§ 1507.1)

... agencies of the Federal Government shall comply with these regulations. It is the intent of these regulations to allow each agency flexibility in adapting its implementing procedures ... to the requirements of other applicable laws.

Agency Capability to Comply (§ 1507.2)

Each agency shall be capable (in terms of personnel and other resources) of complying with the requirements ...

Consider the following case where federal money was allocated for construction of a prison reception and medical center. The project was deemed to be a "major federal action" as a result of the federal funding.[13] After preparing a draft EIS, state and federal officials concluded that it would be difficult to fully comply with NEPA's requirements. They decided to "defederalize" the project by reallocating the federal funds to another state project and financing the project with state-only revenue. When challenged, the state argued that NEPA no longer applied because the project would now be financed purely with state funding. The court ruled that circumventing the intent of Congress in such a manner was unacceptable.[14]

While it may be inappropriate to circumvent NEPA in circumstances where the Act already has been triggered, the courts may be more tolerant of a conscious design to avoid triggering NEPA's requirements from the onset.

In one case, the court concluded that bad faith is indeed a factor relevant to NEPA cases, but only "once NEPA has been determined to apply to a challenged project."[15] Thus, an agency has not acted in bad faith if a project has been intentionally designed to avoid triggering NEPA in the first place.[16]

1.2.2   WHEN MUST NEPA BE INITIATED?

As described earlier, preparation of an EIS should begin as close as possible to the time in which an agency is either developing a proposal or is presented with a proposal. Agencies must prepare

NEPA documents early enough so as to contribute to the decision-making process but not to use them simply to rationalize or justify decisions already made. This timing requirement has been met if the document is prepared in time to meet the decision deadline, but not so early that it cannot meaningfully contribute to the decision-making process (§ 1502.5).

An EIS must also be initiated so that it can be completed in time to be included in any recommendation or report concerning the proposal. The reader should note that a proposal may exist in actuality even if the agency has not officially declared one to exist (§ 1508.23).

1.2.2.1   The Definition of a Proposal

With respect to NEPA, a proposal is defined to exist when the following three conditions are met (§ 1508.23):

- An agency has a goal.
- The agency is actively preparing to make a decision on one or more alternative means of accomplishing the goal.
- The effects can be meaningfully evaluated.

The Supreme Court concluded in the case of *Kleepe v. Sierra Club* that agencies are only required to prepare EISs for actual proposals; merely contemplating a project and preparing a support study does not necessarily constitute a proposal.[17] In the case of *Kleepe*, the Sierra Club claimed that a regional EIS had not been prepared by the Department of Interior for approval of mining projects in the Great Plains. The Supreme Court found that an EIS was not necessary because there was no specific plan or proposal for developing coal deposits. The Court concluded that NEPA only requires EISs for proposals. Because there was no actual proposal for developing the region's coal deposits, no EIS was required.

The Court went on to assert that it was not arbitrary for the Secretary of Interior to determine that the independent activities were not related enough to constitute an integrated proposal that required an EIS. However, the Court was careful to note that a single EIS may be required to analyze a situation where several proposals (1) occur within the same region; (2) are closely related in time; and (3) present a cumulative environmental impact.

1.2.3   WHEN MUST NEPA BE COMPLETED?

As viewed by some courts, the point of commitment has been used for determining a point that cannot be exceeded without completion of the NEPA process. The point of commitment has sometimes been viewed as the point at which an irreversible or irretrievable commitment of resources is made. Other courts have interpreted this timing requirement to be the point at which an irreversible action would be made.

This point of commitment has been interpreted to mean a point at which so much time, material, or resources have been utilized that the agency would find it difficult not to go forward with the project. For example, in one case, the Bureau of Land Management (BLM) granted oil and gas leases, before the NEPA process was completed.[18] The leasing recipient was granted priority rights to future exploration and development of oil and gas fields, but was forbidden from occupying or using the leased land without future BLM approval. When challenged, the court found that issuing this lease was not an irreversible or irretrievable action requiring preparation of an EIS because BLM authorization would still need to be obtained to pursue future exploration and development of these lands. Conversely, an EIS would probably have been required if the BLM had issued leases which permitted exploration and development.

1.2.4   COMBINING AND INTEGRATING OTHER REQUIREMENTS

Selected citations providing direction to integrate or combine NEPA with other documents and planning processes are provided in Table 1.4.

TABLE 1.4

Direction for Integrating and Combining NEPA with Other Agency Planning

Integrate the requirements of NEPA with other planning and environmental review procedures … (§ 1500.2[c])

Agencies shall *integrate* the NEPA process with other planning … (§ 1501.2)

Identify other environmental review and consultation requirements … prepare other required analyses and studies concurrently with, and *integrated with*, the environmental impact statement … (§ 1501.7[a][6])

Hold an early scoping meeting or meetings which may be *integrated* with any other early planning meeting … (§ 1501.7[b][4])

… prepare draft environmental impact statements concurrently with and *integrated* with environmental impact analyses and related surveys and studies … (§ 1502.25[a])

The [ROD] may be *integrated* into any other record prepared by the agency … (§ 1505.2)

… proposals for legislation … shall be *integrated* with the legislative process of the Congress (§1506.8[a])

Any environmental document in compliance with NEPA may be *combined* with any other agency document … (§ 1506.4)

TABLE 1.5

Actions Subject to NEPA's Requirements

Proposals for Legislation (§ 1506.8[a])

The NEPA process for proposals for legislation (§ 1508.17) … shall be integrated with the legislative process of Congress

Major Federal Actions (§ 1508.18)

Actions include the circumstances where the responsible officials fail to act and that failure is reviewable by the courts or administrative tribunals under the Administrative Procedures Act or other applicable law as agency action (§ 1508.18):

(a) Actions include new and continuing activities, including projects and programs financed entirely or partly financed, assisted, conducted, regulated, or approved by federal agencies; new or revised agency rules, regulations, plans, policies or procedures; and legislative proposals.

(b) Federal actions tend to fall into one of the following categories:

    (1) Adoption of official policy such as rules, regulations, … treaties and international conventions or agreements … formal documents establishing policies which will result in or substantially alter agency programs.

    (2) Adoption of formal plans … which guide … uses of federal resources.

    (3) Adoption of programs, such a group of concerted actions to implement a specific policy or plan …

    (4) Approval of specific projects such as construction or management activities in a defined geographic area … actions approved by permit or other regulatory decision as well as federal and federally assisted activities.

Certain limited categories of actions abroad:

For the global commons (such as oceans or Antarctica), an Executive Order provides that environmental impact statements will be prepared for all major federal actions having significant environmental effects.[19]

Detailed direction for integrating NEPA with other planning and permit requirements is provided in Chapter 4.

1.2.5 ACTIONS SUBJECT TO NEPA

NEPA specifically applies to a broad and comprehensive range of federal actions, including policies, programs, projects, enacting some forms of legislation, promulgating regulations, and international agreements. Table 1.5 lists some selected regulatory citations related to the application of NEPA.

1.3 NEPA LITIGATION AND JUDICIAL REVIEW

The companion book, *The NEPA Planning Process*, provides an overview of the process for bringing a suit under NEPA.[1] For an in-depth legal review, the reader is also directed to Mandelker's book, *NEPA Law and Litigation*.[20]

1.3.1 STATUTE OF LIMITATIONS

There is no explicit statute of limitations under NEPA. That is, there is no specific period within which a judicial challenge must be filed. However, some courts apply the statute of limitations for civil suits brought against the United States under the APA, which is 6 years. At least one court, however, has noted that applying this APA limitation would be a "blind application of a statute … and would result in illogical and capricious administration of an important environmental statute."[21]

Some agency authorizing statutes place restrictions on challenges that may be brought against an agency's actions. These restrictions might be viewed as the functional equivalent of an NEPA statute of limitation. For example, in one challenge to a decision by the Secretary of the Interior to take land into trust for an Indian tribe, the action had to be brought within 30 days. As another example, certain actions challenging an FAA Order must be brought within a period of 60 days. Critics counter that such time periods are often far too short to allow a proper public review, let alone sufficient time to prepare legal action.

1.3.2 ADMINISTRATIVE RECORD AND JUDICIAL REVIEW STANDARD

Judicial review under NEPA is based on review of the administrative record prepared by the agency. Barring exceptional circumstances, a judge normally does not accept testimony or receive evidence beyond that already provided in the administrative record. The judge decides whether the agency has taken a sufficiently "hard look" in terms of reviewing the environmental consequences of its proposal.

In the case of *Kleepe v. Sierra Club*, the Supreme Court concluded that NEPA does not contemplate that a court should substitute its judgment for that of the agency as to the environmental consequences of agency's actions. But the court does have a duty to ensure that the agency has taken a "hard look" at their environmental consequences.[22] This standard is described in more detail shortly.

1.3.2.1 Arbitrary Capricious Standard

Judges typically lack the technical background necessary for reaching informed determinations regarding significance. Thus, most courts have been reluctant to second guess the competency and professional judgment of an agency's technical experts. Barring a "clear error in judgment," courts have historically deferred judgment to the agency.

With respect to NEPA, a court's review of an agency's actions, conclusions, and findings of fact are governed by the APA. The APA establishes a standard for judicial review of decisions involving NEPA.[23]

A court may find that NEPA was incorrectly implemented if an agency's actions, findings, or conclusions were "arbitrary, capricious, an abuse of discretion, or otherwise not in accordance with law"[24] (emphasis added). The judicial review is expected to be "searching and careful," but the arbitrary and capricious standard is narrow, and the court does not substitute its own judgment for that of the agency.[25]

If an agency opts not to prepare an EIS (i.e., EA, CATX, or nothing at all), it must put forth a "convincing statement of reasons" that explains why the project will not significantly impact the environment.[26] This account proves crucial in a court's determination of whether an agency took the requisite "hard look" (a judicial criterion commonly used by the courts in NEPA cases) at the potential impacts of the proposal.

An EIS must be prepared if substantial questions are raised as to whether an action may cause a significant impact on the human environment. To trigger this requirement, a plaintiff need not show that significant effects will in fact occur, but instead, it may be sufficient to only raise substantial questions as to whether a project may have a significant effect.[27]

1.3.2.2   Recent Case Law Involving the "Hard Look" Standard

The following recent case is included to provide the reader with some insight into how courts have applied the "hard look" doctrine in their review of EISs.

In 2005, the U.S. District Court for the Eastern District of North Carolina ruled that the analysis in a U.S. Navy EIS was inadequate.[28] The Fourth Circuit applied the "hard look" standard in evaluating the adequacy of the EIS. The court's analysis is instructive.

Based on this EIS, the Navy decided to construct a landing field in Washington and Beaufort Counties, North Carolina. The district court issued an injunction barring the Navy from pursuing its final decision. The Navy appealed. On September 7, 2005, the U.S. Court of Appeals for the Fourth Circuit upheld the need for the Navy to prepare a new supplemental EIS.

What Constituted a Hard Look? The appeals court based its decision on the long-standing principle that its role is to determine whether an agency has taken a "hard look" at an action's environmental impacts. The court wrote that "A 'hard look' is necessarily contextual ..." and should be based on "... a holistic view of what the agency has done to assess environmental impact ... The hallmarks of a 'hard look' doctrine are thorough investigation into environmental impacts and forthright acknowledgment of potential environmental harms."[29]

Impact on Birds. The Navy had evaluated five alternative landing field locations in eastern North Carolina in the EIS, including the proposed site in Washington and Beaufort Counties (alternative Site C). Site C is located about 5 miles from the Pocosin Lakes National Wildlife Refuge, and the flight pattern for training exercises would come within 0.2 mile of the Refuge. Plaintiffs focused on the Navy's evaluation of potential impacts on birds (among other issues).

Court's Review of the Analysis of the Selected Site. The Appeals Court found the following inadequacies in the evaluation of Site C in the original EIS:

1. In regard to the Navy's site investigation, the court found that four one-day visits were insufficient to "conduct systematic observations or perform species-specific studies."
2. The Navy contended that the bird–aircraft strike potential at Site C was similar to that at other flight training facilities. However, the court concluded that "this comparative assessment provided only a useful starting point," and further analysis was necessary.
3. The Navy's literature review identified research indicating that snow geese (who winter at the Refuge) "may be especially sensitive to aircraft activity." The court found that the EIS needed to go beyond simply "citing the articles or abstracts that contradict the conclusions reached [i.e., that impacts would be minor]."
4. The Navy relied on an analysis of environmental effects of aircraft overflights at three existing military facilities to draw conclusions about potential impacts at Site C. The court noted differences between circumstances at Site C and the existing facilities and found that the Navy had not provided a proper factual basis for a comparative analysis.

1.3.2.3   Institute for Environmental Conflict Resolution

The U.S. Institute for Environmental Conflict Resolution was established by the 1998 Environmental Policy and Conflict Resolution Act to assist parties in resolving conflicts over environmental issues, which involve federal agencies.[30] While part of the federal government, it provides an independent and neutral arena for federal agencies to work with citizens, state, local, and tribal governments, and private organizations and businesses to reach common ground, rather than through litigation and other adversarial approaches to dispute resolution.[31]

In this capacity, the Institute provides an alternative mechanism for resolving NEPA disputes. The U.S. Institute is also charged with assisting the federal government in the implementation of the substantive policies set forth in Section 101 of NEPA.

1.3.3 PROCEDURAL VERSUS SUBSTANTIVE

In terms of NEPA, the term "procedural" refers to the specific requirements that an agency must comply with. For instance, an agency must prepare an EIS for actions that may have a significant impact on the environment, and using the EIS process, the agency must perform a public scoping process.

"Substantive" aspects, on the other hand, basically refer to concepts such as the ultimate goal(s) of NEPA, that is, protection and enhancement of environmental quality. For example, Section 101 states that the federal government will use all practicable means (i.e., substantive) to "attain the widest range of beneficial uses of the environment without degradation, risk to health or safety, or other undesirable or unintended consequences." NEPA further requires that all practicable means will be used to "assure for all Americans safe, healthful, productive, and esthetically and culturally pleasing surroundings." In simple language, a substantive requirement directs federal agencies to make decisions that promote environmental quality or limit environmental degradation.

1.3.3.1 Case Law

In a precedent-setting case, *Strycker's Bay Neighborhood Council v. Karlen*, the Supreme Court concluded that NEPA does not require an agency to elevate environmental concerns over other admittedly legitimate considerations. Nor do the courts have the power to order a shift in priority. NEPA simply requires that an agency adequately consider the environmental consequences of its decisions.[32]

In another case, *Baltimore Gas and Electric Co. v. Natural Resources Defense Council*, the Supreme Court ruled that NEPA only requires investigation, consideration, and disclosure of environmental impacts. In this case the court concluded that[33]

NEPA places upon an agency the obligation to consider every significant aspect of the environmental impact of a proposed action. It also ensures that the agency will inform the public that it has indeed considered environmental concerns in its decision-making process.

Congress does not require agencies to elevate environmental concerns over other appropriate considerations. Instead it only requires that agencies take a "hard look" at the environmental consequences before carrying out a major action.

The role of the courts is simply to ensure that the agency has adequately considered and disclosed the environmental impact of its actions and that its decision is not arbitrary and capricious. As the court explained, "It is not our task to determine what decision we, as Commissioners, would have reached. Our only task is to determine whether the Commission has considered the relevant factors and articulated a rational connection between the facts found and the choice made."

However, the intent of NEPA has not been fulfilled if an agency prepares an NEPA analysis simply in order to satisfy procedural requirements and then totally disregards this analysis in its decision-making process. An agency must be able to demonstrate that it has seriously considered and weighed the information presented in the analysis against other concerns and factors before making a final decision on the action.

Despite these earlier cases, lingering doubts remained regarding NEPA's procedural versus substantive authority. Most of these doubts would be resolved in 1989.

1.3.3.2   *Robertson v. Methow*

The landmark case of *Robertson v. Methow Valley*, decided in 1989, involved a Forest Service (FS) study that designated a particular national forest location as having high potential for a new ski resort. Methow Recreation had applied for a special use permit to develop and operate such a resort on the site, and the FS had prepared an EIS for this project. Plaintiffs challenged the FS decision to issue a special use permit for a ski resort. In 1989, the Supreme Court found that:[34]

- NEPA does not impose a substantive duty on agencies to mitigate adverse effects or to include in an EIS a fully developed mitigation plan. Although the EIS requirement implements the statute's sweeping goals by ensuring that agencies will take a "hard look" at the consequences and by guaranteeing public dissemination of relevant information, it is well settled that NEPA itself does not impose substantive duties mandating particular results. While other statutes may impose substantive environmental obligations on federal agencies, "NEPA merely prohibits uninformed—rather than unwise—agency action."
- An important EIS requirement is the discussion of steps that can be taken to mitigate adverse effects. Omission of a reasonably complete discussion of possible mitigation measures would undermine the "action-forcing" function of NEPA. However, "there is a fundamental distinction between a requirement that mitigation be discussed in sufficient detail to ensure that environmental consequences have been fairly evaluated … and a substantive requirement that a complete mitigation plan be actually formulated and adopted."
- The CEQ amendment to its regulations eliminating the requirement to perform a "worst case analysis" was valid. The worst case requirement was not a codification of prior NEPA case law. Thus, it was appropriate for the CEQ to eliminate such a requirement.

Therefore, in concluding that a proposal complies with NEPA's requirements even though the action would degrade environmental quality, the Supreme Court appears to have ruled out the possibility of interpreting NEPA to have a substantive mandate in terms of protecting the environment.

Table 1.6 describes the four basic parties that play unique but vital roles in ensuring that NEPA is fully and correctly implemented.

1.4   CLOSING THOUGHTS

The author wishes to propose three corollaries to Newton's famous Three Laws of Motion. These corollaries are noted for more than humor's sake. Each has a subtle, yet profound implication in terms of safeguarding our environment and the manner in which agencies choose to perform their NEPA planning processes. Presented below are Eccleston's Three Laws of the Environmental Movement:[35]

First law of the environmental movement: A top-level commitment to environmental quality tends to continue in the direction of quality, unless acted upon by other countervailing forces; lack of commitment to environmental excellence tends to promote environmental degradation.

Second law of the environmental movement: The force that NEPA brings to bear in protecting the environment is equal to the mass of forethought expended in the planning process, multiplied by the decision-maker's commitment to environmental protection.

Third law of the environmental movement: For every project proponent attempting to circumvent the NEPA process, there is an equal and opposite adversary waiting to litigate.

TABLE 1.6
Four Principal Roles in Implementing NEPA

Party	Responsibility
Federal Agency	The federal agency's role in the NEPA process depends on its expertise and relationship to the proposed undertaking. The proposing agency is responsible for complying with the requirements of NEPA. In some cases, there may be more than one federal agency involved in which there may be joint lead agencies or a lead-cooperating agency arrangement. A federal, state, tribal, or local agency having special expertise with respect to an environmental issue or jurisdiction by law may be a cooperating agency in the NEPA process.
EPA	As described in Chapter 9, the EPA has a unique responsibility. Under Section 309 of the Clean Air Act, EPA is required to review and publicly comment on the environmental impacts of major federal actions including actions that are the subject of EISs. If EPA determines that the action is environmentally unsatisfactory, it is required by Section 309 to refer the matter to CEQ. The EPA Office of Federal activities has been designated as the official recipient in EPA of all EISs prepared by federal agencies.
The Public	The public has an important role in the NEPA process, particularly during scoping, in providing input on what issues should be addressed in an EIS and in commenting on the findings in an agency's NEPA documents. The public may participate in the NEPA process by attending NEPA-related hearings or public meetings and by submitting comments directly to the lead agency. The lead agency must take into consideration all comments received from the public and other parties on NEPA documents during the comment period.
The Courts	The courts provide an important function, as they can review the NEPA process, and if it is found to be inadequate they can prescribe methods for enforcing federal agency compliance.

PROBLEMS

1. What are the three basic levels of NEPA compliance? Briefly describe each one.
2. What is an environmental assessment? What is its principal purpose?
3. What is a finding of no significant impact? What is it used for?
4. What is a categorical exclusion? Why is this concept such important to federal agencies?
5. Briefly describe the environmental impact statement process.
6. With respect to NEPA, what does complying to the "fullest extent possible" mean?
7. Why was the court case of *Calvert Cliffs Coordinating Committee v. Atomic Energy Commission* so important?
8. With respect to NEPA, when is an action considered to be a proposal?
9. With respect to NEPA, what is meant by the phrase "procedural versus substantive"?
10. What is the importance of the case *Robertson v. Methow Valley?*
11. What is the "hard look" doctrine?
12. What does Eccleston's third law of the environmental movement imply?

REFERENCES

1. Eccleston C. H., *The NEPA Planning Process: A Comprehensive Guide with Emphasis on Efficiency*, Chapter 9, John Wiley & Sons, New York, 1999.
2. Eccleston C. H., *Effective Environmental Assessments: How to Manage and Prepare NEPA's EA*, Lewis Publishers, Boca Raton, FL, 2001.
3. Eccleston C. H., *Environmental Impact Statements: A Comprehensive Guide to Project and Strategic Planning*, John Wiley & Sons, New York, 2000.

4. *Calvert Cliffs Coordinating Committee, Inc. v. Atomic Energy Commission*, 449 F.2d 1109 (D.C. Cir. 1971).

5. 10 CFR 1221, Subpart D, 1992.

6. Eccleston C. H., *The NEPA Planning Process: A Comprehensive Guide with Emphasis on Efficiency*, Chapter 5, John Wiley & Sons, New York (www.wiley.com), 1999.

7. Eccleston C. H., *Effective Environmental Assessments: How to Manage and Prepare NEPA's EA*, Lewis Publishers, Boca Raton, FL (www.crcpress.com), 2001.

8. CEQ, The NEPA Task Force Report to the Council on Environmental Quality Modernizing NEPA Implementation, http://ceq.eh.doe.gov/ntf/report/htmltoc.html (accessed March 3, 2007).

9. *Calvert Cliffs Coordinating Committee, Inc. v. Atomic Energy Commission*, 449 F.2d 1109 (D.C. Cir. 1971), *cert. denied*, 404 U.S. 942 (1972).

10. *Calvert Cliffs Coordinating Committee, Inc. v. Atomic Energy Commission*, 449 F.2d 1109, 1111 (D.C. Cir. 1971).

11. Fairfax S. K. and B. T. Andrews, Debate within and debate without: NEPA and the redefinition of the prudent-man rule, *Natural Resources Journal*, 19, 521, 1979.

12. *Flint Ridge Development Co. v. Scenic Rivers Association of Oklahoma*, 426 U.S. 776 (1976).

13. *Ely v. Velde*, 497 F.2d (4th Cir. 1970) (Ely I).

14. *Ely II*, 497 F.2d.

15. *Gettysburg Battlefield Preservation Ass'n. v. Gettysburg College*, 799 F. Supp. 1571, 1578 (M.D. Penn. 1992), *aff'd without opinion*, 989 F.2d 487 (3rd Cir. 1993) (emphasis added).

16. Tuckfield, D. J., Attempts to avoid NEPA: Is it bad faith? National Association of Environmental Professionals Conference Proceedings, Session M5A, 672–679, 1995.

17. *Kleepe v. Sierra Club*, 427 U.S. 390 (1976).

18. *Conner v. Burford*, 836 F.2d 1521 (9th Cir. 1988).

19. CEQ Memorandum 44 FR 42, March 29, 1979, Pursuant to Executive Order 12114.

20. Mandelker D. R., *NEPA Law and Litigation*, Clark Boardman Callaghan, New York, 1998.

21. *Park County Res. Council vs. U.S. Dept. of Agriculture*, 817 F.2d 609, 617 (10th Cir. 1987).

22. *Kleepe v. Sierra Club*, 427 U.S. 390 (1976).

23. Section 10(a), 5 U.S.C. § 702. Also see, *Lujan v. Nat'l. Wildlife Fed'n.*, 497 U.S. 871, 882–883, 111 L. Ed. 2d 695, 110 S. Ct. 3177 (1990).

24. APA, 5 U.S.C. § 706(2)(A).

25. *Citizens to Preserve Overton Park, Inc. v. Volpe*, 401 U.S. 402, 416, 28 L. Ed. 2d 136, 91 S. Ct. 814 (1971), *Overruled on Other Grounds by Califano v. Sanders*, 430 U.S. 99, 105, 51 L. Ed. 2d 192, 97 S. Ct. 980 (1977).

26. *Blue Mountains Biodiversity Project v. Blackwood*, 161 F.3d 1208, 1212 (9th Cir. 1998).

27. *Idaho Sporting Cong. v. Thomas*, 137 F.3d 1146, 1149 (9th Cir. 1998); *Greenpeace Action v. Franklin*, 14 F.3d 1324, 1332 (9th Cir. 1992).

28. Final EIS for the Introduction of the F/A-18E/F (Super Hornet) Aircraft to the East Coast of the United States.

29. *National Audubon Society et al. v. Department of the Navy et al.*, U.S. Court of Appeals for the Fourth Circuit, September 7, 2005.

30. P.L. 105–156.

31. Environmental Policy and Conflict Resolution Act of 1998, 20 U.S.C. §§ 5601–5609 (2000).

32. *Strycker's Bay Neighborhood Council v. Karlen*, 444 U.S. 223 (1980).

33. *Baltimore Gas and Electric Co. v. Natural Resources Defense Council*, 462 U.S. 87 (1983).

34. *Robertson v. Methow Valley Citizens Council*, 490 U.S. 332, 109 S. Ct 1835 (1989) (companion case to *Marsh v. Oregon Natural Resources Council*).

35. Eccleston C. H., *Environmental Impact Statements: A Comprehensive Guide to Project and Strategic Planning*, John Wiley & Sons, New York, 2000, p. 27.

2 Tools, Techniques, and Approaches for Improving and Streamlining NEPA

This chapter describes tools, techniques, and approaches for improving and streamlining National Environmental Policy Act (NEPA) and environmental planning problems. As illustrated by the following story, the success that an agency has in streamlining or improving NEPA is, to a large measure, dependent on how that agency chooses to practice the Act.

2.1 NEPA AND PROCESSIONARY CATERPILLARS

As many biologists will testify, processionary caterpillars meander their way through tree branches, each one with its head fitted snugly against the rear extremity of its predecessor. The long, winding procession thus formed gives the caterpillars their aptly deserved designation.

Intrigued by this behavior, the naturalist Jean-Henri Fabre lured a colony of these creatures onto the rim of a pot. In due time, the caterpillars began to snuggle up to one another, and after forming an interconnected chain eventually began to move around in a large circle. Since the chain had no beginning or end, Fabre expected that the caterpillars would soon tire of this unceasing parade and head off in a new direction. Such was not the case.

Fascinated, Fabre placed a supply of food next to them, but to no avail. Undeterred by the food that was outside the domain of their circle, the caterpillars continued on. To his dismay, apparently propelled by sheer force of habit, the creeping circle continued edging ever onward in the same unceasing circle for 7 days and 7 nights. Finally, exhaustion and starvation did them in. The moral of the story is that they were unable to break convention and venture out beyond their established paradigm.[1]

2.1.1 IMPROVING FEDERAL PLANNING

Today, federal agencies are increasingly being asked to do more with less. Efficiency is vital to the goal of successfully implementing its mission. Thus, to avoid going the way of the processionary caterpillars, agencies must be open to new and more effective paradigms for complying with environmental requirements. This goal requires an institutional framework within which modern methodologies can be applied to improve the planning process. Perhaps most importantly, there must be a top-down willingness to accept modern tools, techniques, and approaches for improving planning processes.

Once innovative approaches have been instituted, practitioners must be trained to use them. For example, the U.S. Army reports that training is a particularly valuable instrument for streamlining NEPA. Professionals are kept abreast on the latest approaches for implementing its requirements.

The following sections have been written to assist agencies in improving their environmental planning and analysis processes.

2.2 SLIDING SCALE AND THE RULE OF REASON

Professional judgment and common sense must be exercised in determining the appropriate scope of an analysis. Because the following two principles provide powerful tools for reducing cost, delays, and the amount of effort expended in implementing NEPA, they will be referred to throughout this chapter.

2.2.1 THE SLIDING-SCALE APPROACH

The sliding-scale approach recognizes that the amount of effort expended in performing an analysis is a function of the particular circumstances involved in each case. A sliding-scale approach is justified based on the following regulatory directions.

> Impacts shall be discussed in proportion to their significance. There shall be only brief discussion of other than significant issues. (40 CFR § 1502.2[b])

> NEPA documents must concentrate on the issues that are truly significant to the action in question, rather than amassing needless detail. (40 CFR § 1500.1[b])

Thus, environmental issues are investigated and regulatory requirements are applied using a degree of effort commensurate with their importance. In other words, the amount of effort expended in investigating or addressing a specific regulatory issue should generally vary with the significance of the potential impact and its importance to the decision-making process.

2.2.2 RULE OF REASON

Professional experience suggests that a strict or unreasonable application of a regulatory requirement may lead to a course of action, decision, or level of effort that is wasteful or even absurd. The Rule of Reason is a mechanism used by the courts to inject reason into the NEPA process. In other words, common sense must be exercised in determining the scope, detail, and attention devoted to issues, alternatives, and impacts considered in the analysis.

2.3 STREAMLINING AN AGENCY'S NEPA PROCESS

In the mid-1990s, a vice president of a contractor working at the U.S. Department of Energy's Hanford Site issued a memorandum listing the top 10 processes responsible for delaying site projects. The site procurement and NEPA processes were both listed in this memorandum as significant problems.

Instead of denying that a problem existed or placing blame elsewhere, a proactive effort was mounted to identify approaches for improving the site-wide NEPA process. The author was assigned to an interdisciplinary team (IDT) to investigate and streamline this process.

Two value engineering (VE) studies were performed. The first study examined the NEPA process in an effort to identify root problems for inefficiencies and develop a list of recommendations for streamlining NEPA. A second study was then performed to determine how NEPA could be better integrated into site planning and decision-making. As it turned out, the site planning process was complicated and was being performed by a number of independent departments. The goal of this second study was twofold: (1) to prevent project delays by ensuring that NEPA was triggered early in the planning process and (2) to improve the effectiveness of NEPA in contributing to the Department of Energy (DOE) decision-making process.

A VE approach that was used to examine the NEPA process is briefly summarized in the next section. As described in the companion book, *Environmental Impact Statements*, it is important to note that VE techniques can also be used as a tool for helping to assess the scope or issues considered in an environmental impact statement (EIS).

Data such as the length of time required to complete NEPA projects were not collected prior to the VE workshops described above. Therefore, there were no statistics available for gauging the level of success these VE workshops attained in terms of streamlining NEPA. However, there was a general consensus among management that the VE efforts led to a substantially streamlined NEPA process and also improved the ability of NEPA to contribute to the early decision-making process. Perhaps most important, as a result of these improvement efforts, NEPA was dropped from the list of 10 biggest problems on the Hanford Site.

An overview of the VE process and how it can be adapted to resolve problems in NEPA are described below.

2.4 VE IMPROVEMENT PROCESS

The Society of American Value Engineers (SAVE) has revised the VE terminology, redefining VE as the Value Methodology for which new standards have been published. However, consistent with common usage in the engineering disciplines, this book will use the more traditional terminology of VE.[2]

A VE workshop can be an intensely focused effort, such that a large amount of work can be accomplished in a very short period. In determining the preliminary scope, the goal is to leave no stone unturned. Typically, a VE facilitator leads an IDT through a number of rigorous procedures designed to break down preconceived and prejudicial notions and to consider new and alternative concepts that might lead to better solutions.[3] The facilitator is responsible for keeping the session focused and for promoting an open and nonhostile atmosphere where prevailing assumptions, mindsets, and paradigms are challenged to identify more optimal solutions. Prejudicial and preconceived notions are openly challenged as the IDT identifies factors and solutions that might otherwise be overlooked. The typical sequence of steps is shown below.

2.4.1 PRELIMINARY REVIEW

As a first step in the typical VE process, a preliminary review or pre-study (e.g., on a specific proposal, program, project, process, or design) may be useful in determining if the potential return on investment is sufficient to justify a formal VE workshop. Emphasis is placed on examining the proposal/activity to determine if potential alternatives show promise for increasing value. If a VE study is deemed to be cost-effective, the pre-study is also used to

1. define the scope, objectives, and expected deliverables;
2. establish logistical elements necessary to ensure a successful result;
3. gather preliminary information; and
4. select the study group's membership.

If a decision is made to proceed into a formal VE workshop, a facilitator is assigned and charged with responsibility for leading the VE team. An IDT is carefully selected to include professionals possessing a diverse range of technical expertise and experience. The team challenges existing paradigms and assumption in an effort to identify new or alternative concepts that may lead to better solutions.[3] As outlined next, the general VE process consists of a sequence of distinct steps or phases.

2.4.2 SEVEN PHASES

A standard VE study normally consists of the following seven distinct phases. Each of these phases, together with an emphasis on how they can be used to improve NEPA, is described in the following sections:

- Phase 1: Information phase
- Phase 2: Functional analysis phase
- Phase 3: Creativity phase
- Phase 4: Evaluation and analysis phase
- Phase 5: Development phase
- Phase 6: Presentation phase
- Phase 7: Post study/Implementation phase

With respect to streamlining the NEPA planning process, not all of the following phases are necessarily pertinent or need to be performed. Professional judgment must be exercised, on a case-by-case basis, in determining those phases that can be most effectively integrated with NEPA planning.

2.4.2.1   Phase 1: Information Phase

Information pertinent to the scope of the proposal (or action under investigation) is collected and disseminated to the members of the VE study. Limitations and constraints that may affect the study results are identified and, if necessary, ranked with an assigned value. For example, information could be shared regarding how plans are formulated and who is responsible for making certain types of decisions.

2.4.2.2   Phase 2: Functional Analysis Phase

A tool referred to as a "function analysis system technique (FAST) diagram" is frequently developed as part of a standard VE workshop. The FAST diagram can be used to critically evaluate "how" versus "why" critical functions are currently performed or would be performed. For example, a FAST diagram might be prepared to question why certain steps in the planning process are carried out. Items that have high potential for added value may be earmarked for more detailed examination in the later study.

However, a FAST diagram does not necessarily generate information useful to improving an agency's NEPA process. Because of the amount of effort necessary to construct a FAST diagram, its utility (with respect to improving the NEPA planning process) should be carefully questioned before an effort is launched to prepare such a diagram.

2.4.2.3   Phase 3: Creativity Phase

Team participants are encouraged to exercise creativity with the objective of identifying potential solutions for solving a particular problem. Methods such as brainstorming are used to generate innovative ideas for more detailed consideration. For example, ideas could be sought for streamlining various aspects of the agency's planning process.

2.4.2.4   Phase 4: Evaluation and Analysis Phase

In this phase, ideas generated during the creativity phase are organized into concepts possessing similar features. These can then be solidified into potential alternatives and ranked using one of a variety of techniques.

One ranking technique that has been used is referred to as "Criteria Weighting Matrix and Evaluation Analysis Ranking." Another ranking technique known as the "Nominal Group Technique" is described in Table 2.1. Options are then evaluated with respect to their advantages and disadvantages. For example, this phase could be used to single out ideas that show the most promise for streamlining the NEPA process.

2.4.2.5   Phase 5: Development Phase

Potential improvement options deemed to have the greatest potential for improving the NEPA process during the analysis step are further evaluated and developed into viable, efficient, and cost-effective options or alternatives.

2.4.2.6   Phase 6: Presentation Phase

Options for improving the NEPA process are documented in a report and presented as a study proposal.

TABLE 2.1
The Nominal Group Technique

Step 1: Generate ideas
The facilitator asks group members to write down the important problems or options for resolving these problems. Team participants write their ideas on note cards.

Step 2: Record ideas
A round-robin approach is used to record the ideas generated in Step 1. This process continues until all items have been recorded. Next, to prevent confusion, the facilitator asks each participant to explain briefly the items submitted. Participants are not allowed to criticize or critique any of the suggestions during this phase.

Step 3: Polling
A potentially long list of ideas must be narrowed down to capture the most salient ideas. Many methods exist for accomplishing this task. One common method is to have each participant write down three ideas considered to be the most important. These techniques are tabulated by the facilitator. A second round is then conducted to prioritize the final list of ideas.

2.4.2.7 Phase 7: Post Study/Implementation Phase

The workshop recommendations are considered and evaluated by the team and management. Tasks are assigned to individuals who are then responsible for implementing the chosen recommendations for improving NEPA.

2.4.3 Using VE in Preparing an EIS

The Office of Management and Budget has directed federal agencies to apply VE in planning *major federal projects* exceeding a cost of $1 million.[4] As a result, both the EIS and the VE are mandated to be performed on major federal projects. However, this raises a question: Is this simply another case where overlapping or redundant requirements have been mandated?

This question can best be answered by examining the underlying purposes of VE and NEPA.

VE provides a tool box of problem-solving techniques for analyzing problems and identifying solutions. However, VE is not a planning process.

In contrast, NEPA provides a comprehensive planning process, but lacks a set of intrinsic tools for effectively implementing its procedural requirements. Thus, VE offers a set of tools that, potentially, can manage various aspects of the EIS planning process. To date, this connection appears to have gone largely unrecognized. An integrated NEPA/VE planning approach for complex projects such as preparing a programmatic EIS may be advantageous as it provides an efficient means for complying with both Council on Environmental Quality (CEQ) and the Office of Management and Budget mandates.

2.4.3.1 Comparison of VE with NEPA

Table 2.2 compares principal characteristics and goals of NEPA with those of VE. While NEPA and VE share strikingly similar goals and requirements, they are not redundant; in fact, not only are NEPA and VE compatible, but also they complement each other. Their commonality in goals provides a foundation in which VE can be used as a tool for increasing the effectiveness of an EIS planning process. However, it should be noted that although VE and NEPA share many common requirements, their underlying goals differ. While VE offers many tools and techniques useful in managing the NEPA process, its orientation is different. The objective of VE is toward identifying and solving root problems that are related to, but are not normally the underlying focus of, an EIS scoping effort. Accordingly, a successful strategy requires prudence in the selection of the appropriate VE techniques that will be useful in preparing specific aspects of an EIS.

TABLE 2.2

Comparison of VE with NEPA

VE	NEPA
VE provides management tools useful in planning.	NEPA is a planning process.
The Office of Management and Budget has mandated that VE be applied to the planning of all major projects.	NEPA is the only federally mandated planning process applicable to all major federal actions.
Uses a systematic and interdisciplinary process to arrive at a better value solution.	Is predicated on use of a "systematic interdisciplinary" approach.[5]
Provides an important tool for assessing planning problems.	"Combines"[6] other federal planning processes during the "early"[7] planning phase.
Based on an unbiased and rigorous process.	Requires the use of a "public,"[8] unbiased,[9] and "rigorous" process.[10]
VE should be applied early enough to assist in decision-making. It is not intended to justify decisions already made.	An EIS must be prepared early enough to serve as an important contribution to decision-making. The EIS is not to be used to rationalize or justify decisions already made.[11]
VE can be used to assess planning requirements (e.g., cost, schedule, environmental).	NEPA allows the consideration of cost and other factors in the analysis and decision-making process.
Necessitates a full and fair analysis of alternatives to improve the value of the final product.	An EIS analysis must provide a "full and fair discussion" of impacts and reasonable alternatives.[12]
Promotes consideration of all possible alternatives.	An EIS must explore and objectively evaluate all reasonable alternatives.[13] Alternatives form the "heart" of the EIS.[14]

2.5   USING VE IN DETERMINING THE PRELIMINARY SCOPE OF AN EIS

This section describes how the author used a modified VE workshop approach in performing a prescoping effort for a complex EIS. A paper previously published by the author describes this approach in more detail.[15]

As described earlier, pursuing a standard VE study is deemed at times to be an ineffective use of scarce resources. A considerable degree of forethought and professional experience is therefore required in devising a cost-effective approach for using a VE approach in scoping an EIS.

Accordingly, emphasis was placed on developing an approach where VE techniques could be used specifically to focus and manage various aspects of the prescoping effort. Once a strategy had been developed, an agenda and workshop mission statement were prepared to provide a tangible map for implementing the strategy. The following process can be modified for use in performing a public scoping process and in identifying, evaluating, and managing other issues in the EIS process.

Results of this prescoping workshop provided an early indication of the issues, problems, and levels of effort that would be required if it was found advisable to prepare an EIS. Determining the preliminary scope also provided a basis for forecasting the cost and resources required in preparing the EIS. Table 2.3 provides a simplified agenda, while Table 2.4 describes the goals and objectives for this prescoping workshop. Table 2.3 should be revised, as necessary, for performing a formal public scoping process.

2.5.1   ASSEMBLING AN IDT

To ensure that all relevant planning factors were captured, emphasis was placed on assembling an IDT of experts. Approximately 10 full-time and 15 part-time members participated. The prescoping workshop ran over four consecutive days.

TABLE 2.3
Integrated EIS/VE Prescoping Agenda

Day 1

1. Welcome and introductions (session guidelines and expectations).
2. Overview of the problem statement, agenda, VE, and EIS process.
3. Information sharing session.
4. Document potential decision-making assumptions.
5. Preparation of a FAST diagram for the no-action alternative.

Day 2

1. Brainstorming session to
   - identify criteria that might affect future decisions; and
   - identify high-level decisions that might need to be addressed by the EIS. Rank decisions in importance.
2. Construction of a DIT (summarized below).

Day 3

1 (a) List, consolidate, and prioritize preliminary scope of facilities, functions, and issues to be considered in the EIS.
1 (b) Based on the key issues, develop a list of facilities and functions that would need to be analyzed in the EIS.
2 (a) De-scoping: determine issues that will not be included in the preliminary scope of the EIS.
2 (b) Document issues that are important but not included in the preliminary scope so that they can be revisited at a later date. Some issues might be flagged as candidates for later tiering.
3. Identify any outstanding issues or concerns.
4. Brainstorm ideas for reducing EIS cost and schedule in order to identify specific actions that can be taken to help streamline the scope and EIS process.

Day 4

1. Wrap-up: Assign actions and prepare for managerial presentation.

TABLE 2.4
Goals and Objectives of the Prescoping Workshop

Goals of the scoping workshop

- Determine preliminary scope of the actions, facilities, and operations requiring evaluation in the EIS.
- Identify methods for streamlining the EIS process.
- Define the scope of decisions that might need to be considered in the future.
- Reduce risk of later surprises (i.e., changes in scope resulting from inadequate communications or planning).
- Provide a basis for determining the cost and schedule of the EIS.

Objectives of the scoping workshop

- Develop a FAST diagram of the no-action alternative.
- Determine the underlying need for taking action. Identify the purpose (objectives) that the proposal will need to meet.
- Construct a DIT to determine specific decisions that might need to be made (described in Chapter 8).
- Identify key issues and concerns.
- Identify factors that might influence the decision-making process.
- Determine issues that will not be included in the scope (de-scoping).
- Document issues that are important but not to be included as part of this scope, so that these issues can be revisited at a later date (i.e., later tiering).

A kick-off meeting was held prior to initiating the VE workshop to discuss its purpose and objectives, and to assign action items. The prescoping workshop mission statement and agenda were distributed at the meeting. Certain members were identified and requested to prepare presentations for the information sharing session to be held on the first day.

2.5.2 INFORMATION PHASE

The prescoping workshop began with a review of the workshop mission statement and an explanation of the process and techniques used. The workshop then proceeded to the information phase, which was used to gather and share information that might have a bearing on determining the preliminary scope. Issues such as other ongoing NEPA activities, inconsistencies, gaps in NEPA documentation, related documents and studies, and other related or planned projects were identified during this phase. Information such as keywords, memories (e.g., ideas and concepts), issues and concerns, and assumptions were captured on flip charts. The facilitator was responsible for keeping the session focused on schedule. Except for cases where unusual circumstances exist, it is recommended that this phase be limited to no more than a few hours.

An exercise was performed to identify key planning assumptions. A brainstorming exercise then followed in which the facilitator challenged the group to identify assumptions important to this planning effort. Assumptions identified during the remaining portion of the workshop were continuously added to the flip chart list.

Similarly, an exercise was performed to identify principal planning documents and studies that might have a bearing on the preparation of the EIS. Capturing such information at this early planning phase is important, not only because it may affect the scope, but also because it could prevent duplication and wasted effort.

A flip chart list was also maintained for capturing special issues and concerns raised throughout the workshop. Such a list should be referred to both in preparing the formal EIS scoping process and in preparing the actual analysis.

2.5.3 DECISION-IDENTIFICATION PHASE

An EIS cannot properly support subsequent decision-making if it does not correctly anticipate and address decisions that might require future consideration. While such an observation appears obvious, it is not uncommon to find that the EIS has been completed only to discover that it does not adequately anticipate the types of decisions that actually needed to be considered. Such discrepancies frequently result from disconnects between the scope of the analysis and the actual decision-making that follows. This observation is particularly true when dealing with complex actions or dynamic circumstances.

For this reason, the prescoping workshop used a systematic approach developed by the author and referred to as decision-based scoping (DBS) and decision-identification tree (DIT), which is in marked contrast to the way most scoping efforts are typically conducted. A DBS approach is especially well suited for large or complex EISs, and in scoping programmatic EISs. The DIT provided a systematic methodology for identifying and mapping potential decisions that might need future consideration by a decision-maker and which therefore need to be addressed in the scope of the EIS. The DBS and DIT are both detailed in Chapter 8.

2.5.4 UNDERLYING NEED AND PURPOSE: DEFINITION

In support of the DBS approach, an effort was mounted to define the underlying need for future action in a succinct manner. Identifying and properly defining the "need" for taking action can be deceptively complicated. The need might at first appear intuitively obvious, yet on closer inspection confusion or differing views as to its true meaning may arise. Note that *Webster's Dictionary* defines "need" as "a want of something requisite, desirable, or useful."

To this end, the group was challenged to identify the underlying need(s). Responses voiced by the participants were recorded on a flip chart by the facilitator. Next, the group reviewed, sorted, and consolidated these responses to develop the required succinct definition of the primary or underlying need. Similarly, an exercise was conducted to determine potential objectives to aid in identifying the underlying purposes. As detailed in the companion text, *Environmental Impact Statements*, precise definition of the underlying need is an important first step to take because it can drive the range of alternatives that will be investigated later.[16,17] Not surprisingly, even a small change in the definition of need can have profound implications on the alternatives that are eventually chosen for analysis. Thus, correctly defining the need at this early stage can substantially improve the effectiveness of the planning process that follows.

Once consensus was reached regarding the need, the IDT was tasked with defining the purpose for taking action. The term purpose should not be confused with the term need. "Purpose" is defined in *Webster's* as a "goal" or "object" to be obtained.

2.5.5 Enhancing Effectiveness of the EIS Process

An effort was also mounted to identify methods and approaches for reducing cost and expediting preparation of an EIS. Brainstorming techniques were used to elicit ideas from participants in an attempt to investigate every conceivable method for improving the efficiency and effectiveness of the potential EIS. The facilitator recorded suggestions offered by the participants on flip charts. Special VE techniques were used to evaluate these suggestions. The evaluation consisted of three distinct stages or rounds.

In the first round, the group reviewed all the suggestions in an attempt to eliminate those that were clearly unreasonable or of little value. Item by item, the facilitator led the IDT through each suggestion in an attempt to identify which of them could be dropped. The rule used in this round was that unanimous consensus had to be obtained before an item could be eliminated.

Because of time constraints, the second and third rounds that followed were actually combined into a single round for this EIS effort. Normally, the second round consists of consolidating and combining suggestions into manageable categories. The third and final round normally involves the use of matrix weighting methods, such as a nominal group technique, in an effort to generate a final list of concepts. On completing this task, the team formulated specific recommendations that could be undertaken to expedite preparation of an EIS. As a final step, success-oriented criteria (e.g., cost and schedule considerations, ease of implementation, and budget constraints) were used to prioritize the recommendations for consideration by decision-makers.

2.5.6 Investigating the No-Action Alternative

Where an EIS is being prepared for a complex proposal or program, the IDT might also want to consider preparing a FAST diagram to assess activities currently being conducted as part of the no-action alternative. The FAST diagram technique is founded on the observation that a rigorous evaluation of functions underlining a particular process provides a basis for evaluating problems and alternatives. The FAST diagram provides a systematic tool for identifying these functions.

Process functions were first identified and described using sentences containing active verbs and proper nouns. Using these functions, a diagram was constructed illustrating "how" and "why" particular functions are conducted. Scanning the diagram from right to left reveals "why" particular functions are performed. Conversely, scanning the diagram from left to right reveals the sequence of "how" functions are conducted. When completed, the sequence of how functions are conducted should be internally consistent with the reason why each function is conducted.

Next, an effort was undertaken to identify methods for improving or optimizing a program function. A FAST diagram can provide a useful tool for understanding the current baseline, identifying functional requirements and relationships, and challenging preconceived assumptions and ideas. Where appropriate, other planning considerations such as resource and infrastructure requirements may also be identified.

As described in Section 2.4.2.2, preparation of a FAST diagram can be resource intensive, consuming a substantial amount of a scoping workshop schedule. And, as previously stated, in some cases a FAST diagram might contribute valuable information, whereas in other cases little or no benefit would be derived from its construction. Lacking a clear and specific objective for its use, practitioners are ill-advised to devote scarce resources constructing a FAST diagram. For this reason, it is well to reiterate that prudence and professional judgment must be exercised in determining the appropriate use and application (if any) of a FAST diagram with respect to scoping.

2.5.7 Eliminating Issues from the Scope

One of the principal objectives of the scoping process is to eliminate unimportant issues. De-scoping is a powerful method for reducing the cost of an EIS and it is a requirement that is often underutilized.[18] Efforts should be focused on eliminating insignificant or irrelevant issues. Properly applied, such a method can substantially reduce time and resources required in preparing an EIS.

Partly for this reason, an effort was mounted to identify candidate issues for elimination during the formal scoping process. Emphasis was placed on documenting issues that were considered important but eligible for deferral from the immediate scope of this EIS. These issues could thus be revisited, if necessary, at a later date (some issues might be flagged as candidates for later tiering).

2.5.8 Identifying Other Related Requirements

Regulatory requirements can profoundly influence cost, schedules, and the direction of future actions. For this reason, it is recommended that an exercise be performed to identify other related environmental and regulatory drivers.

For example, a preliminary matrix might be prepared of potential regulatory requirements versus potential alternatives that eventually would be analyzed. A regulatory compliance matrix provides a "heads-up" so that the agency is not surprised by unanticipated requirements that may emerge later. The stage is now set for presenting the results.

2.5.9 Finalizing the Prescoping Effort

The final phase involved is presenting the results of the prescoping workshop. Focused toward peer review, this phase allows the agency and decision-maker(s) to review the potential scope to determine if issues that have not been addressed remain.

In this case, the general consensus was that construction of the DIT, in conjunction with the VE workshop, had been very effective in identifying what was otherwise a rather enigmatic set of potential issues and actions. With only one exception, the prescoping workshop successfully identified principal issues and actions considered essential for analysis.

It is important to note that at this stage the scope is still preliminary and may evolve once additional information or public scoping comments are received. Still this effort was very effective in providing an early indication of the scope, which was useful in planning the public scoping process, identifying agency consultations, preparing cost estimates and schedules, and developing preliminary engineering and environmental support studies.

Once this step has been completed, the stage is now set for preparing and issuing a notice of intent (NOI), which completes the prescoping phase.

A facilitated workshop approach such as the one described above may also offer a useful tool for effectively managing and performing formal public scoping sessions. Such an approach also has the advantage of ensuring that each participant has an opportunity to provide feedback and voice opinions.

2.6 A STRATEGY FOR EFFICIENTLY IMPLEMENTING A POLLUTION PREVENTION PROGRAM

In 1990, the U.S. Congress passed the Pollution Prevention Act.[19] President Clinton issued an executive order instructing federal agencies to implement pollution prevention (P2) measures in 1993.[20] Similarly, the International Organization for Standardization (ISO) has developed a series of environmental management, auditing, and performance standards known as the ISO 14000 series; the ISO 14000 series speaks to the merits of pollution prevention, particularly from the standpoint of establishing a top-level policy committed to pollution prevention.

The following section describes a systematic strategy developed by the author for effectively integrating P2 objectives with DOE Hanford Site projects and operations. This integrated NEPA/P2 strategy received two Secretary of Energy commendations. The purpose of the following example is to illustrate how a federal facility's NEPA planning process can be used to efficiently implement a P2 program.

Because project requirements, procedures, operations, and the details of each agency's NEPA process vary, the readers should consider this example from the perspective of how it can be adapted to their own specific internal projects and programs.

2.6.1 NEPA Provides an Ideal Implementation Mechanism

With very few exceptions, all proposed federal actions are potentially subject to NEPA's require-ments. For this reason, NEPA provides the ideal mechanism or framework for efficiently reviewing new (proposed) federal actions that may present opportunities to reduce pollution and waste gen-eration. While P2 is not specifically mentioned, its inherent objectives are captured in the purpose, goals, and requirements of NEPA, and its implementing regulations. Significant cost savings may be realized by properly integrating NEPA with P2.

NEPA regulatory direction is consistent with the purpose and goals of an integrated NEPA/P2 strategy (Table 2.5). Table 2.6 shows that NEPA's policy is consistent with that of P2. As indicated

---

TABLE 2.5

NEPA's Regulatory Direction is Consistent with an Integrated Pollution Prevention Act

- Ultimately, of course, it is not better documents but better decisions that count. NEPA's purpose is not to generate paperwork—even excellent paperwork—but to foster excellent action. (§ 1500.1[c])
- The NEPA process is intended to help public officials make decisions that are based on understanding of environmental consequences, and take actions that protect, restore, and enhance the environment. (§ 1500.1[c])
- Use all practicable means, consistent with the requirements of the Act and other essential considerations of national policy, to restore and enhance the quality of the human environment and avoid or minimize any possible adverse effects of their actions upon the quality of the human environment. (§ 1500.2[f])

TABLE 2.6

NEPA's Policy is Consistent with the Pollution Prevention Act

- "...to promote efforts which will prevent or eliminate damage to the environment and biosphere and stimulate the health and welfare of man; to enrich the understanding of the ecological systems and natural resources important to the Nation ... (Sec. 2 [42 USC § 4321])
- use all practicable means [to:]
  - fulfill the responsibilities of each generation as trustee of the environment for succeeding generations;
  - assure for all Americans safe, healthful, productive, and aesthetically and culturally pleasing surroundings;
  - attain the widest range of beneficial uses of the environment without degradation, risk to health or safety, or other undesirable and unintended consequences;
  - achieve a balance between population and resource use which will permit high standards of living and a wide sharing of life's amenities; and
  - enhance the quality of renewable resources and approach the maximum attainable recycling of depletable resources."

(NEPA, Sec. 101 [42 USC § 4331][b], 1969)

TABLE 2.7
Integrating NEPA with Other Environmental Processes

- Integrate the requirements of NEPA with other planning and environmental review procedures required by law or by agency practice so that all such procedures run concurrently rather than consecutively. (§ 1500.2[c])
- Integrating NEPA requirements with other environmental review and consultation requirements. (§ 1500.4[k])
- Combining (combine?) environmental documents with other documents. (§ 1500.5[i])

TABLE 2.8
NEPA Direction for Analyzing Actions such as Pollution Prevention Measures

- Use the NEPA process to identify and assess the reasonable alternatives to proposed actions that will avoid or minimize adverse effects of these actions upon the quality of the human environment. (§ 1500.2[e])
- Study, develop, and describe appropriate alternatives to recommended courses of action in any proposal which involves unresolved conflicts concerning alternative uses of available resources as provided by Section 102(2)(E) of the Act. (§ 1501.2[c])
- To assess the adequacy of compliance with Sec. 102(2)(B) of the Act the statement shall, when a cost-benefit analysis is prepared, discuss the relationship between that analysis and any analyses of unquantified environmental impacts, values, and amenities. (§ 1502.23)
- Rigorously explore and objectively evaluate all reasonable alternatives … (§ 1502.14[a])

in Table 2.7, NEPA is to be combined and integrated with other environmental requirements and processes whenever practical; and, as demonstrated in Table 2.7, NEPA provides both an effective and efficient environment framework for incorporating P2 objectives into new actions.

2.6.2 NEPA Provides a Comprehensive Decision-Making Process

A component essential to NEPA as well as to some aspects of P2 is a requirement to rigorously analyze environmental impacts, alternatives, and mitigation measures. As depicted in Table 2.8, the NEPA regulations provide specific direction for analyzing actions such as P2 measures.

NEPA provides decision-makers with many advantages not found in other environmental processes. Moreover, while NEPA mandates rigorous requirements for performing environmental analyses, it places no substantive restrictions on agencies regarding the final decision that may be reached. NEPA only requires that the final decision has been adequately investigated, such that a decision-maker can make an informed decision regarding the course of action to be taken. NEPA, therefore, allows decision-makers to consider many diverse factors, such as environmental impacts, economics, schedules, technology, risk, and other practical considerations, in reaching a final decision.

Although P2 is often touted as a cost savings measure, some waste reduction actions may actually result in cost increases. In practice, decision-makers may be faced with having to make difficult tradeoffs between the goals of P2, other environmental requirements, scarce resources, cost, and agency requirements. Fortunately, NEPA provides decision-makers with a flexible and analytical planning process for evaluating such tradeoffs and resolving dilemmas in an effort to reach a rationally based and legally defensible decision.

2.6.3 Process for Integrating NEPA and P2

A set of checklists has been prepared by the DOE for assessing a diverse range of activities in terms of potential P2 measures. These checklists have been used by employees to identify measures that

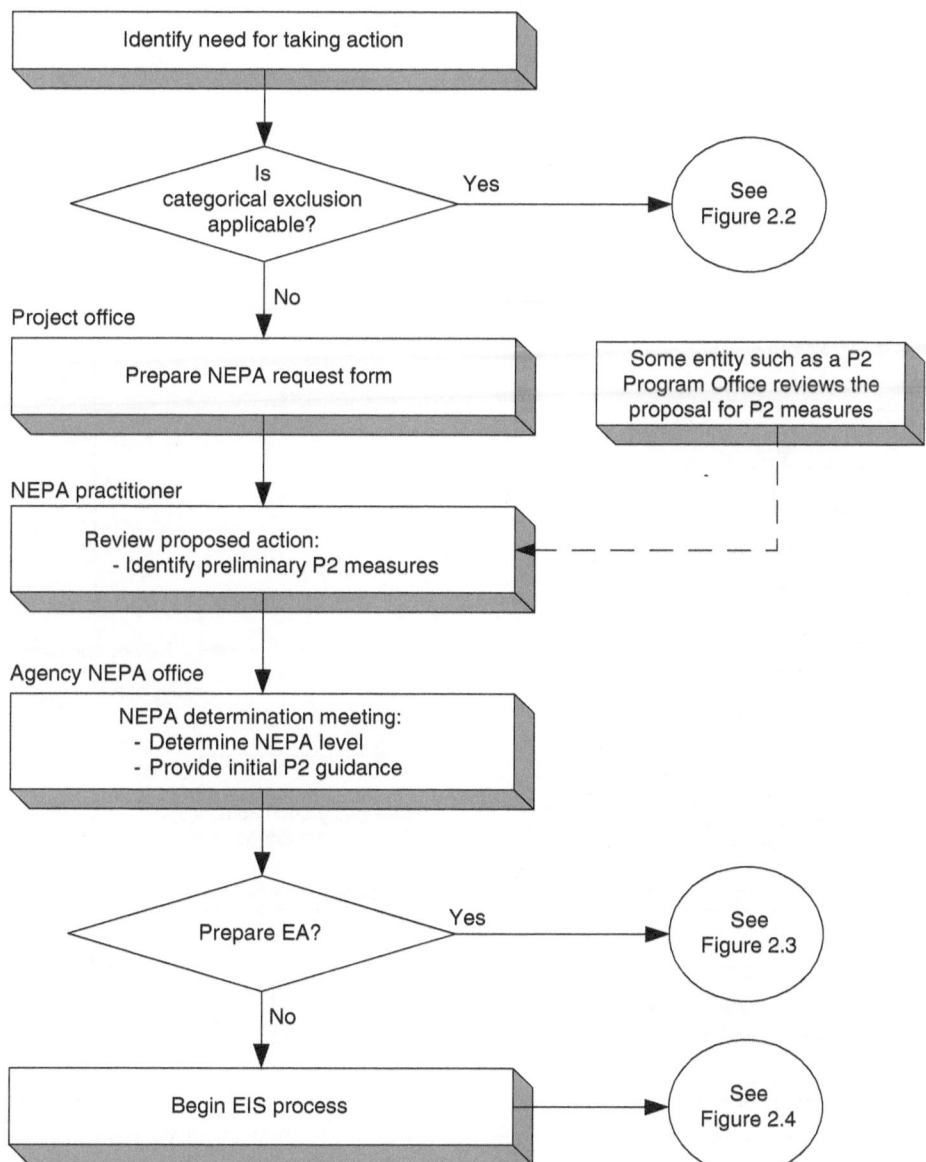

FIGURE 2.1   Proposed process for integrating pollution prevention measures with NEPA.

could be taken to implement specific P2 measures. It is important to note that P2 measures may be applicable to a proposed action even though the action may not result in a significant impact.

Figure 2.1 depicts a high-level integrated strategy developed for the U.S. Department of Energy's Hanford Site. The NEPA process provided the framework for efficiently implementing a comprehensive P2 program. Figures 2.2 through 2.4 describe the details of implementing the integrated NEPA/P2 strategy.[21] The readers should consider how this site-specific strategy could be modified and adopted to other sites or programs.

The following discussion focuses on describing the implementation of a P2 program rather than describing specific aspects of the NEPA process itself. As depicted in Figure 2.1, an integrated NEPA/P2 process begins with identifying a need to take action.

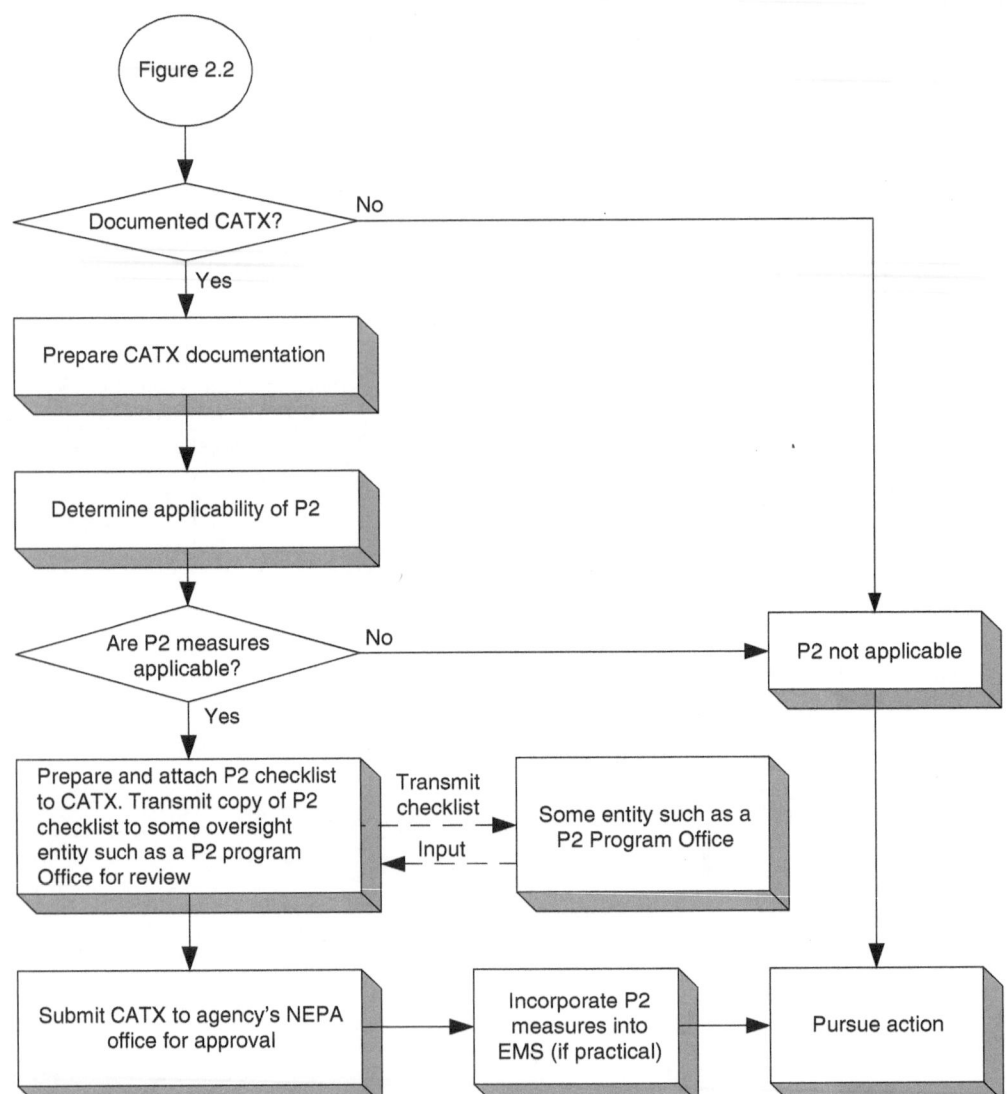

FIGURE 2.2   Process for integrating pollution prevention measures with the CATX process.

2.6.3.1 Categorical Exclusions

Typically, a review is performed to determine if a proposed action can be categorically excluded from further NEPA review. Specific details of a categorical exclusion (CATX) process may change from agency to agency, depending on an installation's internal procedures, requirements, and constraints. Consequently, this section will not dwell on the specific implementation details of the integrated NEPA/P2 process that was developed for the Hanford Site. Instead, readers should consider how a P2 program can be adapted to their specific internal CATX process. A generalized summary of the integrated NEPA/P2 process is described below.

The CATX response is depicted by the diamond labeled "Is categorical exclusion applicable?" in Figure 2.1. If the response is "yes," the reader is directed to Figure 2.2. Figure 2.2 details the process used in determining if and when P2 measures are applicable to an action that is eligible for a CATX.

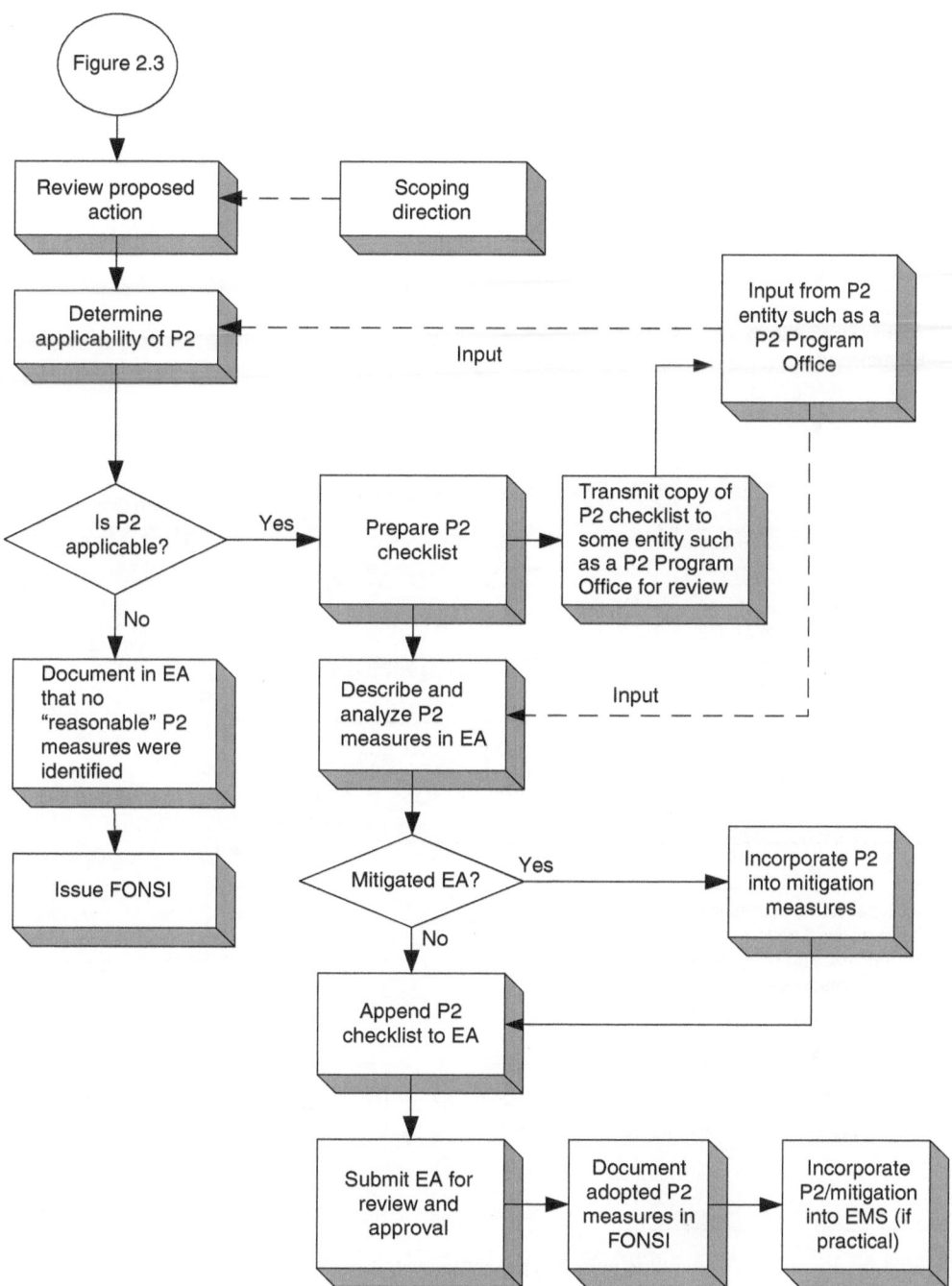

FIGURE 2.3 Process for integrating pollution prevention measures into EAs.

Some agencies recognize two types of CATXs: (1) documented and (2) nondocumented. This step is depicted by the diamond labeled "Documented CATX?" in Figure 2.2. If the CATX requires no documentation ("no" path, diamond labeled "Documented CATX?"), this strategy assumes that the action and its impacts are essentially trivial, and it is therefore not practical to review the action for potential P2 measures. If the CATX requires documentation ("yes" path, diamond labeled

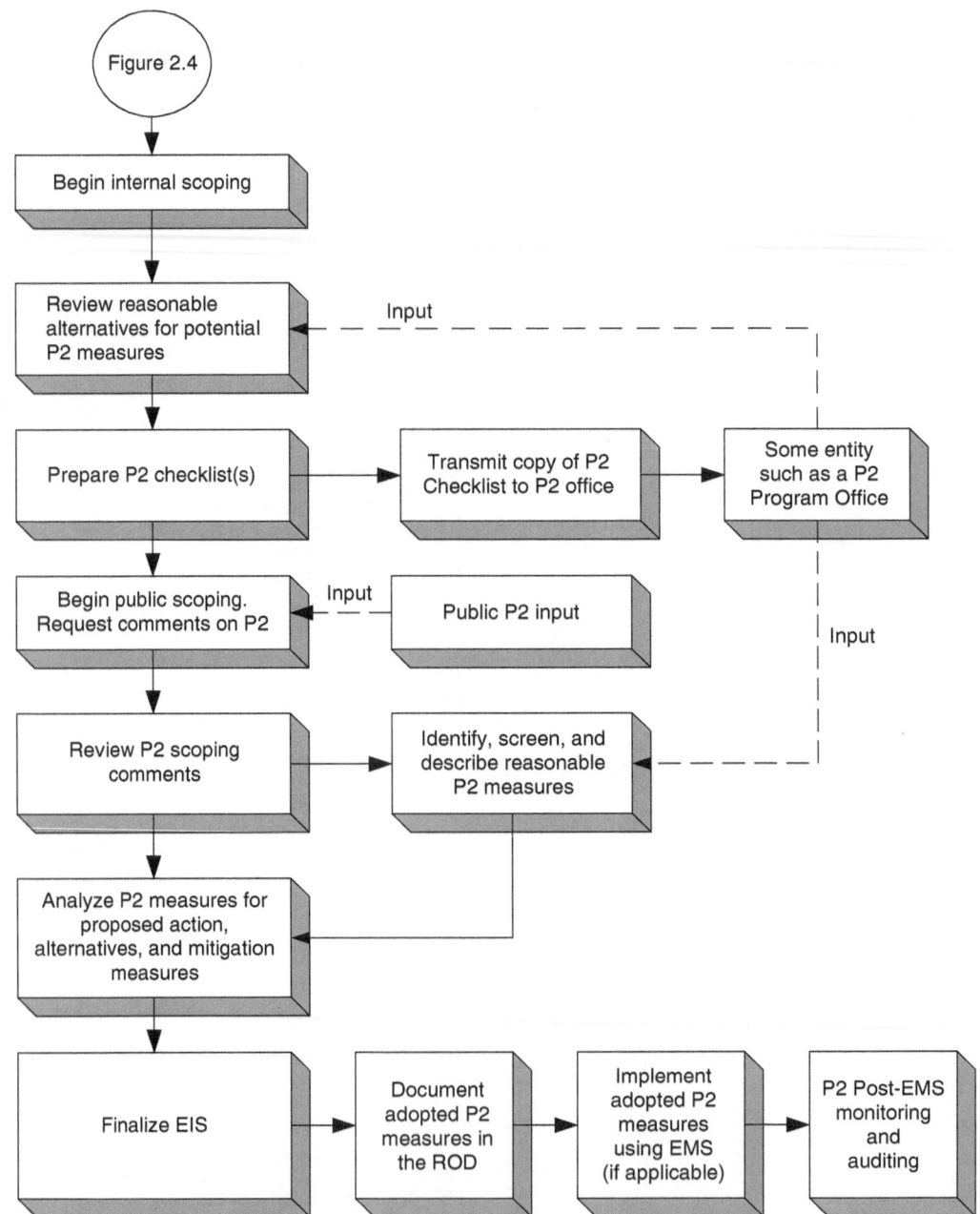

FIGURE 2.4   Process for integrating pollution prevention measures into EISs.

"Documented CATX?"), a review is performed to determine if P2 measures are applicable to the proposed action.

The action is pursued if no practical P2 measures are identified ("no" path, diamond labeled "Are P2 measures applicable?"). If practical P2 measures are identified ("yes" path, diamond labeled "Are P2 measures applicable?"), an appropriate P2 checklist(s) is prepared and attached to the documented CATX for review. If appropriate, the P2 Program Office (or equivalent) provides input and assistance in preparing the P2 checklist. A copy of the completed P2 checklist is transmitted to some responsible entity such as a P2 Program Office. Where practical, any adopted P2 measure could be implemented and monitored through an environmental management system (EMS), such as the one outlined in Section 2.7.

2.6.3.2 Environmental Assessment Process

A different strategy is followed for actions involving preparation of an environmental assessment (EA). This process might be triggered by filling out a NEPA request form (see box labeled "Review proposed action," Figure 2.3). The NEPA request form could be modified to require an initial screening for potential P2 measures; a review is then performed to determine the applicability of P2 measures. The proposed action is reviewed to identify potential measures for achieving NEPA/P2 goals such as reducing or preventing pollution, waste minimization, and energy conservation. Where appropriate, the P2 office assists in identifying potential P2 measures.

If no P2 measures are considered practical ("no" path, diamond labeled "Is P2 applicable?"), this fact is documented in the EA; the EA then continues along the standard NEPA path.

If P2 measures are deemed to be applicable ("yes" path, diamond labeled "Is P2 applicable?"), a P2 checklist(s) should be prepared indicating methods that can be implemented to minimize waste and to prevent pollution. A copy of the completed P2 checklist(s) is transmitted to the P2 office or equivalent.

As appropriate, P2 measures identified in the P2 checklist(s) are incorporated into the description of the proposed action and reasonable alternatives described in the EA. Any applicable P2 measures should then be reviewed to determine their effectiveness.

If a mitigated EA is prepared ("yes" path, diamond labeled "Mitigated EA?"), applicable P2 measures could be incorporated into the mitigation measures.

The P2 checklist(s) is appended to the final EA to be submitted for review and approval. If the EA is eligible for a finding of no significant impact (FONSI), the decision to adopt any applicable P2 measures described in the EA should be documented in it. Where practical, any adopted P2 measures could be chosen through the NEPA process (e.g., FONSI/record of decision [ROD]) and implemented/monitored with an EMS.

2.6.3.3 EIS Process

Where a proposed action involves preparation of an EIS (Figure 2.4), an effort could be performed during the internal scoping process to identify potential P2 measures for achieving NEPA/P2 goals such as reduction or prevention of pollution, waste minimization, and energy conservation. Where appropriate, the P2 office (or equivalent) provides assistance in identifying potential measures.

As indicated in the box labeled "Prepare P2 checklist(s)" in Figure 2.4, a P2 checklist(s) can be filled out indicating potential methods that could be implemented to minimize waste and prevent pollution. A copy of the completed P2 checklist(s) should then be transmitted to the P2 office.

Potential pollution prevention measures identified in the P2 checklist(s), combined with input obtained from the public scoping process, are reviewed to identify specific P2 measures for investigation. Reasonable P2 measures should then be incorporated into the description of the proposed action, and reasonable alternatives described in the EIS. The measures are then analyzed to determine their effectiveness and potential impacts. The analysis should also consider options and costs for implementing P2 measures.

The P2 checklist(s) is archived as part of the administrative record and the completed EIS is submitted for review and approval. The decision to adopt any of the P2 measures investigated in the EIS would be documented in the ROD. Where practical, any adopted P2 measures could be monitored and implemented through an EMS. The EMS could be used in a monitoring and auditing capacity to ensure that the P2 measures are correctly implemented.

2.7 NEPA AND ISO 14001

The effectiveness of NEPA as a planning tool is sometimes diminished because it has not been properly either implemented or integrated into agency's planning process. In 1996, the author began investigating commonalities and similarities which existed between NEPA and the relatively new ISO 14001 EMS. A final report was issued to the President of the National Association of Environmental Professionals (NAEP) in 1997, outlining a process for integrating NEPA with an ISO 14001 EMS, which was published in 1998.*,[22] In 2000, this report was reviewed and approved by the NAEP Board of Directors and issued to the Council on Environmental Quality (CEQ) as a strategy that should be promoted to federal agencies. The following section summarizes this strategy.

Although this strategy is directed at integrating NEPA with an ISO 14001 certified EMS, it can also be integrated with any EMS that is consistent with the ISO 14001 standard. Hence, as used in this chapter, the term EMS is interpreted to mean an EMS consistent with the ISO 14001 series of standards. This strategy has since been generalized to describe a process for integrating any environmental impact assessment (EIA) process with an EMS.[23]

2.7.1 AN ISO 14001 CONSISTENT EMS

The ISO 14000 environmental management standards define specific procedures to assist an organization in complying with applicable statutes, regulations, and environmental requirements; minimizing how the organization's activities negatively affect the environment, and for continually improving on the above. ISO 14000 is similar to the ISO 9000 quality management system in that both focus on the management process used (e.g., how a product is produced) rather than on the ultimate quality of the product itself.

An EMS is part of an organization's management system. The EMS specifies requirements and procedures for implementing the organization's environmental policy. ISO 14001 provides an internationally accepted specification for the EMS. The ISO 14001 EMS has the following five components:

- Detail procedures for defining an environmental policy with a commitment to pollution prevention, continual improvement, and compliance with relevant environmental legislation and regulations.
- Specific procedures for identifying an organization's (products/services/activities) environmental aspects, legal and other responsibilities, and environmental management programs.
- Implementation and operation procedures; includes identification of responsibilities, training and awareness, documentation, and operational controls.
- Developing measurable targets for meeting environmental objectives; verification and corrective action procedures; includes monitoring and measuring performance to meet targets for continual improvement.
- Review by management to ensure that the EMS is addressing changing conditions and information.

Presidential Executive Order 13148 directed federal agencies to develop EMSs by the year 2005.[24] As explained in more detail later, strong parallels exist between the goals and requirements of NEPA and the specifications for implementing an EMS. Because there is a general expectation under an ISO 14001 EMS that action(s) will be taken to improve environmental quality, performance of an integrated NEPA/EMS can provide a mechanism for infusing NEPA's substantive national policy goals into federal decision-making.

Figure 2.5 shows a simplified process for implementing an EMS. The EMS provides a structured system (plan-do-check-act) in which a set of management procedures is used to systematically

---

* Eccleston C.H., " A Conceptual Strategy for Integrating NEPA with an Environmental Management System," issued to the President of the National Association of Environmental Professionals, 1997. Issued to CEQ, 1990/2000.

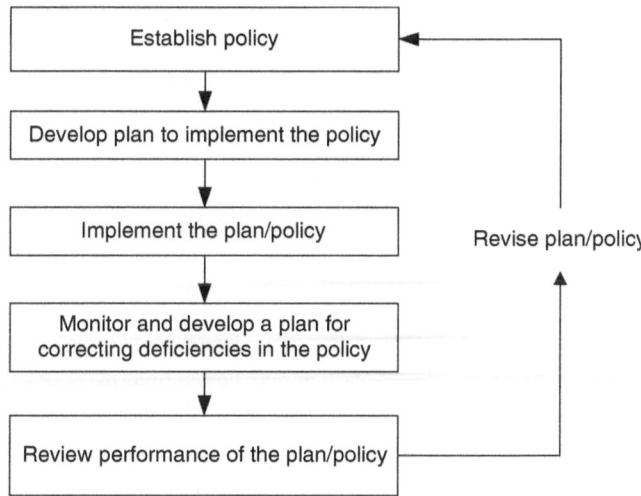

FIGURE 2.5 Simplified overview of a typical EMS.

identify, evaluate, manage, and address environmental issues and performance. The EMS provides organizations with a mechanism which helps ensure that necessary actions are taken to integrate environmental safeguards and compliance into day-to-day operations and long-term planning with respect to an organization's (e.g., including federal facilities) activities, products, and services.

As depicted in Figure 2.5, the EMS employs a rigorous monitoring cycle for managing and continually improving an organization's (federal facility) environmental performance. The continuous improvement cycle is achieved by identifying how the organization's activities affect environment quality. This is accomplished through

- assessing an organization's environmental aspects (described later),
- establishing defined environmental targets and objectives,
- training staff and defining departmental and personal responsibilities,
- cyclical monitoring and implementing corrective actions as necessary to avoid or minimize impacts, and
- instituting procedures and controls for responding to emergencies.

An EMS provides an effective mechanism for tracking, monitoring, and improving the implementation of the agency's selected alternative, including mitigation measures and other commitments established in final NEPA documents; monitoring such commitments can support internal auditing requirements and identification of appropriate corrective actions.

Both NEPA and EMS processes involve a review of activities to identify those with potentially significant environmental impacts and to implement measures to avoid, minimize, or eliminate the causes of adverse environmental impacts. For reasons such as these, an integrated NEPA/EMS methodology can substantially improve an agency's environmental performance and thus environmental quality and may reduce the cost of complying with NEPA and EMS requirements. Some of the key advantages of an integrated NEPA/EMS system are

- incorporating NEPA documents and schedules into EMS objectives and targets;
- archiving the NEPA administrative record and NEPA documents as part of the EMS system for maintaining records;
- integrating NEPA mitigation commitments with other related regulatory requirements and EMS objectives and targets;

- including NEPA impact assessments as part of the process of identifying the environmental aspects under an EMS; and
- monitoring the implementation of the selected NEPA alternative and applicable mitigation and monitoring commitments.

Some agencies have already begun pursuing an integrated NEPA/EMS strategy. For example, new Forest Service planning regulations require 126 Forest Service administrative units to implement EMSs containing procedures for identifying and monitoring environmental conditions. The preamble to the Forest Service regulations declares that:

> … through the implementation of EMS, administrative units will be continually collecting and evaluating the data necessary to create any document that may be required by *NEPA*. This will make the creation of accurate and relevant *NEPA* documents more efficient.[25] (emphasis added)

2.7.2   WHY NEPA AND ISO 14001 COMPLEMENT EACH OTHER

As depicted in Table 2.9, each system has its strengths and weaknesses; moreover, a weakness in one system tends to be offset by the strengths of the other.

Some of the principal characteristics outlined in Table 2.9 are described below.

TABLE 2.9
How an ISO 14001 Consistent EMS and NEPA Complement Each Other

Characteristics	EMS	NEPA
Substantive mandate	Substantive actions are expected to be taken which lead to continual improvement in environmental performance (and thus environmental protection).	Lacks a legally binding substantive mandate to protect the environment.
Goal	ISO's goal is to provide environment protection by identifying impacts. Its continual improvement system is used to reduce these impacts and improve performance.	NEPA's goal is to provide environment protection by ensuring that environmental factors are considered during the early planning process.
Planning function	A planning function is required which provides a system for ensuring that decisions are appropriately implemented. However, ISO 14001 does not prescribe a detailed process for performing the planning function.	A comprehensive environmental planning process is defined, but it lacks an environmental system for ensuring that decisions are properly implemented.
Impact assessment requirements	An investigation of "environmental aspects" is required. However, little specificity is provided regarding the requirements for performing this investigation.	Specifies detailed direction for performing an analysis of direct, indirect, and cumulative impacts.
Life cycle	Details how to perform a life-cycle analysis (ISO 14040 series).	An analysis of reasonably foreseeable impacts over the life cycle of the action is required.
Significance	Lacks detailed direction for interpreting or determining the meaning of significance.	In addition to the context, 10 specific factors are detailed in the NEPA implementing regulations for determining the significance of environmental impacts.

TABLE 2.9 (Continued)

How an ISO 14001 Consistent EMS and NEPA Complement Each Other?

Characteristics	EMS	NEPA
External input	A nonpublic procedure used to record and respond to external parties is required, but ISO 14001 does not prescribe detailed requirements.	Specifies a detailed public participation and formal public scoping process for identifying actions impacts, and alternatives, and for eliminating nonsignificant issues.
Other environmental requirements	A top-level environmental policy is required, including a commitment to prevention of pollution, which is very broadly defined.	CEQ guidance and Executive Orders direct federal agencies to integrate pollution prevention measures, environmental justice, biodiversity, and other considerations with NEPA.
Mitigation	Provides a system that can be used to ensure that mitigation measures are implemented.	Mitigation measures are required to be identified and analyzed, but are not normally required to be chosen or implemented.
Preventive and corrective action	ISO 14001 specifies procedures that can be used to (1) identify circumstances where NEPA commitments or mitigation measures are not being correctly implemented; (2) correcting nonconformities; (3) mitigating their impacts; and (4) developing plans to avoid nonconformities.	A NEPA adaptive management (AM) system can provide an effective process for taking corrective actions, and making mid-course corrections.
Records and documentation	An EMS specifies detailed procedures for controlling and maintaining records needed to demonstrate conformance with the EMS standard.	An EMS can be used for maintaining NEPA records.
Monitoring	Monitoring is mandated as part of the continual improvement cycle.	Encourages (and sometimes requires) post monitoring measures.
Continual improvement	A continual improvement process is a basic concept inherent in an EMS.	CEQ has promoted a cyclical process known as AM.
Audits	ISO 14001 defines internal auditing requirements for periodically assessing conformity to the EMS; the audit results must be presented to management.	NEPA conformance and commitments may be reviewed and audited where such commitments are linked to EMS objectives and targets. An EMS audit provides a means for ensuring that NEPA process and commitments are being correctly implemented.

2.7.2.1 Policy and Plan

The NEPA regulations identify categories of federal activities (i.e., policies, programs, projects, and plans) that are subject to NEPA review. The establishment of federal policies and plans are therefore actions that may be subject to the requirements of NEPA. Thus, federal policies and plans established as part of an EMS may potentially be subject to NEPA requirements, particularly where policies/plans entail significant environmental impacts, issues, or consequences.

While environmental planning is a mandatory element within an EMS, the ISO 14001 standards provide only limited specifications for performing the planning function. For example, specific procedures and requirements with regard to scoping, investigating environmental aspects, defining temporal and spatial bounds, interpreting significance, and other requirements are at present only vaguely inferred or defined.

In contrast, the NEPA regulations provide highly prescriptive direction and requirements for ensuring that an accurate and scientifically defensible analysis is prepared which provides decision-makers with information sufficient to reach an informed decision. Moreover, these requirements are reinforced by nearly four decades of experience gained by agencies through their engagement with diverse missions and environmental issues.

Properly integrated, a combined NEPA/EMS system can provide a synergistic process for planning actions and implementing decisions in a manner that protects and enhances environmental quality, while minimizing cost.

2.7.2.2 Procedural versus Substantive

While an agency must comply with the procedural aspects of NEPA, the courts have largely ruled that an agency is not obligated to select an environmentally preferable alternative or to demonstrate that its decision conforms to the environmental goals (i.e., substantive mandate) established in Section 101 of the Act. Thus, NEPA's contribution is derived not from a substantive mandate to choose an environmentally beneficial alternative but from its procedural requirement, which forces decision-makers to rigorously evaluate and consider the effects of potential actions on the environment, just as they would balance other traditional factors such as cost and schedules.

In contrast, an ISO 14001 consistent EMS involves a general expectation that some type of substantive action(s) will be taken to improve environmental quality. There is at least the expectation that environmentally beneficial actions will be taken, and undertaken in a cyclical process of continual environmental improvement. Thus, an EMS *can* provide at least a limited mechanism for enforcing the substantive environmental mandate that NEPA lacks.

Likewise, NEPA requires analysis of mitigation measures but places no substantive burden on decision-makers to choose or enact them. In contrast, ISO 14001 requires organizations to establish objectives (goals) for improving the environmental quality performance in relation to significant aspects. Similarly, environmental targets are established for measuring and achieving those objectives; targets establish specific and measurable criteria for meeting the objectives.

Achieving these objectives may involve implementing actions similar to NEPA's mitigation measures. Again, NEPA prescribes more rigorous requirements for planning and investigating mitigation measures, while an EMS provides the "teeth" for implementing such measures.

2.7.2.3 Analytical Similarities

Under ISO 14001, the *cause* of an effect is essentially termed an "environmental aspect." Aspects are those specific activities that affect the environment. For example, facility environmental aspects might include production of hazardous waste, wastewater effluents, consumption of fuel, air discharges, traffic, and deliveries or loud plant operation noises. While the environmental aspects must be determined, their environmental consequences or impacts on environmental resources do not necessarily have to be evaluated.

In contrast, the NEPA regulations provide highly prescriptive requirements for ensuring that an accurate and defensible analysis of impacts is performed, which will provide a decision-maker with information that can support informed decision-making. NEPA is more demanding than ISO 14001 as it requires not simply a review of environmental "aspects," but a comprehensive analysis of the actual direct, indirect, and cumulative impacts.

Moreover, as previously stated, the NEPA process is reinforced by nearly four decades of federal experience accumulated by a diverse range of federal agencies, each with its own mission and a spectrum of environmental issues to resolve. From a planning perspective, NEPA provides a more comprehensive and rigorous process for ensuring that environmental impacts are identified, evaluated, and considered before a decision is made to pursue an action.

2.7.2.4 Life Cycle Analysis

As practical, NEPA requires that an analysis be performed over the entire life cycle of an action, including connected actions. Both short- and long-term effects must be considered. To the extent possible, the reasonably foreseeable impacts of future actions must be identified and evaluated. In contrast, the ISO 14040 series describes in detail how a life-cycle analysis could be performed.

2.7.2.5 Public Participation

Public participation is an element essential to the NEPA process. The public scoping process is designed to seek and solicit comments from the public, other agencies, and subject matter experts. Decisions made regarding significance and the choice of alternatives are highly dependent on the concerns of stakeholders. In contrast, the ISO 14001 series provides no requirement for public participation, only a requirement to develop a plan for external communications and inquiries.

Lack of such a requirement can be viewed as a weakness in many parts of the ISO 14001 standard. Thus, NEPA's long experience with public participation and scoping helps to balance the weaknesses of the EMS.

2.7.2.6 Significance

A theme central to both the NEPA and the EMS is the concept of significance. NEPA requires analysis of potentially significant impacts of federal actions. Under ISO 14001, significance is defined vaguely and contains few factors for use in reaching a determination.

In contrast, the concept of significance permeates NEPA's regulatory provisions. The regulations include a detailed definition of significance and in addition to context provide 10 specific factors decision-makers are to use in making such determinations (§ 1508.27). Again, NEPA brings many years of experience to bear on the problem of how best to determine significance.

2.7.2.7 Integration of Environmental Requirements

Federal agencies are instructed to integrate NEPA with other environmental reviews (e.g., regulatory requirements, permits, agreements, project planning, and policies) so that procedures run concurrently rather than consecutively; this requirement reduces duplication of effort, delays in compliance, and also minimizes the overall cost of environmental protection:[25]

- *Identify* other environmental review and consultation requirements … prepare other required analyses and studies *concurrently* with, and *integrated* with the environmental impact statement … (§ 1501.7[a][6], emphasis added)
- Any environmental document in compliance with NEPA may be *combined* with any other agency document … (§ 1506.4, emphasis added)
- *Integrate* the requirements of NEPA with other *planning* and environmental review procedures … (§ 1500.2[c], emphasis added)

The ISO 14001 standard also requires that organizations identify applicable legal and regulatory requirements. The intent is to ensure that organization activities meet applicable legal and regulatory requirements. To this end, an EMS provides a mechanism for efficiently ensuring that regulatory and other requirements are identified and incorporated into an integrated NEPA/EMS process.

2.7.2.8 Integrating Pollution Prevention

The CEQ has issued guidance indicating that, where appropriate, P2 measures are to be coordinated with, and included in the scope of, an NEPA analysis.[26] Some federal agencies have also issued similar directives. The ISO 14000 series speaks to the merits of P2, but mainly from the standpoint of establishing a top-level policy committed to P2.

As described in Section 2.6, ISO 14001 provides a top-down policy for ensuring that P2 is actually incorporated at the operational level, whereas NEPA provides an ideal framework for evaluating and integrating a comprehensive P2 strategy/plan into federal proposals.

2.7.2.9 Preventive and Corrective Action

ISO 14001 requires organizations to establish procedures for taking action to mitigate impacts caused by its activities and for implementing corrective and preventive actions.

Findings and recommendations resulting from EMS monitoring and audits provide the basis for managing corrective/preventive actions. This is similar to NEPA's concept of adaptive management (AM) (described later).

2.7.2.10 Records and Documentation

An ISO 14001 consistent EMS requires that organizations establishing procedures for controlling documents related to the implementation of the EMS. Such records may include environmental aspects (and their corresponding NEPA impacts), regulatory requirements, and monitoring data. This record system can also be used to maintain the NEPA administrative record.

2.7.2.11 Implementation, Training, Monitoring, and Continuous Improvement

Under an ISO 14001 EMS, all personnel whose work could result in a significant environmental impact must receive appropriate training. Such requirements can help ensure that the NEPA decision is correctly and safely implemented.

The NEPA regulations strongly encourage, and in some instances mandate, incorporation of monitoring. Generally, however, the courts have not insisted that agencies incorporate monitoring as part of the NEPA process. In contrast, monitoring is a basic element inherent in an EMS. A properly integrated NEPA/EMS ensures that monitoring is correctly executed.

2.7.2.12 Internal Auditing

ISO 14001 requires periodic internal environmental audits of the EMS. The audit is conducted to verify whether the EMS conforms to the ISO standard and is being properly implemented.

Mitigation and other NEPA-related commitments may be included as part of the EMS audit. Thus, the EMS audit provides another mechanism for ensuring that agency NEPA commitments are appropriately implemented.

2.7.3 STRATEGY FOR INTEGRATING AN EMS WITH NEPA

The NEPA planning process is generally triggered through identification of a need for a proposed action; the process focuses on identifying, assessing, and minimizing impacts of the alternatives to the proposal. In contrast, an EMS addresses the full range of construction and operational activities (i.e., proposals, operations, services) with the intent to continually improve an organization's environmental performance throughout its life cycle.

By establishing applicable environmental objectives and targets, an EMS can help ensure performance and implementation of commitments and mitigation measures through tracking and monitoring programs.

Figure 2.6 illustrates a high-level, general-purpose process developed by the author for integrating an EMS with the NEPA.[22] Figure 2.6 conceptually describes three discrete functions or phases (blocks 2, 3, and 4) that are performed principally using the standard NEPA process:

1. Planning
2. Analyzing potentially significant impacts
3. Decision-making

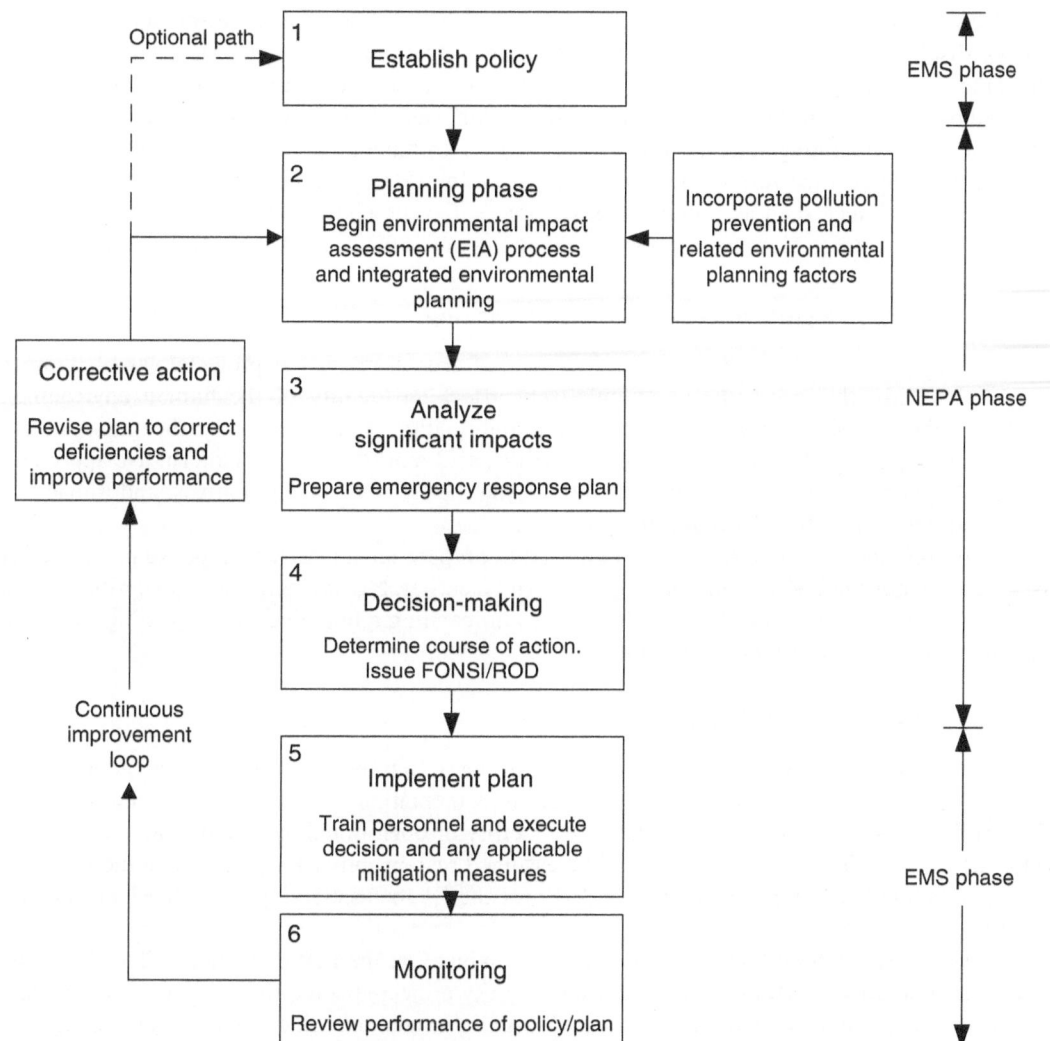

FIGURE 2.6   General-purpose strategy for integrating the NEPA process with an ISO 14001 EMS.

Similarly, the integrated NEPA/EMS system is composed of three additional functions (blocks 1, 5, and 6, as well as a continuous improvement loop) that are performed principally using an EMS process:

1. Establish policy
2. Implement plan/policy
3. Monitoring

2.7.3.1   Policy and Planning Phase

The strategy described in Figure 2.6 is initiated with the establishment of a high-level organizational environmental policy and a commitment to environmental quality (see first block, Figure 2.6).

Next, an effort is mounted to develop a specific plan for implementing the environmental policy (see second block, Figure 2.6); in reality, this might only be performed once.

Drawing on expertise and experience from a diverse array of planning requirements and entities, an interdisciplinary effort may be used in developing an implementation plan (IP). The IP might be prepared for a major federal program or installation. In other cases, an IP might be prepared for a particular facility, operation, or a project-specific action. Here, NEPA provides the interdisciplinary framework for integrating and coordinating all early environmental planning, reviews, and analyses necessary to support the formulation of the IP. Thus, the IP might incorporate an array of related environmental issues such as pollution prevention, safety procedures, habitat management practices, environmental justice, sustainable development, and other related goals or disciplines.

2.7.3.2   Analysis, Significance, and Decision-Making

In the third block of Figure 2.6, NEPA documentation is prepared for projects/plans having the potential to directly, indirectly, or cumulatively affect the quality of the human environment. NEPA's public scoping process is used to obtain public input and separate significant issues from those that are nonsignificant. Consistent with NEPA, different scenarios and alternative approaches are investigated as part of the process for developing the IP. Actions, alternatives, and impacts are evaluated pursuant to NEPA's requirements.

If appropriate, this analysis can also be used to prepare an emergency response plan, thus satisfying an important EMS requirement. As described earlier, NEPA's definition of significance and its 10 significance factors (§ 1508.27[b]) are used in reaching a final decision regarding the IP and actions to be taken (fourth block, Figure 2.6).

2.7.3.3   Implementation Phase

Once a final decision has been reached an effort is begun to implement the decision. A centralized planning function could be used to coordinate implementation of the plan within the respective federal facility(s) and operations. At the lower facility or operational level, environmental compliance officers (ECOs) or equivalents could be assigned responsibility for preparing a more detailed site-specific facility implementation plan (FIP) for the IP, thus satisfying the fifth block shown in Figure 2.6.

The IP provides high-level direction and constraints that the FIP must meet. Thus, individual FIPs could be "tiered" from the IP, providing facility/project-specific direction for implementing results of the planning process through the EMS. A centralized planning office (or equivalent) could be assigned responsibility for approving each FIP to ensure continuity and consistency. Implementation problems and cross-cutting issues could also be elevated to such an office for resolution.

An EMS requires job-appropriate experience/training of affected employees with environmental responsibilities to assure that operational actions (described in the FIPs) are correctly implemented. Defining and tracking the appropriate training requirements could be a centralized EMS function, while performing training is an operations-specific function.

2.7.3.4   Monitoring and Enforcement Phase

A centralized oversight office (or equivalent) could be assigned responsibility for performing reviews and for monitoring facility and operational compliance (sixth block, Figure 2.6). An ECO (or equivalent) could be assigned responsibility for preparing and transmitting input and status reports to the oversight office. Audits could be performed periodically by the oversight office to verify compliance.

Monitoring data are evaluated to verify compliance and effectiveness of the EMS in meeting the established policy and plan. As appropriate, the organizational policy/plan is revised to correct deficiencies (see loop branching to box labeled "Corrective action," Figure 2.6).

Substantial changes could be made at the centralized planning level, while less significant changes might be implemented at the facility level by revising the FIP. Ultimately, the ISO 14001 expectation is that impacts will eventually dissipate, such that the next plan might address performance issues and impacts different from those in the existing plan. Such a process ensures a continuous improvement cycle, which is the hallmark of an EMS, and also promotes CEQ's paradigm of AM.

As described in the next section, an EMS may assist an agency in implementing an integrated NEPA AM system.

2.8 AN INTEGRATED NEPA, AM, AND EMS SYSTEM

One difficulty with the standard NEPA approach is that it tends to be a one-time event, that is, the analysis of environmental impacts is prepared, alternatives and mitigation measures are evaluated, and the decision-maker chooses a course of action. Unfortunately, the NEPA process generally ends at this point.

Whether environmental protection is subsequently achieved depends on the following:

1. The accuracy of the predicted actions and impacts (alternatives and/or mitigation measures) that were evaluated.
2. The agency's commitment to correctly implement the proposal, or perform any monitoring and mitigation measures to which it has committed. However, a major problem is that new information or changing circumstances are difficult to forecast and accommodate in advance.

2.8.1 NEPA AND AM

As described earlier, the President's CEQ is promoting a paradigm referred to as AM.[27] AM provides agencies with a flexible process for actively addressing uncertainty, limits in knowledge, and changing circumstances. The following strategy, published in 2003, presents a process for integrating NEPA with AM and an ISO 14001 EMS.[17,28]

The NEPA implementing regulations contain provisions for continual monitoring and assessment, which are consistent with the process of AM:

CEQ Regulations Section 1505.2(c)—"A monitoring and enforcement program shall be adopted and summarized where applicable for any mitigation."
CEQ Regulations Section 1505.3—"Agencies may provide for monitoring to assure that their decisions are carried out and should do so in important cases."

2.8.1.1 Requirements for Implementing AM

AM is intended to facilitate midcourse corrections in agency actions based on the findings of environmental monitoring. The traditional AM process is characterized by seven key elements:

- Management objectives are revisited on a regular basis and if appropriate revised.
- A model is developed of the system being managed.
- The range of management choices, including mitigation measure, is considered and evaluated.
- The range of potential outcomes is evaluated.
- A monitoring program is developed.
- A continuous improvement mechanism for incorporating lessons learned into future decisions is developed.
- A collaborative structure for stakeholder participation is developed.

Implementing an AM approach necessitates

- ability to identify and adequately analyze the effectiveness of adaptive measures as an alternative or part of an alternative;

TABLE 2.10

The Five-Step AM Process

- *Predict.* Under AM, the standard NEPA process is used to plan actions and evaluate or predict the potential environmental consequences of a proposal (first block, Figure 2.7). Since actions can be later changed to adapt to new information or circumstances, the planning effort, in many instances, can begin at an earlier stage than is normally possible. *Note*: This step must include an analysis of all potential consequences associated with potential changes in the action or mitigation measures.
- *Mitigate.* Alternatives and mitigation measures are considered for mitigating potential impacts (second block, Figure 2.7). Emphasis is placed on implementing alternatives or mitigation measures (evaluated in the "predict" stage) that can effectively address changing circumstances or new information that may be obtained at a later date.
- *Implement.* A decision-maker chooses a course of action and the decision is implemented (third block, Figure 2.7).
- *Monitor.* A monitoring program is adopted (fourth block, Figure 2.7) to review implementation of the proposal to ensure that
  1. the applicable mitigation measures are effective;
  2. the chosen course of action is correctly implemented;
  3. any deviations in either the proposal or its environmental impacts from original predictions are identified and appropriately addressed; and
  4. any significant circumstances or new information that was not considered during the early planning process are identified and appropriately addressed.
- *Adapt.* Actions can be adjusted in this approach to address new circumstances or information acquired through environmental monitoring. The looping arrow (Figure 2.7) illustrates that this is an iterative process. Based on the results of the monitoring program, new mitigation measures may be adopted or changes may be made in the way that the chosen course of action is implemented.

- devising realistic adaptive measures that can effectively respond to a wide range of variations in impacts; these measures should be generally applicable to a range of alternatives whose impacts were analyzed;
- a monitoring method that records environmental effects in a manner that allows practitioners to determine whether adjustments are needed to avoid unexpected impacts;
- approaches for ensuring adequate public involvement; and
- developing adequate performance thresholds or measures for assessing progress and impacts and ensuring the integrity of the NEPA process.

A standard NEPA approach generally involves a process of "predict, mitigate, and implement."[29] As an iterative process, AM involves a discrete five-step process—predict, mitigate, implement, monitor, and adapt.[27] This five-step process is summarized in Table 2.10 and illustrated in Figure 2.7.

2.8.2 POTENTIAL ADVANTAGES

Corrective actions (that have been adequately anticipated and evaluated) can be taken if the agency makes changes in the way the proposal is eventually implemented, or if monitoring detects new information or conditions that are substantially different from those originally anticipated. The goal is to provide decision-makers with a mechanism for effectively addressing uncertainty, and perhaps accepting more uncertainty or risk in the upfront planning process.

The complexity of environmental impacts often makes it difficult to accurately forecast potential impacts with absolute certainty. Unfortunately, agencies are sometimes reluctant to admit that they are not completely confident of their predictions. An AM approach can help address such uncertainties by offering more flexibility for managing overall environmental risks.

Moreover, AM can be useful when an agency is faced with incomplete or unavailable information needed to make reliable predictions. The Regulations (§ 1502.22) address such circumstances; under this section of the Regulations, the NEPA document must disclose information gaps and state

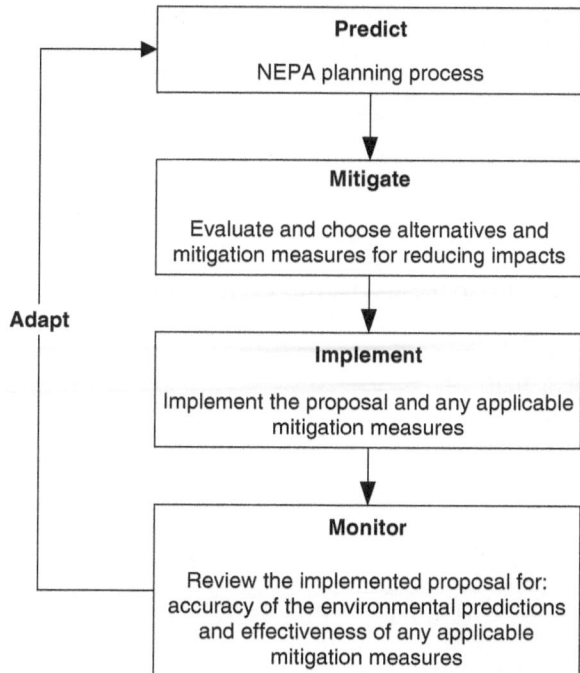

FIGURE 2.7   A typical AM process.

that the evaluation of reasonably foreseeable significant adverse impacts was performed without all relevant information. Practitioners can apply AM to help compensate for this problem by using the monitoring results to more accurately predict and respond to problems associated with uncertainties involving incomplete or unavailable information.

In some, but not all, instances, AM may enhance environmental protection and reduce the overall project-risks and costs. Because of the ability to make midcourse adjustments, costs savings may be achieved where mitigation measures are found to substantially exceed what is later determined necessary to provide environmental protection. Moreover, the ability to make adjustments may prevent the project from being halted to reassess the action and revise the NEPA analysis.

AM, however, is not a mechanism for shortcutting NEPA's regulatory requirements. The predict step must include an analysis of any adaptive changes (and corresponding consequences) that might be made to the chosen course of action. This can make the predict step more complicated, since the analysis must include a standard evaluation of the course of action in addition to potential adaptive changes that might be instituted once the proposal is implemented.

2.8.2.1   NEPA Documents

As appropriate, an agency should consider including AM into its mitigated

- EAs/FONSIs and
- EISs/RODs.

AM can be used to revise the

- method or scope of implementing the proposal,
- timing of proposal implementation, and
- mitigation measures.

Where an AM program is adopted, it is recommended that the AM process should be described and integrated into most or all the chapters of an EA/EIS and should be dealt with comprehensively in an appendix. The AM process should also be outlined in the corresponding FONSI or ROD. At a minimum, the NEPA analysis should describe the

- overall AM approach,
- desired outcome,
- factors used in assessing whether an additional NEPA review is needed,
- how the AM is integrated to the alternatives being considered,
- performance measures that will be used in determining whether the desired outcome is being achieved or an adaptive action is needed,
- monitoring protocol, and
- potential mitigation measures.

When assessing a project to determine if an AM approach is appropriate, the agency should assess the

- potential significance of the impacts and need for providing the flexibility that AM offers,
- complexity of the AM process and potential risks to the project,
- degree to which AM will be accepted by regulators and the public,
- ability to implement performance measures and thresholds,
- ability to evaluate and predict eventual outcomes,
- monitoring criteria and methods that can be employed, and
- cost and agency commitment to implement monitoring and corrective actions.

The author recommends that the agency generally focuses its efforts on integrating AM with the preferred alternative (although AM may also be included to some extent for all alternatives). AM information should also include a discussion of planned monitoring and the decision-making process.

Only occasionally is the "predict, mitigate, implement, monitor, and adapt" methodology of AM fully incorporated into a programmatic analysis. Some agencies use the programmatic analysis to predict, mitigate, and implement; they then apply research methods and monitoring to better understand the environmental linkages and functions that are needed during the adapting and mitigation phase; these successive adaptive actions are modified based on research and monitoring data gathered, and typically require subsequent and tiered NEPA analysis.

The problem with the aforementioned model is that it is just that—a conceptual model. It lacks a rigorous and systematic management system for ensuring that the five-step process is effectively carried out. Enter ISO 14001.

2.8.3 Using an EMS to Implement AM

Strong parallels exist between the objectives of ISO 14001 EMS and AM. An integrated AM/EMS system can provide a synergistic process for implementing decisions in a manner that protects environmental quality while minimizing risk, cost, and delays. Although the following process is directed at integrating AM with an ISO 14001 EMS, it might equally be integrated with any EMS which is consistent with the ISO 14001 standards. As used in this section, the term EMS is, therefore, interpreted to mean an ISO 14001 consistent EMS.

2.8.3.1   An Integrated AM/EMS System

Figure 2.8 presents a general-purpose, integrated NEPA/AM/EMS process published by the author, which uses an EMS as a system or mechanism for implementing AM.[28] This integrated process is composed of three basic functions or phases:

1. Integrated policy, planning, analysis and prediction, mitigation and decision-making
2. Project implementation
3. Environmental monitoring and adaptation

Policy. The AM/EMS system is initiated with the step of establishing a high-level organizational environmental policy with a commitment to environmental quality (see first block labeled "Establish environmental policy," Figure 2.8). As described in Section 2.7, establishing a policy with environmentally significant implications is an action potentially subject to NEPA's requirements; thus, this step might be performed as part of the NEPA planning process (§ 1508.18[b]).

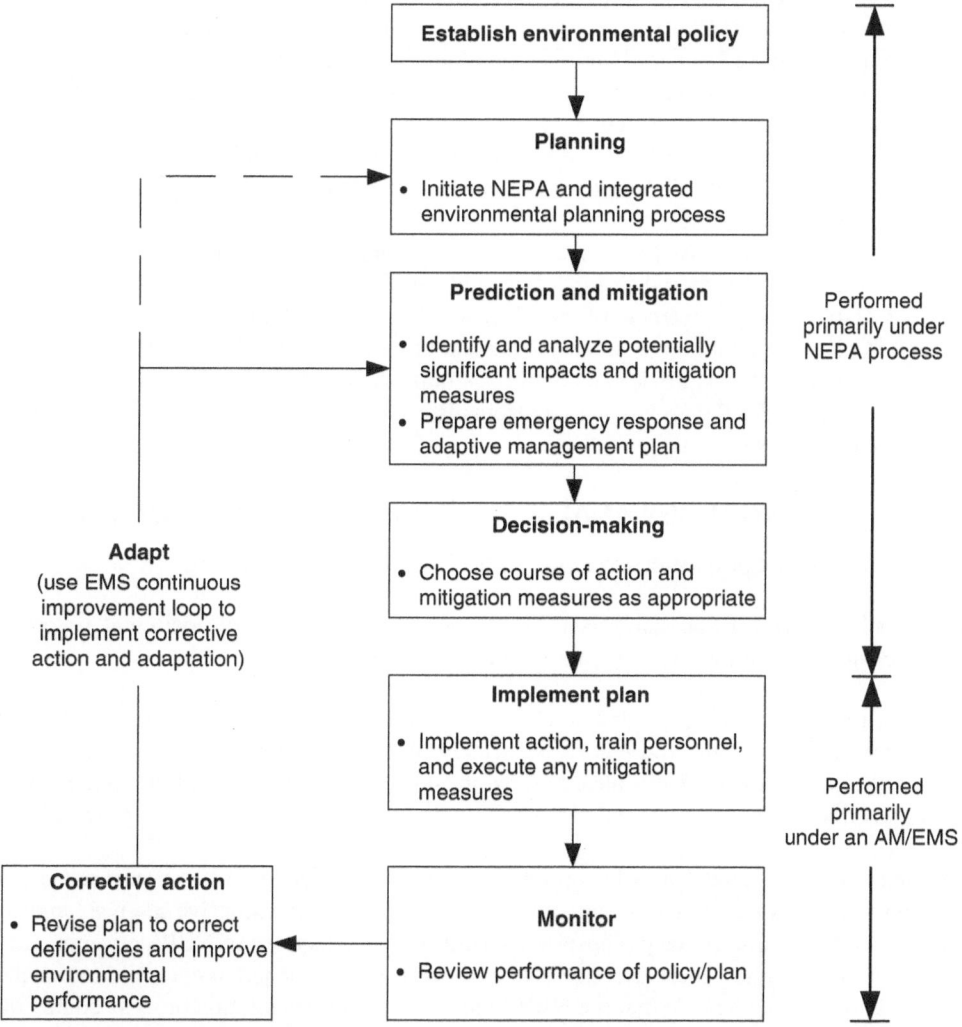

FIGURE 2.8   Conceptual process for integrating NEPA's AM process with an EMS.

Planning. An effort is then mounted to develop a specific plan for implementing the environmental policy (see second block labeled "Planning," Figure 2.8). The NEPA process provides the requirements and framework for performing this planning effort.[30] A proposed action is typically developed, as are reasonable alternatives and mitigation measures. This step must also consider potential changes in the project action and mitigation measures.

Prediction and mitigation. As illustrated by the third block of Figure 2.8 (Prediction and mitigation), the potential impacts of the proposal and alternatives are evaluated pursuant to the requirements of NEPA. This step must also consider the environmental impacts associated with potential changes in the project action or mitigation measures. In addition to the standard NEPA evaluation of the course of action, this assessment must also investigate and analyze potential adaptive changes that might need to be instituted based on the results of the monitoring program. Mitigation measures are also evaluated, as appropriate, for minimizing environmental impacts.

Decision-making. The decision-making phase is performed as part of the standard NEPA process.

Implementation. The EMS provides the management system for ensuring that the agency's plan, final decision, and any applicable mitigation measures are correctly implemented.

Monitoring and corrective action. As depicted in Figure 2.8 (boxes labeled "Monitor" and "Corrective action"), an EMS offers an internationally defined mechanism for monitoring, collecting data, and developing a corrective action plan for adapting the agency's actions to address changing circumstances or new information. Monitoring data are collected and evaluated to confirm compliance with the agency's final decision, and can help verify that the impacts do not significantly exceed projections presented in the NEPA analysis. The readers should note that any corrective actions that are substantially different for the chosen alternative must have been considered and evaluated in the NEPA analyses.

Adapt. As part of an iterative process, a corrective action plan may be prepared so that the organizational policy/plan can be revised to correct any deficiencies (see loop labeled "Adapt," Figure 2.8). If this is a long-term project, the goal of the EMS should be to eventually dissipate the impacts such that a revised plan can address any impacts substantially different from those in the existing plan. Thus, this integrated process provides a continuous improvement cycle, which is the hallmark of an EMS, while AM provides the agency with a mechanism for efficiently managing new information or changing circumstances.

2.9   DETERMINING WHEN ANALYSIS CONTAINS SUFFICIENT DETAIL

The Regulations do not provide criteria of specific direction for determining:

> How much discussion and analysis must be presented in a NEPA document to provide sufficient coverage for an agency to pursue a proposed action?

To date, the courts have provided some generic guidance regarding this issue, but generally this guidance has been vague and has varied widely. While the courts have suggested basic characteristics of inadequacy, they still have not, by-and-large, provided detailed guidance that allows one to determine if the information contained in an analysis is sufficient to provide adequate coverage.

In some cases, an aspect of a proposed action may be considered to be covered by merely mentioning that the specific action will take place. In others, an extensive analysis and discussion may be considered necessary. Thus, in determining the adequacy of NEPA coverage, analysts and decision-makers alike must exercise a considerable degree of judgment. Without such judgment, an argument can be made that the NEPA analysis and documentation process could continue *ad infinitum.*

2.9.1 Dilemma

To reduce risk of a document being found inadequate, agencies frequently prepare NEPA documents containing a level of detail much greater than is reasonably necessary to provide adequate coverage. The phrase *encyclopedia mania* has been used in referring to the production of massive multi volume, often undecipherable NEPA documents. Such overskill can significantly increase the cost of the NEPA analysis and may result in project delays. An inordinate amount of time and resources are often consumed in reviewing NEPA documentation to determine if the amount of discussion and analysis does in fact provide sufficient coverage.

Expectations regarding the adequacy of coverage may reflect an individual's responsibility, professional experience, personal bias, and technical training. It is unlikely that two decision-makers will agree completely on the amount of discussion needed to provide coverage sufficient to allow the agency to implement a proposed action.

The outcome of such reviews is problematic: Even if two decision-makers agree that a particular discussion does not provide adequate coverage they may still disagree on the degree of analysis and discussion needed. Proponents of a project may believe that an NEPA analysis provides sufficient coverage for a proposed project, while critics may argue that the analysis needs to be taken to increasingly detailed levels.

Lacking definitive guidance, decision-makers and critics may point to a universe of potential factors that can be used to defend a position that an action is or is not adequately covered. Assertions are often based on ambiguous opinions that can neither be proved nor disproved.

Ultimately, lack of definitive guidance can also result in project delays, inconsistencies in the treatment of NEPA documents, and risk that a project may be challenged for inadequate coverage.

2.9.2 Criteria for Developing the Sufficiency Test Tool

To help with these problems, the author has developed a peer-reviewed general-purpose tool, called the *Sufficiency-Test Tool*, which is shown in Figure 2.9.[31] Consistent with the rule of reason, this tool is based on the four tests established in Table 2.11. These tests are based on the premise that no useful purpose is served in providing additional details if the information would not in some way improve the decision-making process. Each test also reflects specific provisions that are mandated in the Regulations. The specific provisions are cited for each of the tests.

2.9.3 The Sufficiency Test Tool

The Sufficiency Test Tool is used in determining if the existing discussion of a specific issue or topic is sufficient. If more discussion is deemed necessary, the tool may assist analysts in determining how much additional analysis is needed.

Once the document has been prepared, the Sufficiency Test may be used by reviewers in determining if particular topics or issues are sufficiently covered. During the comment incorporation process, the tool may assist analysts in determining if a comment is relevant and the amount of discussion that may be necessary.

Moreover, the Sufficiency Test may assist decision-makers in determining if the amount of details, discussion, and analysis is sufficient to support a final decision to pursue a proposed action. Finally, if the agency is challenged, the tool might provide a basis for determining if the agency's analysis is or is not in fact sufficient.

2.9.4 Applying the Sufficiency Test Tool

The Sufficiency Test can be applied to an array of different NEPA documentation requirements. Specifically, the tool can be used in determining if there is sufficient discussion of the proposed

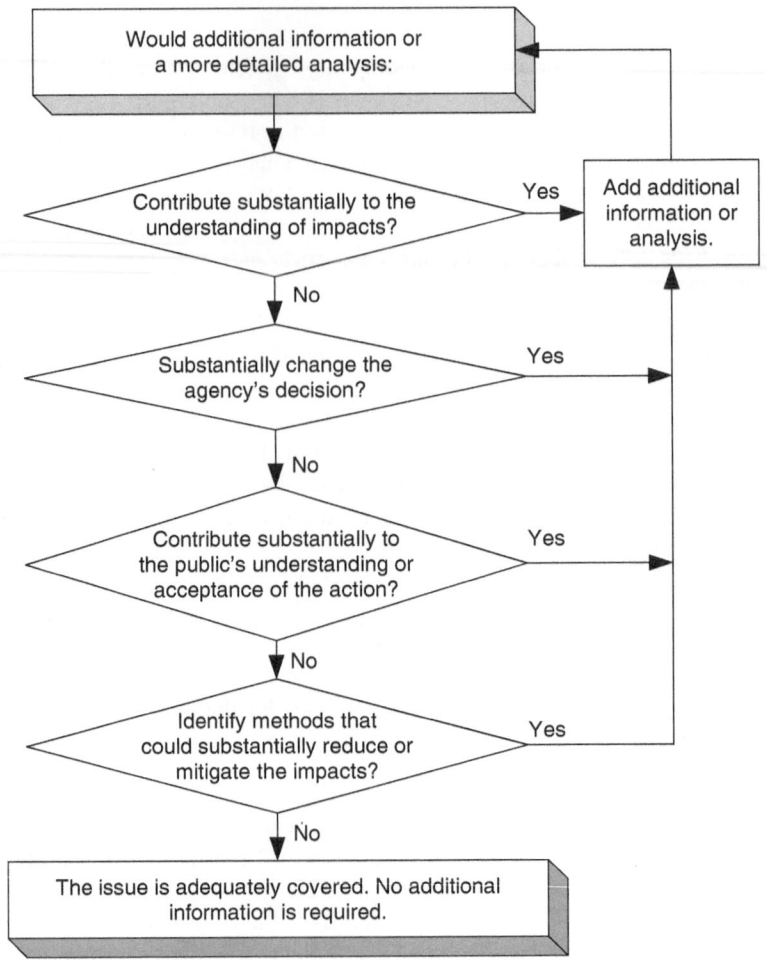

FIGURE 2.9 Sufficiency Test Tool. The tool provides a systematic methodology for determining if the discussion of a topic or issue is sufficient.

TABLE 2.11
Tests for Determining if the Discussion and Analysis are Sufficient to Provide Adequate NEPA Coverage

- Contribute substantially to the understanding of environmental impacts? A more detailed analysis is justified if it provides information that could substantially improve the agency's understanding of the impacts. (§ 1500.1[c], § 1502.1)
- Substantially change the agency's decision? Specifically, would the discussion allow the agency/public "to concentrate on the issues that are truly significant to the action in question" or would it simply amount to an exercise in "amassing needless detail"? A more detailed analysis is justified if, based on environmental factors, it could substantially affect the agency's decision. (§ 1500.1[c], § 1500.4[c], [f], and [l])
- Contribute substantially to the public's understanding or acceptance/rejection of the proposed action/alternatives? NEPA is a public process; a more detailed analysis is warranted if it could either positively or negatively substantially affect the public's assessment of the action, or if it could substantially affect the public's ability to provide comments or input into the decision-making process. (§ 1500.2[b] and (d), § 1500.4[f], § 1501.4[b], § 1502.1, § 1500.1[b])
- Result in new measures that could substantially reduce the impacts associated with the project? A more detailed analysis is justified if it may result in identifying mitigation measures that could substantially reduce adverse impacts. (§ 1500.2[f], § 1502.1, § 1502.14[f], § 1500.1[c])

action and alternatives, analysis of environmental impacts, discussion of the environment affected, and the appropriateness of introductory and background materials.

Application of the tool begins by reviewing a particular topic or issue that either is discussed or will be discussed. The review is conducted by examining each of the four tests with respect to the following question: "Would a more detailed analysis … ?" A "no" answer to all four tests supports a decision that the issue or analysis has been sufficiently covered. There is little justification for preparing a more detailed analysis unless the answer to one or more of these questions is "yes."

A "yes" answer to any test supports a decision that a more detailed discussion or analysis is warranted. Once the additional analysis has been included, this process is repeated; that is, the tool is used again to determine if the analysis is now sufficient. This process is continued until a "yes" response is no longer invoked.

2.9.4.1   Advantages and Limitations

Admittedly, the Sufficiency Test Tool does not completely eliminate subjectivity. After all, decision-making by its very nature is subjective. Although this tool does not completely eliminate such subjectivity, it significantly reduces it because the universe of potential considerations that can be used to justify a particular position is essentially reduced to four narrowly defined tests. Under this tool, decision-makers and critics alike are expected to provide specific and rational arguments, justifying why, in their view, the analysis does or does not meet one or more of the four tests.

As stated earlier, use of the tool should generally be restricted to determining issues concerning "How much information is enough?" The tool is not intended to be used to address NEPA compliance issues such as determining the range of the reasonable alternatives, whether a proposed action has been improperly segmented, or if a connected action has been appropriately considered.

Nor is the test designed to resolve comments related to contradictions, inconsistencies, or either data or grammatical errors. Likewise, it is inappropriate to use the tool to avoid addressing issues or topics that are necessary to comply with the basic requirements of NEPA. Instead, the tool is designed to assist decision-makers in determining if there is technical justification for either

- addressing a particular issue or topic, or
- expanding on the existing discussion of a specific issue or topic that has already been addressed.

Why the Sufficiency Test Tool is defensible. No discussion of this approach would be complete without addressing the question of defensibility: Can an agency legitimately adopt such a tool to determine if an analysis provides sufficient coverage to pursue an action? As we will see, agencies not only have the prerogative to adopt such methodologies but are, in fact, actually encouraged by the Regulations to do so.

Agencies have wide latitude to adopt methodologies. A review of the Regulations reveals that agencies have been given not only the right but the responsibility to develop specific procedures, techniques, and methodologies for implementing the NEPA process. The Regulations were specifically designed to provide agencies with flexibility to tailor the NEPA process to their own needs. Consistent with this philosophy, the Regulations place only minimal restrictions on how agencies must implement the NEPA process. To underscore this philosophy, consider the following directions the Regulations provide:

- Agencies shall "utilize a systematic" approach.[32]
- It is the intent of these Regulations to allow each agency flexibility in adapting its implementing procedures …[33]
- Agencies shall "Identify methods and procedures … to insure that presently unquantified environmental amenities and values may be given appropriate consideration."[34]

Mandelker, author of *NEPA Law and Litigation*, writes

> Because a consensus is usually lacking on the state of the art in environmental methodology, the courts have usually accepted the methodology used by an agency in analyzing environmental impacts. They put the burden of proof on plaintiffs to prove that the methodology was unacceptable...These decisions reflect the usual judicial willingness to uphold the agency when the evidence shows that there is only a disagreement among experts.[35]

Admittedly, this tool is not an impact assessment methodology for analyzing impacts. However, it does provide a tool for assisting agencies in determining if the description of an impact assessment is sufficient. The readers should also note that this tool has been used by more than one agency in resolving legal disagreements involving the sufficiency of descriptions in NEPA analyses.

Determinations are routinely made without a systematic basis. Decision-makers are routinely involved in making determinations regarding the adequacy of coverage. Lacking a formal and systematic procedure, determinations are often made in an inconsistent and contradictory manner.

The sufficiency test does not promote any degree of decision-making beyond the level that is already exercised by agencies on a routine basis. Instead, it simply substitutes a rigorous and systematic approach for the haphazard way in which determinations are currently made. If agencies have authority to make determinations without a formal basis or methodology, certainly a systematic approach is equally defensible. Acceptance of this tool amounts to nothing more than substituting a more formal and systematic procedure for the informal and inconsistent process that is typically used.

2.10   A 10-STEP FORMULA FOR IMPROVING THE NEPA PROCESS

The author offers the following 10-step plan for improving NEPA. Some of the following recommendations would require congressional approval and/or funding, and could be overseen by the President's CEQ.

2.10.1   RECOMMENDATIONS

1. *Streamlining measure.* Each federal agency should be required to establish an internal system for reviewing the effectiveness of its respective NEPA planning process. This system should include establishing a plan for identifying and addressing inefficiencies. All findings should be collected, evaluated by independent experts, and published for governmental agencies, the public, and congressional review, as well as for review by independent entities.

   Congress should also consider having the Government Accounting Office (GAO) or independent consultants perform periodic reviews to identify problems and provide recommendations for improving NEPA implementation. All federal agencies should also be required to begin tracking metrics such as cost, preparation time, readability, and document length. These figures should be reported to the national repository (see recommendation #4) for review by governmental agencies, the public, and Congress.

2. *Education.* Federal agencies should be required to ensure that their environmental staff, contractor personnel, and decision-makers have obtained appropriate environmental training and education. Such education should include recognized management and decision-making tools, techniques, and approaches for performing more streamlined planning analyses.

3. *Annual report.* Originally, an annual report was required to be prepared under NEPA's Title II, Section 201. This requirement was terminated effective on May 15, 2000, pursuant to Section 3003 of Public Law 104-66, the Federal Reports Elimination and Sunset Act.

   The CEQ should be required to reestablish the annual report on the state of the nation's environment and receive the funding to accomplish this. As part of this report, the CEQ should include statistics comparing each federal agency's NEPA process in terms of metrics such as average page length, preparation time, and cost (see recommendation #1).

The report should also summarize each agency's efforts to integrate and streamline its NEPA process. This effort could also document measures that have been taken to adequately train professional NEPA staff.

The CEQ should also issue updated guidance to reflect advances in computer technology and geographic information system (GIS) capability.

4. *National repository.* Currently, there is no uniform method for tracking or evaluating agency NEPA efficiency or streamlining measures. A national repository should be established to allow public online access to all (nonclassified or nonsensitive) EAs and EISs. The effectiveness of streamlining measures cannot be determined without a uniform measurement system. Such a repository would allow the Congress, for the first time, to track the effectiveness and status of NEPA documents.

5. *NEPA journal.* A federal NEPA journal should be established for sharing information and lessons learned among the agencies. Agencies should use this journal for educating federal decision-makers and NEPA practitioners, and for disseminating information concerning new tools, techniques, and methodologies for streamlining the NEPA process.

A national NEPA conference could also be sponsored, allowing all federal agencies to share lessons learned and advances in NEPA practice.

6. *Strategic environmental assessments.* Increasingly, the international community is embracing the concept of preparing strategic environmental assessments (SEAs) in formulating programmatic environmental strategies for combating a host of national and global environmental threats, ranging from developing sustainable energy paths to dealing with water scarcity. While SEAs may play a significant role as a tool for developing national environmental strategies and policies, the acronym, let alone the concept, is barely acknowledged in the United States. This is an important and emerging discipline where the United States is clearly lagging behind those it once mentored.

As a flexible and versatile planning tool, NEPA contains all the requisite elements necessary for preparing SEAs capable of addressing strategic issues of both national and international concern. With respect to the goal of sustainability, the advantages of an SEA are particularly appealing. What appears to be lacking is the awareness and commitment to prepare SEAs.

For example, a NEPA SEA could be prepared for devising a U.S. sustainable energy strategy and policy. This would not only be instrumental in developing a scientifically based strategy for shifting the United States away from vulnerable oil supplies, but it might also go a long way toward reducing global environmental impacts. The SEA would provide an objective and scientific forum for debating strategies, obtaining public input, and achieving national consensus. Thus, a NEPA SEA would afford citizens and interest groups alike a role in actively shaping the direction of future U.S. energy policy. A programmatic EIS could then be prepared and tiered from the SEA for developing specific national program(s) that would implement the high-level strategy established in the SEA.

7. *Environmental management systems.* As described earlier, whereas NEPA provides a process for planning projects, an EMS provides a mechanism for ensuring that the final decision is actually managed and implemented according to the plan. Each system has advantages and disadvantages. Interestingly, many of the weaknesses of an EMS are offset by the strengths inherent in the NEPA process, and vice versa.[22] On the international front, virtually any EIA process can also be integrated with an ISO 14001 EMS.[23] Properly integrated, both processes promote a synergistic system. Agency officials should actively seek to identify projects, programs, and sites where an integrated system could enhance environmental protection/performance.

8. *Substantive versus procedural.* The courts, perhaps correctly so, have interpreted NEPA to be a procedural requirement that all federal agencies must comply with. However, as largely viewed by the courts, NEPA does not place a substantive burden on decision-makers to choose a course of action based on environmental merits. Thus, while NEPA requires evaluation of reasonably foreseeable alternatives and impacts, there is generally

no substantive requirement that a decision-maker base a decision solely on the results of the impact assessment. So how can the link between NEPA's procedural impact assessment requirement and the final decision-making process be strengthened without detrimentally affecting the optimum course of action given nonenvironmental considerations?

Pursuant to NEPA, the federal agency must identify the agency's preferred alternative in an EIS and the environmentally preferable alternative in the ROD. Thus, one approach for mending the linkage problem described above involves amending either NEPA or its implementing regulations to require decision-makers to select the environmentally preferable alternative, *unless* they provide a thorough rational in the ROD as to why other factors outweigh environmental concerns.

While such a requirement would not prevent decision-makers from choosing environmentally undesirable alternatives (based on political, cost, schedule, or other factors), it would place a greater burden on giving serious consideration to environmentally preferable choices. Moreover, it could place greater public scrutiny on the decision-making process and provide greater federal accountability to the public.

9. *Registration/Certification.* Physicians, professional engineers, and accountants are generally required to be licensed or certified where their work may affect or endanger the public. So must architects and teachers. Drivers are all required to pass a test and obtain a license before they can legally drive a vehicle.

Yet, there is no similar requirement regarding the professional qualifications necessary to prepare an environmental analysis, even though the consequences and ramifications might be global and perhaps even catastrophic in nature. It is not uncommon to find that analysts with little or no scientific expertise have been assigned responsibility for preparing complex analyses. Minimum qualifications and a national certification program should be established for NEPA practitioners and decision-makers. At a minimum, environmental practitioners, certified under a nationally recognized program, should be required to review and certify as to the accuracy and objectivity of NEPA documents.

10. *Strengthening NEPA/Regulations.* Currently, federal agencies and contractors prepare NEPA analyses. Yet, the federal government also generally plays the role of project sponsor, advocate, or enabler, which some critics have long charged is a conflict of interest. Under NEPA, the "lead agency" sponsoring, authorizing, or enabling a proposal is given direct responsibility for overseeing and preparing the NEPA analysis.

Consequently, critics have charged that some NEPA documents have been clouded by internal agency conflicts of interest, and consequently such analyses may be less than the objective. To further the goal of impartiality, NEPA's implementing regulations (or even the Act itself) might be amended to require that NEPA analyses be prepared and/or reviewed by outside independent entities or even other impartial agencies (in addition to the EPA).

Another important action that could be taken to promote environmental security beyond U.S. shores would involve amending the NEPA implementing regulations (or the Act itself) to strengthen requirements to consider transboundary effects, particularly with respect to impacts on the global commons.

2.10.2 ADDITIONAL RECOMMENDATIONS FOR IMPROVING NEPA

Hansen and Wolff provide 10 additional recommendations for increasing the effectiveness of NEPA.[36]

1. Integrate the NEPA process with other environmental compliance and review procedures. The first recommendation provides an opportunity to save time, money, and paperwork.
2. Accelerate the decision time for determining the appropriate level of NEPA documentation.

NEPA project managers frequently consume time and resources while awaiting agency decisions on whether or how to comply with NEPA. Use of internal scoping (see point 3 below) and an early determination of whether an EIS is required can avoid potential problems.

3. Conduct early and thorough internal NEPA document scoping.

Internal agency scoping should not be confused with public scoping and should be completed before document preparation starts. The agency cannot adequately prepare for a public scoping process when it has not done its own internal homework.

4. Organize and implement public scoping processes that are participatory rather than confrontational.

Public controversy is an inherent part of many federal projects, but its effects can be mitigated if the public and other agencies feel they are being given the opportunity to really participate. An approach that is gaining increasing acceptance involves allowing participants to form working groups based on major NEPA issues.

5. Maintain an up-to-date compendium of environmental baseline information.

Environmental baseline reports can significantly decrease the time and cost of NEPA document preparation, and can help avoid reinventing the wheel for the affected environment sections in EAs and EISs. Standardizing this information and focusing on what is important can help eliminate encyclopedic discussions and unnecessary details.

6. Prepare more broad-scope umbrella EAs and EISs that can be used for tiering.

Programmatic or site-wide NEPA documents can be used in effectively supporting tiering of narrower, more project-specific analyses.

7. Prepare an annotated outline as a road map for EA or EIS preparation.

An annotated outline can provide important guidance to authors on the desired contents of each section or subsection of the document, the recommended approach to the topic, and data gaps that need to be filled. An outline can be organized in a tabular format with four columns:

a. Outline element (table of contents)
b. Target number of pages for each element
c. Authors responsible
d. Contents and data needs

8. Decrease the length and complexity of highly technical portions of NEPA documents.

Technical data that are highly technical in nature should be presented in a succinct, understandable manner and interpreted for the benefit of both the general public and sophisticated readers. Place detailed technical data in an appendix or in a separate document incorporated by reference.

9. Increase and systematize NEPA compliance outreach, training, and organizational support.

Training in the philosophy, purpose, legal requirements, and methods of NEPA compliance is imperative for everyone involved in the NEPA process.

10. Work diligently to prepare better organized, shorter, more readable NEPA documents.

NEPA documents that are understandable permit greater public participation, increase credibility and support, and can reduce appeals and litigation. Project managers and NEPA professionals must focus at least as much attention on the organization and writing of documents as on their technical content.

PROBLEMS

1. As a class project, develop an integrated NEPA/ISO 14001 EMS for implementing a hypothetical national federal program. Specify who and how officials would execute the integrated process. Outline how federal actions would be identified and evaluated as part of the NEPA process, and how monitoring and mitigation measures would be executed through the ISO 14001 EMS.

2. As a class project, develop an integrated NEPA/P2 program for a hypothetical federal installation. Specify who and how officials would execute the integrated process. Outline how federal actions would be identified and evaluated as part of the NEPA process, and how monitoring and mitigation measures would be executed under the ISO 14001 EMS.
3. With respect to NEPA, explain the "sliding scale" approach.
4. With respect to NEPA, what is the difference between "need" and "purpose."
5. What is "de-scoping"?
6. Explain the purpose of the Pollution Prevention Act and how it relates to NEPA.
7. What is an ISO 14001 EMS? Briefly explain why it can complement the NEPA process.
8. How does the "planning function" within NEPA compare with that of ISO 14001 EMS?
9. What is adaptive management? How is it related to an ISO 14001 EMS?
10. Assume that a Department of Justice wants to construct a 20,000 ft² office building in downtown Chicago. The decision-maker states that "I don't necessarily have to know the color of the building, but I do want to have enough detail to know the number and placement of the doors on the building." Using your personal judgment and the Sufficiency Test, determine if such a detailed description is justified. Justify your answer. Now assume the same problem, but this time with a nuclear materials storage facility.
11. Assume an EA was prepared for a proposed action that involved demolition of an existing building and construction of a replacement of a small office center in a major city. The proposed action would result in the relocation of approximately 20 workers a mile away. During internal review, a review comment is received stating that the EA should quantitatively address socioeconomic and environmental justice impacts, including impacts on highway traffic associated with the relocation of the 20 workers; the comment also states that the EA should assess how a decline in spending, resulting from the relocation of these people, would affect surrounding business; the EA should also address how businesses would be affected in the area to where the employees would be relocated. An accurate assessment of this comment would require a substantial amount of additional time and cost; sidewalk surveys might need to be conducted to determine existing spending patterns. Using your personal judgment and the sufficiency test, determine if the comment is technically justified? How much discussion and analysis do you believe is reasonable? Defend your answer.

REFERENCES

1. Author unknown.
2. SAVE International, *Value Methodology Standard*, May 1997.
3. Miles L. D., *Techniques of Value Analysis and Engineering*, Second Edition, McGraw-Hill, New York, 1972.
4. Office of Management and Budget Circular No. A-131.
5. 40 CFR 1501.2 and 1507.2.
6. 40 CFR 1500.2(c), 1500.5, 1501.2, 1501.2(a), 1501.7(a)(6) and (b)(4), 1502.25(a), 1505.2, and 1506.4.
7. 40 CFR 1500.5(a), 1500.5(f), 1501.1(a), and 1501.2.
8. 40 CFR 1500.1(b), 1500.2(b) and (d), and 1501.4(b).
9. 40 CFR 1502.2(g) and 1502.5.
10. 40 CFR 1502.14.
11. 40 CFR 1502.5.
12. 40 CFR 1502.1.
13. 40 CFR 1502.14(a).
14. 40 CFR 1500.2(e), 1502.1, 1502.14(a) and (c).
15. Eccleston C. H., Applying value engineering and modern assessment tools in managing NEPA: Improving effectiveness of the NEPA scoping and planning process, *Environmental Regulations and Permitting*, Winter 1998, pp. 53–63.
16. Schmidt O. L., The statement of purpose and need defines the range of alternatives in environmental documents, 18 *Environmental Law* 371–81, 1988.

17. Eccleston C. H., *Environmental Impact Statements: A Comprehensive Guide to Project and Strategic Planning*, John Wiley & Sons, New York, 2000.
18. 40 CFR 1500.4(g) and 1501.7(a)(3).
19. Pollution Prevention Act of 1990, 42 USC § 13101 et seq. (1990).
20. Executive Order 12856, Federal Compliance with Right-To-Know Laws and Pollution Prevention Requirements, 58 FR 41901 (August 3, 1993).
21. Eccleston C. H., A strategy for efficiently implementing pollution prevention program through the NEPA planning process, *Journal of Environmental Quality Management*, 15(3), pp. 31–41, Spring 2006.
22. Eccleston C. H., A strategy for integrating NEPA with an EMS and ISO-14000, *Environmental Quality Management Journal,* Spring 1998.
23. Eccleston C. H. and R. B. Smythe, Integrating environmental impact assessment with environmental management systems, *Environmental Quality Management Journal* (Lead Journal Article), 11(4), Summer 2002.
24. Executive Order 13148, Greening the Government through Leadership in Environmental Management, April 21, 2000.
25. CEQ, Regulations for Implementing the Procedural Provisions of the National Environmental Policy Act, 40 Code of Federal Regulations, Pts. 1500–1508, 1978.
26. CEQ, Guidance on Pollution Prevention and the National Environmental Policy Act, published at 58 FR 6478, January 29, 1993.
27. CEQ, The National Environmental Policy Act: A Study of Its Effectiveness after Twenty-Five Years, 1997.
28. Eccleston C. H., Integrating NEPA's concept of adaptive management with an ISO 14000 consistent EMS, *Environmental Quality Management Journal*, 12(3), Spring 2003, pp. 59–67.
29. Eccleston C. H., *The NEPA Planning Process: A Comprehensive Guide with Emphasis on Efficiency*, John Wiley & Sons, New York, 1999, pp. 81–82 and p. 241.
30. 40 CFR 1508.18(b).
31. Eccleston C. H., NEPA: Determining when an analysis contains sufficient detail to provide adequate NEPA coverage for a proposed action, *Federal Facilities Environmental Journal*, Summer 1995.
32. 40 CFR 1507.2(a).
33. 40 CFR 1507.1.
34. 40 CFR 1507.2(b).
35. Mandelker, D. R., *NEPA Law and Litigation*, Chapter 10, Clark Boardman Callaghan, Deerfield, IL, 1993.
36. DOE, National Environmental Policy Act, *Learned Lessons,* March 1, 2001; Issue No. 26.

3 NEPA Streamlining Provisions

Critics have sometimes charged that the National Environmental Policy Act (NEPA) is an inefficient process. Inefficiency is not indigenous to NEPA. To the contrary, the root causes of any inefficiency lie principally with the approaches used and lack of, or failure to adopt, effective tools, techniques, and approaches for implementing NEPA's requirements.

In drafting the NEPA regulations (Regulations), the Council on Environmental Quality (CEQ) placed considerable emphasis on providing direction and methods for promoting efficiency (see Chapter 2). Yet, experience clearly indicates that much of this direction often goes unheeded. In the zest to prepare bullet-proof documents, insufficient attention is often given to guidance pertaining to efficiency.

This chapter presents a systematic description of CEQ requirements, which provides direction for efficiently implementing the NEPA process. Of special interest is a streamlining procedure (interim action provision) described toward the end of this chapter, which allows some actions to proceed in advance of completing NEPA.

3.1 MANAGING, ANALYZING, AND PREPARING NEPA DOCUMENTS

Regulatory direction for streamlining NEPA is categorized in the following three tables:

- Table 3.1: efficiency in managing NEPA
- Table 3.2: efficiency in performing the environmental analysis
- Table 3.3: efficiency in preparing environmental assessments (EAs) and environmental impact statements (EISs)

The primary source of regulatory direction on promoting efficiency is contained in sections § 1500.4 (reducing paperwork) and § 1500.5 (reducing delay). Inspection of Table 3.1 indicates many of these efficiency themes recur throughout the Regulations. Two of the most notable efficiency citations are to

Implement procedures to make the NEPA process more useful to decisionmakers and the public; to reduce paperwork and the accumulation of extraneous background data; and to emphasize real environmental issues and alternatives. Environmental impact statements shall be concise, clear, and to the point and shall be supported by evidence that agencies have made the necessary analysis. (§ 1500.2[b])

Integrate the requirements of NEPA with other planning and environmental review procedures required by law or agency practice so that all such procedures run concurrently rather than consecutively. (§ 1500.2[c])

One of the most effective tools for promoting efficiency is simply to exclude a proposed action from the requirement to prepare an EIS by

- categorically excluding the action, or
- issuing a finding of no significant impact (FONSI) for the action.

TABLE 3.1

Provisions for Promoting Efficiency in Managing the NEPA Process

Apply NEPA Early in the Process (§ 1501.2)
(b) Concurrent circulation and review
(d) Provide for actions by non-federal applications
(d) Earliest possible time (also see § 1501.6[b][1], § 1502.5)

Reducing Delay (§ 1500.5)
(a) Integrating NEPA into early planning (§ 1501.1, § 1501.2)
(b) Inter-agency cooperation (§ 1501.6)
(c) Swift resolution of agency disputes (§ 1501.5)
(d) Early scoping of real issues (§ 1501.7)
(e) Time limits (§ 1501.1[e], § 1501.7[b][2], § 1501.8)
(f) Early preparation (§ 1502.5)
(h) Eliminating duplication with other agency review (§ 1506.3)
(j) Accelerated procedures for legislation (§ 1506.8)
(k) Using categorical exclusions (§ 1508.4)
(l) Using finding of no significant impact (§ 1508.13)

NEPA Policy (§ 1500.2)
(b) Make NEPA process more useful—reduce accumulation of extraneous background data
(c) Integrate NEPA with other environmental planning and review ... so that all such procedures run concurrently rather than consecutively

Purpose of NEPA (§ 1500.1[b], [c])
(b) Concentrate on ... truly significant issues ... rather than amassing needless detail
(c) Better decisions—not paperwork

Reducing Paperwork (§ 1500.4)
(g) Use scoping ... to identify significant issues ... deemphasize insignificant issues, narrowing the scope (§ 1501.7)
(i) Tiering (§ 1502.4[d], § 1502.20, § 1508.28)
(k) Integrating with other environmental review (§ 1502.25)
(n) Eliminating duplication with State and local procedures ... adopting environmental documents ... by other agencies (§ 1506.3)
(p) Using categorical exclusions (§ 1508.4)
(q) Using finding of no significant impact (§ 1508.13)

Purpose of Agency Planning (§ 1501.1)
(a) Integrating NEPA into early planning
(b) Cooperative consultation
(c) Swift resolution of agency disputes
(e) Time limits

Cooperating Agencies (§ 1501.6)
(a) The lead agency shall: (1) request the participation of each cooperating agency in the NEPA process at the earliest possible time; (2) use the environmental analysis and proposals of cooperating agencies ... to the maximum extent possible ...
(b) Each cooperating agency shall: (1) participate in the NEPA process at the earliest possible time ...

Scoping (§ 1501.7)
(b) As part of scoping an agency may: ... (2) set time limits ... (3) adopt procedures to combine its environmental assessment process with its scoping process ... (4) hold an early scoping meeting which may be integrated with any other early planning meeting the agency has.

Major Federal Actions Requiring the Preparation of an EIS (§ 1502.4)
(d) Agencies shall as appropriate employ scoping (§ 1501.7), tiering (§ 1502.20), and other methods listed at § 1500.4 and § 1500.5 to relate broad and narrow actions and to avoid duplication and delay.

Timing of EIS (§ 1502.5)
... as close as possible to the time the agency is developing or presented with a proposal ... early enough so that it can serve practically an important contribution to the decisionmaking process and will not be used to rationalize or justify decisions already made.

(Continued)

TABLE 3.1 (Continued)
Provisions for Promoting Efficiency in Managing the NEPA Process

Environmental Review and Consultation Requirements (§ 1502.25)

(a) To the fullest extent possible agencies shall prepare draft environmental impact statements concurrently with and integrated with environmental impact analysis and related surveys and studies required by the Fish and Wild Coordination Act ... and other environmental review laws and executive orders

Interim Actions While EIS is in Progress (§ 1506.1)

(a) Until an agency issues a record of decision ... no action concerning the proposal shall be taken which would:
   (1) Have an adverse environmental impact: or
   (2) Limit the choice of reasonable alternatives.

Elimination of Duplication with State and Local Procedures (§ 1506.2)

(b) Agencies shall cooperate with State and local agencies to the fullest extent possible to reduce duplication between NEPA and comparable State and local requirements.

Adoption (§ 1506.3[a])

An agency may adopt a Federal draft or final environmental impact statements or portion thereof ...

Coordination/Consultation

Emphasizing cooperative consultation among agencies (§ 1501.1[b])
Providing for swift and fair resolution of agency disputes (§ 1501.1[c])
Cooperating Agencies (§ 1501.6)
Environmental Review and Consultation (§ 1502.25)
Reduce duplication with State/local Procedures (§ 1506.2[b], [c])

Concurrently/Integrated

With other planning and environmental review procedures (§ 1500.2[c])
With other review and consultation requirements (§ 1500.4[k])
Integrate into early planning ... to eliminate delay (§ 1501.1[a])
Integrate with other planning ... to avoid delays (§ 1501.2)
Circulation and review with other planning documents (§ 1501.2[b])
Integrate with other required analysis and studies (§ 1501.7[a][6])
With environmental review laws and executive orders (§ 1502.25[a])
Integrate into State and local planning processes (§ 1506.2[d])

TABLE 3.2
Provisions for Promoting Efficiency in the Environmental Analysis

NEPA Policy (§ 1500.2)

Purpose of NEPA (§ 1500.1)

(b) Needless detail

(c) Integrate with other environmental planning and review

Reducing Paperwork (§ 1500.4)

(b) Analytic rather than encyclopedic

(g) Use scoping ... to identify significant issues ... deemphasize insignificant issues ... narrowing the scope

(k) Integrating with other environmental review

Purpose of Agency Planning (§ 1501.1)

(a) Integrating NEPA into early planning

(d) Early stage identification of significant versus non-significant issues ... narrowing the scope

Scoping (§ 1501.7)

(a)(2) Determine scope and significant issues to be analyzed in depth

(a)(3) Identify and eliminate from detailed study the issues which are not significant or which have been covered by prior environmental review ... narrowing the discussion of these issues ... to a brief presentation of why they will not have a significant effect ... or providing a reference to their coverage elsewhere

(a)(6) Identify other environmental review and consultation requirements so the lead and cooperating agencies may prepare other required analysis and studies concurrently with, and integrated with, the environmental impact statement ...

(Continued)

TABLE 3.2 (Continued)
Provisions for Promoting Efficiency in the Environmental Analysis

Purpose of EIS (§ 1502.1)
Agencies shall focus on significant environmental issues and alternatives and shall reduce paperwork and the accumulation of extraneous background data. Statements shall be concise, clear and to the point ...

Environmental Review and Consultation Requirements (§ 1502.25)
... agencies shall prepare draft environmental impact statements concurrently with and integrated with environmental impact analysis and related surveys and studies required by the Fish and Wildlife Coordination Act ... and other environmental review laws and executive orders

Implementation of EIS (§ 1502.2)
(a) Be analytic rather than encyclopedic
(b) Discuss impacts in proportion to their significance

TABLE 3.3
Provisions for Promoting Efficient Preparation of an EIS or EA

Purpose of NEPA (§ 1500.1)
(b) [eliminate] Needless detail

Reducing Paperwork (§ 1500.4)
(a) Length (§ 1502.2[c], § 1501.7[b][1], § 1502.7)
(b) Analytic rather than encyclopedic (§ 1502.2[a])
(c) Briefly discuss non-significant issues (§ 1502.2[b])
(d) Plain language (§ 1502.8)
(e) Clear format (§ 1502.10)
(f) Emphasis on portions of the EIS useful to the public and decision-makers and reducing emphasis on background material
(j) Incorporating by reference (§ 1502.21)
(l) Requiring specificity of comments (§ 1503.3)
(m) Circulating only the changes when minor (§ 1503.4[c])
(o) Combining documents (§ 1506.4)

Reducing Delay (§ 1500.5)
(i) Combining documents (§ 1506.4)

Scoping (§ 1501.7)
(a)(3) Identify and eliminate from detailed study the issues which are not significant or which have been covered by prior environmental review ... narrowing the discussion of these issues ... to a brief presentation of why they will not have a significant effect ... or providing a reference to their coverage elsewhere
(b) As part of scoping an agency may: (1) set page limits.

Purpose of EIS (§ 1502.1)
Focus on significant environmental issues; Reduce paperwork; Reduce accumulation of extraneous background data; Be concise, clear ... to the point

Implementation of EIS (§ 1502.2)
(a) ... be analytic rather than encyclopedic
(b) [Discuss impacts] in proportion to their significance
(c) [EISs] shall be kept concise and shall be no longer than absolutely necessary to comply with NEPA ... Length should vary first with potential environmental problems and then with project size.

Page Limits of EIS (§ 1502.7)
... shall normally be less than 150 pages and for proposals of unusual scope or complexity ... less than 300 pages

Combining Documents: (§ 1506.4)
Any environmental document in compliance with NEPA may be combined with any other agency document to reduce duplication and paperwork.

Affected Environment (§ 1502.15)
... succinctly describe the environment ... no longer than is necessary ... commensurate with the importance of the impact ... shall avoid useless bulk ... verbose descriptions

3.1.1 Integrating NEPA with SEPAs and Other Processes

As just witnessed, integrating NEPA with other processes can be one of the most effective approaches for streamlining NEPA. For example, more than 25 American states have State Environmental Policy Acts (SEPAs), sometimes referred to as "little NEPAs."[1] Some of these SEPAs have environmental impact assessment processes similar to that of NEPA. However, most of these processes are less comprehensive and rigorous than NEPA.

In a few cases, these SEPA processes have procedural or substantive mandates exceeding that of NEPA. For example, the Massachusetts SEPA requires "a finding that all feasible measures have been taken to avoid or minimize the said impact."[2] Similarly, the New York SEPA requires a finding that "adverse environmental effects identified in the EIS will be minimized or avoided."[3] The California Environmental Policy Act (CEQA) also has a similar mandate. These three SEPAs go beyond NEPA in their requirement to protect the environment.

3.2 SPECIFIC STREAMLINING METHODS

Specific provisions and methods that have been successfully used to streamline NEPA implementation are described in the following sections.

3.2.1 Using Scoping to Narrow the Analysis

A common NEPA litigation complaint involves failure to consider one or more significant issues. Consequently, agencies frequently err on the side of a broader scope with a correspondingly higher degree of detail, thereby diminishing the effectiveness of provisions promoting efficiency. Professional judgment must be exercised in balancing the goal of efficiency with that of preparing a defensible analysis. To this end, the Regulations urge practitioners to use the scoping process to promote efficiency by narrowing the scope of the analysis:

> Using the scoping process, not only to identify significant environmental issues deserving of study, but also to deemphasize insignificant issues, narrowing the scope of the environmental impact statement process accordingly. (§ 1500.4[g])
>
> Implement procedures to … reduce paperwork and the accumulation of **extraneous background data.** (§ 1500.2[b])
>
> Most important, NEPA documents must concentrate on the issues that are **truly significan**t to the action in question rather than amassing **needless detail.** (§ 1500.1[b])

The Regulations further instruct agencies to

> Identify and eliminate from detailed study the issues which are not significant or which have been covered by prior environmental review, narrowing the discussion of these issues in the statement to a brief presentation of why they will not have a significant effect on the human environment or providing a reference to their coverage elsewhere. (§ 1501.7[a][3])

3.2.2 Tiering

To promote efficiency by minimizing paperwork, while reducing delays, the CEQ advocates the use of tiering.[4] Specifically, tiering

> … helps the lead agency focus on the issues which are ripe for decision and exclude from consideration issues already decided or not yet ripe. (§ 1508.28[b])

The Regulations provide additional direction for implementing tiering:

> Agencies are encouraged to tier their environmental impact statements to eliminate repetitive discussions of the same issues and to focus on the actual issues ripe for decision at each level of environmental review. (§ 1502.20)
>
> Agencies shall as appropriate employ ... tiering and other methods ... to relate broad and narrow actions and to avoid duplication and delay. (§ 1502.4[d])

Tiering provides a particularly useful approach for implementing a phased approach to planning, especially in programs that are large, complex and have many components. Tiering is particularly useful when it helps an agency to focus on issues "ripe for decision" and exclude from consideration issues already decided or that are not yet ripe for decision (§ 1500.4, reducing paperwork; § 1500.5, reducing delay). Tiering is appropriate when the sequence of documents is from (§ 1508.28[a], [b])

- program, plan, or policy EIS to a program, plan, or policy statement or analysis of lesser scope or to a site-specific statement or analysis; or
- EIS on a specific action at an early stage (such as need and site selection) to a supplement (which is preferred) or a subsequent EIS or analysis at a later stage (such as environmental mitigation). Tiering in such cases is appropriate when it helps the lead agency to focus on the issues that are ripe for decision and exclude from consideration issues already decided or not yet ripe.

For instance, a programmatic EIS could be prepared to evaluate a national wind energy program. The EIS could cover broad decisions that are ripe for decision during the early planning stages of the program. The programmatic analyses might focus on determining what course of action should be taken, while lower-tier analyses could focus on how such a decision should specifically be implemented.

3.2.3   INCORPORATION BY REFERENCE

The Regulations encourage use of incorporating information by reference as a means of reducing the size of an EIS. Under the Regulations, agencies are, in fact, mandated to incorporate existing material into the EIS if this procedure reduces the length of the EIS without impacting either the agency's or public's ability to review the document (§ 1506.3). This method is also appropriate for reducing the length of EAs.[5]

If incorporating by reference is used, the document must cite this material and briefly summarize its content. Referenced material must also be reasonably available for inspection by interested persons during the comment period; a common complaint has been that agencies fail to identify the location where this material may be obtained by the public.[6]

3.2.4   ADOPTION

To promote efficiency, an agency may adopt an EIS prepared for another project or by another agency. The adopting agency may decide to adopt an entire EIS or any portion of it. The following three distinct circumstances are recognized in which an agency may adopt an EIS (§ 1506.3[a]–[c]).[7] These situations involve circumstances where

- an adopting agency has cooperated in the preparation of an EIS with another federal agency. The cooperating agency may adopt the EIS and issue a record of decision (ROD) without recirculating the EIS for scoping or comment;
- A noncooperating agency undertakes an action that is already evaluated in an existing EIS. In this case, the adopting agency must demonstrate that it has reviewed the EIS and determined that it appropriately evaluates the activities that will be undertaken.

The EIS may be adopted as long as its proposed action is essentially the same as that described in the EIS. The adopting agency must recirculate the EIS, but only as a final EIS. If the actions described in the EIS are different from those proposed by the adopting agency, the EIS must also be recirculated as a draft EIS; in such cases, the agency may adopt all or a portion of the EIS by recirculating the document as a draft and then preparing a final EIS.

The CEQ has issued guidance encouraging agencies to use the adoption process for EAs as well as EISs. Agencies are responsible for ensuring that the scope and content of an existing EA adequately covers the proposed action.[8] Since EISs are generally expected to be site specific and the significance of an impact depends on its context, the adopting agency is responsible for reviewing and verifying that the adopted analysis adequately covers the new project or site.

In practice, adoption is not widely used because even small changes or differences in the affected environment can cause the actual impacts to significantly change. Thus, it is difficult to demonstrate that the impacts of a proposal would be essentially the same as those evaluated by an analysis provided in an EIS for another project or area.

3.2.5 PIGGYBACKING

As a means of reducing paperwork and duplication, agencies are encouraged to combine or piggyback NEPA documents with any other agency documents (§ 1500.4[o] and § 1506.4). Documents that are particularly amendable to piggybacking include regulatory documents, urban impact analyses, and final decision or option documents.[9]

For instance, the forest service has used piggybacking to combine their NEPA documents with the forest management plans. The EIS can be used to identify the agencies' preferred alternative, which is detailed in the proposed management plan. Neither document repeats the other. The EIS and the forest management plan cross reference each other. The EIS can basically be viewed as an independent document that incorporates the proposed action by reference.[10]

An EIS must be physically included in or attached to another report and may use the attached report material as supporting information. The EIS portion of the combined document, however, must be self-supporting and capable of standing on its own right as an analytical document that fully informs decision-makers and the public of the environmental effects and reasonable alternatives; the EIS is not to be presented as an outline or short summary that has been attached to the report.

3.3 LIMITATIONS ON ACTIONS AND INTERIM ACTIONS

Over the period in which a NEPA analysis is under way, an agency may need to implement individual projects or program element actions that fall within the scope of an ongoing analysis. Pursuing an action under such instances may constitute a violation of NEPA's requirements cited in Table 3.4. Fortunately, a special streamlining procedure has been established to address this problem. Actions that may legitimately proceed in advance of completing the NEPA review are referred

---

TABLE 3.4

Requirements Limiting Actions That May Be Pursued in Advance of Completing the NEPA Process

"NEPA procedures must insure that environmental information is available to public officials and citizens before decisions are made and before actions are taken." (§ 1500.1[b])

"Agencies shall not commit resources prejudicing selection of alternatives before making a final decision." (§ 1502.2[f])

EISs will not be used for "... justifying decisions already made." (§ 1502.2[g])

to as *interim actions*. Interim action procedures are provided in § 1506.1 of the Regulations: *Limitation on Actions During the NEPA Process*.

It is recommended that an agency consider preparing a separate interim action justification memorandum (IAJM) for individual actions that qualify for status as an interim action. Such a justification provides evidence that the action has been screened against the criteria specified in § 1506.1. As a short memorandum (1–5 pages), it can be appended to later NEPA documents prepared for the specific action.

3.3.1   Eligibility for Status as an Interim Action

An effort should be mounted early in the NEPA process to carefully review potential actions that may need to begin prior to the completion of the NEPA process. In performing such a review, actions should be reviewed to confirm that they actually fall within the scope of the NEPA review. Actions that do not fall within the scope of an ongoing analysis are, of course, not subject to the limitations presented in § 1506.1. Actions that are subject to the limitations (Table 3.4) and are not eligible for status as an interim action should be factored into project's schedule and planning process.

3.3.2   Two Categories of Interim Actions

The Regulations draw a sharp distinction between interim action requirements applicable to a programmatic EIS versus those that pertain to the preparation of a nonprogrammatic EIS (§ 1506.1). Programmatic EISs are discussed in Chapter 8. Table 3.5 specifies criteria and limitations that must be met in pursuing interim actions within each of the aforementioned categories.

The first category of limitations (project-specific) described in Table 3.5 is interpreted to apply only to project-specific reviews.

3.3.2.1   Additional Limitations

As noted in Table 3.4, the Regulations place an additional restriction on actions that supplement the criteria cited in § 1506.1:

> Agencies shall not commit resources prejudicing selection of alternatives before making a final decision (§ 1502.2[f]).

Candidates for interim action status should be reviewed for compliance with this provision, in addition to the limitations provided in Table 3.5. This provision prevents an agency from taking any

---

TABLE 3.5

Criteria and Limitations That Must Be Met in Pursuing an Interim Action

A. Project-Specific EIS (§ 1506.1[a])

Until an agency issues a Record of Decision (ROD), no action concerning the proposal shall be taken which would:

(1) Have an adverse environmental impact; or

(2) Limit the choice of reasonable alternatives.

B. Programmatic EIS (§ 1506.1[c])

While work on a required program environmental impact statement is in progress and the action is not covered by an existing program statement, agencies shall not undertake in the interim any major Federal action covered by the program which may significantly affect the quality of the human environment unless such action:

(1) Is justified independently of the program;

(2) Is itself accompanied by an adequate environmental impact statement; and

(3) Will not prejudice the ultimate decision on the program. Interim action prejudices the ultimate decision on the program when it tends to determine subsequent development or limit alternatives.

---

action that commits resources to such an extent that it would bias the selection of alternatives; it applies to any interim action, regardless of whether the action is project specific or programmatic in nature.

3.3.3 Limitations on Project-Specific Action

The Regulations do not provide specific guidance for determining precisely what constitutes "limit the choice of reasonable alternatives" (Table 3.5). Committing resources that would prejudice selection of alternatives (§ 1502.2[f]) is a circumstance that certainly appears to trigger this criterion. The third interim action criterion pertaining to the programmatic EIS process also appears to shed some light on this question (§ 1506.1[c]). Based on the third criterion of § 1506.1(c), an action would appear to limit the choice of reasonable alternatives if it "tends to determine subsequent development or limit alternatives."

Consistent with the rule of reason, the author suggests that the following two tests be considered in determining if a particular action limits the choice of reasonable alternatives. Specifically, would the action:

1. determine the subsequent development or direction of the proposed action?
2. commit resources that would prejudice selection of alternatives?

3.3.4 Limitations on Programmatic Actions

The reader should note that the limitations presented in § 1506.1(c) are only specifically required for actions taken during the preparation of a required programmatic analysis. This verbiage was carefully chosen to allow individual program actions to proceed while an agency is voluntarily preparing (in furtherance of the NEPA process) a programmatic EIS, not specifically required under the Regulations.[11] The reader is cautioned to note that the precise line separating a required programmatic EIS from one that is not required, but which is prepared voluntarily, can require a considerable degree of professional judgment.

3.3.4.1 Justified Independently of the Program

As denoted by the first subcriterion in Table 3.5 (§ 1506.1[c][1]), an agency should be prepared to demonstrate that the need for the action exists apart from that of the program to which it is related or supports. Such justification might be demonstrated in a number of ways. For example, an agency may be required by law to implement a certain interim action such as an environmental monitoring effort, whether the entire program is implemented or not. An agency, for instance, might be able to demonstrate that a proposed transmission line is consistent with this criterion because it could be used to support other projects regardless of any decisions reached as a result of a programmatic EIS.

3.3.4.2 Accompanied by an Adequate EIS

The second subcriterion cited in § 1506.1(c) states that to be eligible for the status as an interim action, an activity must be reviewed within an existing EIS. Strictly interpreted, this requirement leads to an absurd dilemma with potentially sever repercussions in terms of schedule and cost because it appears to prevent an agency from pursuing an interim action that is covered under a categorical exclusion (CATX) or FONSI. Thus, even if an interim action can be shown to have no significant impact, a strict interpretation leads to the conclusion that it would still have to be reviewed in an EIS.

The U.S. Department of Energy (DOE) has adopted a more reasonable interpretation for resolving this dilemma. A memorandum prepared by Dr. Ziemer indicates that it is questionable that CEQ actually intended this requirement to apply once a determination that a proposed interim action does not require an EIS has been made (i.e., it is eligible for a FONSI or CATX).[12] Even though this criterion is specifically designated for actions requiring an EIS, it can be logically extended to the entire NEPA process (i.e., EIS, EA, and CATX). Thus, Ziemer's memorandum interprets CEQ's intent to mean that an EIS is required for those actions that are not eligible for a FONSI or CATX. In short, the second subcriterion is interpreted to mean adequate NEPA review in lieu of an environmental impact statement.

3.3.4.3   Not Prejudice the Ultimate Decision

As depicted by the third subcriterion, an agency should be prepared to demonstrate that the interim action would not prejudice the ultimate programmatic decision by predetermining subsequent development or limiting alternatives. Such a demonstration may require balancing the size of the interim action against the entire program, as well as considering the potential connected and related actions that may be triggered by implementing the interim action.

3.3.5   ACTIONS THAT ARE NOT GENERALLY PERMISSIBLE

Conceptual design and feasibility studies are examples of actions that are generally considered to be eligible for status as an interim action. While conceptual design studies are generally deemed permissible, proceeding into detailed design is less certain. Design work beyond the conceptual design stage often tends to be extensive and site specific, which has the effect of eliminating alternatives that include the use of other sites. Additionally, the level of resource commitment and the obvious advantage in project schedule relative to the other reasonable alternatives may tend to bias the agency's choice of alternatives in favor of the detailed design.

For example, the DOE's NEPA implementation procedures specify that, with the exception of interim actions, the NEPA review process should normally be completed "… in advance of, and for use in reaching, a decision to proceed with detailed design. …"[13]

Procurement of certain types of equipment and supplies that require a long-lead time may need to be started early in the planning process to avoid later schedule delays. However, such actions may not be eligible for interim action status if they tend to bias the choice of reasonable alternatives.

The interim action provision is frequently interpreted to prohibit many site preparation activities in support of the proposed action. Such activities tend to be prohibited because they tend to bias a final decision in favor of proceeding with the proposed action at that site.

Drilling a test boring might at first appear to be an innocuous activity, having little or no environmental impact. However, activities such as construction of vehicle trails and drilling site preparation might adversely affect archaeological and other resources. Under the interim action clause, this type of activity should be acceptable if it is first accompanied with an archaeological survey, covered under some level of NEPA review, and if these data will be used to provide information for the NEPA analysis.[14]

3.3.6   AWARDING CONTRACTS PRIOR TO COMPLETING NEPA

Actions involving award of contracts require special consideration. The interim action criteria presented in § 1506.1 may prohibit award of a final contract for a proposed action prior to completing NEPA. This is because the award of a final contract may limit the choice of reasonable alternatives or bias a final decision in favor of proceeding with a particular cause of action. However, NEPA's requirements do not necessarily prohibit award of a nonbinding contract that is contingent on first completing NEPA.

3.4   AGENCY MANAGEMENT AND OPERATIONAL CHANGES

Two management and operational changes that can significantly streamline an agency's NEPA planning process are described below.

3.4.1 Revising NEPA Implementation Procedures

Where the Regulations generally provide guidance on what must be done, an agency's implementation procedures should focus on how these requirements are to be effectively implemented. But this is often not the case. One study found that most agency NEPA procedures that were examined simply restated direction already presented in the Regulations.

The following guidance is offered to assist agencies in more effectively revising their NEPA procedures:

- A review should be performed to determine if the agency's procedures provide appropriate mechanisms for governing how NEPA's regulatory requirements are to be implemented.
- Qualified personnel should be consulted to obtain their views on methods that can be utilized in streamlining their NEPA process. This effort may include consulting with the CEQ or other agencies to determine their experiences and solutions to particular problems.
- Legal consul should be assigned the task of reviewing recent developments in NEPA case law. Existing procedures should be revised, and new procedures should be developed to reflect recent developments in the case law.

3.4.2 Delegation

Significant differences exist in the amount of time spent within the NEPA review/approval cycle among various agencies. For example, at one time it was estimated that as much as 40–50% of DOE's NEPA process schedule was consumed in the review/approval cycle. In contrast, the U.S. Bureau of Reclamations estimated that only about 10–30% of the process time was spent in the review/approval cycle.

The U.S. Army has found that downward delegation has substantially streamlined implementation of its NEPA process. However, care must be exercised to ensure that such delegation does not lead to abuse. For example, abuses can be expected if the delegation is directed to such a low level that the decision-maker does not have appropriate background or experience to make informed decisions.

3.5 STATISTICS ON THE NEPA PROCESS

This section presents statistics on NEPA activity, effectiveness, lawsuits, document length, completion time, and costs. Additional data on NEPA statistics can be found in the companion book, *The NEPA Planning Process.*[15]

3.5.1 NEPA Activity

Since the early 1970s, more than 20,000 EISs have been prepared. Preparation of EISs peaked in the early 1970s, when nearly 2000 EISs were prepared a year.[16] Table 3.6 shows the average number

TABLE 3.6
NEPA Document Activity

NEPA Document	Number of Documents Prepared/Year
Programmatic EISs	30–50
Project EISs	250–450
EAs	30,000–50,000
CATX	Unknown

TABLE 3.7
Environmental Impact Statements Filed by Federal Agencies in 1994

Principal Federal Agencies	Percentage
Department of Agriculture	20
Corps of Engineers	10
Department of Interior	18
U.S. Air Force	4
Department of Energy	5
Department of Transportation	24
Other	19

TABLE 3.8
Results of a DOE NEPA Effectiveness Study

NEPA Effectiveness Rating	Respondents (%)
Not effective at all	3
Not very effective	13
Somewhat effective	7
Effective	23
Very effective	23
Highly effective	30

of annual NEPA compliance reviews that the CEQ estimates occur annually.[16] The number of EISs shown in Table 3.6 represents both draft and final EISs, so the number of actual actions represented is approximately half the total number shown.

As indicated in Table 3.6, upwards of 30,000–50,000 federal actions are given limited review in EAs each year. Yet, agencies typically prepare 250–450 EISs (draft, final, and supplemental EISs) annually. Thus, EISs are prepared on perhaps 1% of these federal projects. This means that about 99% of all federal projects are covered under either a CATX or EA. This estimate is consistent with figures reported for the Federal Highway Administration, which because of its mission prepares a relatively high rate of EAs/EISs, and yet in one year reported that it categorically excluded 90% of its projects.[17]

Since the year 2000, there has been an upward trend in the number of EISs filed. In 2004, 597 EISs were filed with EPA, which is the greatest number since 1995.[18] Much of this increase probably reflects a rise in the number of major federal actions taking place across the country. Far from being undesirable phenomena, this increase can actually be viewed as a desirable metric as it reflects a vibrant, healthy, and dynamic economy with an increase in major federal projects.

Table 3.7 shows a breakdown of EIS filed by federal agencies in 1994.[19] The U.S. Forest Service has generally led all other agencies in the number of EISs filed, followed by the U.S. Federal Highway Administration and U.S. Army Corps of Engineers.

3.5.2 NEPA EFFECTIVENESS

Some opponents have criticized the effectiveness of NEPA in planning actions and reaching good decisions. However, such criticism is not substantiated by actual studies. For example, the DOE has published the results of a questionnaire used in determining NEPA effectiveness. The DOE has reported a steady rise in the number of agency participants that responded favorably. As depicted in Table 3.8, more than three-quarters of the respondents reported a rating between effective and highly effective.[20]

3.5.3   Completion Time and Delays

In setting time limits, an agency must ensure that the limit provides sufficient time to adequately complete the NEPA process.[21]

Where an applicant is involved, failure to meet a time limit may provide sufficient grounds for a suit to enforce the original time limit agreed to. In practice, however, it is unlikely that many applicants would bring a suit against an agency that must grant them a permit.

Due to the diverse types of agency missions, controversial issues, and complexity of analysis, the time required to prepare NEPA analyses can vary greatly. It simply takes longer time and requires substantially more effort to evaluate a TransAlaska Pipeline than it does to examine an electrical transmission line in Nebraska.

Delays can result from many different causes such as an overreaction to litigation fears or from an inadequate analysis that results in the repetition of the analysis. Performing the analysis correctly the first time can prevent delays, that is, a stitch in time can save nine.

3.5.3.1   Reasons for Delays

Several recent studies have shown that NEPA is not a principal cause for federal project delays.

Federal Highway Administration Study. Table 3.9 summarizes, in order of priority, the results of a survey performed by the Federal Highway Administration of 55 offices and 89 projects.

The Federal Highway Administration study indicated that environmental reviews compose only about one-quarter of the total time devoted to planning and constructing major highway projects; this time period does not represent a significant commitment of time when one considers that such projects can substantially alter the nation's landscape. Moreover, this study found that where significant delays in highway projects occasionally occur they are generally due to other causes, such as lack of funding, low priority assigned to a project by the sponsoring state transportation agency, or local controversy over the merits of the project.[22]

NRDC Study. A study of 12 agencies was recently performed by the Natural Resources Council of America. The study confirmed that NEPA is not a major cause of project delays. NEPA did not emerge as the principal cause of excessive delays or costs in any of the 12 agencies that were surveyed. However, a number of other requirements were sometimes cited as resulting in lengthy delays in decision-making, which persons *outside* the agencies attributed to NEPA. To the contrary,

TABLE 3.9

Survey Results of Reasons for Project Delays within the Federal Highway Administration

Reason for Delay	Percentage
Lack of project funding	18
Local controversy	16
Low priority	15
Project complexity	13
Agency reviews	8
Change in scope	8
Endangered Species Act (ESA) review	7
Historic preservation review	6
Wetlands review	4
Law suits	3
Hazardous materials	2

NEPA was frequently cited as the means by which a wide range of planning and review requirements were integrated.[23]

Batts and King Study. Results of a survey of 500 NEPA professionals and decision-makers concerning the causes for NEPA delays were recently published. By an overwhelming margin (67% versus 33%) respondents believed it was factors beyond NEPA that delayed a project rather than the NEPA process itself.[24]

Respondents cited four principal reasons as the root cause for schedule delays: (1) changes in the project description, (2) changes or additions to alternatives, (3) litigation, and (4) poor documentation.

A follow-on survey queried approximately 170 NEPA professionals on how they are overcoming these delays. Top-ranked suggestions included obtaining timely reviews and approvals (decision-maker), having an adequate range of alternatives that bracket the proposal, meeting early with stakeholders (litigation), and allowing adequate time for preparation (poor documentation).

Of particular interest, the survey found that, when respondents were offered various suggestions for expediting NEPA, the factor that achieved the strongest agreement was not necessarily directly related to NEPA; having a strong project manager, or advocate, to keep the project moving was strongly favored by 60% of the respondents.

3.5.3.2   Reducing Delays as a Result of Law Suits

Federal courts have dockets to manage and assign priorities. Congressional direction to grant priority to NEPA cases could substantially expedite disposition such cases.[17] For example, California has a legislative requirement granting priority to CEQA cases. It must be noted, however, that the judiciary could be expected to oppose such a measure, preferring to control its own dockets.

3.5.3.3   EA Completion Time

The U.S. DOE is an example of an agency that has a history of producing relatively lengthy and protracted EAs and EISs. However, the DOE is an example of an agency that must struggle with highly complex and controversial issues having national implications. The DOE recently reported for the 12 months that ended September 30, 2005, the median completion time for 26 EAs was 7 months.[25]

The time required to complete the process also tends to vary among agencies. The U.S. Air Force, for example, estimates that an EA typically takes 3–12 months to complete.

3.5.3.4   EIS Completion Time

The DOE recently reported for the 12 months that ended September 30, 2005, the median completion time for five EISs was 32 months; the average was 30 months.[25] The Air Force estimates that it EIS usually takes one or more years to complete.[14]

3.5.4   Document Length

Data on the page length of EAs and EISs are presented below.

3.5.4.1   Environmental Assessments

Table 3.10 presents the results of a CEQ survey regarding the average length of EAs prepared by various agencies.[26] One respondent in this survey indicated that an EA had been prepared that exceeded 1000 pages in length.[27]

The fact that only about one-third of the respondents indicated that their EAs fall with CEQ's guidance of 10–15 pages[28] is an indication that such direction is probably unreasonable.

TABLE 3.10
Average Length of EAs

Length (Pages)	Respondents (%)
1–15	34
16–25	22
26–45	29
46–60	10
61–100	5

3.5.4.2 Environmental Impact Statements

The CEQ regulations state that an EIS shall normally be less than 150 pages and for proposals of unusual scope or complexity shall normally be less than 300 pages. This page length direction applies only to the main body (e.g., paragraphs [d] through [g] of 40 CFR § 1502.10) of an EIS (§ 1502.7); the main body is defined as

- purpose and need for action;
- alternatives (including proposed action);
- affected environment;
- environmental consequences.

3.5.4.3 Congressional Taskforce

A Congressional NEPA taskforce report claimed that the Cambridge Scientific Abstracts noted that in the year 2000, the average final EIS was 742 pages in length.[18] The length cited in the Cambridge Scientific Abstracts includes appendices (which are sometimes lengthy) and other sections that are not part of the main body of the EIS as defined by 40 CFR § 1502.7. Moreover, this figure probably includes programmatic and site-wide EIS, which are special cases, not generally included under the category of typical EISs. Because of their large scope and frequent national implications, programmatic and site-wide EISs often run several thousand pages; they can greatly skew statistics, leading to erroneous conclusions such as the one (i.e., 742 pages) cited in this report.

3.5.4.4 Environmental Protection Agency Study

The Environmental Protection Agency reviewed the document length of 270 final EISs filed during the calendar year of 1996. This study provides a more accurate assessment of compliance than the figure reported in the congressional taskforce report because it assesses the CEQ page length direction in terms of the main body of an EIS (40 CFR § 1502.10).

The results of the study are presented in Table 3.11.[15] The average length (main body) of text was 204 pages not 742 as reported in the taskforce report. Most importantly, more than one-third of the EISs were less than 150 pages long and more than three-quarters fell below the 300 page limit.

At the upper end, 6% of the EISs had more than 500 pages of text. The longest EIS in EPA's study had 1638 pages of text (main body) and the shortest statement was 12 pages in length.

3.5.4.5 Exceptions

A substantial number of EISs exceed CEQ's guidance of 150 pages for typical EISs and 300 pages for EISs of unusual scope or complexity. The EIS for the national supercollider superconductor project presents a good example of just how carried away an agency can become. This EIS was a multi-volume document nearly 8000 pages long. The printing and mailing cost alone amounted to nearly

TABLE 3.11
Average Page Length (Main Body) of EISs

Section	Average Page Length
Purpose and need for action	11
Alternatives	36
Affected environment	66
Environmental consequences	91
Total length (main body)	204

$1.4 million.[27] It should also be noted that this EIS was prepared for an enormous project with national implications and yet was prepared by the DOE in near record time.

3.5.5 COST

NEPA projects vary substantially in size and complexity. Some relatively small proposals can be much more complex or controversial than large projects. Complex environmental issues or impacts of particular concern such as air or water quality and sensitive or endangered species can substantially affect the cost and time required to comply with NEPA. For reasons such as these, an EIS prepared for a dam may be much more controversial, complex, and costly compared with an EIS for a water treatment plant. Conversely, a relatively small bioengineering laboratory might generate far more controversy and concern than a proposal to construct a dam.

The cost of preparing an EIS typically ranges from a couple hundred thousand dollars to several million dollars. In extreme cases, the cost of preparing a very complex and controversial programmatic EIS can cost tens of millions of dollars. Some representative costs are provided in the next section.

3.5.5.1 Department of Energy and Air Force

A senior-level DOE official (which because of its highly complex projects generates among the most expensive NEPA documents) reports that the cost of DOE's NEPA process is always less than 1% and typically 0.1%.[29]

The U.S. Air Force has found that the cost of preparing an EA with accompanying mitigation measures can be over $350,000 for programs at its Space Systems Division, while the cost of preparing an EIS sometimes exceeds $2 million.[30] For complex projects such as construction of an Air Force launch facility, environmental funding may account for less than 1% of the overall project budget. Conversely, environmental funding for smaller projects has in a few cases run as much as 30% of the overall project budget. For example, environmental compliance accounted for nearly $1 million out of a project budget of 2.7 million that the Air Force spent constructing a 45 mile fiber-optic cable. In this project more than 20 archaeological sites were identified, which resulted in field surveys and consultation with Indian Tribes.[31]

3.5.5.2 CEQ Environmental Assessment Study

In 1993, the CEQ published the results of a survey of the EA process used by federal agencies. Table 3.12 records the results of the average cost of EAs among the Federal agencies surveyed. The CEQ survey reported that nearly 30% of the respondents estimated that the average cost of preparing an EA was between $0 and $1500. Approximately 25% of the respondents stated that the average cost ranged between $1500 and $15,000. Near the upper end to the scale, approximately 30% of the agencies estimated that the average cost lay between $15,000 and $50,000. Finally, nearly 15% of the agencies responded that the average cost was between $50,000 and $100,000.[26]

TABLE 3.12
Average Cost of Preparing EAs

Cost ($)	Respondents (%)
0–500	17
500–1500	12
1500–5000	12
5000–15,000	15
15,000–30,000	15
30,000–50,000	15
50,000–100,000	15

TABLE 3.13
Percentage Breakdown of NEPA Documents That Have Been Subject to a Legal Challenge

NEPA Compliance	Percentage
Categorical exclusions	5
Environmental assessment	43
Environmental impact statement	52

Nearly one-quarter of the agencies replied that in extreme cases an EA could cost as much as $50,000–$100,000. In a substantial number of cases, the agencies either partially or totally recover their preparation costs through fee assessments or other charges.

3.5.6 LAWSUITS

NEPA generates a relatively small volume of litigation. Plaintiffs typically file about 100 NEPA lawsuits per year, representing only 0.2% of the 50,000 or so federal actions reported to require an EA or EIS each year. This represents a nearly infinitesimal fraction of the federal lawsuits that take place annually.[16] The following section provides some important statistics regarding NEPA lawsuits.

3.5.6.1 NEPA Documents Challenged

Table 3.13 presents a breakdown of NEPA documents that have been the subject of a legal challenge resulting in a clear winner or loser.[19] As indicated in the table, cases involving an EIS makeup slightly over one-half of all cases filed.

3.5.6.2 Causes for Action

As detailed in Table 3.14, failures to prepare an EIS and claims that a statement was inadequate have been the two leading causes cited in NEPA litigation.[19] However, issues involving cumulative impacts have become increasingly important.

In the relatively small number of NEPA suits brought by the plaintiffs, nearly 40% of the cases simply involved failure of the federal agency to prepare a NEPA document.[19] In other words, the principal reason for suits involved straightforward issues of outright noncompliance.

TABLE 3.14
Causes for Action Filed under NEPA in 1994

Causes for Action	Percentage
No environmental impact statement	24
Inadequate environmental impact statement	31
No environmental assessment	10
Inadequate environmental assessment	22
No supplemental environmental impact statement	5
Other	8

TABLE 3.15
Breakdown of Lawsuits by Plaintiffs in 1994

Plaintiffs	Percentage
Environmental groups	45
Individuals and citizens groups	21
State governments	2
Local governments	8
Business groups	12
Property owners or residents	11
Native American tribes	1

3.5.6.3   Plaintiffs

As depicted in Table 3.15, individuals and environmental and citizens groups have been the most active plaintiffs bringing NEPA suits.[19]

There is a popular misconception that NEPA is a tool used by "wild-eyed" left-wing environmentalists to slow or halt progress. In reality, individuals and citizens groups, state and local governments, business groups, property owners and residents, and Native American tribes make up 55% of the plaintiffs who have brought suits against the federal government on behalf of NEPA.

3.5.6.4   Injunctions

In 2004, 156 NEPA cases were filed, and in only 11 of those cases did a judge grant an injunction. By way of comparison, in 2004, 281,338 civil cases were filed in the U.S. district courts.

During the same year, 998 cases were filed in the district courts involving environmental matters. Of these, 548 involved only private parties (i.e., not the U.S. Government), and of the balance, which involved suits to which the United States was a party, the government was the plaintiff in 171 cases and defendant in 279. Environmental cases therefore represent a miniscule portion of the federal court caseload, and NEPA cases a modest part of even that small fraction.

With respect to NEPA actions and NEPA litigation, taking the average number of NEPA documents filed annually and the 2004 NEPA injunction figures, a 99.97% rate of NEPA actions were successfully completed without injunctions. Even looking at the relatively modest number of NEPA suits filed, judges did not issue an injunction in 93% of the cases.[19]

PROBLEMS

1. Which two agencies produce the most EISs?
2. List five methods for reducing project delays as a result of NEPA?

3. What are the two leading causes cited for NEPA lawsuits?
4. Do the NEPA regulations encourage integrating NEPA with other processes? Cite your evidence.
5. List three methods for reducing project NEPA paperwork?
6. What is meant by the term "tiering"?
7. What is meant by the term "incorporation by reference"?
8. Which type of NEPA document (CATX, EA, or EIS) is subject to the most legal challenges?
9. How many EAs are estimated to be prepared each year?
10. What is an "interim action"?
11. Prior to completing a project-specific EIS, under what conditions can an action related to the EIS proceed?
12. An EIS for a hydroelectric project is under way. The project engineer decides that to shorten the schedule, he will start the long-lead procurement process; he intends to order six specially designed turbines/generators for the project. These systems will cost 15 million of dollars, and because they are specially designed for this project, they will probably have no other use if the agency decides not to go forward with the proposal. Would such a long-lead procurement process constitute a violation of NEPA?

REFERENCES

1. Council on Environmental Quality, The National Environmental Policy Act: A Study of Its Effectiveness after Twenty-Five Years (1997), p. 3.
2. Mass. Gen. Laws ch. 30, § 61.
3. N.Y. Envtl. Conserv. Law, § 8-0109(8).
4. Council on Environmental Quality's (CEQ) 40 CFR 1500.4, Reducing paperwork; 40 CFR 1500.5, Reducing delay.
5. CEQ's 40 Questions, Question 36a.
6. CEQ, Public memorandum titled, "Talking points on CEQ's oversight of agency compliance with the NEPA Regulations," 1980.
7. CEQ, Memorandum: Guidance Regarding NEPA Regulations, 48 Federal Register, 34263, July 28, 1983.
8. CEQ, Memorandum: Guidance Regarding NEPA Regulations, 48 Federal Register, 34263, July 28, 1983; 40 CFR 1506.3(a–c).
9. Preamble to Final CEQ NEPA Regulations, 43 Federal Register 55978, Section 4, November 29, 1978.
10. CEQ's 40 Questions, Question 21.
11. Yost N.C. and Rubin J.W., The National Environmental Policy Act, unpublished.
12. Ziemer P.L., Guidance Regarding Actions That May Proceed During the National Environmental Policy Act (NEPA) Process: Interim Actions, EH-25 Memorandum, July 12, 1992.
13. 10 CFR 1021.210(b).
14. Lillie T.H. and Lindenhofen H.E., Military construction and the environment: NEPA as a blueprint for compliance, Federal Facilities Environmental Journal, Winter 1991/1992.
15. Eccleston C.H., The NEPA Planning Process: A Comprehensive Guide with Emphasis on Efficiency, John Wiley & Sons, New York, 1999, Chapter 9.
16. CEQ, Environmental Quality 1993: The Twenty-Fifth Annual Report, 1994–1995.
17. Yost N.C., Testimony Before the Committee on Resources United States House of Representatives, Hearing on NEPA: Lessons Learned and Next Steps, November 17, 2005.
18. United States House of Representatives Committee on Resources, Task Force on Improving the National Environmental Policy Act and Task Force on Updating the National Environmental Policy Act, Initial Findings and Draft Recommendations, p. 18, December 21, 2005.
19. Reinke C.C. and Robitaile, NEPA litigation 1988–1995: A detailed statically analysis, Proceedings of the 22nd Annual Conference of the National Association of Environmental Professionals, Orlando, FL, 1997, pp. 759–765.
20. DOE, National Environmental Policy Act: Lessons Learned, Quarterly report for first quarter of fiscal year 1996, March 1, 1996.

21. 40 CFR 1501.8.
22. Dreher R., Testimony Before the Committee on Resources United States House of Representatives, 2005.
23. Smythe R. and Isber C., Natural Resources Council of America, NEPA in the Agencies, October 2002.
24. Batts D. and King J., Fast-tracking NEPA within public land management agencies. Presented at the National Association of Environmental Professionals (NAEP) 28th Annual, San Antonio, TX, 2003; Also Presented at 2006 NAEP Annual Conference, Modifications and attitudes on recent changes in federal laws to expedite NEPA compliance—Opportunities for improvement? Albuquerque, NM, April 23–26, 2006.
25. DOE, National Environmental Policy Act, *Learned Lessons*, December 5, 45, 2005, p. 40.
26. Blaug E.A, Use of the environmental assessment by federal agencies in NEPA implementation, *The Environmental Professional*, 15, 1993, pp. 57–65.
27. Bausch C., A simple formula for crafting better NEPA documents, *Federal Facilities Environmental Journal*, Autumn 1992.
28. CEQ, Forty most asked questions concerning CEQ's National Environmental Policy Act Regulations, 46 Federal Register 18026, March 23, 1981.
29. Cohen E.B., *NEPA News* (published by NEPA Watch), No. 4 (January–February 1997).
30. Lillie, T.H. and Lindenhofen, H.E., NEPA as a tool for reducing risk to programs and program managers, *Federal Facilities Environmental Journal*, Spring 1991.
31. Lillie T.H. and Bowman S., NEPA compliance for Air Force space systems division programs, *Federal Facilities Environmental Journal*, 1(4), Winter 1990/1991.

4 Performing a Systematic and Integrated Planning and Analysis Process

Federal agencies are granted broad discretionary authority regarding how they choose to integrate National Environmental Policy Act (NEPA) into their planning processes including determining the appropriate strategies, procedures, and analytical methodologies to be used. For instance, an agency has a great degree of latitude in determining the scope and detail of the issues to be reviewed, the scientific methodology and models to be used in the analysis, and the methods required to insure scientific accuracy. Typically, the burden of proof in demonstrating that an agency has failed to comply in an appropriate manner with NEPA does not lie with the agency but with the party challenging the agency.

This chapter is designed to introduce the reader to the concepts of integrated and systematic analysis and planning. As such, this chapter introduces some of the fundamental environmental statutes, regulations, principles, and other requirements that must commonly be integrated with the NEPA planning process.

4.1 A FLEXIBLE PLANNING PROCESS

As a planning tool, NEPA is particularly flexible. Unlike most of the other environmental statutes, NEPA allows decision-makers to balance other factors such as political considerations, risk, cost, safety, and schedules in reaching a final decision. NEPA requires agencies to account for environmental factors, yet it does not mandate unwieldy performance standards or other burdensome restrictions on project engineers or decision-makers. Relevant regulatory provisions that underscore this flexibility include the following:

- Identify environmental effects and values in adequate detail so they can be *compared to economic and technical analyses* (§ 1501.2[b], emphasis added).
- An agency may *discuss preferences among alternatives based on relevant factors including economic and technical considerations and agency statutory missions.* An agency shall identify and discuss all such factors including any essential considerations of national policy, which were *balanced* by the agency in making its decision and state how those considerations entered into its decision (§ 1505.2[b], emphasis added).
- The "agency's preferred alternative" is the alternative which the agency believes would *fulfill its statutory mission and responsibilities, giving consideration to economic, environmental, technical and other factors.* The concept of the "agency's preferred alternative" is *different* from the "environmentally preferable alternative," although in some cases one alternative may be both.[1]

The Council on Environmental Quality's (CEQ) NEPA regulations (Regulations) establish goals and procedural requirements, but leave the question of how such requirements are to be implemented largely to the discretion of individual agencies; agencies have thus been granted an unusually wide degree of latitude (i.e., opportunity) and flexibility in determining how they choose to discharge their environmental impact statement (EIS) responsibilities. Nonetheless, many agencies have failed to take full advantage of this.

The preparation of an EIS should be welcomed as an opportunity to formulate the agency's preferred course of action, integrating all pertinent decision-making factors and minimizing cost by preventing disconnects and uncoordinated plans. For example, the EIS requirement to perform an alternative analysis can provide a mechanism for identifying more benign courses of action that may minimize or even avoid future permitting procedures, thus reducing costs while expediting project schedules.

Mitigation measures provide a particularly effective tool for reducing future project risks. By incorporating mitigation into early planning, subsequent impacts may be avoided or reduced to the point where later project implementation and permitting requirements can be curtailed and in some cases entirely eliminated.

4.2   FOSTERING PLANNING AND INFORMED DECISION-MAKING

All too often, NEPA is viewed simply as a process for preparing a document but nothing could be further from the truth. While preparing a document is indeed an integral component, preparation of environmental documents, even excellent ones, is not why NEPA was enacted. Rather, first and foremost, NEPA should be viewed as a federal planning and decision-making process.

The real purpose of NEPA is to provide decision-makers and the public with information that promotes informed decision-making. As noted in Table 4.1, the ultimate objective is to ensure that environmental impacts are properly considered before a final decision is made to pursue a given course of action and "not to justify decisions already made" (§ 1502.2[g], § 1502.5).[2]

Managed properly, the preparation of an EIS is simply the final element of what is otherwise a rigorous environmental planning process. An EIS is merely a tool or mechanism and perhaps the final step of the process undertaken by an agency to record the results of a comprehensive planning and decision-making process.

Table 4.2 presents selected regulatory directions pertaining to planning and agency decision-making.

TABLE 4.1
NEPA's Purpose Is to Facilitate Informed Decision-Making

Integrating the NEPA process into early planning to insure appropriate consideration of NEPA's policies and to eliminate delay (§ 1501.1[a]).

Ultimately, of course, it is not better documents but better decisions that count. NEPA's purpose is not to generate paperwork—even excellent paperwork—but to foster excellent action. The NEPA process is intended to help public officials make decisions that are based on understanding of environmental consequence, and take actions that protect, restore, and enhance the environment (§ 1500.1[c]).

An environmental impact statement is more than a disclosure document. It shall be used by Federal officials in conjunction with other relevant material to plan actions and make decisions (§ 1502.1).

The primary purpose of an environmental impact statement is to serve as an action-forcing device to insure that the policies and goals defined in the Act are infused into the ongoing programs and actions of the Federal Government (§ 1502.1).

… analyses shall be circulated and reviewed at the same time as other planning documents (§ 1501.2[b]).

An agency shall commence preparation of an environmental impact statement as close as possible to the time the agency is developing or is presented with a proposal … so that preparation can be completed in time for the final statement to be included in any recommendation or report on the proposal. The statement shall be prepared early enough so that it can serve practically as an important contribution to the decision-making process and will not be used to rationalize or justify decisions already made (§ 1502.5).

TABLE 4.2
Selected Citations Governing an Agency's NEPA Planning Process

Integrating the NEPA process into early planning … (§ 1500.5[a])

Integrate the requirements of NEPA with other planning and environmental review procedures … (§ 1500.2[c])

Agencies shall integrate the NEPA process with other planning at the earliest possible time … (§ 1501.2)

… utilize a systematic, interdisciplinary approach … in planning … which may have an impact on man's environment … (§ 1501.2[a])

Environmental documents and appropriate analyses shall be circulated and reviewed at the same time as other planning documents (§ 1501.2[b])

Agencies may prepare an EA on any action at any time in order to assist agency planning … (§ 1501.3[b])

Hold an early scoping meeting or meetings which may be integrated with any other early planning meeting … (§ 1501.7[b][4])

[The EIS] shall be used by Federal officials in conjunction with other relevant material to plan actions … (§ 1502.1)

Agencies shall prepare statements on broad actions so that they are relevant to policy and are timed to coincide with meaningful points in agency planning … (§ 1502.4[b])

Agencies shall cooperate with State and local agencies to the fullest extent possible … [and] shall … include: (1) Joint planning processes (§ 1506.2[b])

To better integrate EISs into State or local planning processes, statements shall … (§ 1506.2[d])

… utilize ecological information in the planning and development of resource-oriented projects (§ 1507.2[e])

… insure the integrated use of the natural and social sciences and the environmental design arts in planning and in decision-making … (§ 102[2][A] of NEPA, § 1507.2)

TABLE 4.3
Requirements for Conducting an Early and Open NEPA Process

Integrating the NEPA process into early planning (§ 1500.5[a], § 1501.1[a])

Preparing environmental impact statements early in the process (§ 1500.5[f])

… insure that environmental information is available to public officials and citizens before decisions are made and before actions are taken (§ 1500.1[b])

Agencies shall integrate the NEPA process with other planning at the earliest possible time … (§ 1501.2)

There shall be an early and open process for determining the scope of issues to be addressed and for identifying the significant issues … (§ 1501.7)

… the environmental impact statement shall be prepared at the feasibility analysis (go-no go) stage … (§ 1502.5[a]).

… commence preparation of an environmental impact statement as close as possible to the time the agency is developing or is presented with a proposal (§ 1502.5)

… prepared early enough so that it can serve practically as an important contribution to the decision-making process and will not be used to rationalize or justify decisions already made (§ 1502.5)

4.2.1 EARLY AND OPEN PROCESS

Among the broad array of environmental and safety and health requirements, NEPA is virtually unique in that it is an integral part of an agency's early planning process. Achieving this objective is crucial if an agency is truly using NEPA as a planning and decision-making tool, informing decision-makers and the public of the consequences of potential actions. Table 4.3 presents selected citations that elaborate on this requirement.

4.2.2 Public Involvement

The requirement governing public involvement is one of NEPA's most important provisions. This requirement stated in the Act is as follows:

> Copies of such statement [EIS] and the comments and views of the appropriate federal, state and local agencies ... shall be made available to the President, the Council on Environmental Quality, and the public ... (Section 102[2][C] of the NEPA Act).

This provision makes NEPA a "sunshine act," as it requires that agency decision-making, with respect to environmental effects, be open to public input and review. NEPA's public involvement requirements are described in more detail in later chapters.

4.2.3 Determining the Scope

The term "scope" refers to the breadth and content of an NEPA analysis. The concept of scope involves three essential elements:

> Scope consists of the range of actions, alternatives, and impacts to be considered in an environmental impact statement (§ 1508.25).

The terms "actions, alternatives, and impacts" are defined and dissected in Chapter 8. The term "scoping" refers to the process by which an agency determines the range of issues to be considered in an analysis. Specifically

> There shall be an early and open process for determining the scope of issues to be addressed and for identifying the significant issues related to a proposed action. This process shall be termed scoping (§ 1501.7).

Thus, an early and open process is to be performed in determining the scope of an EIS. Agencies are expected to "make diligent efforts to involve the public in preparing and implementing their NEPA procedures (§ 1506.6)."

In one instance, an agency held scoping meetings and publicly advertised the opportunity to participate in the scoping process. Because no one from the public attended the meetings, the agency continued without public input. When challenged, the court ruled that it was not sufficient for an agency simply to provide the public with an opportunity to participate. The agency, in fact, must actively solicit public attention to gain participation.

4.3 SYSTEMATIC AND INTERDISCIPLINARY APPROACH

NEPA's requirement to use a *systematic* and *interdisciplinary* approach is arguably the single most important requisite for ensuring an accurate and comprehensive scientific analysis. This requirement has been viewed by the courts as having a scope and applicability that extends beyond the EIS requirement to include environmental assessments (EAs) as well.[3] Table 4.4 provides guidance for implementing this requirement.

The term "systematic" is interpreted to denote a disciplined process that is performed using a logically ordered and methodical approach. This requirement implies that a methodical step-by-step approach be used in which one stage of the process builds upon previous stages.

The interdisciplinary requirement places a burden on agencies to ensure that the environmental analysis is performed by knowledgeable individuals or experts representing disciplines that may be potentially affected by or are fundamental to a thorough analysis.

There is an important distinction between *interdisciplinary* and *multidisciplinary* specialists. In the context of NEPA, multidisciplinary implies the preparation of an analysis in which specialists

TABLE 4.4

Direction for Conducting a Systematic and Interdisciplinary Approach

... utilize a systematic, interdisciplinary approach which will insure the integrated use of the natural and social sciences and the environmental design arts in planning and in decision-making which may have an impact on man's environment (§ 1501.2[a])

Make available staff support at the lead agency's request to enhance the latter's interdisciplinary capability (§ 1501.6[b][4])

Environmental impact statements shall be prepared using an inter-disciplinary approach. ... The disciplines of the preparers shall be appropriate to the scope and issues identified in the scoping process (§ 1502.6)

Fulfill the requirements ... of the Act to utilize a systematic, interdisciplinary approach ... (§ 1507.2[a])

from various technical disciplines prepare independent sections of that analysis, but do not necessarily interface or communicate with one another. In contrast, an interdisciplinary analysis implies a "team" approach under which specialists from different technical disciplines interface and communicate with one another.

Related to this requirement is the issue of subject matter expertise. In one case, a challenge was made to an EIS, which resulted in approval of a permit from the Corps of Engineers for the construction of a dam that could affect endangered species. The Corps asserted that it had relied on its internal experts in determining that there was no additional evidence indicating that any supplements needed to be added to the EIS. The court rejected this assertion because the administrative record did not demonstrate that the employees involved in the EIS review were sufficiently qualified to address this issue.[4]

4.3.1 ENVIRONMENTAL DESIGN ARTS

As mentioned earlier, agencies are required to

... utilize a systematic, interdisciplinary approach which will insure the integrated use of the natural and social sciences and the *environmental design arts* in planning and in decision-making which may have an impact on man's environment (§ 1501.2[a], emphasis added).

The requirement to integrate environmental design arts is interpreted to mean that disciplines such as architecture and urban planning (e.g., environmental design arts) are to be integrated into the NEPA planning process so that federal actions blend more naturally into their surrounding environments.

4.4 INTEGRATING OTHER LAWS, PERMITS, AND ORDERS

Experience shows that integrating NEPA into an agency's early planning process is perhaps the single most effective means for improving efficiency. Moreover, a CEQ study concluded that one of the principal causes for delays resulted simply from failure to integrate NEPA properly with other planning and environmental requirements (e.g., wetlands studies, cultural resources studies). For this reason, the CEQ recommends that agencies perform an integrated environmental planning process where various requirements are implemented in parallel rather than sequentially.

As part of the scoping process, agencies are expected to identify other related environmental review and consultation requirements. To the fullest extent possible, agencies are directed to prepare and integrate draft EISs concurrently with other required environmental surveys, studies, and laws. Additionally, there are a number of statutory provisions in other laws and regulations requiring integration of environmental reviews or consultations with NEPA.[5]

TABLE 4.5

List of Principal Environmental Statutes and Requirements That Need to Be Integrated with the NEPA Planning Process

Requirements specifically cited in the Regulations

- National Historic Preservation Act of 1966 (16 U.S.C. 470 et seq.).
- Endangered Species Act of 1973 (16 U.S.C. 1531 et seq.).
- Fish and Wildlife Coordination Act (16 U.S.C. 661 et seq.).

Other environmental review laws and executive orders

- Wild and Scenic Rivers Act of 1968 (16 U.S.C. 1271–1287).
- Coastal Zone Management Act of 1972 (16 U.S.C 1451 et seq.).
- Farmland Protection Policy Act of 1981 (7 U.S.C. 4201 et seq.).
- American Indian Religious Freedom Act of 1978 (42 U.S.C. 1996).
- Pollution Prevention Act of 1990 (P.L. 101–508, 6601 et seq.).
- Environmental Justice (Executive Order 12898).
- Protection of Wetlands (Executive Order 11990).
- Floodplain Management (Executive Order 11988).

In furtherance of such requirements, the CEQ recently issued a draft document titled *Collaboration Handbook*.[6] The purpose of this handbook is to assist federal agency NEPA practitioners in expanding the effective use of collaboration as part of the NEPA process. The handbook outlines general principles, presents useful steps throughout the NEPA process, provides information on methods of collaboration, and presents case examples.

As the first part of Table 4.5 denotes, three requirements are specifically listed in the Regulations as requirements to be integrated with the NEPA planning process (§ 1502.25[a]). Additional requirements that also commonly need to be integrated with NEPA are listed in the second part of Table 4.5.

Some of the most important requirements that need to be integrated or understood in terms of NEPA are described in the following sections.

4.4.1 ENVIRONMENTAL QUALITY IMPROVEMENT ACT OF 1970

The Environmental Quality Improvement Act (EQIA) of 1970 is intended to ensure that every federal agency conducting or supporting public works activities affecting the environment implements policies established under existing law.[7] The EQIA is an Act that supplemented NEPA's authority. The EQIA also 'created' the Office of Environmental Quality. The Office of Environmental Quality is more widely known as the Council on Environmental Quality (CEQ). Among other provisions, the EQIA added additional responsibilities to the CEQ. Its director has been tasked with responsibility for assisting and advising the president on federal policies and programs affecting environmental quality. The Office of Environmental Quality reviews the adequacy of existing environmental monitoring and predicting systems and also assists federal agencies in appraising the effectiveness of existing and proposed facilities affecting environmental quality.

4.4.1.1 Executive Order for Protection and Enhancement of Environmental Quality

In 1970, President Nixon issued Executive Order 11514, which stated that the president, with assistance from the CEQ, would lead a national effort to provide leadership in protecting and enhancing the environment for the purpose of sustaining and enriching human life.[8] Federal agencies are directed to meet national environmental goals through their policies, programs, and plans. Agencies should also continually monitor and evaluate their activities to protect and enhance the quality of the environment. Consistent with NEPA, agencies are directed to share information about existing or potential environmental problems with all interested parties, including the public, in order to obtain their views.

4.4.2 FEDERAL LAND POLICY MANAGEMENT ACT

The Bureau of Land Management (BLM) is an agency within the Department of the Interior whose primary mission is to manage public lands, primarily those in western states. For many years, the BLM managed public lands under a number of different, and sometimes conflicting, statutes.

The Federal Land Policy and Management Act (FLPMA) of 1976 established for the first time a single and comprehensive statutory mandate for retaining public lands under federal ownership and for managing those lands for the public. Under the FLPMA, the mission of the BLM was changed to one of multiple uses—a new concept in the 1970s. As a result, the future of the West was forever changed.

This Act recognized the value of America's public lands and provided a framework for managing them in perpetuity for the benefit of present and future generations. This Act requires the use of planning and the establishment of management programs for protecting the quality of scientific, scenic, historical, ecological, environmental, water resource, and archaeological values on public lands.

4.4.2.1 Policy, Authority, and Responsibility

With respect to the FLPMA, "public lands" or "the public domain" refers to all those lands that the United States has acquired from other nations or from Indian tribes and that have not been sold off or set aside as national forests, national parks, military reservations, etc. These lands, which are managed by the BLM, currently total more than 260 million acres (40% of all federally owned land), or 12% of the total land area, of the United States. The FLPMA also established several statements of general policy:

1. *Federal ownership.*

Following a century old policy of using public lands to promote everything from homesteading to the construction of highways and railroads, FLPMA established a new policy; that is, for the most part, remaining public lands would be retained under federal ownership. Although land exchanges and sale of discrete tracts of land are still allowed, the overarching policy is to retain lands under federal ownership.

2. *Multiple use and sustained yield.*

Under FLPMA, the BLM established a planning process similar to that used by other federal agencies. Periodically, the BLM must inventory public lands and their resources and develop resource management plans (RMPs).

The BLM must manage public lands using multiple use and sustainable yield principles similar to those used by the Forest Service in managing national forests. Resources must therefore be used in a multiple use manner that best meets the needs of the American people and future generations.

The BLM must consider the relative value of resources. This does not necessarily mean, however, that the BLM must promote those uses having the greatest potential economic return or greatest unit output or, conversely, those uses that do not impair productivity of the land.

3. *Withdrawal authority.*

Prior to the FLPMA, U.S. presidents often withdrew public lands on their own initiatives from specific uses or from sale. For example, land was sometimes withdrawn to

- prevent mining and oil/gas development, or
- preserve lands for specific uses (e.g., military bases).

Previously, presidents relied on various federal statutes or even on their own implied power to make such withdrawals. Since 1976, FLPMA has delegated this authority to the Secretary of the Interior to withdraw public lands (BLM-managed lands). The secretary can also withdraw other federal lands such as those lying within national forests with the consent of the appropriate department head. Withdrawals are generally limited to a period of 20 years, meaning that Congress or the president must eventually take action to provide permanent protection to withdrawn lands.

Case law. In a 2004 decision, the U.S. Supreme Court held that a land use plan is generally a statement of priorities.[9] Although land use plans guide and restrain actions, they do not prescribe them. Members of the public cannot generally compel an agency to implement discretionary actions. Because the implementation of land use plans is subject to available appropriations, citizens can only compel an agency to take a specific action that it is already required to take.

4. *Planning and public participation.*

The BLM uses its land use planning process to protect and designate uses of public lands and resources. FLPMA requires the BLM to

- consider present and potential uses of public lands,
- apply principles of multiple use and sustained yield management,
- give priority to the designation and protection of areas of critical environmental concern, and
- weigh long-term benefits to the public against other short-term benefits.

When preparing an EA/EIS during an ongoing RMP revision (together with its accompanying EIS), there may be opportunities to consolidate some components of the NEPA process such as cumulative effects analysis and public involvement activities.

After an RMP is approved, any authorizations and approved management actions based on a project-specific EIS (or EA) must specifically be provided in the RMP or be consistent with its terms, conditions, and decisions.

The planning sections of the BLM manual and the BLM planning handbook can be accessed at: http://www.blm.gov/nhp/200/wo210/landuse_hb.pdf.

4.4.3   NATURAL RESOURCE DAMAGE ASSESSMENT

Section 101(16) of the Comprehensive Environmental Response, Compensation, and Liability Act (CERCLA) defines natural resources as

… land, fish, wildlife, biota, air, water, groundwater, drinking water supplies, and other such resources …

An injury to a natural resource is a measurable adverse change in the chemical or physical quality, or viability, of that resource.

Damages can be assessed on the basis of loss or reduction in quantity and quality of natural resource services and represents the dollar value or the economic loss resulting from the injury. Damage assessments are based on the amount of the residual damage (i.e., damages that are not or cannot be addressed by the remedial or corrective action or that result from such actions).

Services are the physical and biological functions performed by natural resources, including their use by humans and their services to other resources and ecosystems. Examples of resource services include habitat, food, recreation, esthetic value, drinking water, flood control, and waste assimilation.

4.4.3.1   Assessment

A natural resource damage assessment (NRDA) is a process whereby a natural resource trustee may pursue compensation on behalf of the public for injury incurred to natural resources. Only designated federal trustees, authorized representatives of an affected state, or an affected Indian tribe can recover economic resource damages. Natural Resource Trustees conduct NRDAs to calculate the monetary cost of restoring injuries incurred to natural resources resulting from, for example, releases of hazardous substances or discharges of oil.

An assessment plan details the scientific and economic methodologies to be used and the specific data to be collected. The preassessment screening process and the assessment plan activities can be coordinated with ongoing investigations such as a CERCLA Remedial Investigation/Feasibility Study (RI/FS) or the Resource Conservation and Recovery Act Facility Investigation/Corrective Measures Study (RFI/CMS).

NEPA. Timely considerations of NRDA issues in NEPA documents can also be of strategic importance because Section 107 of CERCLA excludes liability for damages that result from a discharge or release when

> … the damages are specifically identified as an irreversible and irretrievable commitment of a natural resource in an environmental impact statement or other comparable environmental analysis.

Thus, if potential natural resource damage was identified in a NEPA document as an irreversible and irretrievable commitment before the action was undertaken, the damage is exempt from future economic damage assessments.

Special conditions imposed by an applicable license or permit that authorized the commitment of resources may also be factored into a decision to exclude the release from liability for damages (other conditions may apply with respect to exclusion of liability for damages to the resources of an Indian tribe).

4.4.3.2   Contingent Valuation Method

Research indicates that most people are willing to pay for environmental benefits and the nonuse of resources that otherwise could be lost to future development. However, unless a dollar value is estimated for nonuse resources, it is likely that they will be treated implicitly as having no value. This leads to a quandary: How much are nonmarket environmental resources really worth and how can their real value be estimated? Often, the only option for estimating their value is by asking questions.

One technique for estimating the economic worth or value of environmental resources is known as the contingent valuation method (CVM). The regulations for cost and damage recovery under the federal Superfund program explicitly recognize the use of contingent valuation as a tool for estimating such values.

CVM can be used to estimate both the use and potential nonuse values of a given nonmarket resource. With respect to NEPA, CVM can be used both to evaluate NRDA and to evaluate and compare the cost and benefits of the alternatives.

CVM involves performing a survey in which people are asked how much they are willing to pay for specific environmental services. The method is called contingent valuation, because people are asked questions regarding their willingness to accept compensation contingent on a hypothetical scenario of losing a given resource. For instance, people might be asked the amount of compensation they would be willing to accept to give up a specific environmental resource.

For example, the Snake River RMP EIS used a CVM to identify nonmarket values associated with the public land parcels. In another case, the Bureau of Reclamation prepared an EIS to reevaluate operations of the Glen Canyon Dam; the bureau used CVM to quantify the impact of various dam flow alternatives on recreation and nonuse value resources. CVM was also used in the valuation of damages that resulted from the *Exxon Valdez* oil spill in Alaska in 1989 and for valuation of air quality improvement at the Grand Canyon.

The questionnaire used in one study put interviewees in the position of becoming decision-makers. In response to cost and benefit information supplied to them, interviewees were asked questions about their preferences regarding incinerators involving different levels of pollution control technology. The researchers found that most people were willing to pay more for stricter control technologies if they believed that the health benefits to be gained outweighed the costs.[10]

Contingent valuation is categorized as a "stated preference" technique, because individuals are asked to estimate and state their values, rather than simply inferring values from actual choices, as is the case with "revealed preference" methods. Under this methodology, the value of resources as diverse as hunting and fishing, wilderness experience, water use, and appreciation of scenic and visual resources can be gauged.

Some prominent economists and psychologists question whether CVM can accurately gauge the true values people place on nonuse goods. Moreover, many have been unwilling to accept the results of CVM studies. Practitioners, therefore, are warned to use such methodologies cautiously.

Nevertheless, nonuse values are real and ignoring them can significantly understate total losses since they are frequently substantial. Despite criticism, many investigators working in natural resource and environmental areas have developed and used CVM, and it is currently the only widely used approach for estimating nonuse values.

Simplified five-step approach. While there are countless variations of CVM, the following five-step approach demonstrates how this methodology can be applied to estimate the value of a resource.

1. Determine the specific resources (water, clean air, lack of health risks) to be examined and determine the relevant population to survey. For example, if the resource is a state park, the relevant population may consist of all citizens of that state. For a city, the relevant population might be restricted only to its citizens.
2. Determine a methodology for performing the survey (e-mail, mail, phone, in-person). Other questions involve determining who will be surveyed and how large the sample size will be. In-person interviews are often considered to be the most effective method but also tend to be the most expensive.
3. The most difficult part of the process is designing the survey. The design process is normally performed by first testing it on groups representing the types of people who will ultimately receive the final survey.
4. After the survey has been designed and tested, it is implemented. Preferably, the survey sample (phone, mail, in-person) should be chosen randomly from the relevant population.
5. Once the survey data have been captured, the investigator analyzes and interprets the results.

4.4.4   POLLUTION PREVENTION ACT

In passing the Pollution Prevention Act of 1990, Congress formally established a national policy to prevent or reduce pollution at its source whenever feasible.[11] This Act establishes the following national policy:

> … that pollution should be prevented or reduced at the source whenever feasible; pollution that cannot be prevented should be recycled in an environmentally safe manner whenever feasible; pollution that cannot be prevented or recycled should be treated in an environmentally safe manner whenever feasible; and that disposal or other release into the environment should be employed only as a last resort and should be conducted in an environmentally safe manner.

The Environmental Protection Agency (EPA) defines pollution prevention (P2) as

> … the use of materials, processes, or practices that reduce the use of hazardous materials, energy, water, or other resources and practices that protect natural resources through conservation or more efficient use.

4.4.4.1 Provisions

Major provisions of the Act include

- providing matching funds for state and local P2 programs through a grant program to promote the use of P2 techniques by businesses.
- establishing a P2 strategy outlining an agency's intent to promote source reduction and collect data on source reduction and recycling.
- operating a source reduction clearinghouse. The Pollution Prevention Information Clearinghouse (PPIC) provides telephone references and referrals, distributes EPA documents, and has a collection of P2 references available for interlibrary loan. For more information, see the PPIC Web site or contact PPIC at (202) 566-0799.

4.4.4.2 Executive Order and the CEQ Guidance

In 1993, President Clinton issued an executive order directing federal agencies and their facilities to comply with the provisions of the Emergency Planning and Community Right-to-Know Act (EPCRA) and the P2 Act.[12]

In 1993, the CEQ issued guidance to federal agencies on how to incorporate pollution prevention principles, techniques, and mechanisms into their planning and decision-making processes and to evaluate and report those efforts, as appropriate, in documents pursuant to NEPA.[13]

4.4.5 CLEAN AIR ACT CONFORMITY AND NEPA

The Clean Air Act (CAA) of 1972 is a comprehensive federal law that regulates air emissions.[14] As designated by the EPA, a nonattainment area is one which exceeds the National Ambient Air Quality Standards (NAAQS). A maintenance area is an area that has been redesignated to attainment from nonattainment; a maintenance area must comply with the NAAQS for a period of 20 years.

The 1977 amendments to the CAA also established the Prevention of Significant Deterioration (PSD) regulations for areas that already meet the NAAQS. The PSD regulations are designed to prevent any significant deterioration in air quality below an established baseline level. In this way, pollutant concentrations may remain well within ambient standards.

Under the CAA, federal actions cannot thwart state and local efforts to remedy long-standing air quality problems that threaten public health issues associated with the six criteria air pollutants (i.e., ozone, nitrogen dioxide, sulfur dioxide, particulate matter, carbon monoxide, and lead).

4.4.5.1 General Conformity

General conformity is an environmental review process mandated under the CAA. The conformity review process is intended to ensure that air pollution emissions from federal actions do not contribute to air quality violations. Under the CAA, states are directed to develop state implementation plans (SIPs), which consist of emission reduction strategies for achieving the goals of the NAAQS. Proposed federal actions must uphold a state's strategy as detailed in its SIP.

In performing a general conformity analysis, a federal agency analyzes direct and indirect emissions associated with a proposal. Air quality models are used in determining conformity. The modeling is used to ascertain whether a federal action contributes to any new violation of a standard or increases the frequency or severity of any existing violation. Specifically, the analysis is performed to ensure that the proposal does not

1. delay attainment of required emission reductions,
2. contribute to any new violation of the NAAQS, or
3. increase the frequency or severity of existing violations.

Emissions of each pollutant must be consistent with the requirements and emissions milestones provided in the SIP. If the conformity determination demonstrates that the proposed action would not conform to the SIP, then the federal agency cannot support, license, permit, or approve that action. The following general six-step process is normally followed for evaluating air conformity impact analyses under NEPA:

Step 1: identify (and quantify) the air pollutants
Step 2: describe the existing air quality conditions
Step 3: identify applicable air quality regulatory standards
Step 4: evaluate the air quality impacts
Step 5: assess the significance (using the air quality regulatory standards)
Step 6: identify appropriate mitigation measures

The companion text, *Environmental Impact Statements*, provides a detailed description of the requirements for preparing CAA conformity assessments under the NEPA planning process.

4.4.5.2   Potential Problems with Nonconformance

The following example illustrates the potential for delay from legal challenges as a result of inadequately addressing conformity in NEPA documents. In March 1991, the U.S. Air Force closed Pease Air Force Base in New Hampshire. The Air Force issued a draft EIS on the disposition and reuse of the base in February. It then issued a final EIS in June followed by a record of decision (ROD) containing a conformity determination in August of that year. In March 1992, the Air Force issued a memorandum that updated the conformity determination in light of new information. The Conservation Law Foundation then filed a citizen's suit under Section 304 of the CAA against the air force alleging, in part, that the final EIS was inadequate because it did not contain a conformity analysis. The Federal District Court agreed and directed the air force to prepare a supplemental EIS to address several CAA issues, including conformity. This case suggests that a general conformity compliance demonstration needs to be completed and taken into account in NEPA documentation.

4.4.5.3   EPA Issues

The EPA has frequently raised general conformity concerns when reviewing draft EISs prepared by federal agencies under authority granted in Section 309 of the CAA. Examples of some of these regulatory issues include

- Concerns that a draft EIS for a proposed flood control project did not address air quality mitigation measures that might be necessary under the general conformity rule. The EPA recommended that the final EIS provide additional information concerning conformity with the SIP.[15]
- Objecting to a proposed groundwater storage program based on potentially significant air quality impacts and the lack of a conformity determination.[16]
- Concerns that, for proposed aircraft facilities, air quality mitigation measures required under the conformity rule were conceptual in nature and lacked definitiveness.[17]

4.4.5.4   Documents on the Web

The U.S. Department of Energy (DOE) has issued guidelines for integrating CAA conformity with NEPA.[18] These guidelines provide direction for complying with EPA regulations (40 CFR Part 93, Subpart B) pertaining to emissions of criteria air pollutants that affect designated nonattainment or maintenance areas. This guidance is available on DOE's NEPA Web site at tis.eh.doe.gov/nepa/. Additional resources are available at

1. tis.eh.doe.gov/oepa/guidance/caa/conformbrf.pdf and
2. tis.eh.doe.gov/nepa/tools/guidance/caaguidance.pdf

4.4.6 CO_2 AND GLOBAL WARMING ISSUES IN NEPA DOCUMENTS

While CO_2 is an unregulated air pollutant, the Intergovernmental Panel on Climate Change (IPCC) stated in 2007 that it is "the most important anthropogenic greenhouse gas" and that "most of the observed increase in globally averaged temperatures since the mid-20th century is very likely due to the observed increase in anthropogenic greenhouse gas concentrations."[19]

With the release of the IPCC study, the author believes that the issues of CO_2 and global warming will increasingly need to be addressed in NEPA analyses. A case in point: To further the purposes of NEPA in response to public comments regarding how the DOE had addressed emissions, DOE issued a supplement to the draft EIS for the Gilberton Coal-to-Clean Fuels and Power Project.[20] The draft EIS, issued in December 2005, analyzes DOE's proposed action to provide cost-shared funding (about $100 million of the total project cost of about $612 million) for construction and operation of facilities near Gilberton, Pennsylvania. The proposal was designated to produce 41 MW of electricity as well as steam and waste by-products.[21]

The Natural Resources Defense Council (NRDC) and several other organizations and individuals submitted comments questioning the accuracy of the CO_2 emissions rate in the original draft EIS (832,000 t/year) and requested information on the reported quantity. Upon review, the DOE found that the draft EIS overlooked a concentrated stream (1,450,000 t/year) exiting the gas cleanup system. The supplemental EIS corrected the value reported in the draft for the annual rate of CO_2 emissions, which was understated by a factor of nearly three.

NRDC and other commentors also stated that DOE should explore potential means of mitigating CO_2 emissions such as through geologic carbon sequestration (burial). In response, DOE analyzed sequestration options in Pennsylvania, but concluded in the supplemental that CO_2 sequestration is not a feasible option during the demonstration period.

The NRDC also stated that the analysis of cumulative impacts should be enhanced. The DOE responded by providing both annual rates of emissions and total quantities of CO_2 potentially released during 50 years of commercial operation. In addition, the supplement provides an enhanced analysis of cumulative impacts under several economic scenarios.

4.4.7 ENDANGERED SPECIES ACT

Following in the footsteps of NEPA, Congress passed the Endangered Species Act (ESA) of 1973 to combine and strengthen its predecessors, such as the Endangered Species Preservation Act of 1966. On signing the Act, President Nixon stated that "Nothing is more priceless and more worthy of preservation than the rich array of animal life with which our country has been blessed."

The ESA has been referred to as the "pitbull" of environmental laws, because it has halted many major federal actions such as dams and major highway projects and has also closed entire forests to harvesting. As one of the first and most prominent major cases, an ESA citizen's suit was brought on behalf of the endangered small snail darter fish. The suit was brought against what many considered to be the "pork barrel" Tennessee Valley Authority Tellico Dam on the Little Tennessee River, one of the last wild rivers in the state. In 1975, dam opponents fought successfully to include the darter on the endangered species list and the dam was halted. Politicians tried to outflank the ESA by establishing a "God Squad" to exempt certain federal actions that would jeopardize listed species. Even though the dam was 90% complete, the God Squad failed to prove that the project was worthy of completion. In 1978, the Supreme Court sided with the citizens group. Nevertheless, even though it had failed in its attempt through the God Squad and the Supreme Court, Congress later specifically exempted the Tellico Dam from the ESA, allowing the project to be completed (see Section 4.4.7.4 for an explanation of God Squad).

Critics have charged that the ESA has had a dismal track record in terms of recovery and delisting threatened and endangered (T&E) species. Supporters counter that in reality the Act has been enormously successful. Recovery of species whose numbers are hovering on the brink cannot be assessed in the short term since it can take many years of careful protection before their numbers increase to a point of real recovery (the bald eagle is a case in point). The fact is that since the passage of the ESA, 98–99% of listed species have been preserved from threatened extinction. In fact, many scientists have stated their belief that the United States would have experienced a much larger number of extinctions but for the protective provisions of the ESA.

Effects on threatened or endangered species is an important factor cited in the CEQ Regulations for determining the significance of an impact (§ 1508.27[b][9]).

4.4.7.1   Administration and Purpose

The ESA is administered jointly by the Secretaries of Interior and Commerce. The U.S. Fish and Wildlife Service (FWS) is responsible for terrestrial species, and the National Marine Fisheries Service (NMFS) is responsible for marine species including anadromous species of fish.

The purpose of the ESA is to ensure that federal agencies use their authority to protect T&E species. Section 7 of the ESA is designed to prevent or modify any projects authorized, funded, or carried out by federal agencies that are

likely to jeopardize the continued existence of any endangered species or threatened species, or result in the destruction or adverse modification of critical habitat of such species.

As described below, the ESA forbids any government agency, corporation, or citizen from "taking" (i.e., harming or killing) endangered animals without an *Endangered Species Permit*. Penalties for violating the ESA can be as serious as a $50,000 fine and up to a year in jail.

Where adverse impacts cannot be avoided, state and local governments and private land owners must develop habitat conservation plans in coordination with the U.S. FWS or NMFS to reduce potential impacts between listed species and development activities; these plans must meet the requirements of Section 10 of the Act.

4.4.7.2   Implementing Regulations

Implementing regulations are provided in

- *50 CFR Part 402*: Department of Interior and Department of Commerce procedures for implementing Section 7;
- *50 CFR Parts 450, 451, 452, and 453*: Department of Interior and Department of Commerce rules for applying for ESA exemptions and for Endangered Species Committee consideration of such applications.

4.4.7.3   Categories of Species

The principal categories of species regulated under the ESA are

- *Candidate species*: Plants and animals that have been studied and that the Service has concluded should potentially be proposed for addition to the federal endangered and threatened species list.
- *Threatened species*: An animal or plant species likely to become endangered within the foreseeable future throughout all or a significant portion of its range.
- *Endangered species*: An animal or plant species in danger of extinction throughout all or a significant portion of its range.

The term "critical habitat" refers to those areas that are necessary for the recovery of a species.

Species can be listed through two mechanisms. The first mechanism involves either the FWS or NMFS taking the initiative to list the species directly. The second mechanism involves a petition by an individual or organization to the FWS or NMFS. If the petition for listing is approved, a notice is published in the *Federal Register*.

The Act contains a citizen enforcement clause that allows citizens and scientists to sue the government either to obtain listing for a species with dwindling numbers or to comply with the law.

4.4.7.4 Scope

With the exception of pest insects, all species of plants and animals are potentially eligible for listing as endangered or threatened. Groups with the most listed species (in order) are plants, mammals, birds, fishes, reptiles, and varieties of clams/mussels.

In addition to federal actions, the ESA also affects private land use. Under Sections 9 and 10 of the Act, nonfederal entities (developers, local governments, and private citizens) must not adversely impact, take, or commercially trade threatened or endangered species; criminal penalties apply. This requirement applies even if the nonfederal entity receives no assistance from a federal agency nor has any other involvement with it.

Exemptions can be granted from the ESA if a cabinet-level Endangered Species Committee decides that the benefits of an activity outweigh the benefits of protecting the species. Because it effectively has the power to condemn a species to extinction, this committee has been referred to as the God Squad. Since 1978, the committee has considered only four exemption requests: snail darter fish in Tennessee, spotted owls in Oregon, bald eagles in Maine, and whooping cranes in Nebraska.

4.4.7.5 Section 7: Consultation

As briefly described earlier, Section 7 of the ESA[22] requires all federal agencies to consult the NMFS or the U.S. FWS if they are proposing an action that may affect listed species or their designated habitat. The term "action" is defined broadly to include funding, permitting, and other regulatory actions.

Each federal agency must ensure that any action they authorize, fund, or carry out is not likely to jeopardize the continued existence of a listed species or result in destruction or adverse modification of designated critical habitat. This consultation requirement is commonly referred to as the "Section 7: consultation process" and may involve both informal and formal dialogues, as well as preparing a biological assessment (BA) and obtaining expert agency opinions.

Under Section 7, federal agencies must comply with the following three requirements:

- Perform (if applicable) a formal consultation with the U.S. FWS on potential impacts to species/habitat
- Prepare (if warranted) a BA on such proposals
- Obtain a permit prior to monitoring, capturing, killing, or performing other scientific studies on threatened or endangered species

4.4.7.6 Consultation

Federal agencies must review actions they undertake or support to determine whether they may affect an endangered species or its habitat. If such a review reveals a potential for adverse effects, the federal agency must consult the FWS or NMFS. Consultation is carried out for the purpose of identifying whether a federal action is likely to jeopardize the continued existence of the threatened or endangered species or adversely affect its critical habitat.

A large percentage of proposals having the potential to adversely impact a listed species can be effectively dealt with through informal consultation during the early planning process. The need for further consultation may be avoided if project design changes can be made that would eliminate adverse impacts to listed species. If the FWS or NMFS determines that a proposed action is unlikely to adversely affect a listed species, the potential for project success is greatly increased as no further consultation is normally required.

A formal consultation process is normally initiated if it is determined that the proposal could adversely affect a listed species or its critical habitat.

4.4.7.7   Biological Evaluation and Assessment

Where listed species are unlikely to be adversely affected and formal consultation is not anticipated, a biological evaluation (BE) is prepared, providing the basis for making a determination during informal consultation.

If a designated critical habitat or a threatened or endangered species is in the area of the proposed action, a BA may need to be prepared to evaluate the potential effects of the project on the species or habitat.

The BA may be prepared as part of the agency's NEPA process. In preparing an EA or EIS, alternatives and mitigation measures should be investigated for avoiding or reducing potential impacts to listed species and their critical habitat. The BA describes

- proposed project;
- project area;
- proposed management activities;
- the listed species that may occur in the project area (including past surveys for such species);
- how the project may affect listed species or critical habitat (direct, indirect, and cumulative effects); and
- measures to avoid, reduce, or eliminate adverse effects.

Within 45 days after concluding formal consultation, the federal agency normally issues a biological opinion. The biological opinion is a document stating the opinion of FWS or The National Oceanic and Atmospheric Administration (NOAA) Fisheries as to whether the impacts (including cumulative impacts) of the federal proposal are likely to jeopardize the continued existence of a listed species or result in destruction or adverse modification of a critical habitat. The majority of biological opinions allow for the proposed action to be undertaken subject to certain conditions; the biological opinion may also recommend "reasonable and prudent alternatives" to the proposed action to avoid jeopardizing or adversely modifying critical habitat.

If there are no feasible alternatives, the lead agency may apply for an exemption with the Endangered Species Committee.

4.4.7.8   Section 9

Under Section 9 of ESA, it is illegal to take any endangered species. The term "take" includes the killing, harming, harassing, or capturing of a threatened or endangered species. This requirement also safeguards a critical habitat.

An endangered species permit is an authorization issued by FWS or NOAA Fisheries under authority of Section 10 allowing an action to go forward that would otherwise be prohibited under Section 9.

4.4.7.9 Section 10

Section 10 of the ESA lays out the guidelines under which a permit may be issued, authorizing otherwise prohibited activities such as the taking of endangered or threatened species.

- Section 10(a)(1)(A): Allows for permits for the taking of threatened or endangered species for scientific purposes or for purposes of enhancement of propagation or survival.
- Section 10(a)(1)(B): Allows for permits for the incidental taking of threatened or endangered species.

4.4.7.10 Invasive Species and Executive Order 13112

Executive Order 13112 has recently been issued, applying to federal agency actions that may affect the status of invasive species.[23] This Executive Order specifically applies to species not native to a particular ecosystem "... whose introduction does or is likely to cause economic or environmental harm or harm to human health." Agencies are prohibited from authorizing or funding actions that may contribute to the introduction or spread of invasive species. Under this order, federal agencies are directed to

1. monitor the populations of invasive species,
2. prevent the introduction of invasive species,
3. detect and respond quickly to control the spread of invasive species populations,
4. provide for restoration of native species and habitat conditions where invasions have occurred,
5. promote public education, and
6. research and develop technologies to control the introduction of invasive species.

4.4.8 SECTION 404 OF THE CLEAN WATER ACT

In 1899, Congress passed the Rivers and Harbors Act which defined "navigable waters" of the United States as "those waters that are subject to the ebb and flow of the tides and/or are presently used, or have been used in the past, or may be susceptible to use to transport interstate or foreign commerce."

The Clean Water Act of 1972 built on the earlier Rivers and Harbors Act, extending the definition of waters of the United States to include tributaries to navigable waters, interstate wetlands, wetlands that could affect interstate or foreign commerce, and wetlands adjacent to other waters of the United States.

A wetland is defined as an area inundated or saturated by surface or groundwater at a frequency and duration sufficient to support, and that under normal circumstances does support, a prevalence of vegetation typically adapted to life in saturated soil conditions.

Effects on wetlands is an important factor cited in the CEQ Regulations for determining the significance of an impact (§ 1508.27[b][3]).

4.4.8.1 Section 404

In 1972, Section 404 of the Clean Water Act established a program to regulate the discharge of dredged and fill material into waters of the United States, including wetlands. The Army Corps of Engineers (Corps) and EPA jointly administer this program. The Corps is responsible for the day-to-day administration and permit review while EPA provides program oversight. In addition, the U.S. FWS, the NMFS, and state resource agencies have important advisory roles.

4.4.8.2 Typical Activities and Exemptions

Activities in U.S. waters regulated under this program include water resource projects (such as dams and levees), conversion of wetlands to uplands for farming and forestry, and fills for development projects such as housing, highways, and airports.

Section 404(f) exempts some activities that would otherwise be regulated under Section 404. Exempted activities include many types of ongoing farming, ranching, and silviculture practices.

4.4.8.3 Basic Requirements

Under the 404 program, no discharge of dredged or fill material is allowed if a practicable alternative exists that is less damaging to the aquatic environment or if the nation's waters would be significantly degraded. A federal permit is required to discharge dredged or fill material into wetlands and other waters of the United States. The program involves three basic requirements:

- Minimizing potential impacts to wetlands
- Taking steps to avoid wetland impacts (where practicable)
- Providing compensation for any remaining, unavoidable impacts through activities to restore or create new wetlands

4.4.8.4 Swampbuster

The Wetland Conservation provision (Swampbuster) of the 1985 and 1990 farm bills requires agricultural producers to protect wetlands on farms they operate if they wish to be eligible for the U.S. Department of Agriculture (USDA) farm program benefits. Producers are ineligible for such benefits if they

1. plant an agricultural commodity on a wetland that was converted by drainage, leveling, or any other means after December 23, 1985; or
2. convert a wetland for the purpose of or to make agricultural commodity production possible after November 28, 1990.

4.4.8.5 Types of Permits

A federal permit is required to discharge dredged or fill material into wetlands and other waters of the United States. The Corps grants two types of 404 permits: general and individual.

General permits are granted on a nationwide, regional, or statewide basis for particular categories of activities that are presumed to cause only minimal adverse environmental impacts. Forty categories of general permits have been established on a nationwide basis to date. Some of these categories simply require notifying the Corps on completing a wetland activity, while others may require submittal of a rigorous preconstruction notification.

An individual permit is usually required for potentially significant impacts. Individual permit applications are evaluated on a case-by-case basis.

As part of the permitting process, the Corps evaluation also includes a review for compliance with NEPA.

4.4.8.6 Permit Limitations and Mitigation

No discharge is normally permitted if it would violate other applicable laws, including state water quality standards, toxic effluent standards, the ESA, or marine sanctuary protections. Nor can the discharge contribute to a significant degradation of wetlands by adversely impacting wildlife, ecosystem integrity, or social amenities such as esthetics. Even if these conditions are met, the applicant must show that all appropriate and practicable steps will be taken to avoid or reduce any possible adverse impacts of the discharge on wetlands.

Only after impact avoidance or reduction criteria have been satisfied can the Corps consider wetlands compensation (e.g., mitigation). The Corps must strive to achieve a goal of no overall net wetlands loss, meaning a minimum of one-for-one functional replacement with an adequate margin of safety to reflect uncertainties.

4.4.8.7 The 404 Permitting Process

Failure to obtain a permit or comply with the terms of a permit can result in civil and/or criminal penalties. The 404 permitting process involves the following basic steps.

4.4.8.8 Public Notice

The Corps issues a public notice after it has received all permit information. This notice describes the permit application, including the proposed activity, potential environmental impacts, and location. The public notice invites comments within a specified time.

4.4.8.9 Comment Period and Public Hearing

After receiving comments from the public, the application and comments are reviewed by the Corps and other interested federal and state agencies, organizations, and individuals. The Corps determines whether an EIS is necessary. Any group of citizens may request the Corps to conduct a public hearing.

4.4.8.10 Permit Evaluation and Statement of Finding

The Corps evaluates the permit application based on the comments received as well as by its own evaluation. A statement of finding is issued to the public explaining how the permit decision was made.

4.4.9 FLOODPLAIN AND WETLANDS EXECUTIVE ORDERS

On May 24, 1977, President Carter issued Executive Orders 11988 and 11990 which provide for the protection of floodplains and wetlands, respectively.[24,25] Both executive orders require federal agencies to consider the impacts of their actions on floodplains and wetlands through existing review procedures such as NEPA.

A floodplain/wetlands assessment is often required for actions that may impact an area falling under this category. The findings of this assessment should be coordinated with and incorporated into the EIS.

Reference to relevant Flood Insurance Rate Maps, Flood Hazard Boundary Maps, as well as consultation with applicable government agency personnel, can be invaluable in determining if an activity may be located within a floodplain. Wetlands can be identified by consulting

- The Corps of Engineers;
- The U.S. FWS national wetlands inventory;
- Wetlands specialists and federal agency specialists;
- State and local wetland inventory databases, land-use plans, maps, and inventories;
- U.S. Geological Survey topographical maps; and
- USDA, Natural Resources Conservation Service local soil identification maps and databases.

4.4.10 COASTAL ZONE MANAGEMENT

The Coastal Zone Management Act (CZMA) of 1972 established a policy for a national program for the beneficial use, protection, and development of the land and water resources of the nation's coastal zone.[26]

4.4.10.1   Coastal Zone Management Act Consistency Regulations

The NOAA recently revised its CZMA Consistency Regulations.[27] The Act requires that all federally conducted or supported activities affecting the coastal zone, including development projects, be undertaken to the maximum extent practicable in a manner consistent with approved state coastal management programs. Specifically, the revised regulations to the CZMA require[28]

> Each Federal agency activity within or outside the coastal zone that affects any land or water use or natural resource of the coastal zone shall be carried out in a manner which is *consistent* to the maximum extent practicable with the enforceable policies of approved State [coastal zone] management programs. (Emphasis added)

4.4.10.2   Consistency Determination

Under the revised regulations, any federal agency activity (regardless of location) is subject to the consistency requirement if that activity will affect any natural resources, land uses, or water uses in the coastal zone. Known as the effects test, this provision requires an agency to consider all reasonably foreseeable direct and indirect effects on any coastal use or resource. The federal agency and the state coastal zone agency may agree to exclude proposals with environmentally beneficial effects on the coastal zone from further review, either on a case-by-case basis or as a category. A federal agency may request state concurrence that certain categories of actions with *de-minimis* coastal zone effects be exempt from further state review.

Under the revised regulations, a federal agency must determine whether its proposed activity has reasonably foreseeable coastal effects. If there are such effects, the agency provides a consistency determination, a report that describes how the proposal is consistent with a state coastal zone management program.

If the agency believes that there are no reasonably foreseeable coastal effects, it can issue a negative determination (i.e., that there are no coastal zone impacts).[29]

4.4.10.3   NEPA and Project Planning

If a negative determination is not required, then the federal agency does not need to notify the state CZMA agency. A consistency determination or negative determination can be provided in any manner that meets the regulation's requirements. Federal agencies may choose, but are not required, to address consistency requirements in NEPA documents. If a federal agency chooses to include its consistency determination or negative determination in an NEPA document, the EA or EIS must include the information needed to support the determination.

To facilitate efficient compliance with all regulatory requirements, practitioners should consider early in the project planning phase whether a proposed action has reasonably foreseeable effects on any land or water uses or natural resources in the coastal zone.

If the proposal has reasonably foreseeable coastal effects, practitioners should coordinate early on with the applicable state(s) coastal zone management agency, in part to help determine whether the agency should integrate a CZMA consistency review with NEPA for the proposal and also to facilitate the state review.

Additional information can be found at: www.nos.noaa.gov/programs/ocrm.html.

4.4.11   Wild and Scenic Rivers Act

The Wild and Scenic Rivers Act of 1968 created the National Wild and Scenic River System, established to protect the environmental values of free-flowing streams from degradation by impacting activities, including water resources projects.[30] The system is administered jointly by the U.S. Forest Service and the National Park Service.

The Wild and Scenic Rivers Act established the policy that certain rivers[31]

- possess outstandingly remarkable scenic, recreational, geologic, fish and wildlife, historic, cultural, or other similar values;
- shall be preserved in free-flowing condition; and
- shall be protected.

This Act both identifies specific river reaches for designation as wild or scenic and provides criteria for use in classifying additional river reaches.[32]

Effects on wild and scenic rivers is an important factor cited in the CEQ Regulations for determining significance of an impact (§ 1508.27[b][3]). The terms "wild" and "scenic" river areas are defined subsequently.

4.4.11.1 Wild River Areas

Those rivers or sections of rivers that are free from impoundments and generally inaccessible except by trail, that have watersheds or shorelines in an essentially primitive state and unpolluted waters. These rivers represent the vestiges of primitive America.

4.4.11.2 Scenic River Areas

Those rivers or sections of rivers that are free from impoundments, with shorelines or watersheds that are still largely in a primitive state, and with shorelines that are largely undeveloped. These rivers may be accessible in places by roads.

4.4.12 Fish and Wildlife Coordination Act

To minimize adverse impacts of proposed actions on fish and wildlife resources and habitat, the Fish and Wildlife Coordination Act of 1934 (last amended in 1965) requires that federal agencies consult government agencies (U.S. FWS, NMFS and State wildlife agencies) regarding activities that affect, control, or modify waters of any stream or bodies of water.[33] It also requires that justifiable means and measures be used in modifying plans to protect fish and wildlife in these waters.

This consultation is generally incorporated into the process of complying with Section 404 of the Clean Water Act, NEPA, or other federal permit, license, or review requirements.

4.4.13 National Historic Preservation Act

In 1966, Congress enacted the National Historic Preservation Act (NHPA) which announces a national policy of encouraging preservation of prehistoric and historic resources.[34] While the NHPA does not mandate preservation of such resources, it directs federal agencies to consider the impact of their actions on historic properties. NHPA also encourages state and local preservation programs.

With only one exception there is no federal requirement affecting the ability of a private property owner to make changes to a building, including its demolition. However, if federal money or a federal permitting process is involved, Section 106 of NHPA is invoked, which requires an assessment of the impact.

The NHPA also establishes an Advisory Council on Historic Preservation (ACHP), now an independent federal agency composed of 20 experts and local government representatives. The purpose of this council is to ensure that private citizens, local communities, and other concerned parties have a forum for discussion on the ways federal decisions impact historic properties. The council has promulgated regulations for implementing the NHPA.[35]

4.4.13.1   State Historic Preservation Officer

Under Section 101(b) of NHPA, each state is responsible for assigning a state historic preservation officer (SHPO) to administer the state's historic preservation program. The term "SHPO," however, usually refers informally to the state historic preservation office. The professional staff of each office has expertise in history, archaeology, and historic preservation. SHPO's duties include

- administering the state's National Register of Historic Places (NRHP) program and
- maintaining the inventory of state archaeological and historical sites, and historic buildings and structures.

4.4.13.2   National Register of Historic Places

As explained below, Section 106 of the NHPA regulations direct federal agencies to take into account the effects of its actions on sites eligible for or already listed in the NRHP. The NRHP is the official list of the nation's cultural resources that have been deemed historically significant and worthy of preservation. As of 2004, the list included 78,000 entries, including many iconic examples of American culture, history, engineering, and architecture.

Any property listed or eligible for listing in the NRHP is considered to be historic. Such properties may include archaeological and historical sites, historic buildings or structures, and objects (e.g., monuments). The NRHP also includes artifacts, records, and material remains related to these properties.

The NRHP is maintained by the National Park Service. Any individual or party can nominate an item or site for inclusion in the NRHP, although historians are often employed to perform this work. If the concerned SHPO approves the nomination, it is then passed to the state's historic preservation advisory board for approval. From there the nomination is sent to the National Park Service, which has final authority to approve or deny its eligibility for inclusion.

Under the CEQ Regulations, an impact on an historic or cultural resource is a factor to be considered in assessing significance (§ 1508.27[b][3]).

4.4.13.3   Eligibility Criteria

The following criteria are used in determining if a property is historically or culturally significant and therefore eligible for listing in the National Register if it

- is associated with events that have made a significant contribution to the broad patterns of our history;
- is associated with the lives of persons who have been significant in the past history of the nation;
- embodies distinctive characteristics of a type, period, or method of construction, or represents the work of a master, or possesses high artistic values, or represents a significant and distinguishable entity whose components may lack individual distinction; or
- has yielded, or may be likely to yield, information important in reconstructing events in either prehistoric or historic times.

4.4.13.4   Section 106

The Section 106 review refers to the federal process for ensuring that historic properties are considered during federal project planning and implementation. The historic preservation review process mandated by Section 106 is outlined in the ACHP regulations.[35] The ACHP administers this review process with assistance from the relevant SHPO.

Section 106 originally applied only to properties actually listed in the NRHP. However, in 1976, Congress extended its provisions to properties not yet listed, but that nevertheless meet the NRHP eligibility criteria.

Under Section 106, each federal agency must comply with two principal requirements prior to carrying out, approving financial assistance to, or issuing a permit for a project that may affect properties either listed or eligible for listing in the NRHP. Specifically, the agency must

1. consider the impact of the project on historic properties and
2. seek the council's comments on the project.

It should be noted that the Section 106 process is purely advisory in nature. As detailed below, the end product is normally a memorandum of agreement (MOA) in which the parties involved agree to a particular plan. An MOA often recommends "document and destroy" in which the historic resource is first photographed and documented, and then demolished.

4.4.13.5 Section 106 Review Process

The SHPO should be consulted to determine the existence of any known resources, either already in or eligible for inclusion in the NRHP.

The federal agency involved in the proposed project is responsible for initiating and completing the Section 106 review process. The four steps of the Section 106 review process are[35]

1. initiate the Section 106 review;
2. identify historic properties within the project's area of potential effects;
3. assess adverse effects on historic properties; and
4. resolve adverse effects.

The agency needs to assume that the potential for historic properties exists until the identification step of the review process has been completed.

The regulations outlining the Section 106 review define the "area of potential effects" to be the "geographical area within which an undertaking may directly or indirectly cause alterations in the character or use of historic properties, if any such properties exist."[35]

4.4.13.6 Consultation

The heart of the Section 106 review is the consultation process, which frequently takes the form of discussions between the agency and the SHPO.

The responsible federal agency first determines whether the undertaking is a type of activity that could either affect historic properties included in the NRHP or that meets the eligibility criteria for the NRHP. If so, it must identify the appropriate SHPO to consult. Federal agencies can also authorize applicants for federal grants, licenses, or permits to initiate consultation with the SHPO, but the agency remains legally responsible for the findings and determinations.

If the agency, in consultation with the SHPO, concludes that the historic property will not be affected, then the agency may proceed with the proposed action subject to any conditions that have been agreed.

The agency, however, must begin consultation to seek ways to avoid, minimize, or mitigate the adverse effects if the parties

- find that there is an adverse effect, or
- cannot agree and the ACHP determines within 15 days that there is a potential for adverse effects.

During consultation, these parties attempt to reach agreement on measures to avoid or mitigate the adverse effects of the agency's undertaking on historic resources. If the parties agree, they generally execute an

- MOA, or a
- programmatic agreement (PA) (if an entire program or a complex staged project is involved).

The MOA outlines those measures that the agency has agreed to take to avoid, minimize, or mitigate the adverse effects. In some cases, the consulting parties may agree that no practical measures are possible.

A PA is a tool by which a federal agency program or other large undertaking will comply with the Section 106 review process by an alternative method. This method is tailored to the needs of the agency. It should be emphasized that PAs are generally for agency-wide agreements and are generally used for repetitive or widespread actions.

NEPA Relationship to Section 106. While Section 106 is a completely separate authority from NEPA, the coordination of studies and documents prepared under Section 106 with those prepared under NEPA is strongly encouraged. The section directing how federal agencies coordinate the Section 106 process with NEPA has been revised to clarify what actions a federal agency must take in making a binding commitment to avoid, minimize, or mitigate adverse effects on historic properties.[36] The binding commitment is satisfied when

1. the commitment is made in the ROD (where an EIS is prepared) or in an MOA as specified in the regulations.
2. the Council on Historic Preservation has commented and the agency has responded to those comments, again as specified in the regulations.

As appropriate, alternatives and mitigation measures should be investigated as part of the NEPA process for avoiding or reducing potential impacts. Analysis of cultural resources is performed by a cultural resource specialist in coordination with any potentially affected Indian tribes. Mitigation measures must be considered where an impact to a cultural resource cannot be avoided.

A project that is considered a categorical exclusion under NEPA is not exempted from review under Section 106. However, projects can be exempted from Section 106 review after consulting with SHPO.

4.4.13.7   Changes to Section 106

In 1999, the Council on Historic Preservation revised its regulations implementing Section 106.[37] The revised regulations implement the 1992 NHPA amendments and streamline the previous regulations. The revised regulations are available at www.achp.gov/

A new section (36 CFR 800.8) allows agencies to comply with Section 106 requirements within the NEPA process. Under 36 CFR 800.8, an agency may use an EA/EIS to comply with Section 106 in lieu of the procedures set forth in 36 CFR 800.3 and 36 CFR 800.6, provided the agency notifies the public and the council and meets certain established standards.

State and local governments, Indian tribes, and the public are also afforded greater ability to become more directly involved in federal activities affecting historic properties. The revised regulations provide for a tribal historic preservation officer (THPO) to substitute for the SHPO when the tribal official has assumed the responsibilities of the SHPO over tribal lands.

4.4.13.8   Archaeological Resource Protection Act

The Archaeological Resource Protection Act (ARPA) requires a permit from the U.S. Department of Interior for the excavation or removal of archaeological resources from public or Native American lands. Criminal penalties are established for the illegal excavation or removal of archaeological items.

4.4.13.9   Archaeological and Historic Preservation Act

The Archaeological and Historic Preservation Act (AHPA) of 1974, as amended, requires preservation of historic and archaeological data that may be placed in jeopardy as a result of federal actions.

4.4.13.10   Indian Religious Freedom Act

The Indian Religious Freedom Act et seq. establishes a national policy protecting the right of Native Americans to exercise their traditional religious and ceremonial rites.[38] Consultation with Native American tribes is required where actions may infringe on religious rites or ceremonial sites.

4.4.13.11   Native American Graves Protection and Repatriation Act

The Native American Graves Protection and Repatriation Act protects Native American graves, human remains and funerary objects, and ensures repatriation of these items in cases where they have been moved to other locations.[39]

4.4.14   FARMLAND AND PROTECTION POLICY ACT

In 1981, as a result of a substantial decrease in the amount of open farmland, Congress enacted the Farmland Protection Policy Act (FPPA) as part of the Agriculture and Food Act (final rules published in 1994).[40] In the statement of purpose, federal programs that contribute to the unnecessary and irreversible conversion of farmland to nonagricultural uses will be minimized. It follows that federal programs, as practicable, shall be administered in a manner compatible with state and local government and private programs and policies to protect farmland. The terms "prime" and "unique" farmland are defined subsequently.

4.4.14.1   Prime Farmland

Land that has the best combination of physical and chemical characteristics for producing food, feed, fiber, forage, oilseed, and other agricultural crops with minimum inputs of fuel, commercial fertilizer, pesticides, and labor, and without intolerable soil erosion.[41]

4.4.14.2   Unique Farmland

Land other than prime farmland that is used for the production of specific high-value food and fiber crops such as citrus, tree nuts, olives, cranberries, fruits, and certain vegetables.[42]

4.4.14.3   Integration with NEPA

The effect on prime farmlands is an important factor in determining the significance of an impact (§ 1508.27[b][3]). In 1990, the CEQ issued a memorandum to assist agencies in analyzing such impacts.[43]

4.4.15   ENVIRONMENTAL JUSTICE

Environmental Justice (EJ) shot into the national spotlight in 1982 when approximately 500 demonstrators gathered in Warren County, North Carolina, to protest the siting of a polychlorinated biphenyl landfill in a predominately low-income African-American community. An investigation initiated by the General Accounting Office (GAO) in 1983 found that three out of four major hazardous waste landfills in the South were located in minority and low-income communities.

Additionally, in 1983, the United Church of Christ's Commission for Racial Justice released a report of a nationwide study that revealed a national pattern of disproportionate location of commercial hazardous waste facilities in minority communities.

Awareness of EJ was further increased when two major environmental conferences were held in the early 1990s. As a result, the EPA created the Office of Environmental Equity in 1992, which was renamed the Office of Environmental Justice (OEJ) in 1994.

4.4.15.1   Executive Order

On February 11, 1994, President Clinton issued Executive Order 12898.[44] This executive order requires that each federal agency identify and address, as appropriate, any disproportionately high and adverse human health or environmental effects resulting from its programs as well as from its policies and activities with regard to minority and low-income populations. The Executive Branch agencies were also directed to develop plans for carrying out the order.

Four methods are outlined in the executive order for considering EJ under NEPA:

- Federal agencies should analyze the environmental effects, including human health, economic, and social effects of federal actions, and also effects on minority populations, low-income populations, and Indian tribes, when such analysis is required by NEPA.
- Federal agencies must provide opportunities for effective community participation in the NEPA process.
- NEPA reviews (such as the EPA's review under Section 309 of the CAA) must ensure that the lead agency in charge of preparing the required NEPA analyses and documentation has appropriately analyzed environmental effects on minority populations, low-income populations, or Indian tribes, including human health, social, and economic effects.
- Mitigation measures identified as part of a finding of no significant impact (FONSI) or ROD should, whenever feasible, address significant and adverse environmental effects of proposed federal actions on minority populations, low-income populations, and Indian tribes.

4.4.15.2   CEQ Guidance

EJ concerns may arise from impacts on the natural and physical environment, such as human health or ecological impacts on minority populations, low-income populations, and Indian tribes, or from related social or economic impacts. The CEQ has issued guidance for implementing the goals of EJ under NEPA.[45] CEQ's guidance outlines six principles that should be addressed in the course of NEPA review to ensure consideration of EJ. These are paraphrased as follows:

- Investigate human demographics in affected areas and determine whether any communities are characterized by low-income levels or high-minority composition. If so, determine whether there is potentially a disproportionately high and adverse effect on such populations.
- Implement effective public participation strategies that seek to overcome linguistic, cultural, institutional, geographic, and other barriers.
- If Indian tribes are involved, make sure that interactions with tribes are consistent with the government-to-government relationship between the United States and tribal governments.
- Assure early and meaningful community representation in the NEPA process.
- Investigate direct, indirect, and cumulative EJ effects on communities.
- Recognize that the cultural, social, occupational, historical, and economic characteristics of a minority or low-income community may intensify the environmental effects on it, as such populations may be more sensitive to these effects and less resilient in adapting to them.

4.4.15.3   Analysis

When determining whether human health or environmental effects are disproportionately high and adverse, agencies are to consider the following factors to the extent practicable:[45]

Disproportionately high and adverse human health effects. Whether the

1. health effects (e.g., risks and rates) are significant or above generally accepted norms. Adverse health effects may include bodily impairment, infirmity, illness, or death.
2. risk or rate of hazard exposure of a minority or low-income population, or Indian tribe to an environmental hazard is significant and is likely to appreciably exceed the risk or rate to the general population or other appropriate comparison group.
3. health effects occur in a minority or low-income population, or Indian tribe affected by cumulative or multiple adverse exposures from environmental hazards.

Disproportionately high and adverse environmental effects. Whether

1. there would be an impact on the natural or physical environment that significantly and adversely affects a minority or low-income population, or Indian tribe. Such effects may include ecological, cultural, human health, economic, or social impacts (when socioeconomic effects are interrelated to impacts on the natural or physical environment) on minority or low-income communities, or Indian tribes.
2. environmental effects are significant and may have an adverse impact on minority or low-income populations, or Indian tribes that are likely to appreciably exceed those on the general population or other appropriate comparison group.
3. environmental effects would occur to a minority or low-income population, or Indian tribe affected by cumulative or multiple adverse exposures from environmental hazards.

Census data can be obtained from: U.S. Bureau of the Census, Population Information, Washington, DC 20233-0001. Web site: http://www.census.gov/; phone: (301) 457-2422.

4.5   ENVIRONMENTAL IMPACT ASSESSMENT

The process of performing an environmental impact assessment (EIA) is depicted in the author's Action-Impact Model illustrated in Figure 4.1. This model illustrates a general-purpose procedure for assessing potential impacts. This model is intended to strike a balance between providing an approach that is rigorous, yet is simple enough to provide a practical process for evaluating environmental impacts.

4.5.1   Actions

An impact is triggered by an action(s). The definition of a federal action is as follows:

"Actions" include new and continuing activities including projects and programs entirely or partly financed, assisted, conducted, regulated or approved by federal agencies; new or revised agency rules, regulations, plans, policies, or procedures and legislative proposals (§ 1508.18[a]).

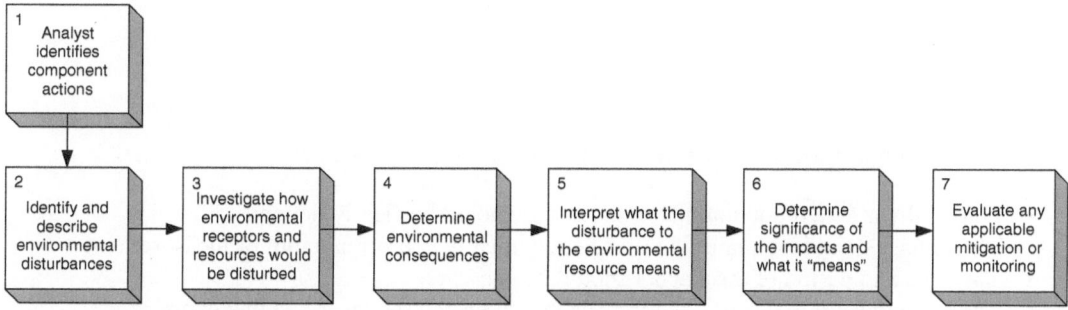

FIGURE 4.1   Action-impact model for investigating environmental impacts.

The definition of scope recognizes three types of alternative action:

- No action alternative
- Other reasonable courses of action (including proposed action)
- Mitigation measures (not included as part of the proposed action)

Potential actions may be viewed as having several stages of development, one of which is a proposal:

> "Proposal" exists at that stage in the development of an action when an agency subject to the Act has a goal and is actively preparing to make a decision on one or more alternative means of accomplishing that goal and the effects can be meaningfully evaluated (§ 1508.23).

A proposal normally consists of a set of discrete component actions. For example, construction of a dam actually involves many component actions: site preparation, construction of access roads, construction of offices and infrastructure facilities (transmission lines, water and sewage facilities), and construction of the actual dam, not to mention the actual operation of the complex.

All component actions related to the proposal need to be identified and adequately evaluated (Figure 4.1, box #1). Chapter 8 provides additional details on the subject of actions.

4.5.2 ENVIRONMENTAL DISTURBANCES

Most actions produce environmental disturbances (e.g., noise, disruption of flora or fauna, air emissions, effluents, vibrations, solid waste, contaminates) that must be identified and defined by the analyst, and described in sufficient detail to support a subsequent analysis of their effect on environmental resources (see Figure 4.1, box #2).

4.5.3 RECEPTORS AND RESOURCES

Environmental disturbances change or perturb one or more receptors (i.e., air or water quality, cultural resources, wildlife, habitat, human health or safety), often referred to as environmental resources or receptors (see Figure 4.1, box #3).

4.5.4 IMPACT ANALYSIS

The analyst must evaluate and determine the impact on the "human environment." The analysis is directed at determining how a given environmental disturbance would affect (change or perturb) receptors/environmental resources (see Figure 4.1, box #4). This analysis is conducted on a resource-by-resource basis.

The result is a set of consequences (i.e., environmental effects or impacts). As detailed in Chapter 9, the terms "impacts" and "effects" are synonymous (§ 1508.8[b]). The CEQ recognizes three distinct types of impacts (§ 1508.25[c]):

- Direct
- Indirect
- Cumulative

Ideally, an impact forecast is characterized by a number of features:[46]

- Potentially affected populations or resources should be clearly defined
- The impact should be quantified using a technically appropriate unit of measurement
- The time and period of the impact should be specified
- The potential significance of the impact should be clearly described and explained
- If appropriate, the likelihood or probability of that impact should be provided

TABLE 4.6

Items That Must Be Addressed in the Analysis of Environmental Consequences

1. Direct effects and their significance.
2. Indirect effects and their significance.
3. Possible conflicts between the proposed action and the objectives of federal, regional, state and local (and in the case of a reservation, Indian tribe) land use plans, policies and controls for the area concerned.
4. The environmental effects of alternatives including the proposed action.
5. Energy requirements and conservation potential of various alternatives and mitigation measures.
6. Natural or depletable resource requirements and conservation potential of various alternatives and mitigation measures.
7. Urban quality, historic and cultural resources, and the design of the built environment, including the reuse and conservation potential of various alternatives and mitigation measures.
8. Means to mitigate adverse environmental impacts.

The term "effects" is defined to include the following attributes:

... ecological (such as the effects on natural resources and on the components, structures, and functioning of affected ecosystems), aesthetic, historic, cultural, economic, social or health, whether direct, indirect, or cumulative (§ 1508.8).

As denoted in Table 4.6, eight requirements must be addressed in the environmental consequences section of an EIS (§ 1502.16).

The analyst interprets what the disturbance to the environmental resource means (see Figure 4.1, box #4).

4.5.5 SIGNIFICANCE

The environmental consequences may be considered either significant or nonsignificant according to specific factors presented in the Regulations (§ 1508.27). The analysis describes the consequences in a manner that allows a decision-maker to reach a decision regarding their significance or nonsignificance (see Figure 4.1, box #6). The reader is directed to Chapter 6 for a more detailed description of the concept of significance. The companion book, *Effective Environmental Assessments*, provides a systematic general-purpose procedure for assessing significance.[47]

4.5.6 MITIGATION AND MONITORING

An agency may choose to mitigate potentially significant impacts (see Figure 4.1, box #7). The reader is referred to Chapter 4 in the companion text *Environmental Impact Statements* for a more detailed treatment of this subject.[48]

The Regulations recognize five distinct types of mitigation measures (§ 1508.20):

a. Avoiding the impact altogether by not taking a certain action or parts of an action
b. Minimizing impacts by limiting the degree or magnitude of the action and its implementation
c. Rectifying the impact by repairing, rehabilitating, or restoring the affected environment
d. Reducing or eliminating the impact over time by preservation and maintenance operations during the life of the action
e. Compensating for the impact by replacing or providing substitute resources or environments

Mitigation measures can include avoiding or minimizing the impacts of an action, repairing the effects of impacts that do occur, and compensating for impacts by replacing or substituting resources that have been damaged (§ 1508.20). If a decision is made to mitigate the impacts to the point of nonsignificance, it is recommended that a mitigation action plan (MAP) be prepared as an integral part of implementing the proposal.

4.5.6.1 Monitoring

A monitoring program (or postmonitoring program as it is sometimes called) and an enforcement plan (as part of the MAP) should be adopted and summarized for any mitigation measures that are chosen (§ 1505.2[c]). It is important to note that specific performance standards need to be established for assessing the effectiveness of mitigation measures.

A monitoring program is also an important step for ensuring that environmental predictions are not exceeded. Agencies are also responsible for making the results of relevant monitoring available to the public. Postmonitoring is particularly useful for ensuring that

- mitigation measures are adequately implemented,
- environmental standards are met, and
- no impacts are encountered that are substantially different from those originally forecasted.

4.5.7 Impact Assessment Methodologies

Many impact assessment tools and methodologies were developed beginning in the early 1970s and continue to be improved upon. Although numerous methods have been developed, many of these assessment methods are rarely used. With the exception of Geographical Information Systems, development of new and more sophisticated methods has generally not received high priority. Some widely used assessment tools and methods include but not limited to

- checklists,
- matrices,
- networks,
- carrying capacity,
- ecosystem analyses, and
- cost–benefit assessments.

4.5.7.1 Environmental Checklists

A checklist can be viewed as either a single column or simplified overview of a proposed action, with only a coarse characterization of the type and magnitude of potential environmental impacts provided. A simplified example is depicted in Table 4.7.

TABLE 4.7
Simple Checklist for Assessing Impacts on Geological Resources

Item	Short Term	Long Term
Surface water hydrology	X	
Surface water quality		X
Groundwater hydrology		
Groundwater quality		
Soils disturbances		X
Land excavations		X
Esthetics		X

While checklists can provide helpful tools, they cannot take into consideration all specific resources or disturbances, or other environmental attributes that may be encountered, particularly on large or complex projects. Their strength lies in the fact that they can support a review of an array of impacts in a systematic way that is relatively easy to use; they can provide analysts with a tool for reducing the likelihood that important impacts will be overlooked.

However, the simplicity of a checklist can also be its undoing. Their use frequently discourages critical thinking (i.e., tunnel vision) and may provide a false sense of a complete assessment; an incomplete checklist can result in a flawed analysis in which important impacts are overlooked. Thus, checklists are often ineffective either because of their incompleteness or because they contain so many irrelevant impacts that they essentially become too large and unwieldy to be of practical use. These disadvantages can sometimes be compensated for by developing checklists tailored to specialized types of projects (constructing bridges, roads, or power plants) where specific actions and impacts tend to be encountered.

Various types of checklists have been developed to provide analysts with a systematic framework so that they do not overlook important environmental considerations. Beyond the standard checklist (Table 4.7), three additional types of checklists are in wide use:

- *Descriptive checklists*—Provide lists of environmental parameters, including information on impact identification and assessment that can assist analysts in identifying relevant impacts.
- *Questionnaire checklists*—Provide a series of questions relating to the impact of a project; these checklists are particularly useful for less experienced practitioners.
- *Weighting (scaling) checklists*—The most complex type of checklist; this method uses weighting factors to assess unquantifiable and intangible impacts using a common scoring and weighting scale.

To address cumulative impacts, checklists need to incorporate all of the activities associated with the proposal, as well as past, present, and reasonably foreseeable future actions. Figure 4.2 shows a hypothetical checklist for identifying potential cumulative effects of a highway project.[49] This table could be viewed as a hybrid between a checklist and an environmental matrix.

4.5.7.2 Matrices

Developed by Luna Leopold in 1971 to assess environmental impacts for the U.S. Geological Survey, matrices are among the most popular and widely used impact identification method.[50] Matrices are an extension of the environmental checklist, in as much as that they can be viewed as two-dimensional checklists, with project activities listed on one axis and potentially affected environmental resources or similar attributes on the other; they allow analysts to assess the importance of individual interactions between activities and resources. An example of a simplified matrix is depicted in Figure 4.3.[51] An "X" indicates a direct impact, whereas an "O" depicts an indirect effect.

Analysts may elect to simply note the presence or absence of an impact in the matrix cells. A more powerful approach, however, is to score impacts based on factors such as importance, magnitude, or probability of occurrence. For example, the values entered into each cell may represent quantified air emissions or number of disturbed acres. Because of the difficulty in quantifying many environmental attributes, matrices often simply use a subjective or relative score. Weighting schemes are also common.

The *Leopold Matrix* is perhaps the most widely used matrix method. It is a project action/environmental resource matrix using measures of potential magnitude and significance to score each impact component. Leopold's original matrix comprised a large grid of 100 potential project actions along the horizontal axis versus 88 environmental attributes on the vertical axis. The matrix only identifies direct impacts and does not indicate factors such as the timing or duration of impacts. Two values are entered into each relevant cell of the matrix: individual cells are filled in to indicate the beneficial or detrimental magnitude (from +10 to −10) and the level of importance (1–10) of the

Potential impact area	Proposed action			Past action	Other present actions	Future actions	Cumulative impact
	Construction	Operation	Mitigation				
Topography and soils	**	*		*			**
Water quality	**	**	+	*		*	***
Air quality		**		*			**
Aquatic resources	**	**	+	*		*	**
Terrestrial resources	*	*		*			**
Land use	*	***		*		*	***
Esthetics	**	***	+	*			**
Public services	*	+				+	+
Community structure		*			*		*
Others							

Key: *Low adverse effect   **Moderate adverse effect   *** High adverse effect
        + Beneficial effect

FIGURE 4.2   A hypothetical environmental checklist for identifying cumulative effects of a highway project. (Courtesy of CEQ.)

Affected habitat	Impoundment	Channelization	Grazing	Logging	Surface mining	Row crop agriculture	Groundwater mining	Flow diversion	Flow augmentation	Urbanization	Irrigation	Hydropeaking
Sediment yield	X		X	X	X	X				X	X	O
Water yield	O		X	X	X	X	X	X	X	X	X	
Channel morphology	X	X	X	X	X	X			O	X	X	O
Substrate characteristics	X	X	X	X	X	X			O	X	X	X
Cover	O	X	X	X	X	X				X	X	O
Timing of flows	X						O	O	O	O	X	X
Magnitude of peak flows	X	O	X	X	X	X				X		X
Magnitude of low flows	X		O	O	O	O	X	X	X	X	X	X
Thermal regime	X		O	O	O	X	O	X	O	X	X	
Water quality	O		X	X	X	X		O	O	X	X	

FIGURE 4.3   Simplified example of a matrix.

impact for each activity against each environmental attribute. A written explanation provides the rationale for the assignment of values.

Matrices are generally superior to checklists because they relate actions to environmental attributes. They also have the advantage of depicting environmental impact data in a simple format. The Leopold matrix attempts to quantify impacts, but in the absence of a systematic scoring system,

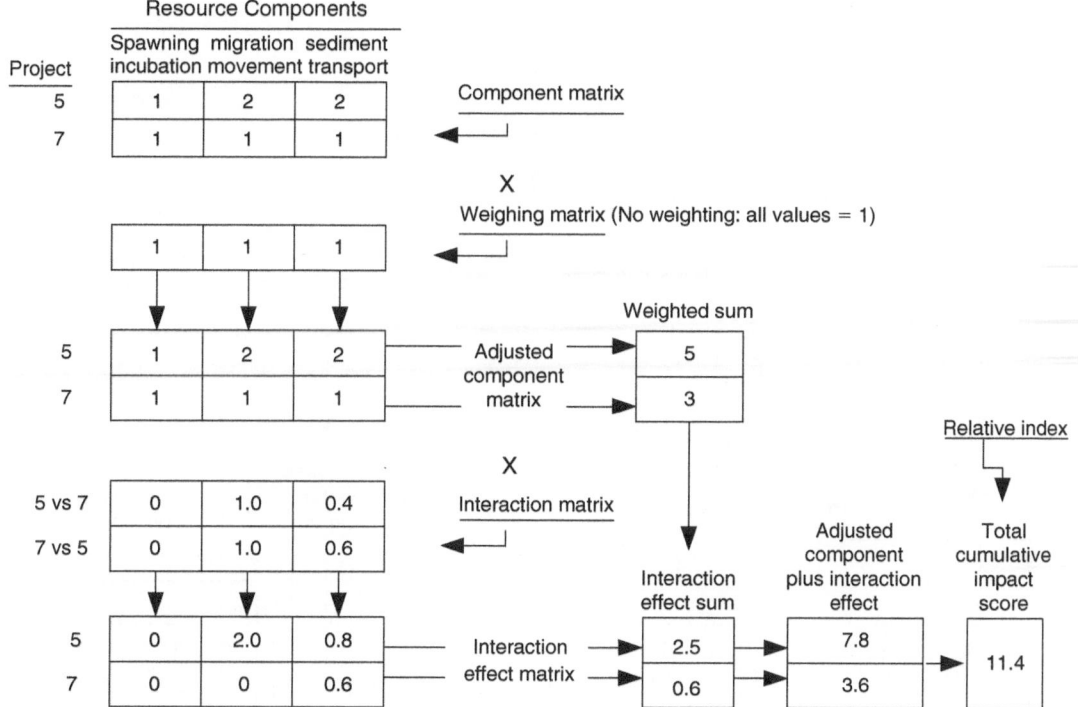

FIGURE 4.4 Example of a matrix involving a cumulative impact computation using matrix algebra to link three resource components and two projects. (Courtesy of CEQ.)

determination of the significance of the impact is ultimately subjective. An additional disadvantage is that impacts can be double counted.

If numerical data have been acquired, a matrix can provide a mathematical tool for (multiplying values in individual cells using matrix algebra) evaluating combined or synergistic effects and, in particular, interactive and cumulative impacts of multiple actions on individual environmental resources. While principally used in socioeconomic assessments, they are increasingly being used to assess physical impacts on environmental resources.

Figure 4.4 shows the cluster impact assessment procedure (CIAP) developed by the Federal Energy Regulatory Commission (FERC) to assess the cumulative impacts of small hydroelectric facilities within a single watershed. The CIAP uses a matrix for each resource (e.g., salmon), consisting of relative effect ratings based on a scale of 1–5, arranged by resource component (salmon, migration, spawning habitat). Each resource matrix contains a summary column that represents the total cumulative impact score across various components for each project.[52]

4.5.7.3 Networks

A network or a system diagram provides a systematic and comprehensive method for identifying potential impacts through sequential cause/effect linkages. The concept of using a network diagram to progressively trace cause-and-effect relationships was pioneered by Sorensen in 1971.[53]

Networks frequently provide the best tool for identifying cause-and-effect relationships, particularly with respect to the assessment of cumulative impacts. Cumulative effects are identified when multiple disturbances affect the same resource.

A network analysis proceeds in only one direction (forward). In contrast, a system diagram allows loops or feedback from one part of the system to another. System diagrams provide a superior

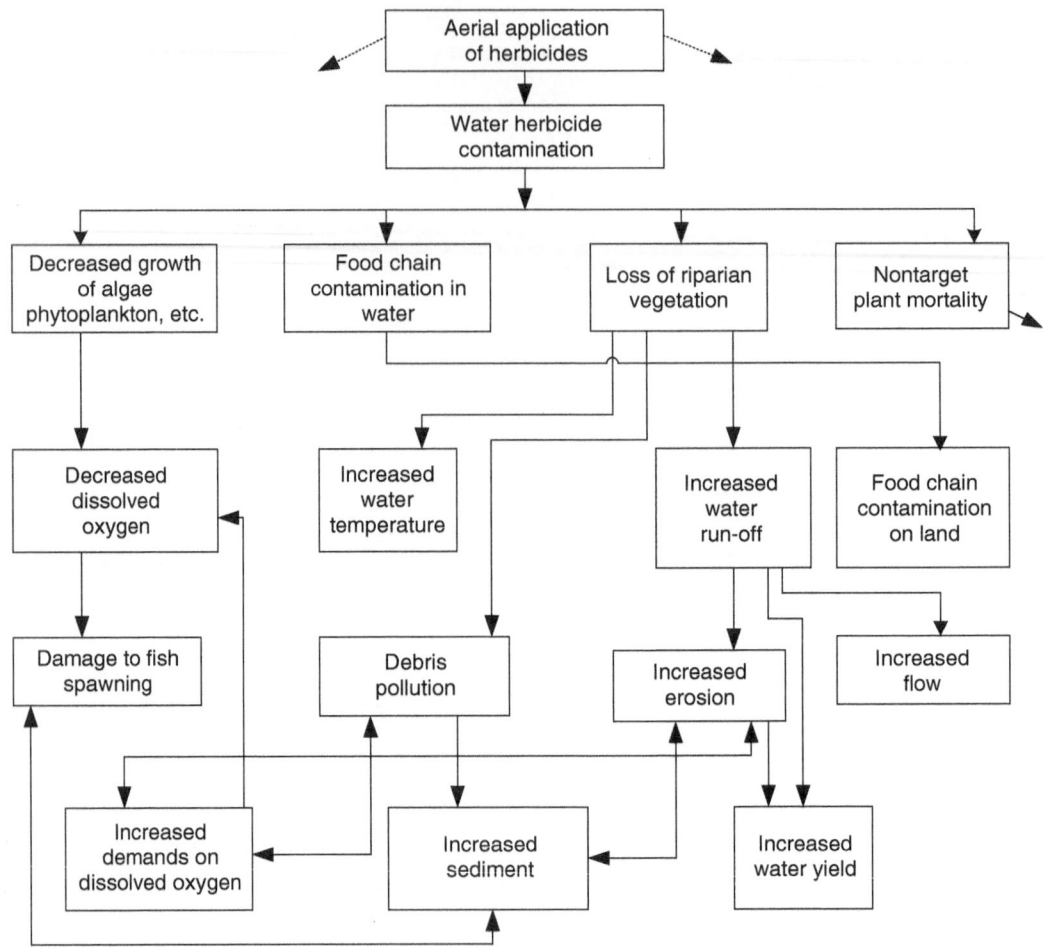

FIGURE 4.5 A systems diagram showing the indirect cumulative impact of aerial herbicide application on an aquatic system. (Courtesy of CEQ.)

mechanism for illustrating interrelationships. Computerized expert systems can be used to facilitate network analyses.

Figure 4.5 illustrates the indirect effects of a single activity resulting in a cumulative impact on a single resource.[54] Specifically, this system diagram shows how fish spawning has been degraded as a result of aerial application of herbicides through five different pathways resulting in low dissolved oxygen concentration and high sediment stress. The low oxygen level is caused by decreased plankton growth and increased oxygen consumption from debris pollution and erosion. Increased sediment is also caused by increased erosion and debris pollution following loss of riparian vegetation.

4.5.7.4   Carrying Capacity Analysis

Carrying capacity analysis is based on the fact that many environmental and socioeconomic systems have inherent limits or threshold levels. Carrying capacity is defined as the threshold of stress below which populations and ecosystem functions can be sustained. The system is considered to be unsustainable if the carrying capacity is exceeded. For example, a reservoir can only supply a finite amount of water to users, a road system can only efficiently accommodate a certain level of traffic, and a rangeland can only sustain a certain number of deer.

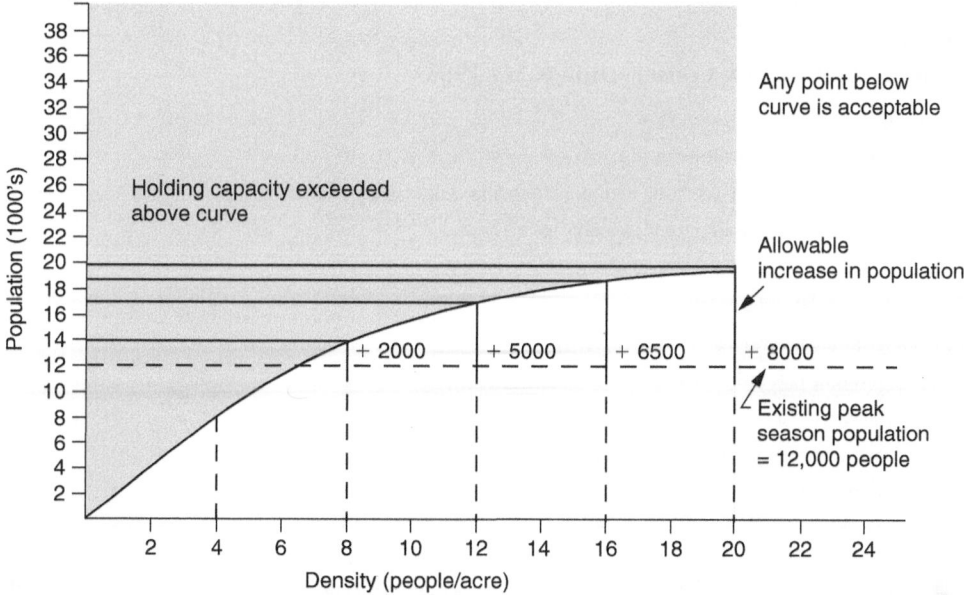

FIGURE 4.6 Carrying capacity analysis for Sanibel Island, Florida. The analysis depicts the allowable population versus runoff assimilation capacity. (Courtesy of CEQ.)

Figure 4.6 illustrates the combinations of population size versus population density that are possible without exceeding the carrying capacity of an interior wetland to assimilate water runoff from a developed area.[55]

With respect to an assessment of cumulative impacts, carrying capacity can be used to identify thresholds for the resources and systems and provide mechanisms for monitoring increases in incremental resource usage. The analysis begins with the identification of potential limiting factors (e.g., grazing land in a pasture). Mathematical equations are developed which quantify and describe the capacity of the resource in terms of these limiting factors. This approach can be used to estimate the effect that a given project would have on the remaining resource capacity. The reader is referred to Section 13.1 for additional description of carrying capacity.

4.5.7.5 Ecosystem Analyses

NEPA analyses have traditionally tended to provide independent, stand-alone assessments of such environmental resources as air quality, hydrology, wildlife, and habitats. While separate or segmented assessments tend to be more straightforward and easier to perform, they also tend to obscure interdependencies and interrelationships, particularly with respect to the understanding of cumulative impacts; proper utilization of an ecosystem approach can be critical to the success of a CIA.

Recognition of the interconnectedness of these "separate" disciplines has facilitated development of more comprehensive methodologies such as ecosystem and watershed management approaches. An ecosystem approach involves consideration of the full scope of ecological resources and their interrelationships. The CEQ has issued a report providing guidance on how to perform and integrate an ecosystems approach into NEPA analyses.[56,57] Table 4.8 describes some basic biodiversity principles that are key to performing a comprehensive and successful ecosystem analysis.[56]

An ecosystem approach can provide broad regional perspective and holistic analyses that addresses the following cumulative impact principles:

- *Focus on the resource or ecosystem*: Ecosystem analyses address biodiversity considerations and incorporate use of ecological condition indicators in assessing impacts.

TABLE 4.8

Principles of Biodiversity Conservation (CEQ 1993)

1. Take a "big picture" or ecosystem view.

2. Protect communities and ecosystems.

3. Minimize fragmentation. Promote the natural pattern and connectivity of habitats.

4. Promote native species. Avoid introducing non-native species.

5. Protect rare and ecologically important species.

6. Protect unique or sensitive environments.

7. Maintain or mimic natural ecosystem processes.

8. Maintain or mimic naturally occurring structural diversity.

9. Protect genetic diversity.

10. Restore ecosystems, communities, and species.

11. Monitor for biodiversity impacts. Acknowledge uncertainty. Be flexible.

- *Address resource or ecosystem sustainability*: Ecosystem management approaches specifically address the interactions and processes necessary to sustain the composition, function, and structure of an ecosystem.
- *Use natural boundaries*: Ecosystem analysis uses ecological boundaries or regions (watersheds, basins) in considering ecosystem functioning and addressing issues such as habitat segmentation.

The ecosystem approach involves three basic principles:

1. Take a big picture approach when assessing the ecosystem
2. Use a diverse suite of biological indicators in assessing impacts
3. Address interactions among ecological components that are needed to maintain the functioning of the ecosystem

A widely cited example of how an ecosystem approach provided a unifying concept for successfully assessing cumulative impacts involved an EIS prepared in 1993 by the U.S. Forest Service and BLM. Panels of experts were assembled to determine the likelihood of maintaining viable population of a suite of species based on available habitat limitations. Addressing the entire ecosystem involved performing a comprehensive, integrated assessment of a wide array of individual but interdependent species, habitats, and ecosystems (terrestrial forest ecosystems, aquatic species, aquatic ecosystems, and aquatic and riparian-dependent species).

4.5.7.6   Cost–Benefit Analysis

A cost–benefit analysis (CBA) involves evaluating and weighing the total expected costs of an action against the total expected benefits in order to determine the best (or most profitable) course of action. Analysts attempt to put all relevant costs and benefits on a common comparative footing. The process generally involves making either monetary or nonmonetary calculations of cost versus expected return or benefit. Costs can be characterized either in terms of dollars or attributes such as losses, sacrifices, or detrimental impacts. The reader is referred to Section 13.3 for an additional description of the CBA process.

Traditionally, most CBA have focused on assessing the monetary costs of very large public and private sector projects. While the focus has been on assessing monetary costs, such analyses are also used to qualitatively access intangible considerations such as a project's benefits to society versus resulting environmental damage.

In many cases, a value must be placed on an environmental resource or even human life, which can generate substantial public or scientific controversy. For example, consider the following issue: Is the reduced risk worth the cost of widening or straightening a dangerous curve on a mountainous roadway if the monetary cost is less than the implicit value of the property damage, injuries, or deaths that could be averted?

Accuracy: Estimating the potential costs, benefits, and reasonable assumptions is often fraught with technical complexities. Assumptions used in a CBA are frequently very controversial as various project proponents and adversaries may each argue that certain costs or assumptions should be included or excluded from a study. The outcome of a CBA should be treated with caution as their accuracy has been brought into question in recent years. The accuracy of such studies depends on how precisely the costs and benefits have been estimated. One peer-reviewed CBA study considered the accuracy of monetary cost estimates for transportation infrastructure proposals; the study found that actual costs of rail projects averaged 45% higher than originally estimated; for roads, the estimates were 20% higher.[58] With respect to benefits, another peer-reviewed CBA study found that actual rail ridership was on average 51% lower than originally estimated; with respect to roads, it found that for half of all traffic projects, the benefits were wrong by more than 20%.[59] Obviously, such inaccuracies constitute a substantial risk to the planning process itself, because inaccuracies of such magnitude are likely to lead to flawed decisions and botched projects.

The following example from the corporate world graphically illustrates how flawed decisions can be made based on an incomplete or unsound CBA. Due to design flaws, the Ford Pinto was deemed more likely to burst into flames during rear-impact collisions. Based on its own internal CBA, the Ford management decided not to issue a recall. Ford's CBA estimated that based on the number of cars in use and probable accident rate deaths resulting from this design flaw would run approximately $50 million; this was the cost Ford estimated it would pay in legal settlements involving wrongful death lawsuits. The CBA estimated that this cost was substantially below the cost of issuing a recall ($140 million). But Ford failed to consider the costs of resulting negative publicity. Public outcry was so great that in the end Ford was forced to issue a recall and the reputation of the Pinto was severely damaged, leading to a significant loss in sales.

As described above, two types of CBAs, described in the following sections, are frequently included in an EIS: monetary CBAs, which assess and compare proposals in terms of dollar values, and nonmonetary CBAs, which assess and compare benefits versus the environmental costs or risks.

Monetary Cost–Benefit in NEPA Documents: There are many methods for presenting monetary CBAs in an NEPA analysis. Perhaps the most common method is simply to compute the qualitative monetary costs of the project and compare that with the benefits that would be derived from implementing the proposal. Consider the following qualitative example.

Consider an EIS that was prepared to assess the impacts of constructing two nuclear power plants. The alternatives involved constructing two large gas- or coal-fired plants. The CBA explained in qualitative terms that while gas-fired plants are relatively inexpensive to construct, they also expose consumers to the potential for volatile price changes and can leave the U.S. economy vulnerable to international disruptions in fuel supply. In contrast, nuclear power plants are relatively expensive to construct, but the operating costs tend to be stable and dampen the volatility witnessed elsewhere in the electricity market.

Now consider how the same EIS quantified the monetary cost and benefits of the proposed nuclear reactor project on the local economy. In considering the increased stress on housing and social services as a result of an influx in workers, the tax base generated from the proposal was projected to generate revenue of approximately $30,000,000 during initial operations. The CBA further explained that most local residents would consider the large tax payments to be a benefit because it would support the development of infrastructure and might lessen the tax burden on local residents.

With respect to the local economy, the EIS stated that the new construction would result in an increased workforce of approximately 700 employees and with the economic multiplier effect this would translate into about 1700 additional jobs. While the plants would result in some environmental degradation, they would inject about $350,000,000 in total economic output into the regional economy.

The capital costs of building the plant were projected to be about $2000/kW. With a total megawatt capacity of a little over 2000 MW, the total construction cost was estimated to be somewhat over $4 billion.

As mentioned below, nonmonetary CBAs can provide either qualitative or quantitative assessments of the environmental cost and benefits.

Nonmonetary Cost–Benefit in NEPA Documents: The EIS for the same two proposed nuclear power plants also assessed the costs and benefits in terms of nonmonetary environmental

TABLE 4.9

Air Pollutant Comparison Cited in the EIS

Pollutant	Coal Emissions (t/Year)	Gas Emissions (t/Year)	Nuclear Emissions (t/Year)
Nitrogen oxides	1800	550	0
Sulfur dioxide	5600	160	0
Carbon monoxide	1800	110	0
Particulates less than 10 μm	90	90	0

TABLE 4.10

Summary of Cost–Benefit Assessment for Two Proposed Nuclear Plants

Cost–Benefit Attribute	Description
Costs:	
Esthetics	The two cooling towers would impair the public's view of the countryside. The steam plum would add to this effect.
Housing	The increased workforce would place a small strain on existing housing/rental units and could result in a modest increase in housing values and rental rates. Housing values would rise by an additional 2% per year while rental rates would rise by an additional 2.5% annually.
Infrastructure	The increased workforce would place a small strain on the local school system, police force, and road system.
Change in land use	The project would result in the loss of a public park and use of a small lake.
Land area sacrificed	The project would result in the lost use of 600 acres of land previously designated for light industry.
Crime	The influx in workers would be expected to result in a small increase in local crime rate.
Cultural resources	The project would result in demolition of a historic building.
Changes in local city character	The influx in workers would result in a relatively small change in the traditional character of the local community.
Benefits:	
Electricity generated	The plants would generate 19 million MWh of needed electricity per year. This would reduce the risk of future brownouts.
Rate stability	The proposed plants would reduce volatility in electric rates.
Unemployment	Job creation and the increased demand for workers would reduce local unemployment.
Health effects	The proposed action would substantially reduce adverse human health effects as a result of avoiding relatively large air emissions that would be produced by operation of a coal- or gas-fired plant.

amenities and attributes. While the nuclear reactors would result in some adverse environmental impacts, the CBA pointed out that the project would also result in substantial air quality and human health benefits. The CBA pointed that unlike gas- or coal-fired plants, nuclear facilities emit virtually no emissions that contribute to global warming or degradation in air quality. In other words, the proposed project would represent a substantial benefit in terms of emissions avoidance when compared with coal- or gas-fired alternatives. Table 4.9 compares the environmental costs in terms of emissions for the coal- and gas-fired alternatives versus the proposed nuclear plants. Table 4.10 summarizes how some other environmental attributes were qualitatively or quantitatively compared.

4.6 REQUIREMENTS FOR PERFORMING AN ANALYSIS

Some of the most important requirements for performing an analysis are provided in the following sections.

4.6.1 Performing a Rigorous Analysis

Table 4.11 provides direction for ensuring that a rigorous analysis has been performed, and Table 4.12 provides additional direction for performing the analysis.

TABLE 4.11
Performing a Rigorous Analysis

[Prepare] analytic rather than encyclopedic environmental impact statements (§ 1500.4[b]).

… supported by evidence that the agency has made the necessary environmental analyses (§ 1502.1)

Rigorously explore and objectively evaluate all reasonable alternatives (§ 1502.14[a])

Devote substantial treatment to each alternative considered in detail (§ 1502.14[b])

TABLE 4.12
Related Directions for Performing the Analysis

Providing adequate detail

Identify environmental effects and values in *adequate detail* so that they can be compared to economic and technical analysis. (§ 1501.2[b], emphasis added)

Considering risk and probability

For the purposes of this section, "reasonably foreseeable" includes impacts which have *catastrophic consequences even if their probability of occurrence is low,* provided that the analysis is *supported by credible scientific evidence,* is not based on pure conjecture, and is within the rule of reason. (§ 1502.22[b][4], emphasis added)

Cost–benefit analysis

If a *cost-benefit analysis* … is being considered … discuss the relationship between that analysis and any analysis of unquantified environmental impact, values and amenities. (§ 1502.23)

Analysis and supporting data

Agencies shall employ writers of clear prose … to write statements which will be *based upon the analysis and supporting data* from the natural and social sciences and the environmental design arts. (§ 1502.8, emphasis added)

TABLE 4.13

Requirements for Performing a Scientific and Accurate Analysis

The information must be of *high quality. Accurate scientific analysis* … [is] essential to implementing NEPA. (§ 1500.1[b], emphasis added)

Accurate scientific analysis … are essential to implementing NEPA. (§ 1500.1[b], emphasis added)

… the analysis is supported by *credible scientific evidence.* (§ 1502.22[b][4], emphasis added)

Agencies shall insure the *professional integrity*, including *scientific integrity* … They shall identify any *methodologies* used … (§ 1502.24, emphasis added)

Identify *methods and procedures* required by section 102(2)(b) to insure that presently unquantified environmental amenities and values may be given appropriate consideration. (§ 1507.2[b], emphasis added)

4.6.2 METHODOLOGY

Table 4.13 provides direction for ensuring that the analysis is scientifically credible, accurate, and of high quality.

The importance of these requirements is underscored by the following court case. The U.S. Court of Appeals held that the Forest Service had misinterpreted an earlier study that resulted in evaluating an average market timber demand that was nearly double what the original study had projected.[60] Thus, the court found the EIS to be misleading because it presented data indicating a public market demand twice the actual value. Moreover, this incorrect interpretation resulted in an EIS that failed to consider an adequate range of alternatives.

4.6.3 FAIR AND OBJECTIVE

Failure to prepare the analysis in an open and unbiased manner can cast doubt on the reputation and integrity of the agency and may provide opponents with a legitimate basis for challenging its final decision. Some of the most important requirements for ensuring a fair and objective analysis are

- … provide *full* and *fair* discussion of significant environmental impacts and shall inform decision-makers and the public of the reasonable alternatives … (§ 1502.1, emphasis added)
- Environmental impact statements shall serve as the means of assessing the environmental impact of proposed agency actions, *rather than justifying decisions* already made. (§ 1502.2[f], emphasis added)
- The statement … will *not be used to rationalize or justify* decisions already made. (§ 1502.5, emphasis added)
- … *objectively evaluate* all reasonable alternatives … (§ 1502.14[a], emphasis added)
- The agency shall … *disclose and discuss* … *all major points of view* on the environmental impacts of the alternatives including the proposed action (§ 1502.9[a], emphasis added)

4.6.4 INVESTIGATING REASONABLE ALTERNATIVES

According to the Regulations, the "heart" of an EIS is the section on alternatives. Selected citations pertinent to this requirement are presented in Table 4.14.

4.6.4.1 "Would" versus "Will"

A common misconception is that the word "would" has the connotation of "future tense." The word "would" actually denotes a conditional statement, or more accurately, a future conditional

TABLE 4.14

Direction for Analyzing the Reasonable Alternatives

Use the NEPA process to identify and assess the *reasonable alternatives to proposed actions* that will avoid or minimize adverse effects of these actions upon the quality of the human environment. (§ 1500.2[e], emphasis added)

Study, develop, and describe appropriate alternatives to recommended courses of action in any proposal which involves unresolved conflicts concerning alternative uses of available resources as provided by section 102(2)(E) of the Act. (§ 1501.2[c], emphasis added)

The ... environmental impact statement shall ... shall inform decision-makers and the public of the *reasonable alternatives* ... (§ 1502.1, emphasis added)

The *range of alternatives* discussed in environmental impact statements *shall encompass those to be considered by the ultimate agency decision-maker.* (§ 1502.2[e], emphasis added)

[The alternatives] section is the *heart* of the environmental impact statement. (§ 1502.14, emphasis added)

Rigorously explore and objectively evaluate all *reasonable alternatives* ... (§ 1502.14[a], emphasis added)

Develop and evaluate alternatives not previously given serious consideration by the agency. (§ 1503.4[a][2], emphasis added)

statement. Thus in the context of NEPA, it implies some level of uncertainty, indecision, or doubt about whether a future action/impact will actually occur. In contrast, the word "will" is a present tense and its use implies that an action/impact will definitely take place.

As just witnessed, every effort is to be exercised in avoiding even the appearance of partiality. Thus, to clearly indicate that no decision has yet been made, discussions of potential actions should be written using the conditional future tense, as if the action/impact could or might take place. The word "would" satisfies the constraint to write in terms of a future conditional statement.

This practice is based on NEPA case law where numerous agencies have lost cases based on the court's conclusion that the analysis lacked or appeared to lack objectivity, or did not give fair and equitable consideration to alternatives. One of the points that has been argued by plaintiffs is that such analyses were flawed because they were written as if the decision had in *de facto* been decided, that is, the proposed action "will" be implemented.

For example, this statement is generally considered inappropriate: "The project will result in a 10% increase in SO_2 emissions." A more impartial statement would read, "The *proposed* action *would* result in a 10% increase in SO_2 emissions." The words "proposed" and "would" clearly indicate to the reader that a final decision has not been made.

For this reason, when describing the environmental impacts or alternatives (including the proposed action), use of nonconditional words such as "will" should be avoided as much as possible. Within the Department of Energy and a number of other agencies, when addressing potential actions, alternatives, or impacts, almost all use of verbiage such as "will" has been eliminated.

However, it is acceptable to use words such as "will" as long as their use is premised on making the entire statement conditional. For example, the following conditional statement is generally considered acceptable: "*If* the *proposed* action is implemented, it *will* result in a 10% increase in SO_2 emissions." The disadvantage is that it is usually shorter to simply write "would" than to add the extra verbiage necessary to make a statement a conditional one.

An exception to this general guidance involves circumstance where a "will" statement describes something that would take place regardless of which alternative (including the proposed action) is implemented. For example, premonitoring activities may need to be preformed for the proposed action and alternatives (including no-action), and therefore could fit under this exception.

4.6.5   WRITING DOCUMENTS IN PLAIN ENGLISH

To provide the public with usable information, the Regulations require that agencies "employ writers of clear prose" and prepare EISs written in "plain language" so that they can be clearly understood by decision-makers and the general public (§ 1502.5, § 1502.8).

The author predicts that this requirement will be the focus of increasing attention and litigation in the foreseeable future. In one case, a court invalidated a Forest Service EIS for the suppression and eradication of gypsy moths on grounds that it was not readable. The EIS was highly technical, using phrases such as "can cause maternal toxicity" instead of "can kill fetus" and "acute lethal dose" in lieu of "deadly dose." The court concluded that an agency has a duty to provide the public with comprehensive information regarding environmental consequences of a proposed action and to do so in a readily understandable manner.[61]

4.6.5.1   Reasonable Man Standard

Some courts have applied a principle known as the *Reasonable Man Standard* (sometimes referred to as the *Clapham Bus Test*) in determining if documents such as an EIS can be understood by the general public. This principle has its roots in old English common law. The English courts used the Clapham Bus Test in determining if a law could be understood by the general public. A common law was considered to be comprehensible if an average man riding a Clapham bus in London could read and understand it.

Additional court directions for preparing EISs that are understandable include:

- "... in language understandable to the general public and at the same time contain sufficient technical and scientific data to alert specialists to particular problems within their expertise."[62]
- a document is to be prepared "so that it translates technical information into language that can be understood by its 'intended readership [including] interested members of the public.' "[63]
- "... *readily understandable by government decision-makers and by interested non-professional laymen likely to be affected by actions taken under the EIS.*" (emphasis added)[64]
- "... written so as to be *readily understandable* by governmental decision-makers and by *interested* non-professional laypersons likely to be affected by actions taken under the EIS."[64]
- "... clear, concise, easily readable form so as to provide a reasonably intelligent non-professional an understanding of the environmental impacts."[65]

The phrase "readily understandable" requires special note. This direction is interpreted to mean that an individual of average intelligence should not need to spend an inordinate amount of time reviewing background information or trying to decipher concepts. Also note use of the phrase, "interested non-professional laypersons" in this same reference. The term "interested" has been interpreted to mean a nonprofessional layman, interested enough in the issues to have done a minimal amount of background research.

4.6.6   INCOMPLETE OR UNAVAILABLE INFORMATION

When an agency is evaluating reasonably foreseeable significant adverse effects in an EIS and there is incomplete or unavailable information, the agency must make clear that such information is lacking (§ 1502.22):

- If the incomplete information relevant to reasonably foreseeable significant adverse impacts is essential to a reasoned choice among alternatives and the overall costs of obtaining it are not exorbitant, the agency must include the information in the EIS.

- If information relevant to reasonably foreseeable significant adverse impacts cannot be obtained because the overall costs of obtaining it are exorbitant or the means to obtain it are unknown, the agency must include an evaluation of such impacts based upon theoretical approaches or research methods generally accepted by the scientific community (§ 1502.22[b][4]).

4.6.6.1   Classified Information

NEPA documents that address classified proposals may be safeguarded and restricted from public dissemination in accordance with agencies' own regulations applicable to classified information. These documents may be organized so that classified portions can be included as annexes, in order that the unclassified portions can be made available to the public (§ 1507.3).

4.6.6.2   Case Law

In one case, a plaintiff sued to compel preparation of an EIS for alleged plans to store nuclear weapons in a proposed facility in Hawaii. The Navy had indeed prepared an EA for construction of a storage facility that was capable of storing nuclear weapons but any plans for storing nuclear weapons at those facilities were classified.

When challenged, the District Court concluded that NEPA applied to the Navy's actions, but given the national security provisions of the Atomic Energy Act and the Navy's own regulations, the Navy had complied with NEPA to the fullest extent possible. The Court of Appeals disagreed and required the agency to prepare and release a "Hypothetical EIS" with regard to operation of a facility capable of storing nuclear weapons.

On appeal, the Supreme Court overturned the decision, finding that a "hypothetical" EIS was not mandated by any statutory or regulatory provisions. The Court concluded that the Navy must consider environmental consequences in its decision-making process, even if it is unable to meet NEPA's public disclosure goals by virtue of Freedom of Information Act exemption for national security purposes.[66] Chapter 8 discusses different methodologies for handling classified information.

PROBLEMS

1. A decision-maker decides to dredge a deeper channel into a nearby river bottom. A construction contractor has been contacted and told to begin the dredging operation 90 days from now. Because the requirements of NEPA must be met before construction activities begin, he next contacts a local consultant to prepare an EA. He notifies the NEPA contractor that the EA must be completed within 90 days, as any longer period will delay the project. Does this process comply with NEPA's regulatory requirements? Justify your answer.

2. The decision-maker described in problem 1 knows that the dredging project will be controversial. To reduce the level of controversy, he decides not to notify the public and to prepare the EA "internally" so as to not "arouse undue emotions." Does this project strategy comply with NEPA's regulatory requirements? Justify your answer.

3. The decision-maker described in problem 1 understands that he must describe the actions and impacts of the proposed dredging project. However, he does not wish to "put the spot light" on alternatives that he believes to be "a nuisance" and with which he has little experience. He notifies the NEPA contractor that they are to prepare an EA which only describes the proposed dredging operation and not to waste any time investigating "screwy" alternatives. Does this project strategy comply with NEPA's regulatory requirements? Justify your answer.

4. In an attempt to save time and money, the NEPA contractor informs the decision-maker described in problem 1 that the proposal involves expertise with wetlands, endangered species, and potential flooding issues. In an attempt to save time and money, the decision-maker decides to have his mechanical engineer who recently graduated from college prepare the EA. Would such a decision comply with NEPA's regulatory requirements? Justify your answer.

5. With respect to NEPA, what does the term "systematic" process mean?
6. Briefly explain what is involved with the Clean Air Act conformity.
7. What is the State Historic Preservation Officer?
8. Does the Endangered Species Act have a *substantive* mandate to protect endangered species? Does the National Historic Preservation Act have a substantive mandate to preserve historic sites?
9. What does Executive Order 12898 state?
10. One of the component actions involved in preparing an oil well drilling proposal involves constructing an access road to the well site. List some of the potential environmental disturbances associated with such a proposal.

REFERENCES

1. CEQ, Forty most asked questions concerning CEQ's National Environmental Policy Act Regulations (40 CFR 1500–1508), *Federal Register* 46(55), 18026–18038, March 23, 1981, Question Number 4a.
2. *Jones v. District of Columbia Redevelopment Land Agency*, 1974, 499 F.2d 502.
3. *Environmental Law Handbook*, Government Institutes, Inc., 10th ed., Chapter 10.
4. *Hughes River Watershed Conservancy v. Glickman*, 81 F.3d 437 (4th Cir. April 12, 1996).
5. Mandelker D. R., *NEPA Law and Litigation*, Clark Boardman Callaghan, New York, 1992.
6. CEQ, *Collaboration Handbook*, September 22, 2006.
7. Environmental Quality Improvement Act of 1970.
8. E.O. 11514, Protection and Enhancement of Environmental Quality, March 5, 1970.
9. *Norton v. SUWA* (June 2004).
10. Rabl A. et al., Impact Assessment and Authorization Procedure for Installations with Major Environmental Risks, Final Report, 1999.
11. Pollution Prevention Act, 42 U.S.C. 13101, 1990.
12. Executive Order 12856, *Federal Compliance with Right-to-Know Laws and Pollution Prevention Requirements* (published August 6, 1993, 58 FR 41981).
13. CEQ, Pollution Prevention and the National Environmental Policy Act, January 12, 1993.
14. 42 U.S.C. 7401 et seq.
15. 65 FR 57336; September 22, 2000.
16. 65 FR 11574; March 3, 2000.
17. 63 FR 12466; March 13, 1998.
18. DOE, *Clean Air Act General Conformity Requirements and the National Environmental Policy Act Process*, April 21, 2000.
19. www.ipcc-wg1.ucar.edu.
20. Supplement to the Draft EIS for the Gilberton Coal-to-Clean Fuels and Power Project (DOE/EIS-0357D-S1) in early January 2007.
21. DOE, *NEPA Lessons Learned*, March 1, 2007, Issue No. 50.
22. 16 U.S.C. Section 1536(a)(2).
23. Executive Order 13112, *Invasive Species*, February 3, 1999; published in *Federal Register*, 64 FR 6183, February 8, 1999.
24. Executive Order 11988, Floodplain Management, May 24, 1977.
25. EO 11990, Protection of Wetlands, May 24, 1977.
26. 16 U.S.C. 1451–1464.
27. 65 FR 77123–77175. December 8, 2000.
28. 16 U.S.C. 1456(c)(1).
29. 15 CFR 930.35.
30. 16 U.S.C. 1271–1287.
31. 16 U.S.C. 1271.
32. 16 U.S.C. 1272.
33. 16 U.S.C. 661.
34. 16 U.S.C. § 470 et seq.
35. 36 CFR Part 800, Protection of Historic Properties, 2001.
36. 36 CFR 800.8(c)(4).

37. 36 CFR Part 800, 64 FR 27044, May 18, 1999.
38. 42 U.S.C. § 1996.
39. 25 U.S.C. § 3001 et seq.
40. PL 97–98; 7 U.S.C. 4201 et seq.
41. 7 U.S.C. 4201(c)(1)(A).
42. 7 U.S.C. 4201(c)(1)(B).
43. CEQ, memorandum of August 11, 1980, Analysis of Impacts on Prime or Unique Agricultural Lands in Implementing the National Environmental Policy Act.
44. Executive Order 12898, Federal Actions to Address Environmental Justice in Minority Populations and Low-Income Populations, 59 FR 7629, 1994.
45. CEQ, *Environmental Justice: Guidance Under the National Environmental Policy Act*, December 10, 1997.
46. Culhane P. J. et al., The precision an accuracy or U.S. Environmental Impact Statements, Environmental Monitoring and Assessment, P217–238, 1987.
47. Eccleston C. H., *Effective Environmental Assessments: How to Manage and Prepare NEPA's EA*, Lewis Publishers, Boca Raton, FL, 2001.
48. Eccleston C. H., *Environmental Impact Statements: A Comprehensive Guide to Project and Strategic Planning*, John Wiley & Sons Inc, New York, 2000.
49. CEQ, Considering Cumulative Effects Under the National Environmental Policy Act, p. A-8, January 1997 (figure is after Bisset 1983).
50. Leopold, L. B. et al., *A Procedure for Evaluating Environmental Impact*. United States Geological Survey, Geological Survey Circular No. 645, Washington, D.C., 1971.
51. USGS, www.mesc.usgs.gov/products/Publications/3910/chapter2.html, after Bovee, K. D., T. J. Newcomb, and T. G. Coon. 1994. Relations between habitat variability and population dynamics of bass in the Huron River, Michigan. National Biological Survey Biological Report 21, 63 pp.
52. CEQ, Considering Cumulative Effects Under the National Environmental Policy Act, p. A-11, January 1997 (taken from FERC. 1985. Procedures for Assessing hydropower Projects Clustered in River Basins (request for comments) *Federal Register* 50, 3385–3403.
53. Sorensen, J. C., 1971, A Framework of Identification and Control of Resource Degradation and Conflict in the Multiple Use of the Coastal Zone, University of California, Berkeley.
54. CEQ, Considering Cumulative Effects Under the National Environmental Policy Act, p. A-16, January 1997 (figure is after Bisset 1983).
55. CEQ, Considering Cumulative Effects Under the National Environmental Policy Act, p. A-34, January 1997 (from Clark J. 1976. The Sanibel Island Report: Formulation of a Comprehensive Plan Based on Natural Systems. Conservation Foundation, Washington, D.C.).
56. CEQ, Incorporating Biodiversity Considerations into Environmental Impact Analyses under the National Environmental Policy Act. 29 pp, 1993.
57. Interagency Ecosystem Management Task Force, The Ecosystem Approach: Healthy Ecosystems and Sustainable Economies, vol. I, Overview, Washington, D.C., 1995.
58. Flyvbjerg, B., Holm, M. K. S., and Buhl, S. L, Underestimating costs in public works projects: error or lie? *Journal of the American Planning Association*, 68(3), Summer 2002.
59. Flyvbjerg, B., Holm, M. K. S., and Buhl, S. L, How (In)accurate are demand forecasts in public works projects? *Journal of the American Planning Association*, 71(2), Spring 2005.
60. *Natural Resources Defense Council v. U.S. Forest Service*, No. 04-35868, 35 ELR 20160 (9th Cir. Aug. 5, 2005).
61. *Oregon Environmental Council v. Kunzman*, 614 F. Supp. 657 (D. Ore. 1985).
62. *Alabama ex rel. Baxley v. Corps of Eng'rs of the United States Army*, 411 F. Supp. 1261, 1267 (N.D. Ala. 1976).
63. *Natural Resources Defense Council, Inc. v. United States Nuclear Regulatory Comm'n*, 222 U.S. App. D.C. 9, 685 F.2d 459, 487 n.149 (D.C. Cir. 1982).
64. *Oregon Envtl. Council v. Kunzman*, United States District Court for the District of Oregon, 636 F. Supp 632, 1986.
65. *Baltimore Gas & Elec. Co. v. NRDC*, 462 U.S. 87, 76 L. Ed. 2d 437, 103 S. Ct. 2246 (1983).
66. *Weinberger v. Catholic Action of Hawaii/Peace Education Project*, 454 U.S. 139, 102 S.Ct. 1917 (1981).

5 Exemptions and Categorical Exclusions

Actions, including categorical exclusions (CATXs), that are exempt from the requirements of the National Environmental Policy Act (NEPA) are described in this chapter. This chapter begins with a comprehensive description of the CATX process and is followed by a discussion on the types of actions that are generally excluded from the requirements of NEPA.

5.1 CATEGORICAL EXCLUSIONS

CATXs, briefly described in Chapter 2, were not fully defined. Because of their clear importance to efficiency, this section has been prepared to provide the reader with a thorough understanding of the entire CATX process. A CATX is defined as

> … a category of actions which do not individually or cumulatively have a significant effect on the human environment and which have been found to have no such effect in procedures adopted by a Federal agency in implementation of these regulations (§ 1507.3) and for which, therefore, neither an environmental assessment nor an environmental impact statement is required. An agency may decide in its procedures or otherwise, to prepare environmental assessments for the reasons stated in § 1508.9 even though it is not required to do so. Any procedures under this section shall provide for extraordinary circumstances in which a normally excluded action may have a significant environmental effect (§ 1508.4).

Since the impacts have already been determined to be nonsignificant, neither an environmental assessment (EA) nor an environmental impact statement (EIS) is required to be prepared for those actions qualifying for a CATX. Moreover, each agency is required to adopt implementing procedures that include specific criteria for, and identification of, typical classes of action that

> (ii) … normally do not require either an environmental impact statement or an environmental assessment (categorical exclusions).[1]

The Council on Environmental Quality (CEQ) recently published draft guidance on establishing, revising, and using CATXs.[2] This guidance is intended to assist federal agencies in improving and modernizing their administration of CATXs under NEPA. The draft guidance recommends procedures and approaches for (1) establishing and revising CATXs; (2) involving the public; (3) documenting development, revision, and use of CATXs; and (4) periodically reviewing CATXs.

Additional direction can be found in § 1500.4[p], § 1500.5[k], § 1501.4[a], § 1507.3[b], and § 1508.4. Although CATXs are one of the most important streamlining tools available to the NEPA practitioner, experience indicates their application is often underutilized.

As far as NEPA's requirements are concerned, once an action has been categorically excluded, the agency is free to pursue the action. The reader should note, however, that while a CATX satisfies NEPA's requirements, it does not necessarily satisfy the requirements of other environmental statutes. Some representative examples of CATXs promulgated by the Department of Energy (DOE) are shown below.[3]

- B2.1 Modifications of an existing structure to enhance workplace habitability (including, but not limited to, improvements to lighting, radiation shielding, or heating, ventilating, air conditioning and its instrumentation, and noise reduction).

- B3.1 (a) Geological, geophysical (such as gravity, magnetic, electrical, seismic, and radar), geochemical, and engineering surveys and mapping, including the establishment of survey marks.
- B3.2 Aviation activities for survey, monitoring, or security purposes that comply with Federal Aviation Administration (FAA) regulations.
- B4.7 Adding fiber optic cable to transmission structures or burying fiber optic cable in existing transmission line rights-of-way.

5.1.1   EXTRAORDINARY CIRCUMSTANCES

Occasionally, an action may be technically eligible for a CATX, but because of unusual circumstances, its potential impacts may be significant. The NEPA regulations (Regulations) provides for such "extraordinary circumstances." Agencies are mandated to make provisions for situations (§ 1508.4) under which an EA or EIS must be prepared even though the proposed action technically falls within the scope of a CATX.

The Regulations, however, do not define what constitutes extraordinary circumstances. Consistent with the rule of reason, the author suggests that this requirement be interpreted to mean

unique or unusual conditions in which a decision-maker, or responsible party, cannot conclude clearly and quickly that the action would result in an insignificant impact.

The author suggests some basic factors that may be useful in devising specific criteria for defining extraordinary circumstances:

- Unresolved conflicts concerning alternate uses of available resources within the meaning of NEPA (§ 102[2][E])
- Actions that may involve effects that are highly controversial (§ 1508.28[b][4])
- Proposals connected to other actions with potentially significant impacts (§ 1508.25[a][1])
- Actions with effects that are uncertain, unique, or involve unknown risks (§ 1508.27[b][5])
- Actions that establish a precedent for future actions with significant effects (§ 1508.27[b][6])

Again, consistent with the rule of reason, the author suggests some additional circumstances where the application of a CATX may be inappropriate:

- Where it has the potential to affect an undisturbed area
- When it is outside the size or scope of activities that would normally be excluded under a CATX
- When it involves the use of hazardous, toxic, or radioactive chemicals that could harm the environment
- When it involves unproven activities or technology
- When it could affect archaeological or cultural resource sites, ecologically critical areas or habitats, sensitive and endangered species, wild and scenic rivers, wetlands, floodplains, or coastal zones

5.1.2   ADOPTING CATXS

Agencies are encouraged by the CEQ to review their list of exclusions periodically and where appropriate update them. The CEQ also encourages agencies to list out examples of activities that fall within these categories.[4]

Agencies are expected to involve the CEQ in the review of proposed CATXs during the draft stage. A formal system should be established for screening proposed actions against CATX eligibility criteria and extraordinary circumstances.

One approach for identifying potential CATXs involves reviewing EAs that have previously resulted in findings of no significant impact (FONSI). Conversely, CATXs should be revoked or revised if experience indicates a history of inappropriate use as follows:

- A scope that is excessively broad
- A potential for significant impacts, either individually or cumulatively

As one example, prior to the approval of a medication, the Food and Drug Administration (FDA) used to prepare an EA for each new drug brought to the market. Yet, after many years of practice, only one case had required the preparation of an EIS. Many years of experience demonstrated a general lack of significant environmental impacts related to the approval of new drugs. Based on this record, the FDA established a firm basis to broaden its CATXs for future drug approvals.[5]

The CEQ reviews and approves the adoption of CATXs to verify that they are consistent with NEPA's regulatory requirements. Because NEPA is a public process, proposed CATXs are published prior to their formal adoption, so that the public is given an opportunity to review and comment on them. As the final step in the adoption process, all CATXs must be published in the *Federal Register*.[4]

5.1.3 APPLYING CATXS

As depicted in Figure 5.1, proposed actions should be carefully reviewed to ensure that they fall appropriately within the scope of a CATX.[6] Agencies might wish to consider the use of a sliding-scale approach in reviewing an action's eligibility for a CATX. Based on such an approach, activities considered routine or trivial in nature may receive only a cursory review, while more atypical or complex activities are scrutinized with an increasing level of attention.

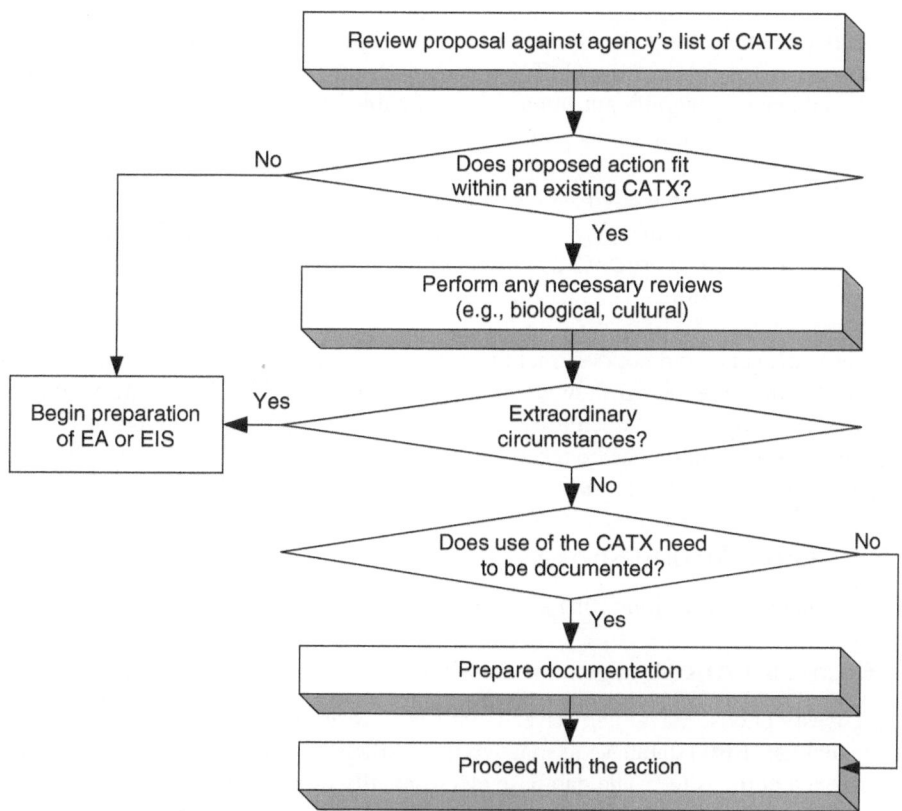

FIGURE 5.1   Generalized process for categorically excluding actions.

If appropriate, a review is performed to confirm that the proposal does not trigger any of the 10 significant criteria (§ 1508.27[b]) listed in the Regulations. Care should also be taken to ensure that the context (e.g., historic or cultural resource, or ecologically sensitive area) in which the action would occur is not overlooked (§ 1508.27[a]). Candidate actions with individually insignificant impacts should also be reviewed to ensure that they are not connected or related to other actions with potentially significant impacts (§ 1508.25[a]). Finally, the review should ensure that no extraordinary circumstances are involved.

Case law. The courts have tended to give agencies broad discretion over the interpretation and application of their CATXs. For example, in 1985 the City of Alexandria, Virginia, challenged the Federal Highway Administration over its decision in favor of a CATX for a highway ramp project.[7] The city claimed that the project involved significant impacts. The court ruled that the agency's interpretation of its own CATXs should be given greater weight than the interpretations of either the court or the city. The court went on to explain that it would defer to the agency's interpretation of its own regulations as long as the agency followed its procedures appropriately.

However, in a second case, a court found that an agency's existing record indicated that a proposed CATX for increasing motorized boat traffic on a wild and scenic river had the potential to impact turtles and salmon and to cause conflicts among user groups.[8]

In a third case, the FAA made a decision allowing an airline to schedule passenger service to LaGuardia from a general aviation airport located in the vicinity of numerous historic parks. The FAA concluded that an additional 10 roundtrip flights per day would have a *de minimus* environmental impact and that a CATX was appropriate. The court concluded that the FAA's failure to prepare an EA or to consult with historic preservation agencies was a harmless error.[9]

5.1.4 DOCUMENTING CATXS

Some agencies document the execution of their CATX processes to demonstrate that they were indeed reviewed, that the use of the CATX was appropriate, and that no unforeseen factors were present that could cause a significant impact. For example, the Forest Service has used a decision memo, the DOE has used environmental checklists, and the Bureau of Reclamation has used a categorical exclusion checklist (CXC) consisting of a 1–2 page narrative of the action and a set of questions regarding its possible environmental impacts.

One expert stated that, in his view, they amounted to "documenting the fact that an action does not have to be documented." Former CEQ chairman Alan Hill stated that, in his view, "… an agency should rarely need to document the fact that an activity is categorically excluded from the NEPA process."[10]

In one case, an agency did not document its reliance on a CATX. The court noted that no documentation existed in the record to show that the agency made the CATX determination at the time the action was approved.[11] As depicted by the third decision-making diamond in Figure 5.1, the author suggests that such practices be reserved for important or potentially controversial situations and for circumstances specified in the agency's NEPA implementation procedures.

5.1.5 STREAMLINING RECOMMENDATIONS

Methods that can be used to streamline an agency's CATX process are described below.

5.1.5.1 Electronic CATX Management Database

The DOE's Westinghouse Savannah River Company has developed an electronic environmental evaluation checklist (EEC) database to manage in an efficient manner the large volume of CATXs that are generated at this site.[12] The database electronically transfers the CATX request form from the initial preparer (usually the project engineer) to the NEPA practitioner responsible for reviewing the request.

The electronic data flow mimics the hardcopy workflow originally developed for the site. The electronic system totally replaces the original hardcopy paperwork that once took weeks to route through the review and approval process. Barton Marcy, a manager at Westinghouse, has reported that a CATX request was processed and approved in a time as short as an hour. The system is fully searchable and provides a graphical, real-time checklist status so that it can be quickly determined if the request for a project has been approved. A monthly activity report that includes a cover letter to the DOE NEPA Compliance Officer is automatically generated by the database at the end of each month.

Figure 5.2 shows one design for such a system, depicting a system that could be designed to allow a requestor (e.g., project engineer) to log into the database and fill out a short request form

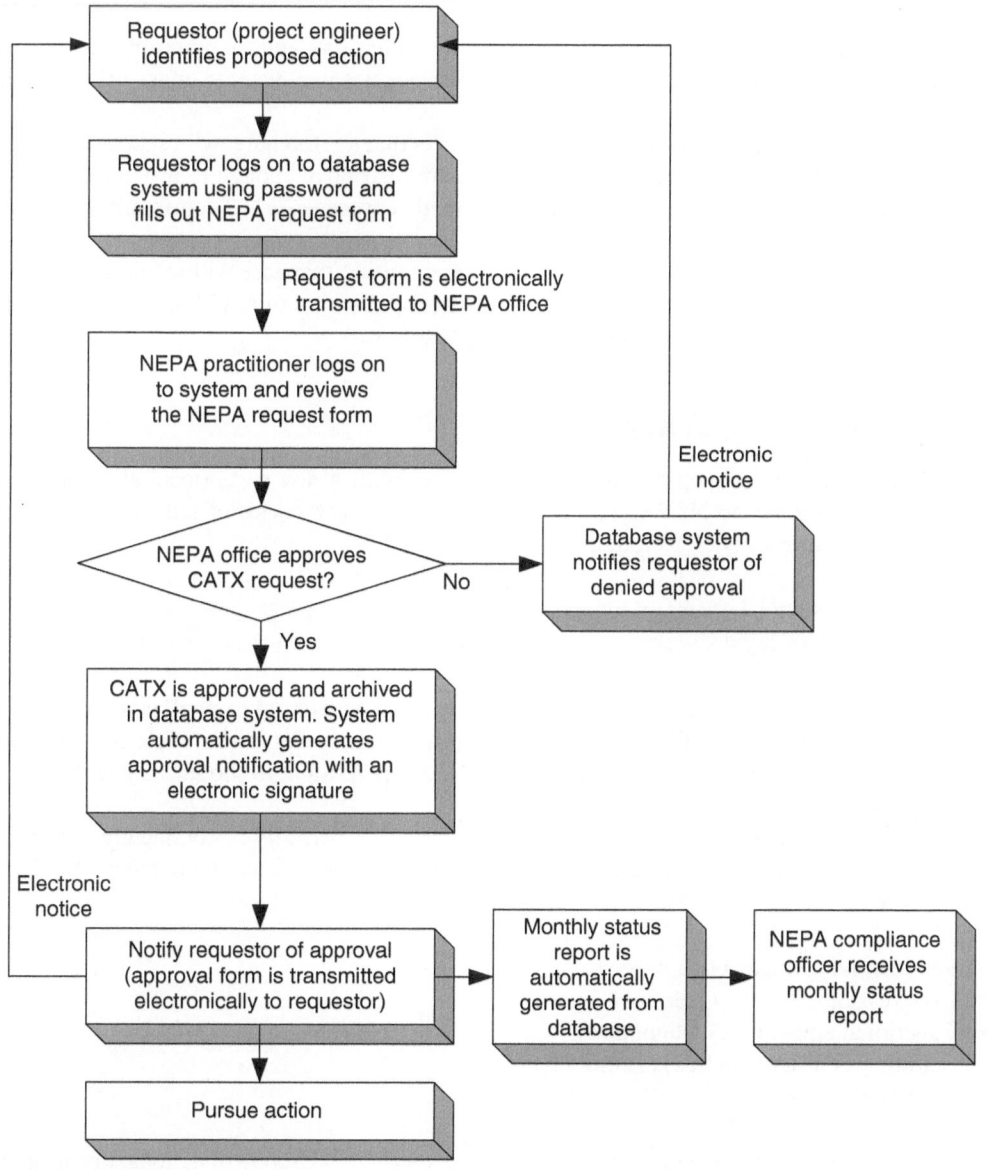

FIGURE 5.2 Example of an electronic database system for automating an agency's categorical exclusion process.

with information such as the nature of the action or even a proposed CATX that the action might fit. On receiving the request, the NEPA practitioner would then use the system, which includes electronic signatures, to process and approve it automatically. A password is used to authenticate the signer's identity and to place both the requestor's and approver's electronic signatures on the approval form. Electronic signatures can be scanned as pictures of actual signatures.

5.1.5.2   Additional Recommendations

Additional recommendations are provided below for assisting agencies in streamlining their CATX processes.

- Practice of documenting CATXs by some agencies should be either eliminated or restricted to actions that are important or potentially controversial.
- Agency officials should periodically review their missions and, as appropriate, expand their supportive lists of exclusions. For example, agencies may want to consider reviewing their lists of CATXs at least every 5 years.
- Agencies have been criticized for adopting CATXs that are too narrow in scope. Prudence should be exercised in developing CATXs that are restrictive enough to prohibit activities that may be significant, yet broad enough to provide coverage for the community of activities that are nonsignificant.
- Agencies should prepare lists of examples of activities that fall within the scope of a CATX. Such examples can clarify the appropriate application of CATXs, reducing potential abuses that may result from their inappropriate use.

5.2   EXEMPTIONS FROM NEPA

NEPA's mandate is both sweeping and comprehensive. With a few exceptions, all proposals for federal action are subject to NEPA's requirements. Only two categories of federal activities were originally exempted from NEPA's requirements (§ 1508.18[a]):

1. Funding assistance by general revenue bonds
2. Operation of the legal system

A limited number of additional circumstances deemed to be exempt from all or some of NEPA's requirements have been identified. Some of these exemptions are noted in Table 5.1. To respond effectively to such situations, federal officials and NEPA practitioners should be keenly aware of these exemptions.

Five categories of actions normally considered to be either partially or completely exempt from NEPA are described below. For additional information, the reader is directed to papers by Schmidt and Swenson.[17]

- Presidential and executive office exemptions
- Congressional (explicit statutory) exemptions
- Functional equivalency exemptions
- Statutory conflict (implicit) exemptions
- Emergency situations

In practice, use of these exemptions is limited to a very narrow scope of actions. Prudence must be exercised to ensure that an action legitimately qualifies for exclusion. Some exemptions involve complexities beyond the scope of this book.

TABLE 5.1
Exemptions from the Requirements of NEPA

Executive Order Exemptions
Federal Actions Abroad

For foreign countries when their environments are significantly affected by major federal actions, agency procedures are to provide for the preparation of environmental review documents in the following situations .... Where the environmental effects of federal actions are within foreign countries, agencies have flexibility under the Executive Order to prepare either concise environmental reviews of the issues involved or to undertake bilateral or multilateral environmental studies. Environmental impact statements will not be required in these circumstances.[13]

For Nuclear Activities Abroad

Unless not required ... the Office of Export and Import Control shall promptly arrange for the preparation of an appropriate environmental document ....[14]

Statutory Exemptions
Clean Air Act Exemption for the EPA

No action taken under the Clean Air Act shall be deemed a major federal action significantly affecting the quality of the human environment within the meaning of the National Environmental Policy Act.[15]

Clean Water Act Exemption for EPA

Except for the provision of Federal assistance for the purpose of assisting the construction of publicly owned treatment works ... and the issuance of a permit ... for the discharge of any pollutant by a new source ... no action by the Administrator taken pursuant to this Act shall be deemed a major federal action significantly affecting the quality of the human environment within the meaning of the National Environmental Policy Act.[16]

Regulatory Exemptions
General Revenue Sharing (§ 1508.18[a])

Actions do not include funding assistance solely in the form of general revenue sharing funds, distributed under the State and Local Fiscal Assistance Act of 1972, 31 USC 1221 et seq., with no Federal agency control over the subsequent use of such funds.

Judicial and Administrative Enforcement Actions

Actions do not include bringing judicial or administrative civil or criminal enforcement actions (§ 1508.18[a]).

Inconsistency of NEPA with Other Statutory Requirements

Parts § 1500–1508 of this title provide regulations applicable to and binding on all *Federal* agencies ... except where compliance would be inconsistent with other statutory requirements ... (§ 1500.3).

The phrase "to the fullest extent possible" in Section 102 means that each agency of the Federal Government shall comply ... unless existing law applicable to the agency's operation expressly prohibits or makes compliance impossible (§ 1500.6).

Legislative Proposals

Preparation of a legislative environmental impact statement shall conform to the requirements of these regulations except as follows: (1) there need not be a scoping process; (2) the legislative statement shall be prepared in the same manner as a draft statement, but shall be considered the "detailed statement" required by statute (§ 1506.6).

Timing Requirements

An exception to the rules on timing may be made in the case of an agency decision which is subject to a formal internal appeal (§ 1506.10[b]).

Emergencies

Where emergency circumstances make it necessary to take an action with significant environmental impact without observing the provisions of these regulations, the Federal agency taking the action should consult with the Council about alternative arrangements. Agencies and the Council will limit such arrangements to actions necessary to control the immediate impacts of the emergency. Other actions remain subject to NEPA review (§ 1506.11).

Inconsistency of Agency NEPA Procedures with Statutory Requirements

Agency procedures shall comply with these regulations except where compliance would be inconsistent with statutory requirements ... (§ 1507.3[b]).

(Continued)

TABLE 5.1   (Continued)

Exemptions from the Requirements of NEPA

Classified Actions—National Security

Agency procedures may include specific criteria for providing limited exceptions ... for classified proposals. They are
  proposed actions which are specifically authorized ... to be kept secret in the interest of national defense or foreign
  policy .... Environmental assessments and environmental impact statements which address classified proposals may be
  safeguarded and restricted from public dissemination in accordance with agencies' own regulations applicable to
  classified information. These documents may be organized so that the classified portions can be included in annexes,
  in order that the unclassified portions can be made available to the public (§ 1507.3[c]).

When Effects Are Only Economic or Social

This means that economic and social effects are not intended by themselves to require the preparation of an environmental
  impact statement (§ 1508.14).

5.2.1   PRESIDENTIAL EXEMPTIONS

As specified in Section 102 of the Act, NEPA applies to agencies of the federal government. The
president is not a federal agency. Thus, decisions made directly by the president (i.e., not generated
by a federal agency) are exempt from NEPA. Consistent with this interpretation, the term federal
agency does not include the Congress, the judiciary, or the president. The Regulations also exclude
"performance of staff functions for the President in his Executive Office" (§ 1508.12).

Court cases involving this exemption have been split. Hence, there is some question concerning its precise scope and validity. Furthermore, some activities involving environmental impacts in
foreign countries (including nuclear activities abroad) are exempted under an Executive Order.

5.2.2   EXPLICIT CONGRESSIONAL EXEMPTIONS

Congress has the authority to exempt specific actions and legislation from NEPA's requirements.
For example, the Energy Supply and Coordination Act of 1974 granted an explicit statutory exemption to the Environmental Protection Agency (EPA) from complying with NEPA when the agency
undertook actions pursuant to the Clean Air Act (CAA) (see Table 5.1). Specifically, this act stated,
"No action taken under the Clean Air Act shall be deemed a major federal action significantly
affecting the quality of the human environment."[18]

Similarly, certain actions taken by the EPA under the Clean Water Act are exempt from NEPA.[19]
Swenson reports several examples and states:

> In addition there are a number of emergency powers given to EPA by various environmental statutes
> that are intended to allow EPA to respond quickly, without formal regulatory findings to various threats
> to the environment which will not wait for the normal process of regulation.[20]

A number of statutory exemptions applying to agency programs other than those of the EPA are
not shown in Table 5.1. Mandelker, who provides a partial list of such exemptions, reports:

> Other environmental protection programs administered by EPA do not contain express exemptions
> from NEPA. Whether these programs are exempted depends on whether a court determines that their
> environmental decision-making procedures are functionally equivalent to NEPA's.[21]

The Disaster Relief Act allows the president to declare an emergency situation so that immediate assistance can be provided. This act exempts a number of emergency relief activities from
NEPA. For instance, the repair and restoration of federal facilities that existed prior to the disaster
are exempt in cases where they are limited to restoring these facilities to the state.

Congress has also exempted from NEPA review controversial projects such as the Alaska pipeline, the San Antonio freeway, and the logging in the Pacific Northwest.[21] In recent years, Congress has increasingly exempted various aspects of other projects and programs from NEPA.

For example, in 2002, the Bush administration proposed a "Healthy Forests Initiative" that exempted loggers from the NEPA process in certain fire-prone federal forests. Instead of filing individual NEPA statements for each concerned forest, the government would issue only one all-embracing large-scale forest thinning plan.

More exemptions are in the works. Section 2055 of the proposed Energy Policy Act of 2005 (H.R. 6) would waive public participation and environmental review under NEPA for many oil and gas drilling activities. Sections 1808 and 2014 of H.R. 6 would allow oil and gas companies to perform their own NEPA analyses and would reimburse the companies for doing so. This clause offers no criteria for ensuring that such analyses would be unbiased and objective.

Title V of H.R. 6 would remove the application of federal laws such as NEPA and the National Historic Preservation Act from energy development decisions on tribal lands. Section 1702 of H.R. 6 limits the evaluation of alternatives to just two: the alternative proposed by the industry and a "no-action" alternative.

5.2.3 FUNCTIONAL EQUIVALENCY EXEMPTIONS

In certain instances, the courts have upheld a doctrine known as functional equivalency. This concept does not appear in any statute, regulation, or executive order. The functional equivalency doctrine is based on an argument referred to as statutory redundancy, originally advanced by the EPA with respect to its statutory mission. Under this argument, certain instances exist where NEPA's requirements are essentially redundant when considered in conjunction with other applicable environmental statutes. That is, since other environmental statutes are essentially the functional equivalent of NEPA, the Act does not apply. Thus, the functional equivalency doctrine implies that

> … A statute is so compatible with the goals of NEPA that an EIS is not needed to ensure protection of the environment.[20]

To say the least, the issue of functional equivalence has been controversial. Its legal foundation has been supported by the courts in some cases but rejected in others. Where courts have ruled in favor of the functional equivalency argument, case law indicates that three criteria have been established for its applicability:[22]

1. Agency's organic statute must provide "substantive and procedural standards that ensure full and adequate consideration of environmental issues."[23]
2. Agency must afford public participation before a final alternative is selected.[24]
3. Action must be undertaken by an agency engaged primarily in the examination of environmental issues.[25]

5.2.4 THE EPA

The courts have generally found that the EPA's activities in furtherance of various environmental statutes are the functional equivalent of compliance with NEPA and that the agency is therefore not required to comply with NEPA in those circumstances. The following cases have concluded that, as a result of functional equivalency, the EPA did not have to comply with NEPA for actions under

- Clean Air Act;[26]
- Ocean Dumping Act;[24]
- Federal Insecticide, Fungicide, and Rodenticide Act;[27]
- Resource Conservation and Recovery Act;[28] and
- Safe Drinking Water Act.[29]

To date, however, the courts have generally declined to apply functional equivalency to any agency other than the EPA, including departments that have substantial environmental responsibilities. For example, one court declined to grant functional equivalency to the National Marine Fisheries Service:[30]

> The mere fact that an agency has been given the role of implementing an environmental statute is insufficient to invoke the functional equivalency exception.

The question of functional equivalency exemptions has been raised most notably with respect to two major environmental laws administered by the EPA:

- Comprehensive Environmental Restoration, Compensation, and Liability Act (CERCLA)
- Resource Conservation and Recovery Act (RCRA)

5.2.5   FUNCTIONAL EQUIVALENCE WITH CERCLA

Before discussing the applicability of functional equivalency to NEPA, it is instructive to summarize some basic differences between the requirements of the NEPA and CERCLA processes. For example, analysis of an affected environment is interpreted much more broadly under NEPA than CERCLA and may extend well beyond the boundaries of a contaminated site. Moreover, NEPA tends to require consideration of a much wider range of alternatives than does CERCLA. Under CERCLA, the public is afforded an opportunity to comment on the selected alternative for remediation. However, unlike NEPA, CERCLA does not require extensive public participation throughout the process. Moreover, CERCLA does not require indirect or cumulative impacts to be addressed.[22]

Application of the functional equivalency doctrine to the cleanup of hazardous waste sites under CERCLA remains somewhat unclear and controversial. The EPA has argued that functional equivalency extends to its actions under CERCLA, since its basic or organic mission is to protect the environment. Thus, NEPA's goals are inherent in the activities performed as part of that mission. Some courts have upheld EPA's claim, granting it a *de facto* exemption from NEPA.[31]

While the EPA has maintained that functional equivalency extends to its own activities, it has generally held that such equivalency does not extend to remediation activities performed by other federal agencies. The CEQ has also maintained that functional equivalency does not extend to agencies other than the EPA. As viewed by the CEQ, a dual NEPA/CERCLA process enhances environmental protection and provides the public with an opportunity to more fully participate in the decision-making process.[32]

Opinions among various agencies have been mixed. The Department of Justice has issued an opinion generally supporting extension of the functional equivalency doctrine to other agencies involved in CERCLA cleanup actions at their facilities. In the past, some agencies such as the DOE have pursued a middle course. For example, rather than depending on functional equivalency, DOE has integrated NEPA values with CERCLA documents. Different departments within the DOD have been split on the issue.

Generally, however, the courts have been reluctant to extend functional equivalency to agencies other than the EPA, even in instances where these other agencies have had significant environmental responsibilities.

5.2.6   FUNCTIONAL EQUIVALENCE WITH RCRA

It has been argued that issuing RCRA permits for the treatment and storage of hazardous waste and for disposal facilities is the functional equivalent of NEPA. Such exemption is based on the fact that the RCRA environmental review performed as part of the permitting process is functionally equivalent to that of NEPA. Swenson reports

> The EPA administrator made a finding on the record that the RCRA permit process was the functional equivalent of the NEPA process, so that an EIS was not needed. The court agreed stating that "most circuits have already recognized ... that an agency need not comply with NEPA where the agency is engaged primarily in an examination of environmental questions and where the agency's organic legislation mandates specific procedures for considering the environment that are functional equivalents of the impact process.[20]

Hansen further clarifies this concept:[33]

> Currently, there is no CEQ or other formal agency guidance available on the subject of functional equivalency. However, some 45 court decisions have granted EPA a *de-facto* "exemption" from NEPA requirements when that agency's substantive and procedural requirements qualify as the "functional equivalent of NEPA". The rationale used for this "exemption" is that EPA's *sole responsibility*, as mandated by Congress and federal statutes is protection of the environment.

Thus, once again, in practice this exemption applies only to the EPA when granting an RCRA permit to a federal agency. It does not extend to another federal agency's actions pursuant to qualifying for or executing actions under the permit. Thus, while the EPA is exempted from the requirement to comply with NEPA in granting the permit, the petitioning agency may still be required to comply with the Act.

5.2.7 IMPLICIT STATUTORY CONFLICTS

As used in this section, the term implicit refers to the fact that certain exemptions that are implied or interpreted to exist have not been explicitly exempted by Congress. This category involves implicit conflicts between NEPA's requirements and statutory responsibilities.

The Supreme Court has ruled that where an agency's duties, mandated under its charter, give rise to a clear and fundamental conflict with NEPA, the latter must yield.[34] Some federal agencies have successfully argued that their legislation conferred statutory authority which is so encompassing that certain actions are not subject to NEPA.[35] The courts have typically taken a very conservative position in their interpretation of such circumstances.[36]

Examples involving implicit exemptions have usually involved scheduling conflicts. One example involves the passage of the 1993 Emergency Petroleum Allocation Act. Congress intended to pass this legislation as quickly as possible. Implementing regulations were to be issued within 15 days of passage and to become effective 15 days after their publication. A suit was brought claiming that an EIS was needed before these regulations could be adopted. The court concluded that an EIS could not be completed within this timeframe and therefore it must be implied that Congress did not intend for NEPA to apply to the adoption of these Regulations.[37]

Other cases have involved conflicts in the processes used for formulating decisions. For example, the Endangered Species Act provides a set of decision-making factors to be used in listing endangered species and does not allow NEPA factors to be taken into consideration.[38]

5.2.7.1 Ministerial

An agency is mandated to perform a ministerial action at the insistence of some authority such as Congress and has no discretion as to whether or not it will take place. An action required by Congress for which an agency has no choice or administrative discretion is normally exempt from NEPA.[39]

One example involves the siting of a high-level radioactive waste repository at Yucca Mountain. When challenged, the court concluded

> [Since] Congress has directed the secretary to proceed with site characterization at Yucca Mountain and nowhere else, site characterization must go forward .... Congress directed the characterization of Yucca Mountain, and the Secretary no longer has discretion over alternative sites—therefore the Secretary had no obligation under NEPA.[40]

In certain events, some aspects of a ministerial action may still be subject to NEPA, particularly where Congress mandates that an action will take place but leaves with the agency discretion as to how or where the action will be implemented. For instance, Congress may direct an agency to construct an experimental energy generation plant at a particular site, but leave the specifics of this project to the discretion of the responsible agency. Although the directive to construct the plant at a specific site may be considered a ministerial action not subject to review under NEPA, an EIS may still be required to investigate alternative designs and modes of operation. Thus, while a no-action alternative may not be required (as the final decision has already been mandated by Congress), an EIS might still be necessary to consider alternative designs and construction and operational issues.

5.2.8 EMERGENCY SITUATIONS

Special procedures prescribed in the Regulations are to be followed in the advent of an emergency:

> Where emergency circumstances make it necessary to take an action with significant environmental impact without observing the provisions of these regulations, the Federal agency taking the action should consult with the Council about alternative arrangements. Agencies and the Council will limit such arrangements to actions necessary to control the immediate impacts of the emergency. Other actions remain subject to NEPA review (§ 1506.11).

Alternative arrangements have run the gamut from taking no additional measures to preparing a memorandum describing the event and outlining the mitigation measures to be taken and to preparing an EIS while the action is actually under way.

Former general council for the CEQ, Nicholas Yost, has stated that in drafting this provision such events were viewed as falling within the realm of "acts of God."[19] The current general council for the CEQ has reported that the Council approves only about one in every three requests for an exemption. Taken over the long term, since 1978 only one or two exemptions have been authorized per year;[41] however, there is some evidence that this number may be increasing.

If time is of the essence, consultation may occur after an action has already been taken. In adopting the Regulations, the CEQ made it clear that it expected such consultation to take place as soon as feasible but not necessarily before an action is taken.[42] The determination of an emergency is often made via a telephone consultation with the concerned agency and then documented by the CEQ in a memorandum.[20]

5.2.8.1 CEQ Guidance

The CEQ offers the following guidance in making emergency alternative arrangements:[43]

1. Do not delay immediate actions necessary to secure lives and safety of citizens to consult, but consult CEQ as soon as feasible. The CEQ will contact your headquarters' NEPA contacts in the event you are unable to reach them (see http://ceq.eh.doe.gov/nepa/contacts.cfm).
2. Alternative arrangements take the place of an environmental impact statement and only apply to federal actions with significant environmental impacts. Lesser actions may be

subject to agency NEPA procedures. Agency NEPA personnel should be contacted regarding agency-specific definitions of significant actions and actions that are categorically excluded.

3. Alternative arrangements for compliance with NEPA may be subject to judicial review. Alternative arrangements do not waive the requirement to comply with NEPA, but establish an alternative means for compliance.

4. Alternative arrangements are limited to the actions necessary to control the immediate impacts of the emergency. They will be developed, based upon the specific facts and circumstances, during the consultation with CEQ.

5. Courts afford CEQ substantial deference regarding its determination of emergency alternative arrangements. Alternative arrangements have been unsuccessfully challenged three times (including Westover, Massachusetts, overflights for Desert Storm training). Once the alternative arrangements are established, CEQ will provide documentation spelling out the alternative arrangements and the considerations on which they are based.

6. Factors to be addressed when crafting alternative arrangements are nature and scope of the emergency, actions necessary to control the immediate impacts of the emergency, potential adverse effects of the proposed action, components of the NEPA process that can be followed and provide value to decision-making (e.g., coordination with affected agencies and the public), duration of the emergency, and potential mitigation measures.

5.2.8.2   Examples of Past Emergencies

The Department of Agriculture received an exemption in 1990 to spray pesticide over residential areas in southern California to control a Mediterranean fruit-fly infestation.

A second example involved a dam located near Yakima, Washington that was deemed to be structurally unstable. Officials voiced concern that it could suffer catastrophic failure. The CEQ agreed to an exemption allowing the dam to be breached, provided that the agency complied with NEPA prior to its reconstruction.

Exemptions were granted for operations Desert Shield and Desert Storm during the Gulf War of 1990–1991. While the war was fought in the Persian Gulf, it involved logistical operations carried out within U.S. boundaries related to the mobilization of men and supplies. One exemption was requested and granted to conduct activities such as changing aircraft over flight patterns and operations, and another was granted in order to carry out an operation that tested the ability to deactivate mines. These exemptions only applied to activities where the DOD did not have sufficient time to comply with NEPA.[20] In granting these exemptions, the CEQ outlined alternative procedures the DOD had to carry out to satisfy NEPA's requirements. These included preparing a memorandum describing the action and any significant impacts that might occur and outlining mitigation measures for reducing such impacts.

Another example involving operations Desert Shield and Desert Storm occurred in September 1990 when the Air Force began to fly C-5A transport planes on a 24-h schedule (including night flights). The CEQ determined that the Middle East situation constituted an emergency such that the Air Force could conduct the flights on an emergency basis (40 CFR § 1506.11).

Plaintiffs challenged both CEQ's authority to allow such arrangements and the applicability of the regulation to the situation at the air base.[44] The court first noted that Section 102 of NEPA requires compliance to the fullest extent possible, indicating that an EIS is not mandatory in all circumstances. For this reason, the court upheld the CEQ's authority to issue the emergency regulation. The court also held that the decision by the CEQ and the Air Force designating the situation as an emergency response was reasonable, given the military's operational and scheduling difficulties.

PROBLEMS

1. Assume an agency has an established CATX for the construction and operation of a small storage facility. Mr. Smith, a federal project engineer, wants to use this CATX to construct a small storage shed across the street from a church for the storage of dynamite and other construction explosives. Do you believe this would be an appropriate application of this CATX? Explain your answer.

2. An agency has an established CATX for the construction and operation of small water wells. However, the senior hydrologist states that the groundwater level has already sustained a cumulatively significant drawdown. Do you believe this would be an appropriate application of this CATX? Explain your answer.

3. Are actions performed directly by the president subject to NEPA? Explain. What about actions taken by a federal judge? Explain.

4. Briefly explain what is meant by functional equivalency?

5. What is a ministerial action?

6. Assume that a river basin is approaching flood stage. A cognizant engineer has stated that if the river continues to raise it could cause the levy system to fail. In his opinion, the levy needs to be reinforced immediately to prevent catastrophic failure. However, Mr. Brown, a regulatory analyst has stated that the NEPA planning process must first be completed. Do you believe Smith is correct? Explain your answer.

REFERENCES

1. 40 CFR § 1507.3[b][2][ii].
2. CEQ, *Establishing, Revising, and Using Categorical Exclusions under the National Environmental Policy Act* for public review, 71 FR 54816, September 19, 2006.
3. 10 CFR 1221, Subpart D, 1992.
4. Council on Environmental Quality, *Memorandum: Guidance Regarding NEPA Regulations*, 48 *Federal Register*, 34263, July 28, 1983.
5. Yost N. C., Testimony before the Committee on Resources United States House of Representatives, *Hearing on NEPA: Lessons Learned and Next Steps*, November 17, 2005.
6. Lillie T. H. and Lindenhofen H. E., NEPA as a tool for reducing risk to programs and program managers, *Federal Facilities Environmental Journal*, Spring 1991.
7. *City of Alexandria v. Federal Highway Administration*, 756 F.2d 1014 (4th Cir. 1985).
8. *Riverhawks v. Zepeda*, unpublished (D. Ore. 2002).
9. *Save Our Heritage, Inc. v. Federal Aviation Administration*, 269 F.3d 49 (1st Cir. 2001).
10. Hill A. A., Former Chairman of the Council on Environmental Quality, remarks made before the Sixth Annual Environmental Review Conference, Environmental Protection Agency, Region IV, Atlanta, GA, October 21, 1982.
11. *State of California v. Norton*, 311 F.3d 1162 (9th Cir. 2002).
12. Barton C. M., Jr., Donald E. G., John R. S., John J. M., and Clayton B. S., Streamlining the NEPA process at the Savannah River Site. Paper presented at a NEPA symposium of the National Association of Environmental Professionals (date unknown).
13. CEQ Memorandum 44 FR 42, 3/29/79, pursuant to Executive Order 12114.
14. Unified Procedures Applicable to Major Federal Actions Relating to Nuclear Activities Subject to Executive Order 12114-44 FR 220, November 13, 1979.
15. Energy Supply and Coordination Act, 15 USC 793[c][1].
16. USC 1371[c][1].
17. Swenson R. T., Desert storm, desert flood: A guide to emergency and other exemptions from NEPA and other environmental laws, *Federal Facilities Environmental Journal*, Spring 1991; Schmidt O.L., Eight Good Reasons not to Prepare an EIS: Eight Thresholds to the Preparation of an EIS, The NEPA Compliance Course, Executive Enterprises Inc.
18. Energy Supply and Environmental Coordination Act of 1974. 15 USC § 793(c)(1).
19. 33 USC § 1371(c)(1).

20. Swenson R. T., Desert Storm, desert flood: a guide to emergency and other exemptions from NEPA and other environmental laws, *Federal Facilities Environmental Journal*, Spring 1991.   .

21. Mandelker D. R., *NEPA Law and Litigation*, Section 5.03(1), Clark Boardman Callaghan, New York, 1998.

22. Memorandum from Bear D., General Council, Council on Environmental Quality, to E. D. Elliott, Assistant Administrator and General Council, Environmental Protection Agency titled *Applicability of the National Environmental Policy Act to Superfund Actions at Federal Facilities*, August 1, 1990.

23. *Environmental Defense Fund v. Environmental Protection Agency*, 489 F.2d 1247, 1257 (D.C. Cir. 1973).

24. *Maryland v. Train*, 415 F. Supp. 116, 122 (D. Md. 1976).

25. *Warren County v. North Carolina*, 528 F. Supp. 276, 286 (E.D.N.C. 1981).

26. *Getty Oil Co. v. Ruckelshaus*, 467 F.2d 349 (3rd Cir. 1972), *cert. denied*, 409 U.S. 1125 (1973).

27. *Merrell v. Thomas*, 807 F.2d 776 (9th Cir. 1986), *cert. denied*, 108 S.Ct. 145 (1987).

28. *Alabamians for a Clean Environment v. EPA*, 871 F.2d 1548 (11th Cir. 1989) and *Alabama ex rel. Siegelman v. EPA*, 911 F.2d 499 (11th Cir. 1990).

29. *Western Nebraska Resources Council v. EPA*, 943 F.2d 867 (8th Cir. 1991).

30. *Texas Committee on Natural Resources v. Bergland*, 573 F.2d 201, 208 (5th Cir. 1978); *Jones v. Gordon*, 621 F. Supp. 7, 13 (D. Alaska 1985), *aff'd in part, rev'd in part*, 792 F.2d 821 (9th Cir. 1986).

31. Bear D., The Role of the National Environmental Policy Act in Promoting Pollution Prevention, Global Pollution Prevention Conference '91 Conference, Washington, D.C., 1991.

32. Swartz L. L., Memorandum on Applicability of NEPA to Federal Agency Actions under CERCLA, Council on Environmental Quality, July 30, 1990.

33. Hansen R. P. and Theodore A. W., NEPA/CERCLA/RCRA Integration: Policy Versus Practice, Current and Future Priorities for Environmental Management, National Association of Environmental Professionals, 18 Annual Conference Proceedings, May 24–26, 1993.

34. *Flint Ridge Development Co. v. Scenic Rivers Association*, 426 U.S.C. 776 (1976).

35. Mandelker D. R., *NEPA Law and Litigation*, Section 5.03(5)(b), Clark Boardman Callaghan, New York, 1998.

36. *Jones v. Gordon*, 792 F.2d 821, 826 (9th Cir. 1986).

37. *Gulf Oil Corporation v. Simon*, 373 F. Supp 1102 (D.C. 1974), affirmed, 502 F.2d 1154 (Emerg. Ct. App. 1974).

38. *Pacific Legal Foundation v. Andrus*, 657 F.2d 829 (6th Cir. 1981).

39. *South Dakota v. Andrus*, 614 F.2d 1190, 1193 (8th Cir. 1980).

40. *Nevada v. Watkins*, 943 F.2d 1080 (9th Cir. 1991).

41. Bear D., General Council, CEQ, Personal communications.

42. CEQ, *Preamble to Final CEQ NEPA Regulations*, 43 *Federal Register* 55978, November 29, 1978, Comments on § 1506.11.

43. Council on Environmental Quality, Memorandum for Federal NEPA Contacts, Emergency Actions and NEPA, Attachment #1, Emergency Alternative Arrangements Under the National Environmental Policy Act, September 8, 2005.

44. *Valley Citizens for a Safe Environment v. Vest* (D. Mass 1991).

6 The Threshold Question: When Is an EIS Required?

Indisputably, the single and most commonly cited provision of the National Environmental Policy Act (NEPA) involves the requirement to prepare a detailed statement:

> ... on proposals for legislation and other major federal actions significantly affecting the quality of the human environment ...[1]

Often referred to as the threshold question of significance, the importance of thoroughly understanding this requirement cannot be overstated as it determines whether an environmental assessment (EA) or an environmental impact statement (EIS) will be required for a particular action. Schedules, budgets, and the success of entire federal projects can rest on conclusions drawn from a review of this requirement.

The threshold question is predicated on a number of key components or criteria, and each of these must be met before the requirement is triggered as a whole. Each of these criteria, therefore, needs to be thoroughly understood before an informed decision can be made regarding the need to prepare an EIS or otherwise comply with NEPA. It should be noted that these criteria were not specifically defined in the Act. Instead, this task was left to the drafters of the NEPA implementing regulations (Regulations) and to the courts. A considerable amount of professional experience is frequently required in determining precisely what circumstances will trigger each one of these criteria.

Table 6.1 breaks the threshold requirement into its discrete criteria. Each criterion is cross-referenced according to where it is defined in the Regulations. These criteria are dissected and examined in detail in the following sections.

6.1 DETAILED STATEMENT

NEPA uses the phrase "detailed statement" in referring to the document that must be prepared for major federal actions significantly affecting the quality of the human environment. The Regulations use the term "environmental impact statement" or "statement" as a synonym for detailed statement.

6.2 PROPOSALS

The requirement to prepare an EIS pertains to proposals for legislation and other major federal actions (Table 6.1). The term proposals might at first appear straightforward, yet numerous challenges have centered on the precise meaning of this term. Three factors have been established by the Council on Environmental Quality (CEQ) for determining when a plan has matured to the stage where it can be considered an actual proposal (§ 1508.23):

- A federal agency has a goal
- The agency is actively preparing to make a decision on one or more alternative means of accomplishing the goal
- The effects can be meaningfully evaluated

TABLE 6.1
The Threshold Requirement

Key Definitions	Reference Location
Detailed statement	§ 1508.11
On proposals	§ 1508.23
For legislation and	§ 1508.17
Other major federal actions	§ 1508.18
Significantly	§ 1508.27
Affecting	§ 1508.3 and 1508.8
The quality of the human environment	§ 1508.14

Based on these three criteria it is clear that a proposal may exist, although the agency has not officially declared one to exist. As depicted by the third criterion, lack of ripeness is not to be mis-construed as a license to move forward, ignoring the requirements of NEPA. Agencies are expected to schedule proposals early enough in the planning process so that an EIS may be completed on time for it to be included in any recommendation or report on the proposal (§ 1502.5, § 1508.23).

6.3   LEGISLATION

The term "legislation" includes (§ 1508.17)

> … a bill or legislative proposal to Congress developed by or with the significant cooperation and support of a federal agency, but does not include requests for appropriations.

The test for "significant cooperation" hinges on whether the proposal is in fact predominantly that of a federal agency as opposed to another source. Legislative proposals include requests for ratification of treaties. These proposals are subject to special requirements described in § 1506.8 of the Regulations. Only the agency with primary responsibility for the subject matter involved is required to prepare a legislative EIS. However, drafting legislation does not, by itself, constitute significant cooperation.

6.4   THE TERM "MAJOR"

The courts have not completely agreed on the definition of the term "major." One of the early his-toric cases in NEPA involved plaintiffs who sued to enjoin (stop) timber sales until the Forest Ser-vice completed an EIS for the management of the area. The Forest Service argued that the phrase "major federal actions significantly affecting the quality of the human environment" created two tests: (1) determining first whether there is a major federal action and (2) determining whether the impact of that action on the environment is significant.

The Forest Service argued that the timber sales were not "major" federal actions.[2] The court concluded that the requirement "major federal actions significantly affecting the quality of the human environment" involved only a single criterion and that was sufficient to trigger an EIS since the two criteria were interwoven. As the court viewed it, "To separate the consideration of the mag-nitude of federal action from its impact on the environment … would [make it] possible to speak of a 'minor federal action significantly affecting the quality of the human environment,' and to hold NEPA inapplicable to such an action … the activities of federal agencies cannot be isolated from their impact on the environment."

However, a few courts have interpreted the term major to be a separate criterion independent of the term significantly, as used in Section 102 of the Act. In such cases, the courts have generally interpreted "major" to be an indicator of either the size or complexity of a project. Factors such as

TABLE 6.2
Examples of Major and Minor Actions

Examples of Actions Held to Be Major	Examples of Actions Held to Be Minor
• A $14 million bridge with 60% federal funding • Conversion of a large federally subsidized housing project with a major change in its use • A 66-mile water channel project costing $1.5 million with $700,000 of federal funding	• A replacement bridge • Demolition if a historic building for which $25,000 of federal funding was committed • Retrofitting 20,000 railway cars to meet federal safety regulations • Transferring a small group of employees and one of the agency functions from an agency field office • Minor traffic improvements

funding levels, allocation of resources, and degree of planning have all been used as indicators for determining if a particular project is considered a major action.

Mandelker has identified examples of activities which the courts have held to be major or minor (see Table 6.2).[3] The reader is cautioned that such examples do not necessarily reflect their significance.

The CEQ, as well as most courts, has taken the position that the term "major" is interpreted to reinforce the term "significantly" but does not have a meaning independent of it (§ 1508.18). Under this interpretation, the actual size or complexity of a project has little bearing in determining if it is an action that may significantly impact the environment.[4]

This second interpretation stems from the view that if an action results in a significant impact, the action is essentially a major action.[5] Such a position avoids potential dilemmas that may arise when an EIS is required for a minor action that results in a significant impact. It also avoids dilemmas where environmental impacts of a major federal action are deemed to be nonsignificant.

6.5 THE TERM "FEDERAL AGENCY"

As defined by the Regulations, the term federal agency includes all agencies of the federal government. It does not include "... the Congress, the Judiciary, or the President, including the performance of staff functions for the President in his Executive Office" (§ 1508.12).

The meaning of "federal" might at first appear to be relatively straightforward. Yet, in some instances, actions undertaken by a nonfederal agency may still be subject to the requirements of NEPA (also referred to as the small federal handle). In recent years, an effort has been under way to privatize many facilities and operations that traditionally have been operated and carried out by federal agencies. This effort has raised many issues with respect to NEPA compliance.

In some circumstances, what would otherwise be considered a nonfederal action may be federalized with respect to NEPA. For example, in one case, a federal agency entered into a contract with a private entity to provide power for a large private project. The federal agency agreed to build a transmission line and to supply power to the private party sponsoring the project. The agency claimed that an EIS was not required since this was a private action. The court ruled that the contract with the private entity had essentially federalized the entire project for the purposes of NEPA. This federalization was of such an extent that the agency was ordered to prepare an EIS to evaluate the impacts of the private plant in addition to those of the transmission line.[6]

6.5.1 FACTORS THAT MAY FEDERALIZE AN ACTION

Three principal factors have been used by the courts in determining if federal agency involvement has made federalized what would otherwise be considered a nonfederal action.[7]

If the involvement is

- supported by a federal contract, grant, loan, or other financial assistance;
- enabled through a federal lease, license, permit, or other entitlement;[8] and
- caused federally.

6.5.1.1 Federal Support by Contract, Grant, Loan, or Financial Assistance

Actions supported by federal payment for services rendered can be viewed as requiring NEPA review.[9] In cases where federal funding has subjected a state or private project to the requirements of NEPA, the funding has been considered not only to be generally active, as opposed to a passive deferral of payment, but also programmatic, in the sense of being provided primarily to further a policy goal of the funding agency.[10] Normally, a substantial percentage or amount of federal funding is necessary to trigger an NEPA review.

Some nonfederal actions that are funded from federal general revenue have been considered to be federal if a federal agency governs how the funds are used. As with federal payment for services rendered, NEPA is required when massive federal financial assistance has been given to a state or private project.[11] Accordingly, the federal government becomes accountable under NEPA for its actions.[12]

Conversely, nonfederal actions have not been federalized when indirect funding seemed marginal at most and where federal officials had no decision-making role.[13] For example, federal participation in a beetle eradication project in California was not sufficient to trigger NEPA compliance, although three federal officials were part of an eight-member board that made recommendations to the state on eradicating a Japanese beetle pest infestation. The court reasoned that the eradication project was not federally funded because the traveling expenses of the participating federal officials were paid by the state, and the eradication project was a state project. The court considered that for the duration of the board meetings, the salaries of the participating federal officials were not reimbursed by the state but still supported a ruling that NEPA did not apply.

6.5.1.2 Enablement by Permit, Lease, License, or Entitlement

When a federal agency has discretion in its enabling decision to consider environmental consequences and when that decision forms the legal predicate for another party's impact on the environment, preparation of NEPA documentation is warranted because the agency has substantially contributed to the environmental impact.[14] The Regulations reinforce the concept that enablement involves the execution of a required federal action that enables a private party to pursue an action. The definition of a major federal action includes granting of permits or other regulatory decisions as well as federal and federally assisted activities. Federal actions that amount to less than a legal precondition are noticeably omitted from this definition.[15]

Enablement is demonstrated by a case involving the Department of Agriculture that was required to prepare an EIS to approve logging operations by a private company. Evidence demonstrated that the federal agency had a responsibility greater than a ministerial act of approval. Not only had it extended logging contracts and modified other contracts, but also it had a financial interest in the lumber acquired. The actions of the federal agency enabled the logging operations in the area to be undertaken, but most significantly, the federal agency was legally obligated by contract to give its approval to the project before it could proceed.[16]

In some cases, federal approval of a private party's project, where that approval is not required for the project to go forward, has not constituted a federal action.[17] However, a nonfederal entity may create a federal action if it consents to federal regulation or grants to a federal agency the ability to control the outcome of the proposed project.[18] A distinguishing characteristic of federal involvement is the ability to influence or control the outcome of a nonfederal project in some material respect.

6.5.1.3 Federal Control

A federal action such as federal approval of a lease, license, permit, or other entitlement that enables a private or state action to take place may be subject to NEPA. In such circumstances, overt federal agency action in furtherance of the nonfederal project is 'federalized' for the purposes of NEPA.[19] For example, the Pueblo Indians leased restricted Indian lands to a development company, and the Bureau of Land Management (BLM) approved this lease.[20] Emphasizing Congress' concern for environmental protection, the court held that BLM approval constituted a major federal action, although the federal government neither initiated the lease nor participated in it financially.[21]

6.5.1.4 Continuing Agency Involvement That Is Modified or Terminated

Some courts also consider whether there is continuing agency involvement in a challenged project such that termination or modification of the agency involvement would terminate or significantly impact the project. Since NEPA only requires federal agencies (not states or private parties) to consider the environmental impacts of their proposed actions, nonfederal actions must sufficiently involve a federal action before it is subject to NEPA.[22]

6.5.1.5 Causation

A nonfederal action may be federalized if the nonfederal action would not otherwise take place were it not for specific actions undertaken by a federal agency. That is, "but for" the federal action, the nonfederal action would not occur. These "but for" actions, by themselves, do not necessarily trigger the requirements of NEPA. Rather, the federal action must also be substantially interrelated to the otherwise nonfederal action.[23]

6.5.2 BASIS FOR A GENERAL-PURPOSE TOOL

Table 6.3 summarizes the case law criteria described above for determining when a nonfederal project becomes federalized for the purposes of NEPA. These criteria provide the basis for the general-purpose tool presented in Figure 6.1, which can be used by decision-makers in determining whether a nonfederal action has been federalized for the purposes of NEPA.[7]

6.5.3 GENERAL-PURPOSE TOOL FOR DETERMINING WHEN NONFEDERAL ACTIONS BECOME FEDERALIZED

Consistent with the rule of reason, the logic diagram presented in Figure 6.1 is based on the criteria established in Table 6.3. The tool, developed by the author and an environmental lawyer, is specifically designed to provide practitioners and decision-makers with a rigorous and

---

TABLE 6.3

Criteria for Determining When Nonfederal Entities May Become Federalized

- Would the nonfederal action involve a substantial degree of financial support by way of a federal contract, grant, loan, or other financial assistance?
- Would the nonfederal action be enabled through a federal lease, license, permit, or other entitlement?
- Would the nonfederal action involve a substantial degree of federal control?
- Is there continuing federal involvement in a nonfederal action to such an extent that termination or modification of this involvement would terminate or significantly impact the nonfederal project?
- Is the federal action substantially interrelated with a nonfederal action to such an extent that "but for" the federal action, the nonfederal action would not take place?

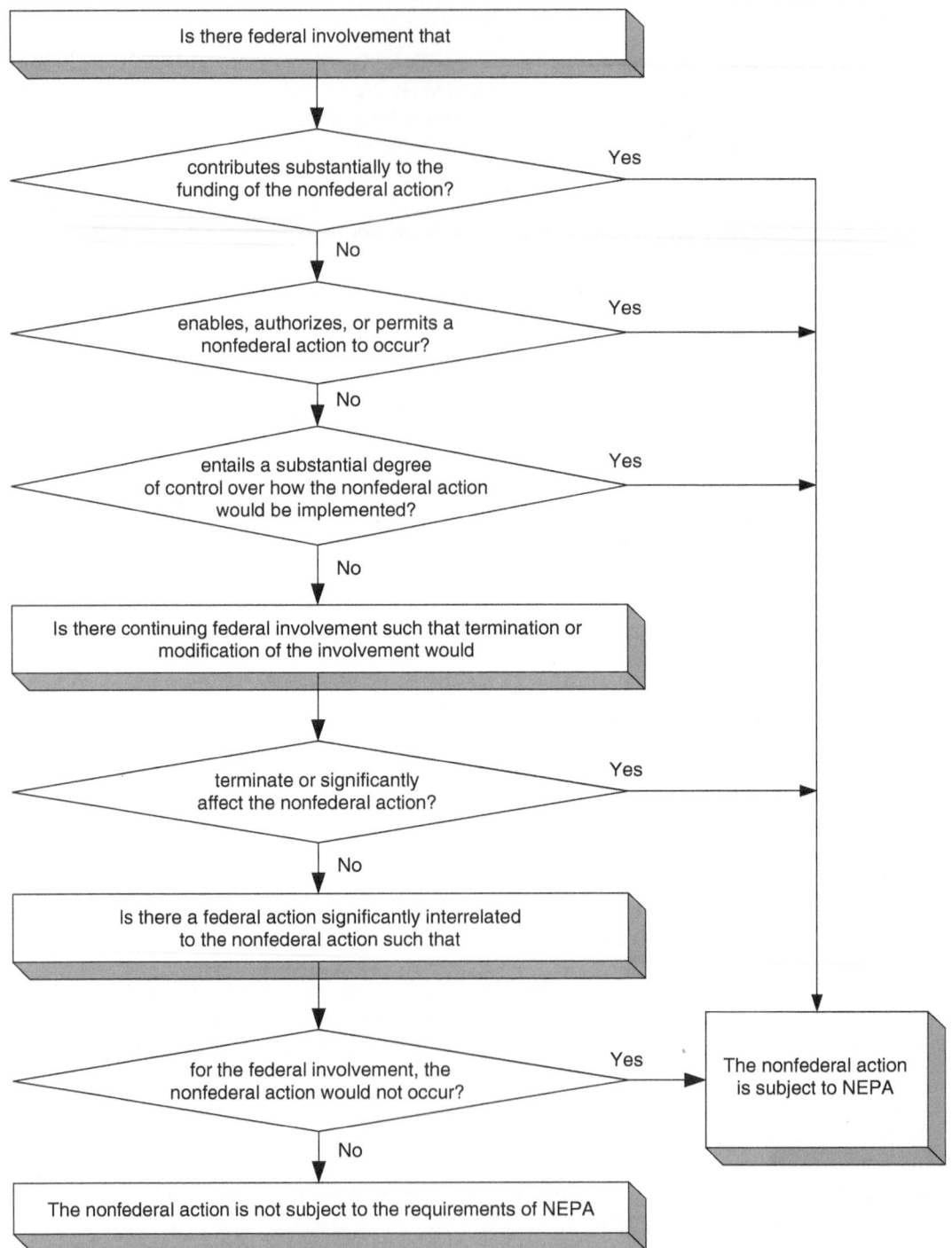

FIGURE 6.1   Tool for determining when a nonfederal action is subject to NEPA's requirements.

systematic procedure to determine if a nonfederal action has become federalized, triggering NEPA's requirements.

Although this tool does not totally eliminate the subjectivity inherent in making NEPA determinations, it provides a valuable technique for substantially reducing subjectivity. It does not promote

TABLE 6.4
Four Categories of Federal Actions

- *Adoption of a Policy.* This category involves actions such as issuing rules, regulations, and interpretations. Adoption of a policy may also involve entering into treaties and international conventions or agreements, or issuing formal documents establishing an agency's policies that will result in or substantially alter agency programs.
- *Adoption of Formal Plans.* Adoption of formal plans may include activities such as issuing official documents that guide or prescribe alternative uses of federal resources and provide the basis for future agency actions.
- *Adoption of Programs.* This category includes actions such as adopting a group of concerted actions to implement a specific policy or plan. Systematic and connected agency decisions allocating agency resources to implement a specific statutory program or executive directive also fall under the heading of agency programs.
- *Approval of Specific Projects.* Adoption of specific projects may include actions such as construction or management activities located in a defined geographic area. Projects also include actions such as approving permits and other regulatory decisions as well as federal and federally assisted activities.

any degree of decision-making beyond the level already exercised in making such determinations in the first place; instead, it provides decision-makers with a rigorous, systematic, and defensible approach for reaching such determinations.

Because this tool is intended to be used as a general-purpose decision-making tool, it may not cover every conceivable condition. Technical aspects of the case law summarized earlier should thus be considered in responding to each of the tests.*

6.5.3.1 Using the Tool

Begin at the top of Figure 6.1 by answering the first question: "Is there federal involvement that contributes to the funding of the non federal action?" If the response is no, the decision-maker continues down through the remaining tests. A "no" answer to all of the tests supports a decision that the nonfederal action is not subject to the requirements of NEPA. A "yes" answer to any single test is sufficient to support a decision that the nonfederal action is subject to the requirements of NEPA.

6.6 ACTIONS

As described in more detail in Chapter 9, actions include "… projects and programs entirely or partly financed, assisted, conducted, regulated, or approved by federal agencies; new or revised agency rules, regulations, plans, policies, or procedures; and legislative proposals" (§ 1508.18[a]). They also include activities that are regulated, assisted by, or require the approval of a federal agency. These actions include both new and continuing activities.[24] A brief description of each of these categories is presented in Table 6.4. Taken together, these categories are sometimes referred to as the "4 Ps."

The Regulations specifically call out only two circumstances where federal actions are not subject to the requirements of NEPA. For the purposes of NEPA (§ 1508.18), these were

- funding assistance solely in the form of general revenue sharing funds, distributed under the State and Local Fiscal Assistance Act of 1972, with no federal agency control over the subsequent use of such funds (*note*: this caveat is no longer applicable);
- bringing judicial or administrative, civil or criminal enforcement actions.

---

* Case law varies among various judicial circuits and courts. Specific questions should be referred to legal counsel.

6.6.1 Inaction

Does NEPA apply to nonactions? This question was addressed in a case where a federal agency had the capability to inhibit a nonfederal action but did not exercise this authority.[25] When challenged, the court concluded that NEPA does not apply where "… an agency has done nothing more than fail to prevent the other party's action from occurring …"

In general, an action is not subject to the requirements of NEPA when a federal agency has an option to act and decides not to do so. However, where a federal agency has a mandatory responsibility to act but fails to do so, this failure to act may constitute an action subject to NEPA.[26]

6.6.2 Applicability of NEPA to International Actions

Federal actions conducted outside the borders of the United States are referred to as extraterritorial actions. Common examples of extraterritorial actions include federal assistance in the construction of highways or dams and licenses for the export of nuclear fuel. The applicability of NEPA to extraterritorial actions is complex and has been the subject of ongoing controversy.

It involves the consideration of two separate issues. Specifically, does NEPA extend to (1) the global commons or (2) the domain of foreign nations? The global commons is generally understood to include such portions of the earth as the oceans, Antarctica, and the upper atmosphere that are understood to be held in common by all nations.

The Act does not place either explicit or implicit limits on the applicability of NEPA to activities conducted outside U.S. borders.[27] For this reason, NEPA has been interpreted by some to extend over U.S. involvement in international actions. This interpretation has its basis partly in the term human environment that is used in Section 102 of NEPA. As used in this context, the term human environment does not appear to limit the requirements of NEPA to the geographical borders of the United States. Moreover, Section 102(2)(f) places specific responsibilities on federal agencies to

> recognize the worldwide and long-range character of environmental problems and, where consistent with the foreign policy of the United States, lend appropriate support to initiatives, resolutions, and programs designed to maximize international cooperation in anticipating and preventing a decline in the quality of mankind's world environment.

Professor Lynton Keith Caldwell, the principal father of NEPA, has indicated that applicability of the Act was intended to include federal actions outside the boundaries of the United States.[28] Because of potential ramifications on U.S. foreign policy, some agencies, including the U.S. State Department, have raised objections to this interpretation and have been reluctant to apply NEPA to activities beyond U.S. borders.[29]

6.6.2.1 Executive Order

In addressing this issue, President Carter issued an executive order providing direction for applying NEPA to extraterritorial actions.[30] This order, drafted in consultation with the CEQ and the U.S. State Department, discusses the scope of NEPA's applicability to international activities and provides direction and procedures for implementing its requirements.

In the author's opinion, this order is confusing and poorly crafted. Surprisingly, it focuses on the preparation of EAs, defining only two cases where an EIS is applicable to extraterritorial actions.[31] Under this order, preparation of an EIS is required for major federal actions significantly affecting the environment of the global commons. However, this order exempts an EIS from having to address impacts on the environment of a foreign nation.

6.6.2.2 Transboundary Effects

The CEQ has developed guidance for assessing the effects of proposals within the United States as well as its territories and possessions that may have transboundary effects and affect another country's environment.[32] While this guidance was developed primarily in the context of negotiations undertaken with the governments of Mexico and Canada to develop an agreement on transboundary environmental impact assessment in North America, the guidance pertains to all federal agency actions that are normally subject to NEPA, whether covered by an international agreement or not.

CEQ's guidance does not expand the range of actions to which NEPA currently applies, nor does it apply to so-called extraterritorial actions (i.e., U.S. actions that occur in another country or otherwise outside the jurisdiction of the United States). Instead, it pertains only to those proposed actions currently covered by NEPA that would take place within the United States and its territories.

This guidance is consistent with long-standing principles of international law. Since the Trail Smelter Arbitration of 1905, it has been a customary law that no nation may undertake acts on its territory that will harm the territory of another state. Moreover, this rule of customary law has been recognized as binding in Principle 21 of the Stockholm Declaration on the Human Environment) and Principle 2 of the 1992 Rio Declaration on Environment and Development. Under these provisions, states have the duty to give notice (including preparation of environmental impact assessments) to others to avert potential harm from the actions they take. Assessing transboundary impacts of federal agency actions that occur in the United States is therefore an appropriate step toward implementing those principles.

NEPA case law has reinforced the need to analyze impacts regardless of geographic boundaries within the United States and has also assumed that NEPA requires analysis of federal actions that take place entirely outside the United States but could have environmental effects within the United States. Courts that have addressed impacts across U.S. borders have assumed that the same rule of law applies in a transboundary context.

Under CEQ's guidance, agencies are consequently expected to include analysis of reasonably foreseeable transboundary effects in EAs or EISs prepared for federal actions undertaken within the United States.

6.6.2.3   Case Law

As the courts have been less than definitive in their rulings, the issue of extraterritorial actions is somewhat confusing. What some courts appear to be suggesting is that NEPA is applicable to actions occurring in the global commons but not to actions taken within the borders of other sovereign nations. Even if this is the intent, few EISs either rigorously consider or evaluate extraterritorial actions or transboundary impacts, or effects on the global commons. Six different cases are described below. It is recommended that the reader consult with legal counsel in determining to what extent transboundary issues may need to be addressed in NEPA analysis.

U.S. Naval Bases in Japan. In 1993, a district court found that NEPA did not apply to U.S. naval operations at three bases in Japan. The court ruled that an EIS was unnecessary because plausible assertions were made that the preparation of an EIS would have impact on the U.S. foreign policy. In the court's view, foreign policy interests outweighed the benefits to be gained from preparing an EIS.[33]

Naval Activities in Exclusive Economic Zones. The navy's littoral warfare advanced development program (LWAD) involved testing experimental technologies, including active sonar at sea. The scientific community is generally in agreement that high-intensity underwater sounds such as those generated by active sonar can adversely affect whales, dolphins, and other marine life.

Most of these tests were conducted on the high seas or within the U.S. exclusive economic zone (EEZ). The EEZ is a zone extending seaward from the boundary of the territorial sea out to a distance of 200 miles. The navy prepared an overseas environmental assessment for every sea test and in each case concluded that the impacts were insignificant.

A plaintiff sued seeking to enjoin (stop) the navy from conducting further sea tests until the navy completed a programmatic NEPA document for the LWAD program.[34] The navy argued that because some of the tests take place in international waters, NEPA does not apply to activities under the program.

The court found that the presumption against the extraterritorial application of U.S. laws did not apply because the planning for the LWAD program occurred entirely within the boundaries of the United States. In the eyes of the court, the federal activity regulated by NEPA is the decision-making process of the agencies, not the underlying project. Because the decision-making process surrounding the approval of sea tests occurred within the United States, the application of NEPA to the LWAD sea tests was not an extraterritorial action.

The court distinguished this suit from others with different rulings by concluding that the rationale in other cases for finding that NEPA did not apply to particular actions was that its application would either have important foreign policy implications or would demonstrate a lack of respect for another nation's sovereignty.

Furthermore, the court reasoned that regarding natural resource conservation and management, "the United States does have substantial, if not exclusive, legislative control of the EEZ." As a result, the court held "that NEPA applies to federal actions which may affect the environment in the EEZ."

Johnston Atoll. In 1990, a court examined the extraterritorial applicability of NEPA to the removal, transportation, and destruction of chemical weapons stored in the Federal Republic of Germany. Under an international agreement, the Department of the Army undertook a joint plan with the West German Army to remove the weapons and to transport them to Johnston Atoll, a U.S. territory in the Pacific Ocean, for treatment and disposal.

The U.S. Army prepared two separate EISs, one for the disposal of the weapons stockpile stored in Germany and the other for construction, operation, and treatment of an incinerator located on Johnston Atoll. Pursuant to Executive Order No. 12114, the army also prepared a global commons EA which analyzed the impacts of the munitions shipment from Germany to Johnston Atoll. However, no NEPA analysis was prepared to evaluate the movement of the munitions within Germany.

Plaintiffs filed a suit against the U.S. Army to prevent the movement of the munitions to Johnston Atoll on the grounds that the U.S. Army had failed to prepare a comprehensive EIS covering all aspects of transportation and disposal of the German stockpile.[35]

The district court concluded that "it is not convinced that NEPA applies extraterritorially to the movement of munitions in Germany or their transoceanic shipment to Johnston Atoll." While the court recognized that "the language of NEPA indicates that Congress was concerned with the global environment and the worldwide character of environmental problems," it reasoned that actions taken under NEPA "should be taken 'consistent with the foreign policy of the United States.'" In the court's words, "Congress intended to encourage federal agencies to consider the global impact of domestic actions and may have intended under certain circumstances for NEPA to apply extraterritorially."

Notwithstanding, the court concluded that NEPA did not apply to actions taken within Germany. In reaching this decision, the court wrote that it "... must take into consideration the foreign policy implications of applying NEPA within a foreign nation's borders to affect decisions made by the President in a purely foreign policy matter." Further, the court reasoned that imposing a requirement to assess environmental impacts on actions within Germany would "... encroach on the jurisdiction of Germany to implement a political decision which necessarily involved a delicate balancing of risks to the environment and the public, and the ultimate goal of expeditiously ridding West Germany of obsolete chemical munitions."

With respect to the transoceanic phase of the action, the U.S. Army had prepared an EA pursuant to Executive Order No. 12114. On this point, the court wrote that it could not "... conclude, as defendants would suggest, that Executive Order 12114 preempts application of NEPA to *all* federal agency actions taken outside the United States...." However, in these particular circumstances, the court was persuaded that NEPA did not require the U.S. Army to consider the global commons portion of the action in the same EIS that covered the Johnston Atoll facility.

McMurdo Station in Antarctica. Plaintiffs challenged the National Science Foundation's (NSF) plan to incinerate waste at McMurdo Station in Antarctica. Essentially, they argued that NEPA applies extraterritorially and thus, the NSF should have prepared an EIS.[36]

In this case, the court overturned an earlier decision which had held that despite NEPA's broad mandates there was no clear congressional intent that NEPA should apply beyond U.S. borders and that NEPA, therefore, did not apply to NSF's decision to build an incinerator in Antarctica. The court held instead that application of NEPA to federal actions is not limited to actions occurring or having effects within U.S. borders. Rather, NEPA is designed "to control the decision-making process ... not the substance of agency decisions" that takes place almost exclusively within the United States. Thus, the court held that NEPA did apply to NSF actions in Antarctica.

Imperial-Mexicali Transmission Lines. The U.S. District Court for the Southern District of California decided in favor of the Department of Energy (DOE) and the BLM in a suit brought by the Border Power Plant Working Group. On November 30, 2006, the court found that the EIS for the Imperial-Mexicali transmission lines was adequate and that the agencies had not violated the Clean Air Act (CAA) by failing to prepare a conformity determination (see Chapter 4).[37] At issue were permits for transmission lines to carry electricity into the United States from two new power plants in Mexico. The DOE issued permits for transmission lines at the U.S.–Mexico border. The BLM issued permits for the power lines to cross land it manages in California. The plaintiff alleged that DOE/BLM violated the CAA by failing to prepare a conformity determination.

The government contended that (1) the conformity determination is not required for the emissions from the power plants because these emissions occur in Mexico and not in the Imperial County nonattainment area and (2) issuance of the presidential permits for the cross-border transmission lines is a foreign affairs function exempt from the conformity requirements.

In its ruling, the court found that the DOE did not have to consider emissions from outside Imperial County in a conformity determination. Regarding the second point, the court found that the DOE did not need to consider emissions from the power plants in Mexico, sources that are permitted and regulated by a foreign government. However, the court disagreed with the DOE's claim that it was exempt from the requirements because issuance of the permits for the transmission lines in the United States is a foreign affairs function.

The plaintiff had also alleged that the EIS failed to ensure the scientific accuracy of information in the consideration of alternative cooling technologies. However, the court viewed the challenges to the treatment of alternatives as "a battle of experts," in which "an agency must have discretion to rely on the reasonable opinion of its own qualified experts." The court refused to "flyspeck" minor technicalities in the EIS in the light of its "comprehensive discussion of the proposed actions and their environmental impacts."

Court Finds Transboundary Impacts in Mexico Do Not Need to Be Considered. The U.S. District Court for the District of Nevada recently ruled that a supplemental EIS is not needed for a proposal where it found the potential environmental impacts to be too speculative and beyond the U.S. control.[38] This ruling stems from a challenge filed in 2005 to the Bureau of Reclamation's final authorization of the All-American Canal Lining Project. The 80-mile-long canal carries water from the Colorado River in Arizona to southern California. Seepage from the unlined canal reduced the amount of water available to users in California but contributes to recharge of an aquifer that underlies the Mexicali Valley in Mexico.

The Bureau completed an EIS in 1994 and decided to line the canal, thereby reducing seepage and providing more irrigation water to users in California. A decade later, however, this project still had not commenced. In January 2006, the Bureau issued a supplemental information report concluding that no substantial change, or significant new information or circumstances, existed that required preparation of a supplemental EIS.

Plaintiffs charged, among other things, that the Bureau violated NEPA by not preparing a supplemental EIS to address alleged significant new information regarding

- a wetland in Mexico and its value as habitat for an endangered species;
- socioeconomic impacts in Mexicali, Mexico, and across the border in the United States; and
- potential impacts to the Salton Sea, a 376-square-mile lake located in a southern California desert ecosystem.

The court divided the plaintiffs' allegations into ones dealing with impacts in Mexico and those with effects within the United States. With respect to transboundary impacts, the court concluded that "… because the impacts in Mexico are beyond agency control and their impacts within the United States are too speculative, NEPA's 'rule of reason' did not require …" the Bureau to prepare a supplemental EIS.

The court's review of allegations related to domestic impacts centered on the Bureau of Reclamation's 2006 Supplemental Information Report. In this case, the court found that the analysis in the report was sufficient and concluded that a supplemental EIS was not required.

6.7 SIGNIFICANCE

Arguably, the concept of significance is the single most complex, elusive concept in NEPA. Probably no other concept has elicited as much confusion or litigation. A thorough understanding of this concept is essential because it establishes the threshold between the relatively simple task of preparing an EA and that of the much more involved process of preparing an EIS.

The term "insignificant" implies that the magnitude of an impact is zero. In contrast, the term "nonsignificant" implies that while the impact is not significant, it may still have some measurable environmental effect. Because most environmental disturbances are either not zero, or cannot be mitigated to the point of zero, many practitioners prefer the term "nonsignificant" over "insignificant."

Experts, let alone the public, often disagree on the significance or nonsignificance of an impact. To a certain extent, the interpretation of significance is in the eye of the beholder.

6.7.1 THE ROLE OF THE COURTS IN DETERMINING SIGNIFICANCE

Unfortunately, guidance from the courts has been so narrowly defined that it generally lacks applicability to situations beyond very restricted circumstances. In many instances, the courts have done little more than redefining significance in terms of other equally enigmatic concepts or wording. For example, the courts have variously defined significantly to mean "not trivial," "appreciable," "important," and "momentous."

In an early precedent-setting case, plaintiffs challenged an EA prepared for the construction of a jail and other facilities in New York City.[39] The EA described a number of environmental impacts and concluded that the project was not an action having a significant environmental impact. In reviewing this case, the court concluded the following:

1. Most major federal actions, no matter how limited in scope, have some adverse effect on the human environment. Congress could have decided that every major federal action should be the subject of an EIS. However, by adding "significantly," Congress raised the bar, beyond that which might simply be required for any major federal action.
2. CEQ guidelines suggest that an EIS should be prepared where the impacts are controversial, referring not to the amount of political or public opposition, but to where there is a substantial technical dispute as to the size, nature, or effect of the major federal action.
3. In deciding whether a major federal action would significantly affect the environment, an agency should be required to review the proposed action in light of the extent to which it

would cause adverse environmental effects in excess of those created by existing uses in the area as well as the cumulative harm that results.

4. Agencies must develop a reviewable environmental administrative record for the purposes of a threshold determination under Section 102(2) (C).

6.7.2 REGULATORY DEFINITION

Section 1508.27 of the Regulations states that the intensity as well as the context in which an impact would take place must be considered in making a determination of significance. Beyond this, the Regulations provide little substantive direction for making such determinations.

6.7.2.1 Context

Experience has shown that decision-makers sometimes focus an inordinate amount of attention on intensity, sometimes to the extent that the context of a proposed action is excluded. The requirement to consider context acknowledges that the setting or location of an environmental disturbance can have an important bearing on conclusions regarding significance. Specifically, the Regulations state

> ... the significance of an action must be analyzed in several contexts such as society as a whole (human, national), the affected region, the affected interests, and the locality. Significance varies with the setting of the proposed action. For instance, in the case of a site-specific action, significance would usually depend upon the effects in the locale rather than in the world as a whole. Both short- and long-term effects are relevant (§ 1508.27[a]).

A proposed power plant, for example, might have a much greater impact on both the environment and human health if it is located in the middle of a large metropolitan area that already has substantial air quality problems, rather than if it is sited in a more remote area. Similarly, the impacts of a proposed airport might be very significant if it is located near a populated area as opposed to a remote location.

6.7.2.2 Intensity

Intensity is a measure of the degree or severity of an impact. The CEQ has established 10 factors (significance factors) that should be considered in evaluating the intensity (§ 1508.27[b]). As shown in Table 6.5, these factors are to be considered in terms of the context in which the impacts would occur.

An agency cannot necessarily determine that impacts are insignificant simply because the action is considered to be temporary (§ 1508.27[b][7]). Although the impacts directly attributable to a proposed action may not be significant, the impacts of other related actions may result in a determination of significance. It is equally important that one should not try to avoid a determination of significance by segmenting or "breaking a project down into smaller component parts," which individually do not have a significant impact (§ 1508.27[b][7]).

Compliance with Regulatory Standards. An impact is likely to be deemed significant if an applicable environmental standard or requirement is threatened or breached. This reasoning is captured in significance factor #10. For example, the CAA regulations define limits on hazardous air emissions. Accordingly, a hazard treatment plant is likely to be viewed as resulting in a significant impact if it would cause the local air quality to exceed an established prevention of significant deterioration level.

Care should be exercised in assessing significance with respect to local standards (e.g., noise standards) because some local communities have established unrealistic and, in some cases, virtually unattainable standards. This being the case, decision-makers may want to address significance

TABLE 6.5

Intensity Factors for Evaluating Significance

1. Impacts that may be both beneficial and adverse; a significant effect may exist even if the federal agency believes that, on balance, it will be beneficial.
2. The degree to which the proposed action affects public health or safety.
3. Unique characteristics of the geographic area such as proximity to historic or cultural resources, park lands, prime farmlands, wetlands, wild and scenic rivers, or ecologically critical areas.
4. The degree to which the effects on the quality of the human environment are likely to be highly controversial.
5. The degree to which the possible effects on the human environment are highly uncertain or involve unique or unknown risks.
6. The degree to which the action may establish a precedent for future actions with significant effects or represents a decision in principle about a future consideration.
7. Whether the action is related to other actions with individually insignificant but cumulatively significant impacts. Significance exists if it is reasonable to anticipate a cumulatively significant impact on the environment. Significance cannot be avoided by terming an action temporary or by breaking it down into small component parts.
8. The degree to which the action may adversely affect districts, sites, highways, structures, or objects listed in or eligible for listing in the National Register of Historic Places or may cause loss or destruction of significant scientific, cultural, or historical resources.
9. The degree to which the action may adversely affect an endangered or threatened species or its habitat that has been determined to be critical under the Endangered Species Act of 1973.
10. Whether the action threatens a violation of federal, state, or local law or requirements imposed for the protection of the environment.

in terms of either state standards or federal standards. In such instances, it is recommended that the decision-maker carefully document why the state or federal standards were used in lieu of the local standards.

Misconceptions. Decision-makers frequently focus a large, perhaps even excessive, amount of attention on determining whether an action would violate existing regulatory standards (significance factor #10); a common misconception, particularly among project engineers, is that no significant impacts will occur as long as a project complies with all applicable environmental laws and regulations. However, many actions (hydroelectric dams, power stations, federal facility siting) can extract sizable impacts even though the project complies with all applicable laws and regulations. Moreover, even though an action does not violate any existing environmental requirements, the context as well as the remaining nine significance factors are often sufficient to support a determination of significance. For instance, an emission that conforms to applicable CAA standards can still be considered significant should it be regarded as highly controversial (significance factor #6), or by virtue of its proximity to unique characteristics (significance factor #3), or other considerations.

Controversy. Significance depends on the degree to which impacts are considered to be highly controversial (significance factor #4, Table 6.5). This significance factor deserves special mention as its application is widely misunderstood. The courts have ruled that this factor refers primarily to controversy of a technical or scientific nature rather than to mere opposition or controversy of a political nature. As expressed by one court:[40]

> ... cases where a substantial dispute exists as to the size, nature, or effect of the major federal action rather than to the existence of opposition to a use, the effect of which is relatively undisputed.... The suggestion that "controversial" must be equated with neighborhood opposition has also been rejected by others.

Scientific versus Political Controversy. Scientific controversy may involve disagreements over a specific methodology or approach used in analyzing impacts, data used in the study, or interpretations of the environmental impacts. As demonstrated by the following court cases, a NEPA document that fails to disclose and analyze important but differing scientific opinions (e.g., scientific controversy) is defective.

In the case of *Foundation for North American Wild Sheep*, the court concluded that the case involved true scientific controversy because numerous scientists and knowledgeable individuals were highly critical of a particular EA and disputed its findings.[41] However, in the case of *Friends of Endangered Species*, the court disagreed with the plaintiffs, finding that there was nearly unanimous agreement within the scientific community regarding the contents and conclusions presented in the EA.[42]

When faced with conflicting views, agencies have discretion to rely upon their own experts. As one court explained, "When specialists express conflicting views, an agency must have the discretion to rely on the reasonable opinions of its own qualified experts even if, as an original matter, a court might find contrary views more persuasive."[43]

Potential Accidents. The Regulations do not explicitly mention the significance of impacts that could result from potential accidents. To complicate matters, potential accidents that have a very remote chance of occurring can be particularly difficult to assess. However, the author suggests that the following significance factors may be considered when estimating the extent of the impacts that could result from potential accidents and used in assesing the significance of a potential accident:

- The degree to which the proposed action affects public health or safety (§ 1508.27[b][2])
- The degree to which the effects on the quality of the human environment are likely to be highly controversial (§ 1508.27[b][4])
- The degree to which the possible effects on the human environment are highly uncertain or involve unique or unknown risks (§ 1508.27[b][5])
- Whether the action threatens a violation of federal, state, or local law or requirements imposed for the protection of the environment (§ 1508.27[b][10])

The reader is referred to Chapter 10 for additional information on assessing accidents.

Beneficially Significant Impacts. An action may result in a significant environmental impact even if it is believed that, on balance, the effect will be beneficial (§ 1508.8[b], § 1508.27[b][1]). Thus, an action that would result in a significant beneficial impact (with no significant adverse impacts) may still be subject to an EIS. It might at first appear to be unreasonable to require preparation of an EIS for an action that would significantly improve environmental quality. However, it may be difficult if not impossible, to demonstrate that an action would actually result in a significant beneficial impact (with no significant adverse side effects) without first preparing an analysis to review thoroughly the direct, indirect, and cumulative impacts. An action that is substantial enough to significantly improve the environment might also involve hidden or unknown adverse impacts that can be adequately identified only through preparation of a detailed analysis.

Additional Factors. It is important to point out that the significance factors presented in Table 6.5 include some, but not necessarily all, of the factors that might need to be considered in assessing significance. Where appropriate, other relevant factors may also need to be considered before reaching a final decision. Consistent with the rule of reason and regulatory and statutory provisions, the author proposes 10 additional factors that might also be considered in assessing significance (Table 6.6). Some of these factors are closely related to existing significance factors provided in the Regulations while others are relatively novel.

Consideration of the additional significance factors suggested in Table 6.6 would strengthen NEPA and its goal of protecting environmental quality. One of these suggested factors (#4) requires special mention in the next section.

TABLE 6.6
Additional Significance Factors

1. *Multiple Nonsignificant Impacts.* The degree to which a multiple number of different and substantial but individually nonsignificant impacts affect the environment. For example, consider a proposal which results in air emission, waste effluents, degradation of a visual resource, and copious generation of waste. Assume that none of these impacts actually breach an environmental standard (or other CEQ significance factor), and each is individually considered to be nonsignificant and yet, collectively, these four distinct effects could be deemed to constitute an overall significant impact. As this represents a diverse set of impacts, each affecting a different resource, this concept should not be confused with that of a cumulative impact.

2. *Low Magnitude Impact across a Large Spatial or Temporal Domain.* The degree to which the intensity of a normally nonsignificant impact affects a large spatial or temporal domain. For example, consider a proposed 15-mile pipeline that does not significantly harm or affect a species or other environmental resources. However, while its effect on visual resources is considered to be marginal at any particular location, the effect is summated across a long geographic area. Another example could involve a project that will result in a small but long-term (20-year) increase in the ambient noise level. While this increased noise level does not breach any existing noise standards and may not be technically significant in the traditional sense, its long-term contribution might still be deemed to be significant.

3. *Inconsistencies with Existing Land Uses.* The degree to which an action is inconsistent with existing use or with land use policies or plans. This guidance is consistent with existing case law and is grounded in the supposition that if a proposed action is consistent with existing land uses, its adverse impacts are likely to be less significant. In one case, a court used an example of a plan to construct an additional highway in an area already honeycombed with roads and highways. In this case, the court reasoned that construction of an additional highway in such an area probably would have fewer adverse impacts than if it was constructed in a roadless area. This factor also has its basis in language from the NEPA regulations (§ 1502.16[c]) and is related to the concept of "context" (§ 1508.27[a]).

4. *Significant Deterioration.* The degree to which the action would degrade an environmental resource even if would not breach a threshold of significance. For example, an action that does not breach any environmental standard or requirement but which substantially degrades environmental quality might be viewed as significant (see Chapter 9).

5. *Waste Generation.* The degree to which the action would contribute to the production of dangerous or nonhazardous waste. With respect to NEPA, this factor provides consideration of the Pollution Prevention Act of 1990. This factor also has its basis in language from the NEPA (Sec. 101 [42 USC § 4331[b][6]]).

6. *Degrade Visual Resources or Amenities.* The degree to which an action would alter, degrade, or impair visual, natural landscape, cultural, or geological resources, esthetics, or natural amenities. This factor would implement language from the NEPA (Sec. 101 [42 USC § 4331[b]]). It is also an effect that needs to be evaluated (§ 1508.8) but which is not specifically called out as a significance factor in § 1508.27.

7. *Urban Sprawl.* The degree to which the action would contribute to urban sprawl or intrusion into less developed areas. It is vaguely captured in § 1508.27 (a) (6). Effect on urban quality must be evaluated (§ 1502.16) but is not specifically cited as a significance factor in § 1508.27.

8. *Environmental Justice.* The degree to which adverse environmental effects or high risks to human health resulting from an action taken in response to an agency's programs, policies, and activities might disproportionately affect minority and low income populations. This factor has its basis in language from Executive Order 12898. Effects on urban quality are to be evaluated (§ 1502.16[g]) but not actually called as a significance factor in § 1508.27.

9. *Nonrenewable Energy and Natural Resources.* The degree to which an action would consume nonrenewable energy or natural resources. This factor has its basis in language from the NEPA (Sec. 101 [42 USC § 4331[b]], Sec. 102 [42 USC § 4332[c][6]] and § 1502.16[e][f]). This consideration is not explicitly called out as a significance factor in § 1508.27.

10. *The Degree to Which an Action Would Limit the Range of Beneficial Uses of the Environment for Americans or Future Generations of Americans.* This factor has its basis in language from the NEPA (Sec. 101 [42 USC § 4331[a][b][3]]) but is not actually called out as a significance factor in § 1508.27.

6.7.2.3 Significant Departure

An action that does not breach but substantially degrades environmental quality might also be viewed as significant. Consider a proposal for the construction of a federally sponsored plant in an area that is relatively pristine. Assume that the concentration of a certain air pollutant would violate regulatory standards when it reaches a concentration of 200 units. Further, suppose that the current ambient concentration, prior to constructing the facility, is 20 units. The facility would increase the ambient concentration level to 100 units. Even though a value of 100 units is still well below the regulatory limit of 200 units, it is equally true that this action would increase the current concentration by 400%, degrading air quality to half its permissible level. Thus, an action that could significantly change the environmental baseline may be considered significant even if it does not breach an environmental standard or threshold. The author refers to this concept as the significant departure principle, which is described in detail in Section 9.3.

Decision-makers must also consider significance from a cumulative as well as an individual perspective. Regarding the example given above, such a large degradation in air quality might well be viewed as significant by the public if not by the decision-maker. While a value of 100 units may not immediately breach an environmental requirement, it will nevertheless substantially degrade the air quality baseline and increase the chances that air quality standards would be breached at some point in the future.

6.7.3 SIGNIFICANCE AND THE NATIONAL HISTORIC PRESERVATION ACT

Decision-makers need to note that "significance" as used in Section 106 of the National Historic Preservation Act (NHPA) has a very specific meaning. Under the NHPA, a resource is defined to be significant if it meets the eligibility criteria for the National Register of Historic Places and is not disqualified by any criteria considerations. If a cultural resource is not considered significant from a cultural resources perspective within the meaning of Section 106, it would also normally be considered insignificant from the standpoint of NEPA.

6.7.4 DETERMINING A SIGNIFICANCE BASELINE

Determining the appropriate environmental baseline (affected environment) is not always straight-forward. In one case, the DOE restarted a nuclear reactor located at a Savannah River site that had been in cold shutdown for approximately 15 years.[44] Prior to the 15-year shutdown, the reactor had discharged heated coolant water into a nearby creek, severely damaging the surrounding wetlands. Despite the original damage, the wetland had recovered over the 15-year period in which the reactor had not operated. The Natural Resources Defense Council (NRDC) brought a suit claiming that restarting the reactor would cause new damage to the recovered wetlands.

In its defense, the DOE argued that restarting the reactor would cause no further damage to the condition of the wetland as had existed 15 years earlier. Essentially, the agency was arguing that it was not the recovering wetland but rather the previous (damaged) environmental baseline that should be used to measure the impacts of restarting the reactor.

If the impacts were measured against the previous degraded environmental baseline, then restarting the reactor would not contribute any significant impacts beyond those that had existed earlier, and an EIS would not be required. Conversely, if the existing environmental baseline was used, this action would significantly affect the recovered wetlands, requiring the preparation of an EIS. On reviewing the case, the court found that the wetlands had indeed recovered during the intervening years from much of the damage originally suffered. Based on this finding, the court ruled that the environmental impacts should be measured against the current prevailing (recovered) environmental baseline rather than the previous existing baseline.

In a different case, a court reached an opposite conclusion. This case involved a proposal to replace a bridge destroyed during a hurricane 3 months earlier. The bridge was scheduled to be reconstructed within 6 months of its destruction, in the same location, and with a design similar to the bridge that had been destroyed. A citizens' group sued to have an EIS prepared for this action.[45] The court ruled that the previously existing environment that included the bridge and its associated impacts should be used instead of the existing baseline that did not include the bridge. Based on this determination, the court ruled that replacing the bridge would not result in a significant impact relative to the former baseline that had included the bridge. It is left to the reader's professional judgment to determine how these two opinions should be applied.

6.8 AFFECTING

An EIS is required to be prepared for major actions significantly affecting the quality of the human environment. In the past, this term has been misconstrued to mean that an EIS is only required for actions that definitely result in a significant impact. To avoid such confusion, the Regulations define the term "affecting" to mean "will or may have an effect on" (§ 1508.3).

In reviewing this term, it is important to note that not all actions or events necessarily affect the environment. NEPA's requirements are not triggered if the physical environment has not been affected.

An action affects the environment only if it produces a change in one or more of the resources that together form the environment. Conversely, if an action has not produced such a change, neither has it resulted in an environmental impact. Some exceptions may apply particularly if an action basically commits an agency to a future action that would affect the environment or result in an irretrievable commitment of resources. However, a reasonably close connection must exist between the action and its resulting effects on the environment.[46]

6.9 HUMAN ENVIRONMENT

An early historic lawsuit in NEPA involved plaintiffs who sued to enjoin (stop) timber sales until the Forest Service completed an EIS on the management of that area. The court rejected the Forest Service's conclusion that there was no effect on the human environment from the timber sales, because there was no evidence that "human" users of that area would be directly impacted by such timber sales. As the court viewed it, "This appears to be too restrictive a view of what significantly affects the human environment. We think NEPA is concerned with indirect effects as well as direct effects. There has been increasing recognition that man and all other life on this earth may be significantly affected by actions which on the surface appear insignificant."[47]

Accordingly, the term "human environment" is interpreted comprehensively to mean "... the natural and physical environment and the relationship of people with that environment" (§ 1508.14). An argument can be made that in one way or another, some relationship exists between humans and virtually every aspect of the natural and physical environments. Thus, from a practical standpoint, there is little distinction between the terms environment and human environment. For this reason, it is a common practice to use the term environment in lieu of human environment.

The term "environment" is nearly all-encompassing and can be broadly interpreted to include ecological and natural resources as well as the components, structures, and functions of ecosystems. The environment also includes esthetic, historic, health, cultural, economic, and social resources (§ 1508.8).

From the standpoint of human health and safety, the term "human environment" is also interpreted to include the interior of buildings and facilities. The human environment also includes quality of urban life (noise, traffic, crime, and drug trafficking).[48]

6.9.1 THE PHYSICAL ENVIRONMENT

In the case of *Douglas County v. Babbitt*, plaintiffs challenged a decision under the Endangered Species Act (ESA) to designate critical habitat for a threatened or endangered species without complying with NEPA. The court concluded that "NEPA procedures do not apply to federal actions that do nothing to alter the natural physical environment" (§ 1505). To clarify this point, the court held that[49] "If the purpose of NEPA is to protect the physical environment, and the purpose of preparing an EIS is to alert agencies and the public of potential adverse consequences to the land, sea, or air, then an EIS is unnecessary when the action at issue does not alter the natural, untouched physical environment at all."

In contrast to the *Douglas County* case, another suit involved plaintiffs who challenged a similar critical habitat designation that had been made without compliance with NEPA.[50] Here the court specifically referenced and disagreed with the 9th Circuit *Douglas County* decision and held that ESA procedures did not displace NEPA requirements. The court agreed with the plaintiffs' claim that the proposed designation "will prevent continued governmental flood control efforts, thereby significantly affecting nearby farms and ranches, other privately owned land, local economies, and public roadways and bridges." The court characterized these impacts as "immediate and the consequences could be disastrous." The court stated:

> While the protection of species through preservation of habitat may be an environmentally beneficial goal, Secretarial action under ESA is not inevitably beneficial or immune to improvement by compliance with NEPA procedure.... The short- and long-term effects of the proposed governmental action ... are often unknown or, more importantly, initially thought to be beneficial, but after closer analysis determined to be environmentally harmful.

6.9.2 IMPACT AND EFFECT

As described in Chapter 9, three types of impacts are recognized: (1) direct, (2) indirect, and (3) cumulative effects. As discussed earlier, environmental impacts include but are not limited to ecological, esthetic, historic, cultural, health, economic, or social considerations (§ 1508.8[b]). From the standpoint of NEPA, the terms effect and impact are synonymous (§ 1508.8).

With respect to indirect effects, consider the following case involving a proposal to build a freeway interchange that would support privately sponsored, nonfederal development near that interchange. Construction of this interchange would have had a direct influence on future private development. The agency was sued for failing to prepare an EIS that would analyze the indirect impacts that would occur once the interchange was built. The court ruled that the EIS must include an analysis of the proposed nonfederal development that would occur as a result of this interchange.[51]

6.9.3 ENVIRONMENTAL IMPACTS THAT DO NOT HAVE TO BE CONSIDERED

Impacts on economic and social resources are not, by themselves, sufficient to require the preparation of an EIS. Some issues such as noise and congestion clearly affect the physical environment while others such as employment and crime definitely fall within the realm of socioeconomics. However, for proposed actions where an EIS is necessary, the effects on economic or social resources should be addressed when they are interrelated with the natural or physical environment (§ 1508.14).

Religious Practices. While historic and cultural resources are subject to analysis, at least one court has ruled that impacts on religious practices such as those covered under the American Indian Religious Freedom Act are not within the scope of NEPA. While an analysis of impacts on religious and spiritual practices may not be required, impacts on archaeological resources associated with Indian religious practices are subject to the requirements of NEPA.[52]

Psychological Stress. In one case, a court held that an NEPA analysis did not have to address the psychological health and community well-being of people residing near the Three Mile Island nuclear power plant resulting from the restart of the reactor.[53] The court concluded that NEPA does not require an agency to evaluate every impact of its proposed action, but only the impact on the physical environment.

Remote and Speculative Impacts. Environmental impacts do not have to be evaluated if they are determined to be remote or speculative. Fogleman has identified three factors used by the courts in determining if an impact should be considered remote or speculative:[54]

- The level or degree of confidence that the agency has in predicting the impact
- The information available to the agency that provides a basis for describing the impacts in a manner that is meaningful to the decision-maker
- The potential that that the decision-maker will meaningfully consider the effects at a later date without being obligated to continue the action because of past commitments

An action is likely to be considered reasonably foreseeable if it is a logical stepping stone to potential development or accelerates the development of an area. The degree of speculation tends to increase as projected impacts become removed or disconnected from the precipitating action. An additional step may be considered speculative even if the additional step, by itself, is considered reasonably foreseeable.

Consider a case involving the expansion of an airport in Hawaii. In reviewing this case, the court ruled that an increase in tourism resulting from expansion of the airport was reasonably foreseeable. Notwithstanding, the court concluded that the potential increases in Hawaii's permanent population that might occur from increased tourism was over-speculative.

PROBLEMS

1. A federal agency needs to treat hazardous waste to meet land disposal restrictions. In lieu of constructing and operating a hazardous waste treatment facility, the agency proposes to purchase those services from a private party. Specifically, the agency proposes to lease a parcel of land within the federal enclave to a private entity that would then finance, construct, and operate the hazardous waste treatment facility to process the agency's hazardous waste. The agency would have no involvement in the actual waste processing operation. The agency solicits proposals from private entities to supply such services. Is this privatization action subject to a NEPA analysis?

2. A federal agency has a need for additional sanitary treatment capacity and solicits proposals from private entities to supply such services. Further, suppose that a private company responding to the request for proposal already owns the land and a treatment plant, and plans to enlarge the plant's processing capacity regardless of whether or not it is selected to provide the services for the federal agency. The quantity of federal waste water would constitute only a small fraction of the total volume that would be processed by the entity, and existing transfer lines would be modified to move the sewage. In essence, this example is typical of many types of actions that are contracted out for service. Is the nonfederal action (renovating the plant) subject to NEPA?

3. With respect to NEPA, what is a detailed statement?

4. With respect to NEPA, what are the conditions that define a proposal?

5. A federal agency normally prepares an EA for small buildings and research facilities less than 10,000 ft² in size. In this instance, the agency proposes to build a relatively small 8000 ft² laboratory that would be used to perform genetic engineering experiments with potentially unknown risks. The project engineer argues that an EIS is not required because an EIS is only required for major federal actions; since this is not a 'large' action, an EIS is not required. Is his reasoning correct? Explain your answer.

6. A state highway department proposes to fund, design, construct, and manage a 30-mile road. Does NEPA apply to this action? Explain your answer.

7. A private company proposes to build a 20-mile-long gas pipeline. The company would fund, design, construct, and operate the pipeline. No federal permits or other approvals are required. However, the pipeline would cross 10 miles of federally owned property for which it must obtain federal authorization, and the pipeline cannot be built without this authorization. Does NEPA apply to this project? Why or why not?

8. Construction of a 6000-ft^2 office building would partially disrupt or destroy an area of a valley in which early settlers had originally settled. This area is potentially eligible for listing in the National Register of Historic Places. The project engineer maintains that it would be ridiculous to prepare an EIS on such a small building. Is an EIS required? Cite specific reasons to justify your answer.

9. A federal agency proposes a project to eradicate nonnative invasive plants (NNIPs) in an otherwise pristine area. The project would consist of digging up the plants, burning them, and applying approved herbicides to eradicate any remaining traces. The chief biologist concluded that the project would result in a significant beneficial improvement in environmental quality. He further claims that NEPA does not apply since "NEPA only applies to projects that adversely affect the quality of the human environment." Is he correct? Is it possible that an EIS might need to be prepared? Explain.

10. In response to employee health concerns, a project engineer proposes to renovate a federal laboratory by installing safety equipment and a modern air filtration system. Experiments with new chemical processes are being proposed. The engineer maintains that NEPA does not apply because these are safety and health upgrades that do not affect the environment. He also declares that NEPA only pertains to the natural, exterior environment and not to the interior of buildings. Is he correct? Why or why not?

REFERENCES

1. 42 USC § 4332(101)(C).
2. *Minnesota Public Interest Research Group v. Butz*, 498 F.2d 1314 (8th Cir. 1974).
3. Mandelker D. R., *NEPA Law and Litigation*, Clark Boardman Callaghan, New York, 1992.
4. *Colorado River Indian Tribes v. Marsh* 1985.
5. *Guide to the National Environmental Policy Act*, Section 3.2.
6. *Sierra Club v. Hodel*, 544 F.2d 1036, 9 ERC 1449 (9th Cir. 1976).
7. Eccleston C. H. and Williamson B. D., Determining when NEPA applies to nonfederal activities, *Federal Facilities Environmental Journal*, Winter 1997.
8. *National Association for the Advancement of Colored People (NAACP) v. The Medical Center*, 584 F.2d 619, 630 (3rd Cir. 1978).
9. *National Association for the Advancement of Colored People (NAACP) v. the Wilmington Medical Center, Inc.*, 426 F. Supp. 919 (1977) (where a large portion of the nonfederal renovation project would be borne by the federal government through payment for services provided); later proceeding 453 F. Supp. 280 (1978); later proceeding on discrimination claim 453 F. Supp. 330 (1978); and remanded 599 F.2d 1247 (3rd Cir. 1979), on remand 491 F. Supp. 290 (1980); aff'd 657 F.2d 1322 (1981) and later proceeding on attorney fees 530 F. Supp. 1018 (1981), rev'd on other grounds 689 F.2d 1161 (3rd Cir. 1982); *reh. den.* 693 F.2d 22 (3rd Cir. 1982); *cert. den.* 460 U.S. 1052 (1983), 103 S. Ct. 1499 (1983).
10. *Landmark West v. US Postal Service*, 840 F. Supp. 994 (1993).
11. *City of Davis v. Coleman*, 521 F.2d 661 (federal financial participation in the construction of a highway system); *Monroe County v. Volpe*, 472 F.2d 693 (2nd Cir. 1972) (federal financial assistance of 60% of proposed highway cost of which the viaduct section alone would exceed $14 million); *Scherr v. Volpe*, 466 F.2d 1027 (7th Cir. 1972) (federal financial aid for the extensive construction of two-lane conventional highway into a four-lane freeway).
12. *NAACP* at 631.
13. *Sierra Club v. Hodel*, 848 F.2d 1089 (quoting Rogers W., *Environmental Law*, p. 763, 1977) and *Almond Hill School v. US Dept. of Agriculture*, 768 F.2d 1030, 1039 (9th Cir. 1985).
14. *NAACP*, 584 F.2d 619, 633 (3rd Cir. 1978).
15. 40 CFR 1508.18.
16. *Minnesota Public Interest Research Group v. Butz*, 498 F.2d 1413 (8th Cir. 1974).
17. *NJ Dept. of Env'l Protection v. Long Island*, 30 F.3d 403 (3rd Cir. 1994) (where a nonfederal party voluntarily informs a federal agency of its intended activities to ensure that they will comply with law and regulation, and to facilitate the agency's monitoring of the activities for safety purposes, the agency's

review of the plan does not constitute a major federal action); See also, e.g., *National Forest Preservation Group v. Butz*, 485 F.2d 408 (9th Cir. 1973) (federal government exchanged certain park lands for those owned by a private enterprise to exercise impact on the lands); *Davis v. Morton*, 469 F.2d 593 (10th Cir. 1972) (government agency required by law to approve a lease on Indian property in order to have a valid lease was ordered to file an EIS because without its approval the lease would be invalid as a matter of law); *Lathan v. Volpe*, 455 F.2d 1111 (9th Cir. 1971) (DOT ordered to file an EIS where federal approval of highway construction plans was legally required in order to qualify for federal funding).

18. *Environmental Rights Coalition, Inc. v. Austin*, 780 F. Supp. 584 (1991).
19. *Defenders of Wildlife v. Andrus*, 627 F.2d 1238, 1245 (D.C. Cir. 1980) and *NAACP* at 629.
20. *Davis v. Morton*, 469 F.2d 593 (10th Cir. 1972).
21. *Goos v. Interstate Commerce Commission*, 911 F.2d 1283, 1269 (8th Ar. 1990) (where the Eighth Circuit reinforced its threshold applicability decisions from *Ringsred v. Duluth*, 828 F.2d 1305 (8th Cir. 1987). See also *Sierra Club v. Hodel*, 675 F. Supp. 594, 612 (1987).
22. *Gettysburg Battlefield Preservation Ass'n. v. Gettysburg College*, 799 F. Supp. 1571, 1577 (1992) (quoting *Environmental Rights Coalition, Inc. v. Austin*, 780 F. Supp. 584, 1991, where the court resolved that without the requisite involvement in a project by a federal agency, the project simply does not involve a major federal action [necessary to trigger NEPA] no matter how much the project may impact the environment).
23. *Landmark West v. U.S. Postal Service*, 840 F. Supp. 994 (1993).
24. CEQ NEPA Regulations, 40 CFR 1508.18.
25. *Defenders of Wildlife v. Andrus*, 627 F.2d 1238 (D.C. Cir 1980).
26. *Guide to the National Environmental Policy Act*, Section 3.4.
27. Bear D., NEPA at 19: a primer on an "old" law with solutions to new problems, *Environmental Law Reporter—News and Analysis*, February 1989.
28. Caldwell L., personal communications, April 1994.
29. *Environmental Law Handbook*, Government Institutes, Inc., 10th ed., Chapter 10.
30. Executive Order No. 12114; 44 *Fed. Reg.* 1957, 1979.
31. Mandelker D. R., *NEPA Law and Litigation*, 2nd edition, New York, 1992.
32. CEQ, Council on Environmental Quality Guidance on NEPA Analyses for Transboundary Impacts, July 1, 1997.
33. *NEPA Coalition of Japan v. Aspin*, 837 F. Supp. 466 (D. D.C. 1993).
34. *Natural Resources Defense Council v. U.S. Department of the Navy*, unpublished (C.D. Cal. 2002).
35. *Greenpeace USA v. Stone*, 748 F. Supp. 749 (D. Haw. 1990).
36. *Environmental Defense Fund v. Massey*, 986 F.2d. 528 (D.C. Cir. 1993).
37. DOE/EIS-0365, December 2004.
38. *Consejo de Desarrollo Economico de Mexicali, AC, et al. v. U.S. et al.* (Case No.: 05-0870).
39. *Hanley v. Kleindienst*, 471 F.2d 823 (2d Cir. 1972), *cert. denied*, 412 U.S. 908 (1973).
40. *Sierra Club v. Bosworth*, 199 F. Supp. 2d 971 (N.D. Cal. 2002); *League of Wilderness Defenders–Blue Mountains Diversity Project v. Marquis-Brong*, not reported (D.Ore. 2003) and *League of Wilderness Defenders v. Zielinski*, 187 F. Supp. 2d 1263 (D. Ore. 2002).
41. *Foundation for North American Wild Sheep*, 681 F.2d 1182.
42. *Friends of Endangered Species, Inc. v. Jantzen*, 760 F.2d 976 (9th Cir. 1985).
43. *Marsh v. Oregon Natural Resources Council*, 490 U.S. 360, 109 S.Ct. 1851 (1989).
44. Holt et al., CRS Report for Congress: NEPA Compliance at Department of Energy Defense Production Facilities, March 6, 1990.
45. Guide to the National Environmental Policy Act, Section 3.6.
46. *Minnesota Public Interest Research Group v. Butz*, 498 F.2d 1314 (8th Cir. 1974).
47. Guide to the National Environmental Policy Act, Section 3.6.1.
48. *Douglas County v. Babbitt*, 48 F.3d 1495 (9th Cir. 1995), *cert. denied*, 116 S. Ct. 698 (1996).
49. *Catron County Board of Commissioners v. U.S. Fish and Wildlife Service*, 75 F.3d 1429 (10th Cir. 1996).
50. *City of Davis v. Coleman*, 521 F.2d 661 (9th Cir. 1975).
51. *Lockhart v. Kenops* (CA8, 1991) 927 F.2d 1028.
52. *Metropolitan Edison Co. v. People Against Nuclear Energy (PANE)*, 460 U.S. 766, 103 S. Ct. 1556 (1983).
53. Folgman V. M., Guide to National Environmental Policy Act, Section 3.5, 1990.
54. *Life of the Land v. Brinegar*, 485 F.2d 460, 469 (9th Cir. 1973), *cert. denied*, 416 U.S. 961 (1974).

7 Preparing Environmental Assessments

In the early 1970s, federal agencies had no option but to prepare an environmental impact statement (EIS) as many actions could not be immediately excluded as being clearly nonsignificant. Consequently, EISs were frequently prepared only to reach the conclusion that no significant impacts existed. In 1978, the Council on Environmental Quality (CEQ) responded to this problem by creating an environmental assessment (EA), the third level of National Environmental Protection Act (NEPA) compliance designed to provide an efficient mechanism for bridging the gap between the categorical exclusion (CATX) and EIS.

When the CEQ created the EA, it believed that the EIS would still be the principal instrument used for evaluating impacts. Instead, EAs have become the principal instrument used for evaluating impacts. This observation is supported by the fact that, on an average, approximately 100 EAs are prepared for each EIS. Moreover, the CEQ estimates that 30,000–50,000 EAs are prepared each year compared with just 300 to 500 EISs.[1]

This chapter describes the EA process and its documentation requirements. For a more in-depth discussion of the EA process, the reader is directed to the author's companion book, *Effective Environmental Assessments*.[2]

7.1 OVERVIEW

When challenged, an EIS is often easier to defend than an EA because an EA must prove that none of the potential environmental impacts is significant or, if one is, that it can be adequately mitigated. In contrast, there is no such requirement for an EIS.

Additionally, because of its smaller size, a judge is more likely to take personal interest in an EA and actually read through it. Thus, an EA may receive more rigorous judicial review than an EIS. Project advisories have taken note of this fact and have revised their strategies accordingly. As a result, in recent years agencies have witnessed a movement away from challenging EISs as many opponents have refocused their efforts instead on EAs, considering them to be more vulnerable targets.

Most agencies have not remained docile in the face of greater opposition. Evidence suggests that the quality of EAs has generally improved and, consequently, agencies have been become more successful in defending their findings of no significant impact (FONSI).

7.1.1 THE PURPOSE OF AN EA

As illustrated in Figure 1.1, the preparation of an EA is necessary for federal actions that cannot be excluded under a CATX and for which the agency has not prepared an EIS. An EA can serve any one of the following three objectives (§ 1508.9):

1. Briefly provide sufficient evidence and analysis for determining whether to prepare an EIS or a FONSI
2. Aid an agency's compliance with NEPA when no EIS is necessary
3. Facilitate preparation of an EIS when one is deemed necessary

The most important function of an EA is, used as a screening device, to evaluate actions to determine if they may result in a significant impact, requiring the preparation of an EIS. To justify issuing a FONSI, an EA must provide clear and convincing evidence that the proposal would not result in any significant impacts or that any identified significant impacts can be mitigated to the point of nonsignificance.

7.1.2 COMPARISON OF EAs TO EISs

Table 7.1 compares some of the basic traits and differences between EAs and EISs. Three principal reasons why EISs are longer and more complex:

- EIS has more extensive documentation requirements. It also requires more extensive procedures in preparation and issuance.
- EIS public participation (scoping and reviewing the draft EIS) usually raises issues and concerns that require more analysis and explanation.
- EIS must thoroughly describe and analyze the significant impacts (which by definition are not at issue in an EA qualifying for a FONSI) and show how decisions about the action are related to an understanding of these impacts.
- EIS must rigorously explore a range of reasonable alternatives, while an EA typically provides a more cursory review of these alternatives.

TABLE 7.1
Comparison of an EA with an EIS

Environmental Assessment	Environmental Impact Statement
The principal purpose is to determine if the proposed action would result in a significant impact requiring preparation of an EIS.	The question of significance is no longer at stake. The EIS is prepared to support decision-making by identifying and evaluating alternative methods for meeting the purpose and need and mitigating the impacts.
The proposed action can only be pursued if its impacts are insignificant or can be mitigated.	An agency may pursue any analyzed alternative regardless of the resulting impacts.
A substantial change in the project would normally require preparation of a new NEPA document.	An EIS reviews a range of reasonable alternatives. A substantial change is therefore less likely to result in a delay since it may already be covered in one of the alternatives.
	A different alternative may simply be chosen by supplementing the record of decision (ROD).
No restrictions exist regarding who may prepare an EA (federal agency or a contractor).	A party may not prepare an EIS if it has a financial or other interest in the outcome (conflict of interest).
EAs are normally substantially faster and cheaper to prepare.	EIS are typically more complex, lengthy, and expensive.
The analysis normally focuses on the proposed action. Other alternatives are usually given only cursory review.	Substantial treatment must be devoted to each of the reasonable alternatives investigated.
While an EA is a public document, a formal public scoping process is not normally required under the regulations. An EA is not normally required to be publicly circulated for formal review and comment, as is the case for an EIS (however, many agencies now do this).	EISs require a much larger degree of public involvement. A formal public scoping process is required. EISs must be publicly circulated for review and comment.
EAs are not typically supplemented. A substantial change normally requires preparation of a new EA.	EISs may be supplemented if there is a significant change in the action, information, or circumstances.
EAs are often more susceptible to a successful legal challenge.	If challenged, an EIS is frequently easier to defend.
No requirement exists to consider mitigation measures.	Mitigation measures must be analyzed.

7.1.3   Time Periods

Preparation of EAs must be so timed that they can be circulated at the same time as other planning documents (§ 1501.2[b]). The CEQ has advised agencies that an EA process should normally require a maximum of 3 months to complete and generally need substantially lesser time.[3] In practice, however, many agencies report that the total process normally exceeds this time frame. The following section provides a thorough review of the process requirements governing the preparation, review, and approval of an EA and a FONSI.

7.2   PREPARING AND ISSUING THE ASSESSMENT

Figure 7.1 depicts a generalized procedure for preparing an EA. If the agency is uncertain whether an action would result in a significant impact, it may choose first to prepare an EA to determine if the action qualifies for a FONSI. This is the course normally taken, since preparing an EIS requires substantially more effort.

In cases involving an applicant (i.e., a nonfederal entity applying for a federal permit, license, or approval), preparation of the EA must begin early in the process and "no later than immediately after the application is received" by the agency (§ 1502.5[b]).

An EIS must be prepared either by the agency or by a contractor who has signed a disclosure statement indicating that it has "no financial or other interest in the outcome of the project" (§ 1506.5[b]). No such requirement exists for contractors assigned the responsibility of preparing an EA (§ 1506.5[b]). The agency, however, is responsible for evaluating the environmental issues involved in the project and also assumes legal responsibility for their scope and content.

7.2.1   Public Involvement

After a decision is made to prepare an EA, the most appropriate level of public involvement must be determined. The stage is now set for initiating internal (and if applicable, public) scoping (see first box, Figure 7.1).

Consultations with outside authorities and agencies are also initiated, as warranted. This step is important, as revealed by a case in which an agency reported that the NEPA process was particularly useful in helping the state and the Indian tribe concerned to resolve their differences about a proposed action.[4]

While the CEQ NEPA Regulations (Regulations) do not specifically require an agency either to incorporate or respond to public comments in an EA, such practice is highly recommended. Where appropriate, either public hearings or meetings should be conducted. Criteria for determining if these are warranted include (§ 1506.6[c]):

- Circumstances where there is substantial interest in holding a hearing or where an environmental controversy is involved
- A request for a hearing is made by another agency with jurisdiction over the action, supported by reasons why a hearing would be helpful

7.2.1.1   Public Notification

The CEQ has stated that EAs are to be made available to the public and that agencies are to give public notice of their availability (§ 1506.6[b]). The goal should be to notify all interested or affected parties.[5] Repeated failure to notify interested parties or the affected public could be interpreted as a violation of the Regulations.

A combination of methods may be used to notify the public. For instance, appropriate notification for proposals of national interest might involve publishing a notice of availability in the *Federal Register* and national publications as well as mailing the notice to interested national groups.

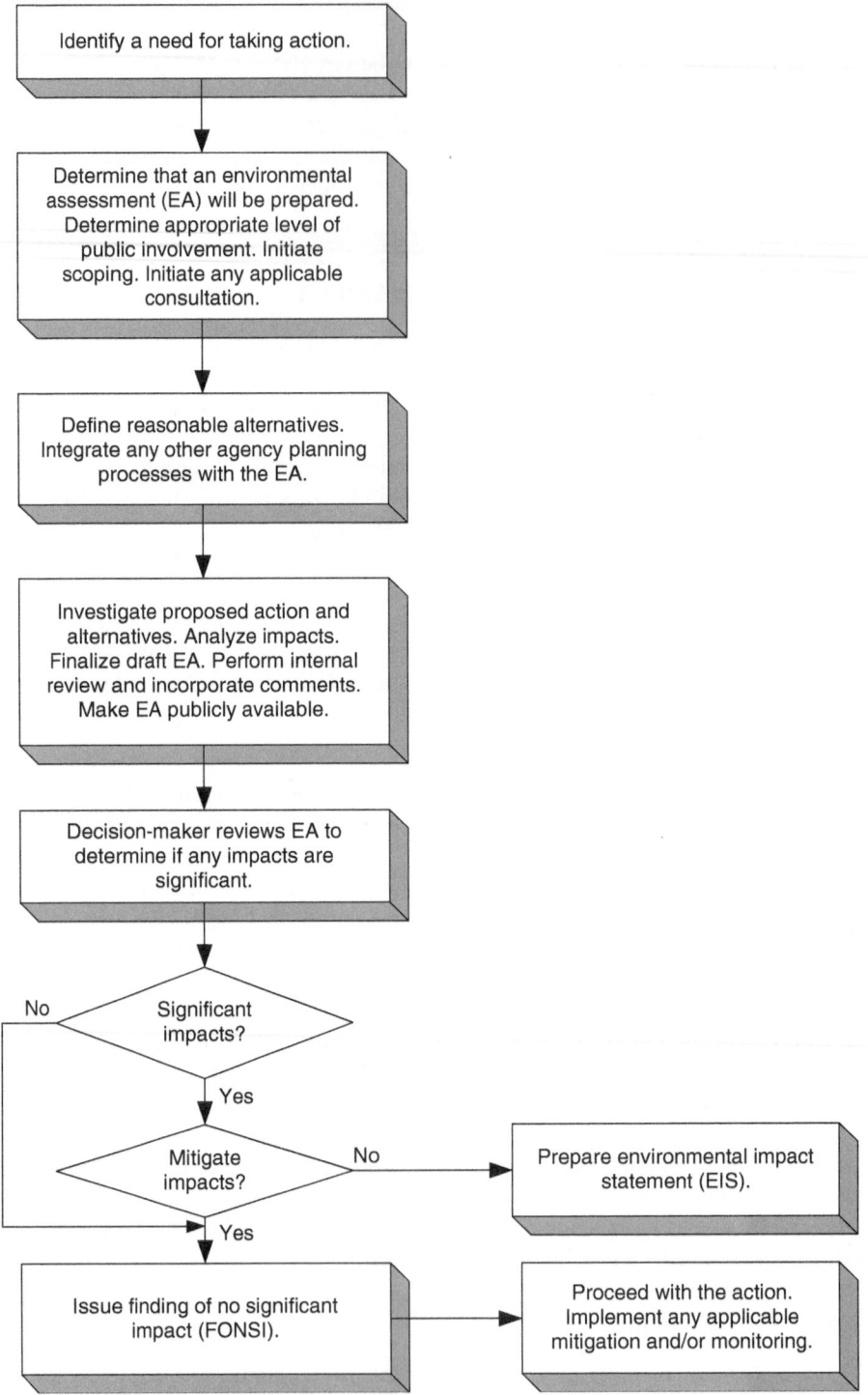

FIGURE 7.1   Typical environmental assessment process.

In other cases, appropriate notification for a site-specific proposal, such as publishing notices in local newspapers, may be sufficient.

7.2.1.2 The Public Review Process

While EAs must be made publicly available, the Regulations require no public comment, review, and incorporation period. Thus, agencies neither are specifically required by the Regulations to respond to public comments nor are EAs and FONSIs required to be filed with the U.S. Environmental Protection Agency (EPA).

From an agency's standpoint, thoroughly involving the public in the review of an EA can be distinctly advantageous. For example, in some situations, courts have ruled that an agency cannot be forced to prepare an EIS if the plaintiff has had ample opportunity to dispute the EA and FONSI process but failed to do so.

7.2.1.3 Consultation

As appropriate, agencies are required to consult with other agencies in preparing their NEPA documents. Such consultation facilitates a more thorough analysis (§ 1501.1, § 1501.2[d], § 1502.25). While the direction provided in § 1502.25 is directed at the preparation of EISs, it is also interpreted to be equally applicable to EAs. This is evidenced by the fact that an EA is required to list "... agencies and persons consulted" (§ 1508.9[b]).

Chapter 2 of the companion text *Environmental Impact Statements* provides additional information that may be of use in promoting public involvement.[6] The EA process is integrated with other planning studies or analyses (see second box, Figure 7.1).

7.2.1.4 Case Law

A U.S. Court of Appeals concluded that the Army Corps had complied with the NEPA regulatory direction to involve the public in preparing an EA. The court held that the agency met the "to the extent practicable" requirement by issuing public notice of the proponent's application, conducting two public hearings, responding to public comments in the EA, and conferring with environmental agencies. The court did not agree with plaintiffs that the agency should have prepared a draft EA for public comment. This case is in contrast to a second case described below in which the agency had a much more limited public involvement process.

In the second case, a district court held that the Forest Service violated NEPA by failing to provide for effective public involvement in the preparation of the EAs for four logging projects. In its defense, the Forest Service argued that issuing a scoping notice and releasing the final EA to the public satisfied the mandatory public involvement requirements. The court noted that while the Regulations do not require the circulation of a draft EA, they require that the public be informed to the extent practicable. The scoping notices contained no analysis of the environmental impacts of the projects. Moreover, they failed to give the public adequate information to effectively participate in the decision-making process.[7]

7.2.2 PERFORMING THE ANALYSIS

The Regulations are surprisingly silent when they come to providing direction for performing an EA analysis. Nonetheless, to ensure that it is adequate in substance and content, the EA must provide an accurate, unbiased, and scientifically based study for determining the significance or nonsignificance of potential impacts. To the extent feasible, an effort should be made to quantify and explain the probable intensity of potential impacts. The results of the investigation should be clearly documented, and the analysis should be based on professionally accepted technical and scientific methodologies.

Although the Regulations do not specifically state that an EA must consider the impacts of connected, similar, or cumulative actions, it is obvious that it cannot adequately make a determination of significance without considering them (§ 1508.27[b][7]). As illustrated by the following case, an analysis should always consider and evaluate such impacts as appropriate.

In one case, a plaintiff challenged the adequacy of an EA prepared for the importation of spent nuclear fuel rods from Taiwan to the United States. The court found the EA to be inadequate, noting that the examination of alternatives was bound by the rule of reason and that the level of analysis should commensurate with the severity of the impacts. The court also found the agency's choice of alternatives and its analysis of the cumulative risks of radiation exposure to be inadequate.[8]

7.2.2.1   The Proposed Action

While an EIS must devote substantial consideration to each of the analyzed alternatives, the focus of attention is predominantly on the proposed action. Typically, reasonable alternatives are only briefly described before being dismissed. In these cases, the EA should clearly explain the reasons for the dismissal of each alternative. Such a practice is justified because the principal purpose of an EA is generally distinctly different from that of an EIS. While the principal purpose of an EIS is to explore reasonable alternatives to proposed actions that could avoid or reduce significant impacts, EAs are normally prepared to determine if an action would result in significant impacts. Thus, attention is normally focused on the proposed action with correspondingly lesser attention devoted to the alternatives. It is recommended that a sliding-scale approach be used in determining the number of alternatives as well as the degree to which the reasonable alternatives should be analyzed.

7.2.2.2   Reserving Significance Findings for the FONSI

The purpose of the EA is to provide decision-makers with facts about potential impacts. Many EAs have been successfully challenged because they either made or appeared to make a determination that the impacts of a proposed action were nonsignificant.[9] For this reason, precautions should be taken to avoid any perception of such judgments made or giving the impression of partiality. Any actual judgment regarding the significance of an impact is reserved for the FONSI.

Consistent with this direction, nonjudgmental terms such as consequential, inconsequential, substantial, large, and small are generally considered acceptable. Conversely, judgmental terms such as significant, nonsignificant, acceptable, and tolerable should be avoided.

7.2.3   Addressing Cumulative Impacts in EAs

While the Regulations require EISs address cumulative impacts, they are silent on this requirement for EAs. For this reason, claims have been made that analyses of cumulative impacts do not need to be addressed in an EA. However, as discussed above, it is reasonably clear that a cumulative impact assessment (CIA) must be performed before an agency can conclusively determine that the impacts of an action are nonsignificant (§ 1508.27[b][7]).

In one case, a court found that while some individual projects had independent utility and thus need not be considered together in the same NEPA document, the EAs prepared for each project did not adequately consider their cumulative impacts as reasonably foreseeable actions.[10]

7.2.3.1   Cumulative Impact Study

A study has been carried out to determine the adequacy of cumulative impact analysis in EAs.[11] This study reviewed 89 EAs prepared by 13 federal agencies that were announced in the *Federal Register* during the first half of 1992. Based on certain criteria, each EA was examined to determine if it adequately addressed cumulative impacts.

TABLE 7.2

Reasons for Inadequate Cumulative Impact Analysis in EAs

Reasons for Inadequate Cumulative Impact Analysis	Percentage (%)
Did not even mention cumulative impacts	61
Concluded that cumulative impacts were insignificant without presenting any analysis of evidence	14
Failed to address cumulative impacts on all resources	21
Failed to adequately address all past, present, and reasonably foreseeable future actions	1
Adequately addressed cumulative impacts	3

To begin with, of the 89 EAs examined, only 35 mentioned the term "cumulative impacts." Of these 35, 13 concluded that the cumulative impacts were insignificant without presenting any analysis or evidence on which to base such a decision. Of the remainder, 19 failed to discuss cumulative impacts for all the resources that would be affected. Of the three remaining EAs, only two correctly identified all the past, present, and reasonably foreseeable future actions in their analyses. Table 7.2 summarizes the results of this study.

7.2.4 THE CILIX METHOD

Preparing a CIA that fully and rigorously satisfies the regulatory requirements set forth in 40 Code of Federation Revolution (CFR) § 1508.7 (see Chapter 9) can be a daunting if not an impractical task, particularly when preparing an EA. As just witnessed, a close examination of CIAs in EAs reveals that such analyses are frequently inadequate or insufficient to demonstrate that a rigorous examination has been undertaken to prove that the cumulative impacts are nonsignificant pursuant to regulatory requirements set forth in 40 CFR § 1508.7.

NEPA is governed by the rule of reason, that is, reason should be applied when a regulatory requirement results in an impractical, irrational, or absurd result. Under some specific circumstances (described below), the author offers a streamlined approach referred to as the Cilix method which provides a reasonable and practical method for demonstrating that a cumulative impact is clearly nonsignificant.[12] This technique can be used to demonstrate nonsignificance even in situations where the cumulative baseline has already sustained a significant impact.

7.2.4.1 Standard Approach

Assessing cumulative impacts under the standard approach may require identifying and assessing a potentially large array of past, present, and reasonably foreseeable future actions. On a resource-by-resource basis, analysts then need to evaluate and 'add' these impacts together to produce a cumulative impact baseline. In turn, the impact of the proposed action is also be 'added' to this baseline. If the baseline has already breached the threshold of significance, then from a strict, absolute standpoint, any contribution beyond that point can be interpreted as significant and an EIS would need to be prepared.

7.2.4.2 Cilix Methodology

The concept behind the Cilix methodology is straightforward:

> If a proposal is deemed eligible for a FONSI (i.e., the impacts are nonsignificant), the CIA need only demonstrate that the cumulative incremental impact is clearly so small as to be negligible or unimportant and therefore 'nonsignificant' in terms of its contribution to the cumulative impact baseline.

Thus, the method is applicable to proposals for which the environmental contribution, when 'added' to the impacts of other past, present, and reasonably foreseeable actions, is so small as to constitute no appreciable increase in the cumulative effect. Because the contribution is negligible, it would not affect, influence, or contribute to any significant change or increase in the cumulative impact on an environmental resource.

With respect to cumulative impacts, in such instances, neither there is a rational justification to forego the proposed action nor a rational justification for preparing an EIS. A FONSI or a CATX can thus provide an appropriate mechanism for NEPA review. The significance factors (40 CFR § 1508.27) should be used in assessing and demonstrating that the incremental impact is trivial, but instead of considering the significance factors from an absolute value (total impact), they can be assessed in terms of the relative change in the impact.

In essence, the Cilix method can be summarized as follows. Rather than preparing what may be highly detailed and complex assessment of a cumulative impact, the Cilix method simply demonstrates that, regardless of what the cumulative baseline is, the incremental impact is too small to have an appreciable effect upon it. Therefore, the cumulative impact is nonsignificant (negative declaration statement), consistent with the purpose of a FONSI.

The beauty of the Cilix method is that one does not necessarily have to undergo an exhaustive analysis of the environmental baseline. Such an analysis is unnecessary since the proposal will have a negligible effect on the baseline, regardless of its state. This is consistent with the purpose of an EA, since the intent of a FONSI is to demonstrate that there will be no significant impact (i.e., negative declaration statement).

7.2.4.3 Restrictions

Application of the Cilix method is restricted to the following two circumstances:

1. Where an area has not been significantly affected, and where the direct and indirect impacts of the proposed action are clearly so small as to have no appreciable incremental effect
2. Where an area has already sustained a significant environmental impact but the direct and indirect impacts of the proposed action are clearly so small as to have no appreciable incremental effect

Application of the Cilix method is invalid where

the direct and indirect impacts (i.e., the incremental contribution) may result in some appreciable incremental effect on the cumulative impact baseline.

As described above, the Cilix method is restricted to cases where the incremental contribution of the proposed action is clearly trivial or unimportant. Moreover, it cannot be used in the rare circumstance where a trivial incremental impact of a proposal could provide the final contribution necessary to breach the threshold of significance.

7.2.4.4 Application Scenarios

The following three hypothetical scenarios are presented to demonstrate the logic behind the Cilix method as well as circumstances under which its application is valid. These cases are provided for illustrative purposes only.

Case 1: Assume a situation in which a particular water resource has a water quality environmental baseline (provided by considering past, present, and reasonably foreseeable future actions) with a fictitious value of 70 units. Further, assume that a federal project adds 40 units to the baseline, resulting in a total impact of 110 units. The legal (significance) limit is 100 units. The cumulative impact of the proposal would therefore breach the threshold of significance. In this case, application of the Cilix method would be invalid as the incremental impact is clearly significant, requiring the preparation of an EIS.

Case 2: Assume the same problem, but this time, using a different set of numbers to depict significance conditions. Again, the threshold of significance (maximum legal limit) is assigned a value of 100 units. But in this case, the cumulative impact baseline has already sustained a significant impact of 110 units. The proposed action would add 0.00002 units, resulting in an impact of 110.00002. Thus, this proposal would add to the significant impact but only infinitesimally so. A strict interpretation of the CIA leads to the conclusion that once the significance threshold has been breached, any additional impact should likewise be considered significant—clearly an impractical result in this case.

Recall that NEPA is governed by the rule of reason. The Cilix method provides a practical solution to this second scenario because it considers such an increase not from an absolute value but instead from its relative contribution, that is, the increase is so small (0.00001%) as to have no appreciable effect on water quality. In this case, the Cilix method can be used to demonstrate that the degradation is clearly so small as to be deemed nonsignificant.

Case 3: Finally, assume the same problem, but this time, using a different set of significance conditions. As with the previous cases, the threshold of significance (maximum legal limit) is 100 units. Again, the environment baseline has already sustained a significant impact accruing to 110 units but in this case the proposed action would add a further 20 units, resulting in a total impact of 130, a relatively large (18%) contribution to the environment baseline. Since the Cilix method is valid only for circumstances in which the environmental contribution is clearly nonsignificant, such a large increase in a value that has already been significantly affected would provide a basis for a rational decision-maker to conclude that the additional units represent a significant increase requiring preparation of an EIS.

7.2.4.5 Advantages

Under the Cilix method, it is not necessary to prepare what might be a highly complicated and expensive analysis of cumulative impacts so long as the incremental impact is so trivial as to be unimportant. What is needed here is the ability to demonstrate that the incremental increase is clearly an unimportant contribution in terms of the cumulative baseline. The Cilix method is therefore invalid in any circumstance where this cannot be done.

Beyond providing a more practical approach to the assessment of cumulative impacts in EAs, the Cilix method has a second important advantage: Even if the cumulative impact baseline of the applicable resource is significantly affected (or could be in the future), a FONSI can still be issued with respect to the CIA as long as the incremental contribution from the proposed action relative to the baseline is so small as to constitute no appreciable environmental change to the resource. This is because, from an environmental quality standpoint, it would make no practical difference whether the proposal is implemented or not, since there would be no substantial cumulative change in the affected environmental resources.

7.2.4.6 Example

The following abbreviated example is provided for illustrative purposes only. This example demonstrates how the Cilix method can be applied to evaluate the cumulative impacts involved in the proposed construction of a federal building in the crowded downtown business center of a large city. The area has already sustained a significant cumulative impact. For instance, virtually all of it has been filled with buildings and paved over with concrete. The streets that run in every direction are crowded with commuter traffic, and there is a high level of associated traffic noise. Because of all the other high-rise construction, the proposed building is practically unviewable from more than a block away.

The cumulative impact descriptions presented below assume that a rigorous analysis of the direct and indirect impacts of the proposed building have already been performed in the EA. In addition to the CEQ's 10 intensity factors for assessing significance, the setting (context) may also need to be considered since the impacts of a building located in a crowded downtown area can be quite different from those of exactly the same structure sited inside a nature preserve.[13]

The example described below considers two different alternatives: (1) no-action alternative and (2) proposed action. It also considers an abbreviated Cilix analysis of cumulative effects on the following environmental resources: (1) visual quality, (2) noise, and (3) traffic congestion.

Cumulative Visual Resource Impacts. The following example demonstrates how the Cilix method can be used to assess cumulative impacts on visual resources within an EA:

Alternative 1 (No-Action): As the no-action alternative would not directly result in any measurable change in visual resources, this alternative would not contribute to any cumulative effect on them.

Alternative 2 (Building): Existing buildings currently surround the proposed site, and future buildings are proposed for the surrounding area. Because of the limited amount of area that would be affected, the proposed construction is considered to be small-to-negligible and would therefore contribute little or no incremental increase in visual impairment when combined with other past, present, and reasonably foreseeable projects. Consequently, there would be no substantial change in cumulative visual resources within the surrounding area.

Cumulative Noise Impacts. The following example demonstrates how the Cilix method can be used in an EA to assess cumulative impacts on noise:

Alternative 1 (No-Action): As the no-action alternative would not directly result in any measurable incremental impact, this alternative would not contribute to any cumulative effect on noise levels.

Alternative 2 (Building): The construction and operational noise level resulting from existing activities is deemed to be high. Construction and operational noise levels associated with reasonably foreseeable projects will add to those already existing. However, the incremental increase in noise level construction will be both temporary and negligible when compared with the total ambient noise level. The long-term incremental increase in operational noise level will be negligible when compared to the total ambient noise level. Because the proposed project would contribute little or no incremental increase in noise when combined with other present and reasonably foreseeable projects, there would be no substantial change in cumulative noise levels within the surrounding area.

Cumulative Traffic Congestion Impacts. The following example demonstrates how the Cilix method can be used in an EA to assess cumulative impacts on traffic congestion:

Alternative 1 (No-Action): As the no-action alternative would not directly result in any measurable change in traffic congestion, this alternative would not contribute to any cumulative effect on traffic levels.

Alternative 2 (Building): The traffic congestion resulting from existing construction and operational activities is deemed to be substantial. Traffic congestion impacts from reasonably foreseeable activities will add to the existing congestion. However, the incremental increase in traffic congestion would be negligible when compared with the total current and future congestion levels. Because the proposed project would contribute little or no incremental increase in traffic when combined with other present and reasonably foreseeable construction projects, there would be no substantial cumulative change in congestion levels.

7.2.4.7   Justification

As shown above, the Cilix method is not only practical to use but justifiable, since it can often provide a more rigorous demonstration to show that an impact is either significant or nonsignificant. As long as the proposal's contribution to the overall impact can be shown to be negligible there is little or no practical justification for performing a full and complicated CIA that requires an in-depth evaluation of all past, present, and reasonably foreseeable future impacts. The Cilix method can therefore provide evidence that no further investigation is warranted even if the environmental resource has already sustained a significant cumulative effect.

Thus, the difference between the standard and Cilix methodologies is that under the standard approach, as defined in 40 CFR § 1508.7, significance is assessed from an absolute perspective, that is, in terms of whether the impact triggers or breaches the threshold level of significance. This can require a complicated assessment of the cumulative impact baseline to which the impact of the proposed action is 'added.' In contrast, under the Cilix method, significance is assessed from a relative perspective, that is, "would the impact significantly change the cumulative impact baseline?"

7.3   ISSUING A FONSI

As in the case of an EA, a FONSI is a public document and must therefore be made publicly available. The Regulations require that the FONSI be made available to the affected public (§ 1501.4[e][1]). However, the Regulations provide little direction as to how this is to be accomplished.

7.3.1   WAITING PERIOD

In most circumstances, a proposed action may be initiated as soon as the FONSI is issued. However, in some limited instances, it must be made publicly available (including at state- and area-wide clearinghouses) for a minimum review period of 30 days before the agency makes its final determination on whether or not to prepare an EIS (§ 1501.4[e][2]). In such circumstances, no action may be taken with respect to the proposed action until this 30-day review period has elapsed. A 30-day review period is required in the following circumstances:

- Proposed action is similar to one normally requiring preparation of an EIS under the agency's implementation procedures.
- Nature of the proposed action is one without precedent.

Additionally, a presidential directive has been issued requiring a FONSI to be made available for a minimum of 30 days when the action affects a wetland or floodplain.[14] In the CEQ's opinion, a 30-day review period is necessary in the following circumstances also:[15]

- Borderline case, such that there is a reasonable argument in favor of preparing an EIS
- Unusual case, a new kind of action, or a precedent-setting case such as the first intrusion of even minor development into a pristine area
- Case in which scientific or public controversy exists over the proposal

7.4   THE EA DOCUMENTATION REQUIREMENTS

The Regulations provide only sparse directions regarding the preparation and content of an EA. For example, the documentation requirement consists of only one paragraph, which rather than being presented in the main body of the Regulations, is relegated to the last section that defines NEPA terms (§ 1508).

TABLE 7.3

Documentation Requirements for an EA

Need for the proposal

Alternatives (as required by Section 102[2][E] of NEPA)

Environmental impacts of the proposed action and alternatives

List of agencies and persons consulted

7.4.1 PAGE LIMITATIONS

While the Regulations do not specify page limitations for EAs, the CEQ advises agencies to keep their length to a maximum of approximately 10 to 15 pages.[16] As is often the case, agencies may be confronted with a "more is better" mindset, resulting in more costly analyses. While the CEQ has maintained that EAs should be limited to the above page length, less than 30%, or 14 out of 41 of the agencies surveyed by the CEQ, indicated that their assessments normally fall within this range.

7.4.2 REQUIREMENTS

As indicated in Table 7.3, the Regulations specify only four documentation requirements that an EA must meet (§ 1508.9). The first three items shown in the table are common to both the EA and the EIS. The fourth item is required only for the EA.

7.4.2.1 Need for the Proposal

According to the CEQ, the section on the "need" for a proposal should briefly describe the following:[17]

- Information that substantiates the need for the project; incorporate by reference, information that is reasonably available to the public. For example, "This agency is preparing to erect a temporary emergency response facility to replace facilities disrupted or destroyed by hurricane Katrina, in order to facilitate rescue and relief efforts in an effort to minimize further death and adverse health conditions and restore communications and power."
- The existing conditions and projected future conditions of the area impacted by the project. For example, "The area(s) in which the temporary facility will be located or relocated is identified in the attached map. This area consists of ..." [add brief description of the environmental state of the area that will be affected by the location and operation of the facility, focusing on those areas that are potentially sensitive; the goal is to show that refueling sites are not on top of aquifers, nesting areas, graves, sacred sites, etc., that is, show the utility and need to identify actual place-based environmental issues rather than compiling laundry lists of environmental resources that are not at issue.]

7.4.2.2 Proposed Action and Alternatives

In the CEQ's opinion, the section describing the need for the proposal should briefly describe the proposed action and any alternatives that meet the purpose(s) of the proposal.[18] The agency has discretion to determine the number of alternatives. The alternatives should focus on the objectives of the purpose and need statement. For example, the need to use existing infrastructure necessary to support a proposed facility is a potential basis for focusing on a discrete number of alternatives.

When there is consensus about the proposed action based on input from interested parties, the agency may consider the proposed action and proceed without consideration of additional alternatives. Otherwise, the EA should describe reasonable alternatives to meet project needs. (NEPA Section 102[2][E]).

7.4.2.3   Environmental Impacts of the Proposed Action and Alternatives

According to the CEQ, this section should[17]

- briefly describe impacts of the proposed action and each alternative. The alternatives must meet the purpose and need. The description should provide enough information to support a determination to either prepare an EIS or find no significant impact;
- concentrate on whether the action would significantly affect the quality of the human environment (40 CFR 1508.27);
- tailor the length of the discussion to the complexity of each issue. Focus on those human and natural environment issues where impacts are of concern (telephone or e-mail consultations, and discussions with local, tribal, or state and federal agencies with appropriate experience or expertise may help focus such discussions); and
- incorporate by reference, data, inventories, and other information or analyses relied upon (hyperlinks in Web-based documents is encouraged). This information must be reasonably available to the public.

The agency may discuss the impacts (direct, indirect and cumulative) of each alternative together in a comparative description or discuss each alternative separately.

7.4.2.4   Agencies and Persons Consulted

According to the CEQ, this section should list the agencies and persons consulted.[17] For example, as appropriate, the EA should include the people, offices, and agencies that were consulted to ensure that the location of the project did not unintentionally cause an adverse impact.

7.4.3   Suggested Outline

Many agencies typically exceed the minimal requirement depicted in Table 7.3. It is essential that a balance be struck between preparing an EA that meets the goals of NEPA, yet is not exorbitant. A generalized outline is suggested in Table 7.4. This outline may need to be tailored to meet the agency's particular mission and any additional requirements cited in its NEPA implementation procedures.

A description of the affected environment (Table 7.4) is important because it provides a baseline of environmental resources that may be affected. This information is important in assessing the potential significance of an impact.

Commonly cited criteria for determining significance involve the assessment of an action for compliance with environmental permits, laws, and regulations (§ 1508.27[10]). For this reason, Section 5.0 of the suggested outline can assist decision-makers in assessing conformance with

---

TABLE 7.4
Suggested Outline for an EA

Title page
Glossary
Executive summary
1.0 Purpose and need for the proposed action
2.0 Description of the proposed action and alternatives
3.0 Description of the affected environment
4.0 Environmental impacts of the proposed action and alternatives
5.0 Applicable environmental permits and regulatory requirements
6.0 List of agencies and persons consulted
7.0 References

TABLE 7.5

Suggested Outline for the Section on Alternatives

2.0 Description of proposed action and alternatives

2.1 Brief introduction:

    2.1.1 Briefly describe process used to identify alternatives

    2.1.2 Briefly describe alternatives considered but not analyzed

2.2 No-action alternative

2.3 Proposed action:

    2.3.1 Location

    2.3.2 Cost and schedule

    2.3.3 Construction activities

    2.3.4 Operations

    2.3.5 Support activities

    2.3.6 Routine maintenance and upgrades

    2.3.7 Project termination and decommissioning activities (if applicable)

    2.3.8 Mitigation measures (if applicable)

2.4 Alternative A

    (Similar to that shown for Section 2.3, although alternatives in an EA are usually covered in less detail and then dismissed)

2.5 Alternative B

    (Similar to that shown for Section 2.3, although alternatives in an EA are usually covered in less detail and then dismissed)

---

applicable permits and regulatory requirements. Since NEPA is an "up front" planning process, this information can also allow the agency to begin planning for future permits that will be required.

7.4.3.1 Suggested Outline for Section on Alternatives

A generalized outline for describing alternatives, which expands upon Section 2.0 in Table 7.4, is suggested in Table 7.5. This outline may need to be tailored to meet the agency's specific mission and circumstances.

The no-action alternative shown in Table 7.5 is not specifically required by the Regulations, but its inclusion is considered to be a good practice and may be required by the courts (e.g., reasonable alternative). If not for any other reason, it should be included because it provides a baseline against which impacts of the proposed action and alternatives can be compared.

7.4.3.2 Pollution Prevention

The CEQ has issued guidance instructing agencies to take every opportunity to incorporate pollution prevention considerations into their early planning and decision-making processes, including EAs and EISs.[17] Where practical, pollution prevention measures should be incorporated as part of the proposed action, its reasonable alternatives, and its mitigation measures.

7.5 THE FONSI

As we have seen, based on the review of an EA, the purpose of a FONSI is to document a decision-maker's determination not to prepare an EIS because a proposed action will not result in a significant impact (§ 1501.4[e]). Specifically, a FONSI is defined as

> … a document by a Federal agency briefly presenting the reasons why an action, not otherwise excluded (§ 1508.4), will not have a significant effect on the human environment and for which an environmental impact statement therefore will not be prepared …

Agencies shoulder the burden of proving that no significant impacts will occur. Decision-makers must carefully weigh the evidence presented in an EA before concluding that a FONSI is appropriate. Because the agency is faced with the burden of proving nonsignificance, all conclusions presented in the FONSI must be directly related to the analysis presented in the EA. A poorly prepared FONSI is vulnerable to a successful challenge. The importance of maintaining an objective analysis prior to making a determination of significance cannot be overemphasized. As a FONSI becomes more defensible, the likelihood that an adversary may attempt a challenge tends to diminish. Moreover, the agency is also more likely to be viewed positively by the public.

Care must be exercised in demonstrating that no significant impacts will occur. In providing evidence sufficient to justify a determination of nonsignificance, the FONSI should demonstrate how both the intensity and the context of the impacts were considered in reaching the determination.

Each FONSI should be tailored specifically to the proposed action under consideration. The FONSI should be prepared in such a way that it clearly demonstrates that the decision-maker responsible for signing the document thoroughly understands the scope of the action and clearly comprehends the implications and potential for producing significant or nonsignificant impacts. Few things can damage an agency's credibility more than a discussion that is so ambiguous and confusing that it leaves the court unable to believe that the officials responsible clearly understood what they were approving.

7.5.1   DOCUMENTATION REQUIREMENTS FOR FONSIs

The Regulations provide only a cursory discussion of the documentation requirements that a FONSI must meet. These requirements are described in Table 7.6 (§ 1508.13, § 1501.7[a][5]).

As indicated by the second bullet (Table 7.6), the FONSI must either include the EA or a summary of it. There should be no question as to what will take place or its implications for the environment. The FONSI should

- include information concerning the scope of the action and where it will take place;
- indicate who has proposed the action, why it was proposed, and when it is scheduled to be carried out; and
- explicitly state that the proposed action would not result in a significant impact based on the analysis presented in the EA. Such a statement is the legal equivalent of stating that an EIS is not required.

According to the CEQ, the finding is to state succinctly the reasons for deciding that the action will have no significant environmental effects. As applicable, it should also indicate which factors were weighted most heavily in making the determination. Moreover, the FONSI may include the EA or a summary, or incorporate it by reference.[19] A checklist for assisting NEPA practitioners in preparing a FONSI is presented in Table 7.7.

TABLE 7.6
Documentation Requirements for FONSI

Brief explanation of the reasons why the action will not have a significant effect
  on the human environment. If the EA is included, this discussion need not
  repeat discussion in the assessment but may incorporate it by reference.
The EA or a summary of it.
Any other environmental documents related to the scope of the proposed action.

TABLE 7.7

Checklist for Preparing a FONSI

Specific Considerations

1. The FONSI clearly demonstrates that the responsible decision-maker thoroughly understands the scope and comprehends the implications of the action?
2. The FONSI has been specifically prepared and tailored to the action in question?
3. The FONSI
   (a) explicitly states that no significant impacts will result from the proposed action and
   (b) demonstrates that both the intensity and context were taken into account in reaching a decision of nonsignificance (§ 1508.27)?
4. The FONSI conclusively demonstrates and explains why the action will not result in any significant environmental impacts (direct, indirect, and cumulative)?
5. The FONSI indicates which factors were weighed most heavily in reaching the determination of nonsignificance?
6. The FONSI explains the scope of the action (e.g., who, what, when, where, why, and how)?
7. The FONSI includes the EA or a summary of it?
8. The conclusions in the FONSI are directly tied to the analysis presented in the EA?
9. The FONSI notes any other environmental documents related to the scope of the action?
10. The FONSI
    (a) describes any mitigation measures that will be adopted and
    (b) any such measures have been designed and customized to address specific impacts?
11. The EA has adequately evaluated the effectiveness of any mitigation measures committed to in the FONSI?
12. Any mitigation measures adopted in the FONSI
    (a) would mitigate any significant impacts to the point of nonsignificance and
    (b) are free from scientific controversy?
13. Funding and technical means exist for implementing any mitigation commitments made?
14. If mitigation measures are adopted as part of the FONSI, a specific monitoring and implementation plan is included to ensure that the mitigation measures are successfully adopted (such a plan is not specifically required under the Regulations but should be performed as part of good NEPA practice)?

7.5.1.1 EA Checklist Format

Some agencies use an environmental checklist or a Leopold matrix as a tool for reviewing environmental issues and to help in determining if significant issue exists that requires detailed analysis. While use of such tools can be helpful, some EAs are issued using a checklist format. In the author's opinion, an EA checklist format generally does not meet the regulatory requirement to use plain language, nor does such a format generally provide a clear description of impacts, or sufficient evidence that the impacts are nonsignificant, or an adequate discussion of alternatives or potential mitigation measures.

Standardized analysis forms are used for noncontroversial projects with relatively small environmental impacts, particularly where there are no conflicts in alternative uses of available resources. On the other hand, standardized analysis forms are generally inappropriate for controversial or high-profile projects, where the project is relatively complex or involves more than simply evaluated impacts, where mitigation is proposed, or where there are potential conflicts involving alternative uses of available resources.

7.5.1.2 Judicial Review of EAs and FONSIs

Case law has established that an agency's decision not to prepare an EIS (i.e., issue a CATX or FONSI) can normally be overturned only if the decision was arbitrary, capricious, or an abuse of discretion. In reviewing an agency's FONSI, a court's responsibility is to ensure that the agency took a "hard look" at the environmental consequences. In reviewing a FONSI, the court determines whether the agency

- identified and investigated relevant issues and areas of environmental concern;
- provided sufficient evidence and made a convincing case that the impacts would be insignificant; and
- convincingly established that any changes in the project or mitigation measures would sufficiently reduce the potential impacts to the point of nonsignificance.

7.6 MITIGATION

The question of mitigated FONSIs has a rather convoluted history. This issue has important ramifications because a mitigated EA can be completed typically in less than half the time it would normally take to complete an EIS for the same action. Early on, questions were raised about the acceptability of mitigating impacts as a means of avoiding the preparation of an EIS.

At one time, the CEQ discouraged the use of mitigated FONSIs.[20] This position was primarily a result of

- EAs receiving less public review than EISs and
- concerns over the lack of appropriate and rigorous requirements for implementing monitoring and mitigation measures.

In recent years, the controversy has subsided as courts have generally accepted the use of mitigated FONSIs. Today, many agencies are issuing them.

7.6.1 TYPES OF MITIGATION

With respect to EAs, the term "impact mitigation" or more simply "mitigation," generally refers to measures used for reducing significant environmental impacts to the point of nonsignificance, so that the project qualifies for a FONSI.

In contrast, the term "impact minimization" or "impact reduction" generally refers to measures used for reducing adverse impacts regardless of whether the impacts are significant or nonsignificant. Agencies are not legally required to identify or implement measures for reducing impacts that are considered to be nonsignificant. However, consistent with Section 101 of NEPA, agencies should seriously consider the environmental merit of mitigating even nonsignificant impacts. The CEQ encourages agencies to include alternatives and impact-reduction methods in an EA, even if the impacts are considered nonsignificant.[21]

Mitigation measures are additional steps that can be taken to mitigate impacts beyond what would normally be part of the proposed action (§ 1508.25[b][3]). In the opinion of the CEQ, actions that are standard engineering practice or required under law or regulation are not normally considered mitigation measures.[22]

Any mitigation measures adopted as part of a FONSI should be carefully considered by the decision-maker since such commitments are legally binding. Steps should be taken to ensure that funds and technical means exist for implementing such mitigation measures.

7.6.2 ADVANTAGES AND DISADVANTAGES

A persuasive argument can be made that mitigated EAs actually result in substantially less impact than might otherwise occur if an EIS was prepared for the same action. An agency is free to choose any EIS alternative (regardless of its impact), so long as it has been analyzed adequately. In contrast, a mitigated FONSI must reduce impacts to the point of nonsignificance; thus, a mitigated FONSI can actually result in greater environmental protection than would occur if an EIS were prepared.

Mitigated EAs may also have disadvantages. For example, an EIS normally provides a more thorough analysis of impacts and alternatives and may provide a more effective tool for combining the entire environmental process into a single, integrated planning exercise. A mitigated EA can also be used maliciously to reduce or even circumvent public scrutiny and the more comprehensive public involvement process normally associated with an EIS.

Perhaps, most significantly, an EIS provides greater protection from opponents desiring to halt a project. Although an EIS involves more time and resources in its initial preparation, it may ultimately save an agency additional time, resources, and political embarrassment in the event of a legal challenge.

7.6.3   ANALYZING THE EFFECTIVENESS OF MITIGATION MEASURES

Sometimes mitigation measures have been proposed without a corresponding investigation of their effectiveness in diminishing impacts to the point of nonsignificance. It is no surprise that some measures may be highly effective in this regard, while others may have little or no effect.

Case law has clearly established that the analysis must take into account the effectiveness of proposed mitigation measures in reducing potential impacts to the point of nonsignificance. NEPA does not make a distinction between beneficial and adverse impacts. Thus, some courts have required an EIS in situations where mitigating adverse impacts would result in significantly improved environmental quality.[23] Currently, this issue has not been resolved definitively by the courts.

7.6.4   LEGAL CRITERIA FOR MITIGATION

As we have seen, the burden of proof lies with the agency to demonstrate that a mitigated action will not result in a significant impact. For this reason, judicial review of mitigated EAs is often more stringent than for either an EIS or a nonmitigated EA.[24] As shown in Table 7.8, the courts appear to impose six criteria or requirements for mitigated FONSIs. The agency should therefore carefully review the EA in question to ensure that mitigation measures are consistent with these criteria.

As indicated in Table 7.8, disagreement or controversy among experts substantially weakens an agency's ability to prove that no significant impacts will occur. Further, an agency cannot rely on future or to-be-determined mitigation, since there is no way to adequately assess the effectiveness of such measures; more "to the point" mitigation measures must be designed to address specific actions and impacts.

7.7   STREAMLINING THE EA PROCESS

The CEQ encourages the use of EAs as a means to streamline the NEPA process (§ 1500.4[p], § 1500.5[K]). Specific methods for accomplishing this goal are described below.

TABLE 7.8
Legal Requirements for Mitigating EAs

Mitigation measures must effectively reduce impacts to the point of nonsignificance.

Effectiveness of measures should be free from scientific controversy.

Measures must be demonstrably effective. Cursory statements regarding the effectiveness of mitigation measures are generally insufficient; an EA must present evidence to support mitigation claims.

Measures should be fully identified and defined prior to filing the FONSI.

Measures should address specific environmental issues and concerns, including cumulative impacts.

Mitigation measures that are vague or general in nature are normally inadequate.

Measures should show that the methods would effectively mitigate the impacts.

A specific monitoring or implementation plan should be included to ensure that mitigation measures are carried out.

7.7.1 Reducing Duplication and Delays

An agency's review and approval cycle should be examined periodically for inefficiencies. For example, to the extent practical, reviews involving more than one entity should be performed in parallel. Agencies should also consider delegating approval to the lowest competent decision-making level within an organization. Value engineering and other methods may prove useful in identifying and rectifying such inefficiencies (see Section 2.3).

A number of agencies have reported significant savings when efforts were made to identify and coordinate NEPA with environmental studies performed by other agencies (§ 1506.2[b][4] and § 1502.5[b]).

7.7.2 Tiering

Agencies are strongly encouraged to use tiering as a means of expediting the NEPA compliance process (§ 1502.20). An EA prepared for an action that is within the scope of a broad EIS need only summarize the issues discussed in the EIS and state where a copy of the EIS can be obtained. The EA can then incorporate discussions presented within the EIS by reference.

Questions have been raised regarding the appropriateness of tiering a new EA from an existing EA. According to the CEQ, tiering one EA from another is inappropriate (i.e., EAs should only be tiered from an EIS). The Regulations do not address this issue.

However, an EA can incorporate another EA by reference. Thus, this issue is essentially a mute point, since the practice of incorporating by reference provides an agency with essentially the same capability as that provided by tiering.

7.7.3 Reducing the Length of Assessments

Factors such as unusual circumstances, the degree to which an action may be controversial, or the complexity of the proposed action all play a substantial role in determining the ultimate length of an EA. For this reason, a sliding-scale approach (see Section 2.2) should be applied in determining the appropriate length and complexity of an EA. But one must also be careful that going the extra mile in some circumstances does not establish an institutional precedent. To prevent setting such a precedent, some agencies have added a "disclaimer," indicating why additional material, which is not required under the Regulations, has been included.

Agencies can also benefit greatly by adhering to the CEQ's guidance and direction for reducing the length of EAs. As just described, incorporating information by reference can be a particularly useful streamlining practice and is highly encouraged by the CEQ.

PROBLEMS

1. A decision-maker reviews an EA for the construction of an electrical transmission line. The EA evaluates 10 potentially significant impacts. Nine of the impacts are later determined to be nonsignificant. However, the 10th issue involves destruction of critical habitat and appears to be potentially significant. Can the decision-maker issue a FONSI if the overwhelming majority of the impacts are nonsignificant but one impact may be potentially significant? If not, what courses of action are open to the decision-maker?
2. What is the estimated ratio of the preparation of EAs to that of EISs?
3. Must an EA consider potential cumulatively significant impacts? Explain your answer.
4. According to the CEQ, how long should it take to prepare an EA?
5. Is an EA a public document?
6. An EIS contractor must sign a statement indicating that it has no financial or other interest in the outcome (conflict of interest provision). Does a similar restriction exist on who may prepare an EA?

7. Under what circumstances must a FONSI be made publicly available for a 30-day review period?
8. What are the regulatory documentation requirements for a FONSI?
9. Is it possible to mitigate potentially significant impacts such that a FONSI can be issued for a proposal?
10. Suppose a proposal involves building a 10,000-ft^2 office building that would house approximately 75 workers in a highly developed 50 Mi^2 military installation that has already sustained a significant cumulative impact. The military installation has over 200 facilities and a staff of 10,000 and is located in a relatively isolated desert environment. Assume that a preliminary investigation indicates that the building would not result in a significant direct or indirect environmental impact. You are assigned responsibility for preparing an EA for the proposed office. Using the Cilix method, prepare a hypothetical description of cumulative impacts on noise and land use resources. You may develop your own criteria and details to support the description of the cumulative impact.

REFERENCES

1. CEQ, *The National Environmental Policy Act: A Study of Its Effectiveness after Twenty-Five Years*, p. 19, January 1997.
2. Eccleston C. H., *Effective Environmental Assessments: How to Manage and Prepare NEPA EAs*, CRC/Lewis Publishers, Boca Raton, FL, 2001.
3. Council on Environmental Quality, Forty Most Asked Question Concerning CEQ's National Environmental Policy Act Regulations (40 CFR 1500–1508), *Federal Register* 46(55), 18026–18038, March 23, 1981, Question Number 35.
4. U.S. DOE, *NEPA Lessons Learned*, Issue No. 20, p. 17, September 1, 1999.
5. CEQ, 40 Questions, Question No. 38.
6. Eccleston C. H., *Environmental Impact Statements: A Comprehensive Guide to Project and Strategic Planning*, John Wiley & Sons, New York, 2000 and Eccleston C. H., *The NEPA Planning Process: A Comprehensive Guide with Emphasis on Efficiency*, John Wiley & Sons, New York, 1999.
7. *Sierra Nevada Forest Protection Campaign v. Weingardt*, Nos. CIV-S-04-2727, -05-0093, 35 ELR 20151 (E.D. Cal. June 30, 2005).
8. *Sierra Club v. Watkins*, 808 F. Supp. 852 (D.D.C. 1991).
9. Davis R., *National Environmental Policy Act*, Presented as part of Government Institutes, Inc. class on Environmental Laws and Regulations, Richland, WA, September 1987.
10. *Native Ecosystems Council v. Dombeck*, 304 F.3d 886 (9th Cir. 2002).
11. McCold L. and Holman J., Presented at the 18th Annual Conference of the National Association of Environmental Professionals, Raleigh, NC, May 24, 1992.
12. Eccleston C. H., The Cilix methodology: a practical methodology for assessing cumulative impacts in environmental assessments, *Journal of Federal Facilities Environmental Management*, pp. 37–44, Winter 2006.
13. 40 Code of Federal Regulations (CFR) 1508.27(b).
14. Executive Orders 11988, 11990.
15. CEQ, 40 Questions. Question No. 37b.
16. CEQ. 40 Questions. Question No. 36a.
17. Council on Environmental Quality, *Memorandum for Federal NEPA Contacts: Emergency Actions and NEPA*, Attachment #2, *Preparing Focused, Concise and Timely Environmental Assessments*, September 8, 2005.
18. Council on Environmental Quality, *Guidance on Pollution Prevention and the National Environmental Policy Act*, 58 FR 6478, January 29, 1993.
19. CEQ, 40 Questions, Question No. 37a.
20. CEQ, 40 Questions.
21. CEQ, 40 Questions. Question No. 39.

22. CEQ, *Public Memorandum: Talking Points on CEQ's Oversight of Agency Compliance with the NEPA Regulations*, 1980.
23. Huber K. D., NEPA: mitigation and the need for an environmental impact statement, *Federal Facilities Journal*, 1(2), Summer 1990; *EDF v. Marsh*, 651 F.2d 983 (5th Cir. 1981); and *National Wildlife Federation (NWF) v. Marsh*, 721 F.2d (11th Cir. 1983).
24. Daniels S. E. and Kelly C. M., Deciding between an EA and an EIS may be a question of mitigation, *Western Journal of Applied Forestry*, 5(4), March 1991.

8 Preparing Environmental Impact Statements

This chapter focuses on providing environmental planners and agency officials with an understanding of the basic step-by-step procedural or process requirements that must be followed in preparing an environmental impact statement (EIS). Emphasis is placed on providing the reader with specific tools and techniques for streamlining the EIS process.

A thorough treatment of all the requirements relevant to the preparation of an EIS is beyond the scope of a single chapter. Thus, for a more detailed discussion of the EIS process including the documentation requirements, the reader is referred to the companion book, *Environmental Impact Statements*.[1]

8.1 OVERVIEW OF FUNDAMENTAL EIS CONCEPTS

The EIS process consists of two distinct phases, which will be described later:

1. Draft EIS
2. Final EIS

The term EIS is a generic term used to describe a statement in either its draft or final stage. The EIS process essentially begins at the time an agency first considers a proposed action that may significantly affect the quality of the human environment.

As we consider the details of the EIS process, the reader should note that the principal purpose of an EIS is to serve as an "action-forcing" device "… to help public officials make decisions that are based on understanding of environmental consequences and take actions that protect … the environment (§ 1500.1[c])."

The following sections describe fundamental concepts underlying the EIS process. These concepts, beginning in Section 8.2, provide the basis for describing the preparation of EISs.

8.1.1 LEAD AND COOPERATING AGENCIES

There are times when an action requiring an EIS may involve two or more federal agencies (§ 1501.6). The agency(ies) with overall responsibility for preparing the EIS is referred to as the lead agency (§ 1501.5, § 1508.16). State and federal agencies may also serve as "joint lead agencies" as long as the group includes at least one federal agency (§ 1501.5[b]).

Any agency(ies) that assists the lead agency in preparing an EIS is referred to as a cooperating agency. These agencies may include any federal agency other than the lead agency having "jurisdiction by law or special expertise with respect to any environmental impact" that will be considered in the EIS. Cooperating agencies may also include state and local agencies, or an Indian tribe if the proposed action may affect a reservation (§ 1508.5).

8.1.1.1 Disputes

Occasionally, disagreements may arise between the lead and cooperating agencies over the scope and content of an EIS. When this happens the agencies are expected to settle any disputes among themselves. For example, an EIS could be deemed technically adequate but still fail to cover sufficiently the needs of a cooperating agency if the lead agency fails to provide a comprehensive scope.[2]

Where both a lead and a cooperating agency believe that the EIS has been adequately prepared, it can be issued even if the two may disagree over the ultimate conclusions or decisions recorded in the record of decision (ROD); in other words, the EIS can be issued as long as the agencies do not disagree over the content of the information or the analysis performed. The lead and cooperating agencies may wish to select their own preferred alternatives and are free to pursue separate courses of action once their respective RODs have been issued.[2] (For additional information, refer to Section 8.4.)

8.1.2   Selecting the Lead Agency

Potential candidates for the position of lead agency must decide among themselves which one will be appointed. Factors used in determining the lead agency designation are listed below in order of descending importance (§ 1501.5[c]):

1. Magnitude of the agency's involvement
2. Project approval/disapproval authority
3. Expertise concerning the action's environmental effects
4. Duration of the agency's involvement
5. Sequence of the agency's involvement

If, after a period of 45 days, the concerned federal agencies cannot agree on a lead agency, a request for determination can be filed with the Council on Environmental Quality (CEQ), which will make the decision (§ 1501.5[e]).

Responsibilities. Responsibility for preparing an EIS lies with the lead agency or with a contractor selected by it. The lead agency may also delegate this responsibility to a cooperating agency (§ 1506.5[c], § 1501.6[b]). However, because Section 102(2)(C) of the Act strongly implies that an EIS can only be prepared by a federal agency, questions arose early on regarding the extent to which a federal agency can delegate its responsibility.

NEPA Amendment. As a result of mounting confusion, Congress amended the National Environmental Policy Act (NEPA) in 1975 by passing Public Law 94-83. This added a new section, 102(2)(D), enacted at the time primarily to facilitate situations where state agencies receive block grants from a federal agency for highway construction. The amended Act allows EISs to be prepared by state agencies or state officials in certain situations where major actions are funded under a federal program or by a grant to a state.[3] However, federal agencies are still responsible for the scope, content, and objectivity of the EIS. The allocation of responsibilities for specific environmental issues should be completed during the scoping stage (§ 1501.7[a][4]).

8.1.2.1   Lead Agency Responsibilities

Once a lead agency has been identified it should begin seeking cooperation and assistance from other federal agencies that have jurisdiction by law or special expertise concerning issues related to the proposal.[4] The lead agency is also expected to request the participation of cooperating agencies at the earliest possible time and must identify by letter or memorandum the other agencies that will undertake cooperating responsibilities (§ 1501.5[c]). A federal agency may also request the lead agency to designate it as a cooperating agency. The lead agency should solicit cooperation from both state and local agencies as well as from affected Indian tribes in cases relating to their lands or sites of religious or cultural significance to the tribe.

The lead agency is ultimately responsible for the content and accuracy of the EIS. As funds permit, it is also expected to underwrite the costs of the activities and analyses that it requests cooperating agencies to perform. In their budget requests, potential lead agencies are expected to request funds sufficient to cover the cost of preparing the EIS (§ 1501.6). Other lead agency responsibilities include (§ 1501.7)

- publishing the notice of intent (NOI) in the *Federal Register*,
- determining the scope of the EIS,
- identifying and eliminating insignificant environmental issues, and
- determining the EIS schedule.

Cooperating Agency's Responsibilities. Under the direction of the lead agency, cooperating agencies are expected to assist it in providing information and preparing the required analysis. Cooperating agencies may be expected to provide staff resources to assist the lead agency in the analysis and to use their own funds to cover activities that the lead agency is unable to finance (§ 1501.6[b][3]).

Occasionally, a cooperating agency may not be able to meet its responsibilities in preparing an EIS. In these cases, the cooperating agency must inform the lead agency in writing that it is unable to meet some of the commitments requested because of other program obligations. A copy of this reply must also be submitted to the CEQ (§ 1501.6[c]).[4]

8.1.2.2 Applicants

Where an action involves an applicant, the preparation of an EIS must begin early in the process and "no later than immediately after an application is received" by the agency (§ 1502.5[b]). If requested by an applicant, the agency must set time limits on the EIS process that also need to be consistent with the purposes of NEPA and other essential considerations of national policy (§ 1501.8[a]).

Taking Action against an Applicant. An agency must notify an applicant promptly that it will take action to ensure compliance with NEPA's requirements in cases where the agency is considering an application from a nonfederal entity and learns that it is about to take an action within the agency's jurisdiction that would (§ 1506.1)

1. result in an adverse impact, or
2. limit the choice of reasonable alternatives.

Information Provided by Applicants. Federal agencies may request an applicant to furnish information that will be used in preparing the EIS (§ 1501.2[d]). In these cases, the agency should outline the information it requires.

An agency may request an applicant to prepare an environmental report or to submit other types of data, but it is also required to carry out an independent verification and evaluation to ensure the accuracy of the data provided (§ 1506.5). An analysis that exhibits "unquestioning acceptance" of a project applicant's statements regarding its objectives may be deemed defective. The agency must also conduct or commission an independent analysis of the alternatives offered by an applicant.[5] Direct responsibility for preparing the actual EIS remains with the federal agency or with the independent contractor selected by it. This is because the EIS conflict of interest provision prevents applicants from preparing the actual EIS themselves.[6]

Case law indicates that federal agencies may charge applicants the cost of preparing the required NEPA documentation for their proposed actions. For example, in one instance, the Bureau of Land Management charged an applicant the cost of preparing NEPA documentation to cover a right-of-way that it had requested.[7] When challenged, the court ruled that the bureau was correct in assessing the charge because the EIS would not have been prepared had the applicant not made the request.

8.1.3 SCHEDULE AND TIMING REQUIREMENTS

The principal steps to follow in preparing a typical EIS are outlined in Figure 8.1. Table 8.1 provides a comprehensive list of all time limits and schedule requirements presented in the NEPA regulations (Regulations) governing various aspects of the EIS process.

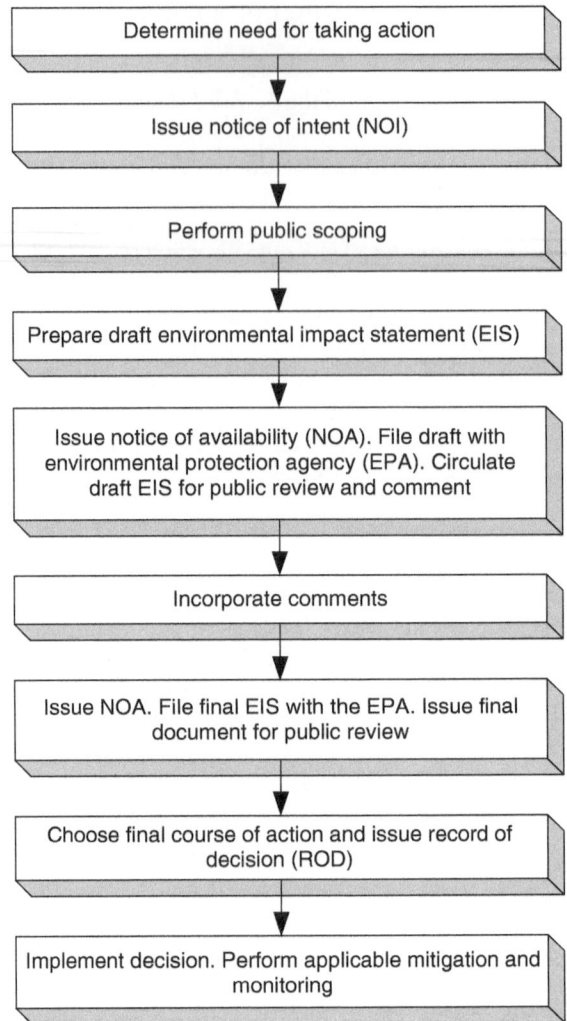

FIGURE 8.1   Principal steps in a typical EIS process.

8.1.3.1   When Should an EIS Begin?

An EIS must be prepared early enough to contribute to the decision-making process. It must not be used to rationalize or justify decisions already made (§ 1508.23, § 1502.5).

An agency must begin preparation of an EIS as close as possible to the time in which the agency is developing or is presented with a proposal, so that the statement can be completed in time to assist the decision-maker in reaching a final decision. Table 8.2 provides specific regulatory directions as to when an EIS should be phased to support typical types of activities (§ 1502.5).

8.1.3.2   How Long Should an EIS Take?

"Documentation procrastination" is a term that has been used in referring to the practice of failing to set a timely schedule for the preparation of an environmental assessment (EA)/EIS. An achievable schedule should be established. As specified in the Regulations, the preparation of an EIS even for large and complex projects should normally require 12 months or less to complete; a programmatic EIS (P-EIS) may require a somewhat longer period. The CEQ believes that such a time frame

TABLE 8.1
Schedule and Time Limits for Preparing an EIS

- Completion of an EIS should normally require less than 1 year.[8]
- If, after a period of 45 days, an agreement cannot be reached regarding the designation of a lead agency, a request for determination can be filed with the CEQ, which will then make that decision (§ 1501.5[e]).
- In cases where an agency circulates a summary of an EIS and receives a timely request for the entire statement and for additional time to comment, the time shall be extended by at least 15 days beyond the minimum period (§ 1502.19[d]).
- In cases where a draft EIS is to be considered at a public hearing, the agency should make the statement available to the public at least 15 days in advance unless the purpose of the hearing is only to provide information pertaining to the draft (§ 1506.6[c][2]).
- Following consultation with the lead agency, the EPA may reduce or extend the periods prescribed in the Regulations. However, if the lead agency does not concur with the extension of time, the EPA may not extend it for more than 30 days (§ 1506.10[d]).
- A proposed action shall be made or recorded under § 1505.2 by a federal agency as follows (§ 1506.10[b]):
  1. 90 days after publication of the NOA for a draft EIS or
  2. 30 days after publication of the NOA for a final EIS.
- Where a final EIS is filed within 90 days following the filing of the draft EIS with the EPA, the minimum 30-day period and the minimum 90-day period may run concurrently. However, subject to § 1506.10[d], agencies shall allow not less than 45 days for comments on draft EISs (§ 1506.10[c]).
- A legislative EIS may be transmitted to Congress up to 30 days later than its accompanying legislative proposal to allow time for completion of an accurate statement that can serve as the basis for public and congressional debate (§ 1506.8[a]).
- In responding to a referral, the CEQ shall take no longer than 60 days to complete actions for resolving the issue (§ 1504.3[g]).
- A referring agency shall deliver its referral to the CEQ no later than twenty-five (25) days after the final EIS has been made available to the EPA, commenting agencies, and the public (§1504.3[b]).
- No later than twenty-five (25) days after a referral has been made to the CEQ, the lead agency may deliver a response to both the CEQ and the referring agency (§ 1504.3[d]).

TABLE 8.2
When an EIS Should Be Phased to Support Agency Activities

Projects directly undertaken by a federal agency: An EIS must be prepared at the feasibility analysis (go/no-go) stage.
  If necessary, the EIS may be supplemented at a later stage when more reliable information is available.
Private applications to the agency: Preparation of the EA/EIS must begin no later than immediately after the application has been received.
Adjudications: A final EIS must be completed before the final staff recommendation is submitted to the decision-makers.
  As appropriate, the EIS may follow preliminary hearings designed to gather information for use in its preparation.
Informal rulemaking: A draft EIS should be prepared in time to accompany the proposed rule.

is well within the planning cycle of most large projects.[8] In practice, however, the EIS process often exceeds these guidelines, indicating such guidance may be unrealistic.

8.1.3.3   What Is the Maximum Page Limit for an EIS?

A final EIS (e.g., paragraphs [d] through [g] of § 1502.10) should normally be less than 150 pages in length. Where a proposal is of unusual scope or complexity, the EIS should normally be less than 300 pages (§ 1502.7).

8.1.4   EIS CONTRACTORS

Due to staffing and expertise limitations outside, consulting firms are often used in preparing EISs, but it should be noted that their use has raised issues involving potential conflicts of interest.

8.1.4.1   EIS Contractors and Conflicts of Interest

To avoid any conflict of interest, an EIS contractor can only be selected by the lead agency or, where appropriate, by a cooperating agency (§ 1506.5[c]). Specifically, the Regulations state

> Contractors shall execute a disclosure statement prepared by the lead agency, or where appropriate the cooperating agency, specifying that they have no financial or other interest in the outcome of the project.

This disclosure requirement is interpreted broadly by the CEQ to mean "... any known benefits other than general enhancement of professional reputation."[9] Under this interpretation, a company is not eligible to prepare an EIS if it has a guarantee of future work. Whereas § 1506.5(c) prohibits private contractors having a vested interest in the outcome from actually preparing an EIS, it does not specifically prevent them from participating in its preparation.

For example, a conflict of interest may exist if a construction company has a guarantee of future design or construction work on a proposed project. A conflict might also include indirect benefits such as a guarantee in which the company would benefit from being awarded the contract to build a new business center following its construction of a new roadway off-ramp. However, the company would probably not be disqualified from preparing an EIS simply because it had been involved in developing initial data or plans for the proposed project, as long as it has no future financial or other interest in the outcome of the proposal. It is also free to submit bids for future work on the project after the EIS has been approved.[10] The reader should note that this EIS disclosure requirement does not specifically apply to the preparation of EAs.

The CEQ has issued an opinion indicating that at times it believes agencies have been overly strict in interpreting this provision. For example, some firms have been excluded from bidding on an EIS contract simply because they had links to a parent company with design or construction capabilities. As interpreted by the CEQ in § 1506.5(c), firms are merely prohibited from preparing an EIS if they have financial or other interests in the outcome of the project during the period in which the EIS is under way.[6]

If a contractor is selected to prepare an EIS, the responsible federal official must provide the necessary guidance and also "participate in [its] preparation." The agency is responsible for independently evaluating and verifying the accuracy of the EIS prior to its approval and assumes full "responsibility for its scope and contents (§ 1506.5[c])."

Preparer versus Participant. A review of case law reveals that failure to comply with the conflict of interest provision can, in and of itself, provide a basis for successfully challenging an EIS.[11] Some confusion centering on the use of the term "preparer" in § 1506.5(c) has developed regarding the applicability of this provision. Neither the Regulations nor the subsequent CEQ guidance has clearly defined this term. The courts have made a distinction between the roles of a "preparer" versus that of a "participant." A preparer is one who translates information into written form or writes a document. It follows that entities may participate (e.g., contribute information) in the preparation of an EIS without being designated as preparers.

The difference between being considered a preparer or a participant largely centers on the degree of discretion one has to accept, revise, or reject information submitted for consideration by a participant. For instance, a contractor not involved in preparing an EIS but engaged in preparing significant support or background papers for it is unlikely to be considered a preparer and would therefore not be subject to the disclosure statement requirement.[12] Similarly, other courts have concluded that the participation of a consultant in the preparation of an EIS and its supporting documentation is not improper, provided the agency takes responsibility and actively participates in preparing the analysis.[13]

Generally, individuals who have not been granted authority to determine the content of an EIS may be given responsibility for formulating proposed responses to EIS comments or for organizing a team to develop comments without being considered preparers. This is because the individuals' roles have been limited to providing information and making recommendations. Conversely, granting entities with decision-making authority over the content of the EIS is likely to trigger the definition of a preparer and would therefore be subject to the conflict of interest requirement.

8.1.5 INTEGRATING OTHER ENVIRONMENTAL REQUIREMENTS

A draft EIS must list all federal permits, licenses, and other entitlements that must be obtained in order to implement the proposal (§ 1502.25[b]). Permit requirements are usually identified through agency consultation.[14] As feasible, the NEPA analysis should be prepared jointly with applicable state or local agencies (§ 1502.5[b]).

During scoping, the lead agency is also required to identify related environmental review and consultation requirements so that the lead and cooperating agencies may prepare other required studies concurrently and integrate them with the EIS (§ 1501.7[a][6], § 1502.25[a]). As described in Chapter 4, some of these requirements include

- Fish and Wildlife Coordination Act (16 U.S.C. 661 et seq.),
- National Historic Preservation Act (NHPA) of 1966 (16 U.S.C. 470 et seq.),
- Endangered Species Act of 1973 (16 U.S.C. 1531 et seq.), and
- Other environmental review laws and executive orders.

The EIS process is to be integrated with other planning and environmental review procedures required by law or agency practice so that all such procedures run concurrently rather than consecutively (§ 1500.2[c], § 1502.25[a]).

The lead agency is expected to consult other agencies that have jurisdiction by law or special expertise in environmental issues related to the preparation of an EIS. This requirement is known as consultation. The term "jurisdiction by law" means having agency authority to approve, veto, or finance all or part of a proposal (§ 1508.15). The term "special expertise" means possessing statutory responsibility, agency mission, or related program experience (§ 1508.26).

The Regulations direct federal agencies to cooperate with state and local agencies in order to reduce duplication between NEPA and comparable state and local environmental requirements. Where these requirements do not conflict with NEPA, federal agencies are directed to prepare an EIS that fulfills both these requirements and those of NEPA (§ 1506.2[c]).

8.1.5.1 Pollution Prevention

As described in Chapter 4, the CEQ has issued guidance instructing agencies to take every opportunity to incorporate pollution prevention (P2) considerations into their early planning and decision-making processes. Where appropriate, documents such as the EA, EIS, and ROD should record the results of this review.[15] The term "pollution prevention" is defined more broadly by the CEQ than by the EPA under the Pollution Prevention Act of 1990.[16]

Agencies are encouraged to include P2 as an issue for review during the scoping process. As appropriate, P2 measures should be included as part of the proposed action, reasonable alternatives, and mitigation measures examined by the agency.

8.1.6 NATIONAL REGISTER OF HISTORIC PLACES

NEPA pronounces a goal to

... preserve important historic, cultural, and natural aspects of our national heritage ... (NEPA Section 101[b][4])

As described in Chapter 4, Section 106 of the NHPA of 1966 regulates federally assisted actions that may affect cultural and historic resources. Consultation with the state historic preservation office (SHPO) may also be required.

Under the NHPA, actions must be reviewed for potential significance where a reasonable possibility exists for them to impact structures or sites having potential cultural, historic, or archaeological value. Specifically, a review must be conducted to determine if a resource is eligible for listing on the National Register of Historic Places. Where applicable, a cultural resources review (CRR) is conducted to identify the potential impacts to such structures and sites.

Programmatic Agreement (PA). Any study of the kinds described above should begin early in the process. In certain instances, it may be impractical and too costly to identify impacts until the detailed design stage or where the activities would extend over a wide geographical area.

One approach for dealing with this contingency might involve entering into a PA with the SHPO, specifying the steps to be taken to minimize impacts to potential resources encountered after the project is under way. Under this approach, the EIS could discuss the PA and reference specific steps that would be followed to avoid or minimize impacts.

8.1.6.1   Endangered Species

Where potential actions may significantly disturb any plants, animals, or their habitats, a "biological resources review" may need to be performed. As described in Chapter 4, the biological resources review should identify those species present in the areas under review. Special consideration is given to species of concern including any flora or fauna listed by state or federal agencies as sensitive, endangered, or threatened. Species that are candidates for listing should also be reviewed. Consultations with the U.S. Fish and Wildlife Service should be conducted, especially where a biological resources review indicates that a listed species or a candidate for listing may be impacted.

8.1.6.2   Floodplain and Wetlands Requirements

As described in Chapter 4, President Carter issued Executive Order 11990 pertaining to the protection of the nation's wetlands.[17] Under this order, federal agencies are required to establish procedures to ensure that adequate consideration is given to protecting wetlands during the planning process.

Wetlands include geomorphic features such as swamps, marshes, and ponds. The specific designation of a wetland area is normally based on three primary characteristics: (1) soil type, (2) frequency of saturation, and (3) types of species present.

The reader should note that an area does not need to be continuously wet to be considered a wetland. An area may be designated a wetland even if it is only occasionally saturated, as long as it is capable of supporting certain types of species.

In that same year, President Carter issued Executive Order 11988 governing management of the nation's floodplains.[18] This order requires federal agencies to establish procedures for ensuring that the effects of proposed federal actions within a floodplain are adequately considered during an agency's planning process. Implementing guidelines were issued in the following year.[19]

A floodplain/wetland assessment is normally required if an activity may potentially impact either of these. The U.S. Corps of Engineers should be contacted when preparing such an assessment.

Criminal liabilities are associated with performing activities without prior authorization that may affect a floodplain or wetland. For this reason a review should be performed early in the EIS process to determine if one or both of these would be affected by a proposed action or one of its alternatives.

8.1.7   Classified Proposals, Emergency Situations, and Periodic Reviews

With respect to preparing an EIS, three circumstances require special mention: classified proposals, emergency situations, and periodic NEPA reviews.

8.1.7.1 Classified Proposals

Agencies are not required to disclose a NEPA document publicly if the information it contains might jeopardize national security. As applicable, an agency's NEPA implementation should include specific procedures for exempting classified information from public disclosure (§ 1507.3[c]). Such procedures apply to NEPA documents that have been properly classified under criteria established by Executive Order or statute in the interest of national defense or foreign policy.

It is important to note that such exemptions do not absolve an agency from its responsibility to prepare and use the EIS in the agency's internal decision-making process.[20] Dissemination of the classified portion can be restricted to appropriate decision-makers and individuals according to requirements that apply to classified information.

Information may be organized in the EIS that separates classified information from unclassified material. Unclassified portions of the document can be made available for public review, whereas the classified portion may be circulated only within an appropriate branch of the agency in accordance with the procedures for handling such material (§ 1507.3).

8.1.7.2 Emergency Situations

As described in Section 5.2, a special provision has been established for situations where emergency circumstances make it necessary for an agency to take actions that do not comply with NEPA's requirements (§ 1506.11). Agencies are expected to consult the CEQ about alternative arrangements. These must be limited to actions considered necessary to respond to the emergency.

8.1.7.3 Guidance on Periodically Reexamining EISs

The CEQ has identified two conditions where EISs more than 5 years old should be reexamined to determine if they are still valid where

- a proposed action has not yet been implemented, or
- the proposed action involves an ongoing action.

If an agency's review indicates that a substantial change has occurred, or if there are significant new circumstances or information relevant to environmental concerns that have bearing on the proposed action or its impacts, a supplemental EIS (S-EIS) must be prepared (§ 1502.9[c]).[21]

At least one court has ruled that an EIS cannot be invalidated simply because it is too old. In this case, an EIS was prepared in 1975 for the construction of a levee by the Army Corps. The construction was a progressive project performed over many years. An EA was initially prepared to cover the first portion of the construction project. This concluded that no significant impacts would result that had not already been evaluated in the EIS. In 1990, the EIS was challenged on the basis that it was too old to continue providing coverage for work that was starting on a new segment of the project. The court determined that there was no reason to prepare a new EIS simply because the existing EIS was old. In other words, as long as the original information and conclusions remain valid, an EIS cannot be invalidated simply because of its age.[22]

8.2 PREPARING THE EIS

The following sections describe the special issues of interest, procedural requirements, and step-by-step process for preparing an EIS.

8.2.1 NOTICE OF INTENT

Following the decision to prepare an EIS and prior to the beginning of the formal scoping process, the lead agency must publish an NOI in the *Federal Register* (§ 1501.7) that describes both

TABLE 8.3

Methods for Notifying Parties Where Effects Are Primarily of Local Interest

- State and area wide clearinghouses
- Consulting with Indian tribes whose lands or sites are of religious and cultural significance
- Local newspapers and other local media
- Interested community organizations and small business associations
- Newsletters and direct mailing to owners and occupants of nearby or affected property
- Posting of notice on and off site in the area where the action is to be located

the proposed action and possible alternatives (§ 1508.22). Its purpose is to inform the public and other interested parties that an EIS will be prepared. In situations where a long lag time may exist between the decision to prepare the EIS and its actual preparation, an agency's implementation procedure may allow it to postpone the publication of the NOI until a date is reached that still provides reasonable time before work begins on the draft EIS (§ 1507.3[e]).

If an agency fails to provide appropriate notice, the lapse may provide sufficient grounds for successfully challenging the EIS. However, a party does not necessarily have sufficient grounds to challenge an agency on this point alone if it has received notice by some other means.[23]

Table 8.3 presents suggested groups to which notices may be directed, and methods of notification. Where an action may be of national concern, notices must be mailed to national organizations that are expected to have an interest in the matter (§ 1506.6[a]).

The U.S. Environmental Protection Agency's Office of Wetlands, Oceans, and Watersheds has issued the guidance document *Community Culture and the Environment: A Guide to Understanding a Sense of Place*.[24] The guide, together with related training, provides tools for working with community groups to protect the environment. Copies of this guide can be obtained from the National Center for Environmental Publications and Information at (513) 489-8190, (800) 490-9198 or by mail to NCEPI, U.S. EPA Publications Clearinghouse, P.O. Box 42419, Cincinnati, OH, 45242.

8.2.1.1 *Federal Register*

The *Federal Register* is the official daily publication for rules, proposed rules, and notices of federal agencies and organizations, as well as executive orders and other presidential documents. It is updated daily by 6 a.m. and is published Monday through Friday, except federal holidays.

Information on the availability of documents, schedule of meetings, and decisions is published in the register. In addition, EPA publishes a list of EISs that they have received from agencies each week and a summary of ratings on EISs that they have just reviewed.

This register is the location where one can find notices from federal agencies regarding their NEPA actions. It can be accessed at http://www.gpoaccess.gov/fr/index.html. The easiest way to pull up notices is to have as much information as possible. Key words such as the name of the agency, location of the action, and date or date ranges of the publication are all helpful in the search.

8.2.2 THE SCOPING PROCESS

It is imperative that an agency accurately determines the proper scope of potential actions and their associated impacts. The term "scoping" is a NEPA expression that describes one major public involvement aspect of the NEPA EIS process (§ 1501.7). A well orchestrated scoping process provides agencies with a particularly effective means for reducing paperwork and delays. The concept of a formal and public scoping process was one of the features that the public most strongly supported during a review of the draft Regulations in 1977.[25]

TABLE 8.4
Principal Goals of Scoping

- Ensure that all problems are identified early in the process and are properly studied
- Identify alternatives that will be examined
- Identify significant issues that need to be analyzed
- Eliminate unimportant issues
- Identify public concerns
- Identify state and local agency requirements such as permits and land use restrictions

---

An agency begins the formal scoping process following publication of the NOI. If the proposal is later canceled, the agency should issue a notice of cancellation in the *Federal Register*.

Each proposed action represents a unique set of circumstances. Not surprisingly, the level of public interest may vary greatly. A sliding-scale approach is recommended in determining the appropriate degree of public participation. This approach recognizes that the degree of public participation varies with the particular circumstances since some proposed actions, particularly those that are controversial, may necessitate more extensive public participation efforts than others.

8.2.2.1 Purpose of Scoping

The purpose of scoping is to solicit input from other agencies and the public so that the analysis can be more clearly focused on issues of genuine concern.

Table 8.4 lists specific goals that the scoping process is intended to accomplish.[6] Attention is often focused on identifying the scope of alternatives, issues, and the impact that will be analyzed. Just as important, however, is the use of this process to de-emphasize insignificant issues so as to narrow the scope of the EIS. The author refers to this task as de-scoping.

There are two situations where the scoping process is not required for the preparation of an EIS (§ 1502.9[c][4], § 1506.8[b][1]):

- S-EIS
- Legislative EISs

It is important to note that commenting on a proposal is not a "vote" on whether the proposed action should take place. Nonetheless, the information provided by the public during the EA/EIS process can influence the decision-makers and their final decisions because NEPA does require that federal decision-makers be informed of the environmental consequences of their decisions.

8.2.2.2 Initiating the Scoping Process

While the Regulations do not prohibit the use of the public scoping process before issuing an NOI, this cannot substitute for the normal scoping process that follows afterward. The only exception to this rule involves cases where scoping is performed before a decision has been made to proceed with an EIS. In these instances the early scoping process may be substituted. This provision only applies if an earlier public notice is issued for the preparation of an EA that clearly indicates that the EA scoping process might be used to substitute for the later EIS scoping process. Once the EIS process begins, however, the NOI must still be issued and it must state that written comments on the scope of alternatives and impacts will still be accepted and considered.[26]

The Regulations do not mandate a specific duration or schedule for performing scoping, nor do agencies, for example, have any obligation to extend a scoping period for public comment beyond the date originally set.[27]

The lead agency must publicly identify any EA or other EIS under preparation that is related to the scope of the EIS being prepared. Other environmental review and consultation requirements must also be identified so that these requirements can be integrated with it (§ 1501.7[a]).

8.2.2.3 Performing the Scoping Process

The steps and measures taken to satisfy the scoping requirement are largely left to the discretion of individual agencies as are the methods used to seek public input.[6] The agency may choose whatever communications methods are considered best for informing the public (whether local, regional, or national) and obtaining input on the proposal. Videoconferencing, public meetings, conference calls, formal hearings, or informal workshops are all legitimate ways to conduct scoping.

Although the CEQ strongly encourages that public scoping meetings or hearings be held, the Regulations do not specifically mandate this practice. Nonetheless, many federal agencies now do require that public meetings or hearings be held. The following factors may indicate a need to conduct a public meeting or hearing (§ 1506.6[c]):

- When there is substantial environmental controversy concerning the proposed action
- When there is substantial interest in holding the hearing
- When a request for a hearing has been made by another agency with jurisdiction over the action, supported by reasons why a hearing will be helpful

8.2.3 Decision-Based Scoping and Decision-Identification Tree

No single approach is universally accepted for determining the scope of an EIS. This is most commonly determined by first identifying the range of activities and/or facilities to be analyzed. Once these have been identified, the scope of impacts can be more accurately determined. Nonetheless, agencies sometimes complete an EIS only to discover that its scope has not adequately addressed all the decisions that eventually will need to be made or considered.

Such discrepancies can often be traced to "disconnects" existing among the scoping process, the subsequent EIS analysis, and later decisions. Clearly, if the scope is not accurately identified at the outset, it is unlikely that the completed EIS analysis will adequately support future decision-making. Compounding this problem is the lack of rigorous tools and techniques available to help in accurately identifying the scope of analysis.

A peer-reviewed approach developed by the author for resolving the aforementioned problem, referred to as decision-based scoping (DBS), is discussed in the next section.[28] The DBS approach is in marked contrast to the way scoping efforts are typically conducted. Figure 8.2 provides a conceptual illustration of the difference between DBS and the more conventional scoping approach.

Under DBS, emphasis is placed first on identifying the potential decisions (decision-points) that may eventually need to be considered by the decision-maker. By doing this, practitioners obtain valuable input that can be instrumental later in making more accurate determinations. Once agreement has been reached on the scope of actions and reasonable alternatives, a sound basis will have been established for identifying the environmental impacts requiring evaluation. Properly implemented, the DBS approach can reduce costs while enhancing the effectiveness of NEPA planning. Application of this approach is not limited to NEPA scoping efforts alone. Indeed, a general-purpose methodology is introduced below that can be applied to a diverse range of other planning applications.

8.2.3.1 Decision-Based Scoping

The DBS approach is especially well suited for identifying the scope of potential decisions and actions that need to be considered in large or complex EISs, particularly in P-EISs. This is especially true in cases where many potential decision-points could be involved, where decision-points might otherwise go unnoticed, or where the agency is unsure about the specific nature of the decisions that might need to be considered.

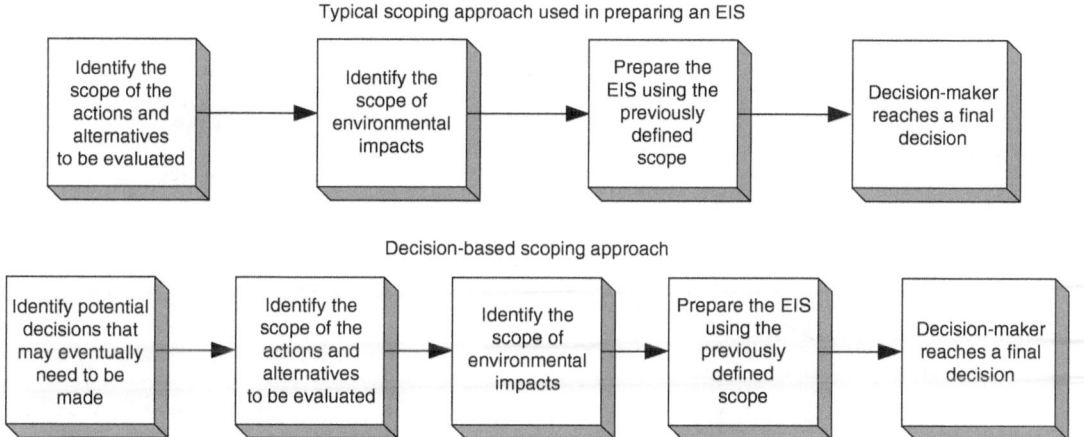

FIGURE 8.2   Comparison of the DBS approach with the approach typically used in preparing an EIS.

A facilitated interdisciplinary workshop can provide an excellent forum for implementing a DBS process. Selected value engineering (VE) techniques (described in Section 2.3) can also be applied to assist practitioners in identifying potential decisions that may eventually need consideration. For example, a VE facilitator could present a project and challenge a group to identify and prioritize all possible decisions that might need to be made or considered by a decision-maker. Described next is an environmental planning tool, developed by the author, referred to as a decision-identification tree (DIT).[28] Consistent with the rule of reason, this tool is intended to assist practitioners in identifying potential decision-points. However, before describing this, a word of caution is in order. The DIT is *not* intended to be used in determining the actual outcome of a given decision. Instead, the DIT provides a tool for *identifying* potential decision-points that may eventually need to be considered so that the EIS can be designed to support them.

The DIT provides an effective tool for identifying the range of decisions and consequently the potential actions that may need to be taken for later analysis. The utility of this tool becomes increasingly more apparent, as the scope or complexity of the EIS planning process expands. By reducing risks associated with uncertainty, this approach can result in long-term cost savings.

8.2.3.2   Constructing the DIT

Figure 8.3 provides a simplified example of a DIT for a hypothetical waste-management program and is shown for illustrative purposes only. A DIT for a different type of project would be markedly different. Professional judgment must be exercised in determining how best to construct a DIT for a given project.

As depicted in Figure 8.3, the box to the left of the vertical dashed line denotes potential factors or decision-points that are outside the scope and control of the EIS decision-making process. Such factors or uncertainties are important to capture, as these might influence potential decisions; thus they are mapped out to the right of the vertical dashed line.

The first step in constructing a DIT involves identifying the principal or key decisions that will or might need to be considered by the decision-maker. This task can be deceptively complicated. A VE technique described in Chapter 2 known as the "nominal group technique" may be helpful in identifying such decision-points.

The horizontal axis of the DIT is referred to as the "will-axis" because it indicates potential decision-points designated in terms of the following question: "Will or might decision-makers be faced with having to consider and make decisions with respect to the following course of action?" Specific decision-points that might need to be considered are denoted by diamond figures along

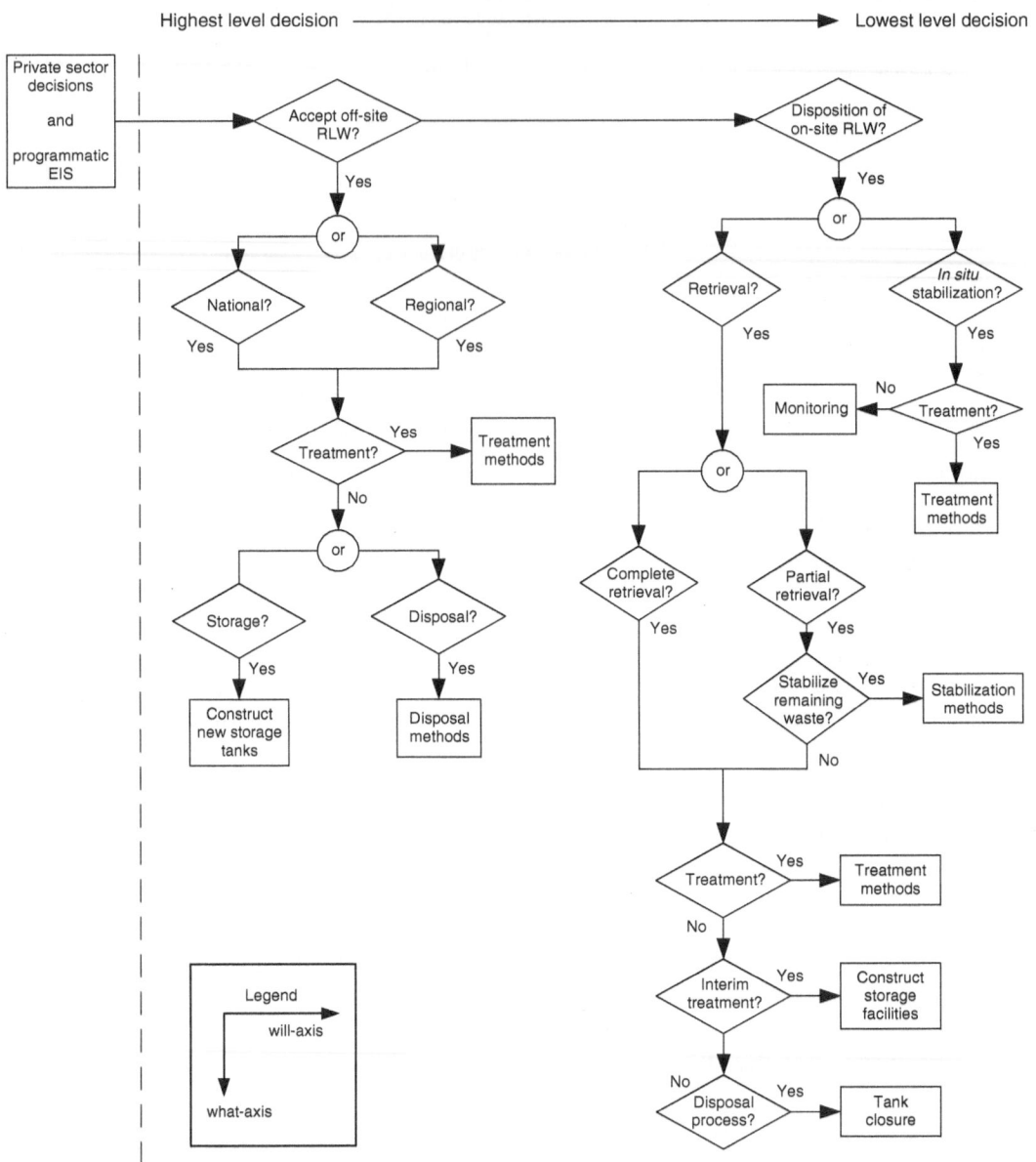

FIGURE 8.3 Example of a DIT for a hypothetical RLW management program.

the will-axis. Highest level (i.e., most significant, fundamental, or important) decisions are drawn on the left side of the will-axis, proceeding progressively toward less important decisions, as one moves toward the right. Decisions along the will-axis are largely independent from one another.

A decision to pursue a given course of action often triggers or spawns subsequent, lower level considerations (i.e., dependent decision-points). Accordingly, the DIT also builds downward along the vertical axis. Where a choice to pursue a given decision would trigger or spawn the need to consider subsequent lower level considerations, the triggered decision-points are mapped downward, from the most important to the least important subdecisions, along the vertical axis. For this reason, the vertical axis is referred to as the "what-axis," because it denotes the question: "What types of decisions will or might decision-makers have to consider if a decision was made to pursue the preceding course of action?" The following example illustrates the value of preparing a DIT.

Example. Consider a hypothetical agency project involving the modernization and possible expansion of a radioactive liquid waste (RLW) management operation at one of its installations. For the purposes of this example, the term "RLW" is used to signify a generic rather than a legally defined waste type. Further assume that a decision regarding future acceptance of offsite RLW is identified as the most fundamental decision that the agency may have to consider. Accordingly, the diamond-shaped icon labeled "Accept offsite RLW?" is entered as the first element along the will-axis (Figure 8.3).

Decision-points denoted along the will-axis are either entirely or largely independent from one another. Thus, if a decision was made to accept offsite RLW, it would not trigger or influence the next highest order decision regarding disposition of existing onsite RLW.

Now consider how pursuing a particular decision can automatically trigger other additional subdecisions. For instance, in this example, assuming a decision was made to accept offsite RLW, the highest level (i.e., most significant or important) decision that would be triggered involves determining the extent to which this waste would be accepted (e.g., accepting the waste on either a restricted regional basis or from across the entire nation). The "or" gate indicates that one but not both of these decisions would need to be made. It is important to note that the DIT does not necessarily follow the standard logic convention used in many other types of logic flowcharts. A "yes" response indicated at the bottom of the two diamonds simply denotes the logic path that would be taken *if* a decision (i.e., yes) was made to accept waste on either a national or regional basis (but not both). A "no" path is unnecessary, as such a decision simply indicates that further decision-making along that path would stop. The next, lower level decision involves the need for potential treatment. Various treatment technologies would need to be evaluated if a decision is eventually made to treat the offsite waste (see box labeled "Treatment methods"). Two potential decision-making paths, labeled "Storage" and "Disposal," branch downward from the diamond labeled "Treatment." A decision to pursue either a storage or a disposal option would require consideration of either "construction of new storage tanks" or potential "disposal methods."

As soon as these decision-points have been completely mapped along the what-axis, this process is repeated, starting back along the top of the will-axis. In this example, a decision regarding disposal of onsite RLW waste is considered to be the next highest level decision-point that would need to be contemplated. The procedure for mapping the lower level decisions along the what-axis is performed in a fashion similar to that just described.

Completing the DBS Effort. As soon as the DIT has been completed, it should be reviewed by a decision-maker who should thoroughly understand the latitude of decision-making ability that will be available based on the scope depicted. Once agreement on this has been obtained, attention should then turn to the more standard task of identifying the scope of facilities, operations, activities, and the specific alternatives needing evaluation. Together, these will provide the scope of the actions and impacts of potential decisions that environmental planners and the EIS must be prepared to support.

8.2.4 Preparing the Draft EIS

Input obtained from the scoping process is used in determining which staff members will participate in the EIS analysis. Their particular disciplines will reflect the scope and issues identified during the scoping process (§ 1502.6). The draft EIS is expected to conform to the scope agreed upon during that process.

An agency has a large degree of discretion in determining the sources of data that will be used in preparing the analysis. For example, in one case, a court agreed with an agency. Instead of collecting new data, the agency had used data from an EIS prepared for a previous lease agreement that had adequately examined the direct and indirect impacts of an oil and gas development plan. The court found that the agency had made a reasoned judgment that the old data were relevant to the new plan and yielded a useful analysis of the possible cumulative impacts.[29] Thus, while an agency must take a "hard look" at environmental consequences, the EIS "need not be exhaustive to the point of discussing all possible details bearing on the proposed action."[30]

On completing the draft EIS, the public must be notified of its availability for review. The EIS, together with comments received from public review of the draft and any support documents made available under the Freedom of Information Act (§ 1506.6[f]), must then also be made available to the public.[31] Where practical, these documents should be provided to the public either free of charge or at a fee that covers only their actual reproduction costs.

8.2.4.1   Issuing the Notice of Availability and the EPA Filing Process

This section elaborates on the earlier discussion of the NOI by describing the specific process of its issuance and its relationship to the EPA 309 review process. The lead agency is responsible for filing five copies of both the draft and final EIS (including appendices) with the EPA office in Washington, D.C., which then files one copy of the EIS with the CEQ (§ 1506.9). In addition to the five copies that are filed with EPA headquarters, agencies should also provide a copy of the EIS directly to the appropriate EPA regional office(s) for review and comment. Material that is incorporated into the EIS by reference is not required to be filed with EPA. Delivery of the EIS to the CEQ satisfies the requirement of making the EIS available to the president of the United States (§ 1504.1[c]).

EPA also should be notified of all situations where an agency has decided to withdraw, delay, or reopen a review period on an EIS. All such notices will be published in the *Federal Register*.

The filing agency (usually the lead agency) should prepare a letter of transmittal to accompany the five copies of the EIS. The letter should identify the name and telephone number of the official responsible for both the distribution, contents of the EIS, and should indicate that the transmittal has been completed. The filing period should not occur before the draft EIS has been transmitted to commenting agencies and made available to the public.

Once EPA has received the EIS, it is stamped with an official filing date and checked for completeness and compliance with § 1502.10. If the EIS is not complete (i.e., if the documents do not contain those elements outlined in § 1502.10 of the Regulations), EPA will contact the lead agency to obtain the omitted information or to resolve any problems prior to the publication of the notice of availability (NOA) in the *Federal Register*.

The EPA Management Information Unit (MIU) is responsible for centralized data management and maintaining the reporting system for the review process. A database known as COMDATE provides a weekly computerized report listing all EISs filed during the previous week (§ 1506.9).

8.2.4.2   Publication of the Notice in the *Federal Register*

A report is prepared by EPA each week listing all EISs filed during the preceding week. This report includes an EIS accession number, EIS status (draft, final, supplemental), date filed with EPA, the agency or bureau that filed the EIS, the state and county of the action that prompted the EIS, the title of the EIS, the date comments are due and the agency contact. The contents of this report are published each Friday under an NOA in the *Federal Register*. Upon publication of the NOA, EPA sends the information in its EIS status report to the CEQ.

All notifications are published in the *Federal Register* on the Friday following the week in which the EIS is filed with the EPA (see Figure 8.4). Thus, each week, the *Federal Register* lists all EISs filed during the preceding week (§ 1506.10[a]).

The EIS filing date is defined as the date EPA publishes the NOA in the *Federal Register*, not the date that the document is transmitted or received by the EPA. Thus, the minimum EIS commenting and waiting periods are calculated from the date the NOA is published in the *Federal Register* (§ 1506.10[b], [c], and [d]). Review periods for draft EISs, draft supplements, and revised draft EISs shall extend 45 calendar days, unless the lead agency extends the prescribed period or a reduction of the period has been granted.

The waiting periods for final EISs shall extend for 30 calendar days from the publication of the NOA, unless the lead agency extends the period or a reduction or extension in the period has been

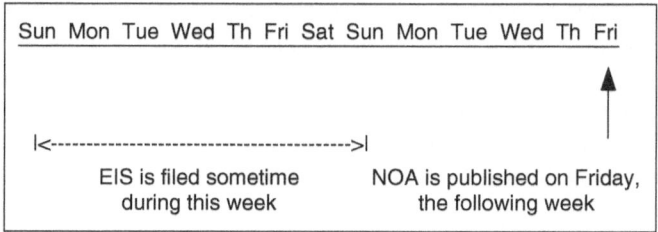

FIGURE 8.4   This figure indicates the sequence in which the NOA is published, relative to the date on which the EIS has been filed with the EPA.

granted. If a calculated time period would end on a nonworking day, the assigned time period will be the next working day (i.e., time periods will not end on weekends or on federal holidays).

It should be noted that there is an exception to the rules of timing (§ 1505.10[b]). An exception may be made where an agency decision is subject to a formal internal appeal. EPA has the authority to both extend and reduce the time periods on draft and final EISs based on the demonstration of "compelling reasons of national policy" (§ 1506.10[d]). The CEQ also has the authority to approve alternative procedures for preparing, circulating, and filing supplemental draft and final EISs (§ 1502.9[c][4]). Agencies sometimes mistakenly publish (either in their EISs or individual notices to the public) a date by which all comments on an EIS are to be received; agencies should ensure that this end date is based on the date of publication of the NOA in the *Federal Register*.

All regulatory timing requirements should be adhered to rigorously. However, an innocent error does not necessarily provide sufficient justification for a successful challenge. For example, in one case, an agency was challenged over an irregularity when a notice was published in the *Federal Register* on the day the EIS was circulated rather than during the following week as specified in the Regulations. The court concluded that this minor violation did not affect the ability of entities to review the statement for 30 days following the publication of the NOA. The court's justification was that this trivial error did not, by itself, constitute sufficient grounds for challenging the EIS.[32]

Prescribed review and waiting periods may be extended by the lead agency. Likewise, the EPA may extend the minimum periods for reasons of national policy in cases where a request has been made by another federal agency. In contrast, if requested by the lead agency, the EPA may also, for reasons of national policy, shorten the prescribed periods. The reader should note that a failure to provide timely comments is not considered sufficient reason, of itself, for extending a prescribed period. If the lead agency does not concur with the extension, the EPA may not extend a prescribed period by more than 30 days. However, when the EPA does reduce or extend any period of time, it is required to notify the CEQ (§ 1506.10[d]).

8.2.4.3   EPA EIS Repository

Filed EISs are retained at the Office of Federal Activities (OFA) for a period of 2 years and are made available to office staff only. After 2 years, these EISs are sent to the National Records Center.

The EPA Library houses a microfiche collection of final EISs filed from 1970 through 1977, and all draft, final, and S-EISs filed from 1978 through 1990. These microfiches are available through interlibrary loans. It can be contacted at

Environmental Protection Agency Library
Headquarters Library
EPA West Building
Constitution Avenue and 14th Street, NW, Room 3340
Washington, DC
202-566-0556

One of the largest collections of EISs is available from Northwestern University's Transportation Library. Nearly all of the EISs issued since 1969 are held here in both draft and final form.

The Cambridge Scientific Abstracts (CSA) is a privately owned information company that publishes abstracts and indexes to scientific and technical research literature for a charge. Detailed abstracts of EISs published from 1987 to the present are available; individual copies of EISs may also be available on special order. CSA can be contacted at

CSA
Edward J. Reid
Editor, EIS: Digest of Environmental Impact Statements
7200 Wisconsin Avenue – Suite 601
Bethesda, MD 20814
301-961-6742

8.2.4.4   Circulating the Draft EIS

As noted above, the completed draft EIS must be circulated to the general public for comment and it must satisfy the requirements of NEPA (§ 1503.1) to the fullest extent possible.

A minimum comment review period of at least 45 days is required for drafts (§ 1506.10[c]). Where one is to be considered at a public hearing, the agency should make it available to the public at least 15 days before that date. However, this requirement does not apply to cases where the purpose of the hearing is simply to provide information for the EIS (§ 1506.6[c][2]).

If, during the first circulation, the draft EIS is found to be so inadequate as to preclude meaningful analysis, the agency must prepare and recirculate a revised draft (§ 1506.10). If appropriate, the agency may limit preparation and circulation only to the portion of the EIS that was determined to be inadequate (§ 1502.9[a]).

Efficient EIS Distribution. Electronic distribution of the Yucca Mountain EIS saved to the Department of Energy (DOE) an estimated $200,000.[33] Rather than distributing paper copies of the entire 5000-page Yucca Mountain final EIS, the DOE instead primarily distributed CD-ROMs and paper copies of the EIS summary. The CD-ROMs contained the entire EIS as well as images of more than 13,000 EIS comments that were not part of the EIS itself. The agency also distributed about 75 paper copies of the entire document to selected federal, state and local agencies, and to other entities that requested a hard copy.

In the initial distribution of about 6200 CD-ROM/paper summary sets, the project manager provided information to recipients concerning how they could request paper copies of the entire document that included an option to call a toll-free telephone number. Each set costs about $7 to produce and distribute. The total production and distribution cost amounted to slightly more than $100,000. If paper copies of the entire EIS had been circulated instead, the cost would have exceeded $300,000.

The DOE waited an extra week before filing the EIS with the EPA so that anyone requiring the complete document could receive it before that date and the subsequent publication of an NOA by the EPA.

8.2.4.5   Circulating a Summary

In some instances, if a draft or final EIS is unusually long, the agency may circulate a summary in lieu of the entire EIS (§ 1500.4[h], § 1502.19). Nevertheless, the entire EIS must still be provided to the following parties:

- Any federal agency having jurisdiction by law or special expertise in the environmental impact as well as federal, state, or local agencies that are authorized to develop and enforce environmental standards
- The project applicant (if any)

- Any person, organization, or agency requesting the entire EIS
- In the case of a final EIS, any person, organization, or agency that submitted substantive comments on the draft

If a summary is circulated and the agency receives a request within a short period of time for the entire statement and for additional time to comment, an extension of at least 15 days beyond the minimum review period must be granted for that requestor (§ 1502.19[d]).

Where a draft has appendices, it may be circulated without them. However, they must be made available upon request (§ 1502.9).

Where changes to the draft are minor, the agency may choose to reduce paperwork by attaching and circulating only the changes made to the draft EIS rather than rewriting and circulating an entirely new document (§ 1500.4[m]).

8.2.5　EPA Review under Section 309 of the CAA

As noted in the Introduction to this book, NEPA was signed into law on New Year's Day of 1970. The CEQ and EPA were likewise established in the same year. The Clean Air Act (CAA), which was enacted to place controls on activities that can degrade air quality, was also passed at the end of 1970.

The CAA contains an unusual provision that appears to lie outside the scope of the purpose for this statute, which is to protect air quality. Section 309 of the CAA directs EPA to review certain proposed actions of other federal agencies subject to NEPA and to make those reviews public. If the lead agency fails to make sufficient revisions and the project remains environmentally unsatisfactory, EPA may refer the matter to the president's CEQ for mediation.

Inclusion of Section 309 within the CAA can be understood in terms of the context of controversial projects that were in the headlines during the early 1970s. One such project, for example, involved the proposed supersonic transport aircraft (SST), which involved controversial environmental ozone issues and eventually became a crucial test of NEPA.

The Department of Transportation (DOT), the lead agency proposing the SST project, chose not to disclose EPA's review comments on the SST EIS before issuing its final decision. Although the later 1978 NEPA regulations would clarify this ambiguity, Congress had a legislative statute on the table in which to make its sentiments known—the CAA. Senator Edmund Muskie who crafted Section 309 (NEPA review requirement) of the CAA stated to the senate in submitting the conference report that once EPA has completed its 309 review, it must make its NEPA comments public and "not when the environmental impact agency decides the public should be informed."[34]

Thus, EPA was granted responsibility for performing the Section 309 EIS review because "it is essential that mission-oriented federal agencies have access to environmental expertise in order to give adequate consideration to environmental factors."[35] Over the years, EPA has reviewed most of the approximately 20,000 draft and final EISs produced since NEPA's enactment.

8.2.5.1　EPA Review Responsibilities

The following section describes EPA's EIS review process for both draft and final EISs. EPA plays a critical role in instilling quality and honesty into the NEPA processes. The EPA is responsible for reviewing, commenting in writing, and rating each EIS. It is also responsible for working with the involved agency to resolve any outstanding issues. With respect to this responsibility, Fogleman writes

Whereas plaintiffs in NEPA lawsuits may be precluded from challenging an agency's substantive decision to proceed with an action, section 309 expressly grants the EPA administrator the power to comment on the substantive decision, to publish the decision, and to refer the matter to CEQ. Once a matter is referred to the CEQ, agencies tend to accept the CEQ's suggestions or to reach an agreement with the EPA. Thus the EPA has considerable power to ensure that environmentally destructive actions do not proceed. The power is increased substantially if the CEQ agrees with the EPA's comments. This power is largely undeveloped.[36]

EPA's Section 309 review authority is delegated to the OFA. OFA maintains a manual prescribing the duties, procedures, and responsibilities for performing the "309 review process," which is described below.[37]

The OFA has developed a set of criteria for rating draft EISs. The rating system provides a basis by which EPA makes recommendations to the lead agency for improving the draft. If the recommended improvements are not made in the final EIS, EPA may refer the final EIS to CEQ.

The OFA is the official recipient of all EISs prepared by federal agencies. The OFAs and its regional counterparts typically review about 450 EISs and some 2000 other actions annually. Besides actions for which an agency may not prepare an EIS, the OFA reviews, to mention a few, activities such as proposed agency regulations and proposals for which an agency has determined that no EIS is needed (whether or not the agency has published a FONSI).

As mentioned earlier, the OFA publishes the NOA in the *Federal Register* for all draft and final EISs, and S-EISs that are filed with the EPA. These notices start the official "clock" for public EIS review/comment period and waiting period. The OFA is required to review and provide comments on the adequacy of the analysis and environmental impacts on every draft or final EIS that is filed.[38] Under Section 309 of the CAA, the EPA is directed to review and publicly comment on the environmental impacts of federal activities, including actions for which EISs are prepared.[38]

EPA program offices are responsible for providing technical assistance and policy guidance on review actions related to their areas of responsibility. Each EPA regional office is responsible for carrying out the review process for the proposed federal actions affecting its region. Each EPA regional office designates a regional environmental review coordinator (ERC) who has overall management responsibility for the review process in that region.

For each EIS, an EPA principal reviewer (PR) is designated to coordinate the 309 review and to prepare the comment letter on the proposed federal action. Associate reviewers (ARs) may also be assigned to this review. The AR is a person designated by the PR to provide technical advice in specific review areas and to provide the views of the office in which the AR is located. To track their status, the EPA maintains a publicly available computer database known as COMDATE.

If the agency's preferred alternative is found to be unacceptable, the EPA is required under law to refer the matter to the CEQ (§ 1504.1[b]). Other agencies may also make similar reviews. The results of these reviews must be made available to the president, the CEQ, and the public (§ 1504.1[c]).

In a few circumstances, if requested by the public or another agency, the EPA may also review categorical exclusions (CATXs) and EAs. Thus, while other agencies may only refer EISs to the CEQ, the EPA may refer these other documents as well.

8.2.5.2   EPA's Rating of the Draft EIS

Upon completing the review of a draft EIS, the PR will rate the statement according to the alphanumeric system described below and include the designated rating in the comment letter. Unless an alternate review period is agreed, EPA's review comments must be provided to the lead agency within the standard 45-day review period that begins with the publication of the NOA.

The purpose of the rating system is to synthesize the level of EPA's overall concern with the proposal and to define the associated follow-up that will be conducted with the lead agency. This review rating is normally focused principally on the preferred alternative identified by the agency in the draft. If, however, a preferred alternative is not identified, or if the preferred alternative has significant problems that could be avoided by selection of another alternative, or if there is reason to believe that the preferred alternative may be changed at a later stage, the reviewer may also rate

TABLE 8.5
EPA's System for Rating the Draft EIS

Environmental Rating	Adequacy
LO (lack of objections)	1 (adequate)
EC (environmental concerns)	2 (insufficient information)
EO (environmental objections)	3 (inadequate)
EU (environmental unsatisfactory)	

individual alternatives. The EPA is also expected to comment on specific mitigation measures as well as any actions that may lead to a possible violation of environmental standards.

To the extent possible, assignments of the alphabetical rating will be based on the overall environmental impact of the proposed action, including those impacts that are not adequately addressed in the draft EIS.

Alphanumeric Rating System. EPA uses an alphanumeric rating system that summarizes their recommendations to the lead agency. This system is used to rate the

- environmental impact of an action and
- adequacy of the draft EIS.

The draft is reviewed to determine the severity of the environmental impacts that would result from the project. Based on the EPA's judgment, the proposal is reviewed and rated according to whether the environmental impacts are considered acceptable or unacceptable. As shown in Table 8.5, the proposal is given an environmental rating according to one of four alphabetical categories: LO, EC, EO, and EU.

Each alphabetic rating is also assigned a numeric rating (i.e., 1, 2, or 3) according to the adequacy of the draft document. This alphanumeric rating system is summarized in Table 8.5.

8.2.6 SCOPE OF DRAFT EPA COMMENTS

The review will include EPA's assessment of the expected environmental impacts of the action. If substantive impacts are identified, an evaluation of the adequacy of the supporting information presented in the EIS with suggestions for additional information that is needed will be described.

For categories EO, EU, or 3, the EPA will ensure that the lead agency is notified of the general EPA concerns prior to the receipt of EPA's comment letter. For categories EU and 3, ERC must attempt to meet the lead agency to discuss EPA's concerns prior to the submission of the comment letter to the lead agency. The purposes of such a meeting are to describe the specific EPA concerns and discuss ways to resolve those concerns to ensure that the EPA review has correctly interpreted the proposal and supporting information, or to become aware of any ongoing lead agency actions that might resolve the EPA concerns.

To ensure the objectivity and independence of the EPA review responsibility, the EPA comment letter itself and the assigned rating are not subject to negotiation and should not be changed on the basis of the meeting unless errors are discovered in EPA's understanding of the issues.

As explained earlier, if after this review the EPA administrator determines that the proposal is unsatisfactory from an environmental standpoint, he or she is required by law to refer the matter to the CEQ (§ 1504.1). Any action such as a referral is only to be taken after every effort has been made to resolve the issue(s) with the agency. This referral process is described in more detail in Section 8.4.

Although the EPA does not have authority to halt a project simply because it could result in unacceptable environmental impacts, use of the alphanumeric rating may well provide political leverage for bringing it into serious question. Additionally, this ranking may become a legal element if the EIS is challenged.

8.2.6.1   EPA Review of the Final EIS

Most EISs have improved in quality by the time they reach their final stages. The final EIS will be checked to determine whether the statement adequately resolves the problems identified in the EPA review of the draft EIS. A detailed review and submission of comments on the final EIS will be performed for those proposals rated EO, EU, or 3 during the draft stage.

The review of the final statement will generally be directed on the major or unresolved issues, focusing on the impacts of the project rather than on the adequacy of the statement. Normally, the scope of review and the comments will be limited to issues raised in EPA's comments on the draft EIS that have not been resolved in the final EIS and any new, potentially significant impacts that have been identified as a result of information made available after publication of the draft EIS.

An alphanumeric rating is not assigned to a final EIS. Instead, the EPA prepares narrative comments in its review that generally focus on the impacts and unresolved issues contained in the document rather than on its adequacy. The final comment letter will not normally be distributed externally until after the lead agency has received their set of comments. As appropriate, the PR will ensure that

1. EPA receives a copy of the ROD;
2. where appropriate, the lead agency has included all agreed upon measures as conditions in grants, permits, or other approvals; and
3. the lead agency has incorporated into the ROD all agreed upon mitigation and other impact reduction measures.

8.2.6.2   EPA Monitoring and Follow-Up

The PR will review the ROD on all final EISs on which the EPA has expressed environmental objections and/or those where the EPA has negotiated mitigation measures or changes in project design. As necessary, the EPA will perform a follow-up to ensure that

1. EPA participates as fully as possible in any post-EIS efforts to assist agency decision-making;
2. any agreed-upon mitigation measures are fully implemented (e.g., permit conditions, operating plan stipulations, etc.); and
3. any agreed-upon mitigation measures are identified in the ROD.

8.2.6.3   Inviting Comments on the Draft EIS

Table 8.6 lists public and private parties from which an agency must obtain or at least request comments (§ 1503.1).

TABLE 8.6

Entities from Which the Lead Agency Must Obtain or Seek Comments

- Any agency that has requested statements on actions of the kind proposed
- Any federal agency that has jurisdiction by law or special expertise with respect to any environmental impact involved, or that has authorization to develop and enforce environmental standards
- Appropriate state and local agencies that are authorized to develop and enforce environmental standards
- Indian tribes, when the effects may be on Indian land or sites of religious or cultural significance to the tribe
- Applicant (if any)
- The public

Any agency with jurisdiction by law, having special expertise or that is authorized to develop and enforce environmental standards, must comment on EISs that fall within its jurisdiction, experience, or authority. An agency may reply simply that it has no comments on the EIS. In other cases, to reduce paperwork, the lead agency should require commentators to be as specific as possible in their remarks (§ 1503.2, § 1503.3).

The DOE prepared a highly controversial EIS for the national Yucca Mountain Geologic Repository concerning the storage of high-level radioactive waste. The DOE performed 21 public hearings for the draft EIS and also established a public comment period of almost 200 days. In response, more than 11,000 comments were received including some 2300 letters and other submittals.[39]

But this does not even come close to the U.S. Forest Service's experience in preparing a highly controversial national EIS for its Roadless Area Conservation Program and related proposed rule to be applied to some 160 national forests and grasslands. Public participation activities in the draft EIS for the Roadless Area Conservation Program included about 450 public scoping meetings and hearings. During its scoping process, the Forest Service received more than 517,000 letters, cards, and other submittals containing well over 1 million comments! Form letters and postcard campaigns accounted for about 481,000 of the submitted items.[39]

Over a 60-day draft EIS public comment period, the Forest Service estimated it received an additional 1 million letters, cards, and other items including about 60,000 individually written letters—6,000 of them from local, state, and federal agencies. The forest service assigned 95 full-time staff to analyze these comments.

8.2.7 PREPARING THE FINAL EIS

All substantive comments received on the draft are to be included as part of the final EIS regardless of whether or not the comment is believed to merit individual discussion. In preparing the document, the agency must consider the comments received both individually and collectively. Comments that are exceptionally lengthy may be summarized (§ 1503.4[b]).

Where an agency does not agree with a comment, it must explain why it considers it to be unwarranted. The agency must respond to all the comments (§ 1502.9), but if two or more comments are very similar, the agency may group them together and prepare a single response. The comments and the agency's responses are often placed in an appendix to the EIS.

An agency is not required to respond at length to vague comments, such as a generic complaint that the EIS is inadequate. However, it must provide an adequate response to specific comments. For example, if a comment indicates that a summary of an air dispersion model was not included in the EIS, the agency may need to prepare and include a summary of that model as part of its response.[40]

If a commentator could have raised the issue of a new alternative during the scoping period but failed to do so, the agency does not necessarily have to investigate this option. In contrast, if the new alternative was only discovered or developed after the scoping period had ended, a supplemental draft EIS may need to be prepared to address it.[41]

The agency must also address any responsible opposing views that were not adequately discussed in the draft EIS (§ 1502.9[b]). Normally, in lieu of a simple response, this should result in changes being made to the document. Potential agency responses include (§ 1503.4[a])

- modifying the alternatives;
- developing and evaluating alternatives that have not yet been seriously considered;
- modifying, improving, or expanding the analyses;
- making corrections to the text; and
- explaining why comments do not warrant further agency response, citing the source, authority, or reasons that support the agency's position.

Once comments from the draft have been resolved and incorporated, the final EIS is reissued to commenting agencies and the public for review.

8.2.7.1   Issuing the Final EIS

As described below, the final EIS is also required to be publicly circulated using a process similar to that for the draft EIS.

8.2.7.2   Procedures for Issuing the Final EIS

Once comments have been incorporated and the final EIS is complete, the agency files the statement with the EPA. Although it is a requirement that the final EIS be recirculated, there is no regulatory requirement for an agency to request or incorporate comments on it, as in the case of a draft.[13] However, during this 30-day review period, the public and other agencies may provide further comments.[42]

The EPA publishes an NOA in the *Federal Register* and is responsible for sending one copy of the final EIS to the CEQ (§ 1506.9). This filing period should not occur before the final EIS has been transmitted to commenting agencies and made available to the public (§ 1506.9).

An entire copy of the final EIS must be circulated to any person, organization, or agency that has submitted substantial comments on it (§ 1502.19). However, where changes are minor, the agency may attach and circulate only the changes to the draft EIS in lieu of rewriting and circulating the entire document (§ 1500.4[m]).

8.2.7.3   Mandatory 30-Day Waiting Period

As explained earlier, the agency must wait for a minimum period of 30 days following the publication of the NOA before making a final decision regarding the proposed action. The 30-day waiting period commences on the date that the NOA is published (§ 1506.10).

This 30-day waiting period is not intended for the purpose of obtaining additional comments but to provide other agencies and the public with adequate time to review the final EIS and take any desired action such as referring the matter to the CEQ or filing a legal suit. No decision regarding the proposed action may be made or recorded until the 30-day waiting period has elapsed and no decision can be taken on the NOA for the draft EIS until at least 90 days following its publication (§ 1506.10[b]).

In instances where the final EIS is filed within 90 days of the draft, the minimum 30-day and 90-day periods may run concurrently. However, agencies may not allow less than 45 days for comments on draft EISs (§ 1506.10[c]). These periods represent the minimum requirements but it is quite typical for agencies to exceed them, particularly in cases that are highly controversial or involve complex issues.

For example, suppose that the comment period for a draft EIS runs for a period of exactly 45 days starting on the day that the EPA publishes the NOA in the *Federal Register*. The agency promptly incorporates the comments into the draft and the EPA publishes the NOA for the final EIS 9 days following the completion of the mandatory 45-day comment period. In this case, the 45-day comment period, in addition to the 9-day comment incorporation period, followed by the mandatory 30-day waiting period, amounts to 84 days, 6 days fewer than the 90-day minimum requirement. The agency must thus wait out these extra days before making a decision. Typically, most EISs exceed the 90-day requirement.

8.2.7.4   Exceptions

These are a few exceptions to the 30-day waiting period. An agency involved in rulemaking under the Administrative Procedures Act or any other statute concerned with the protection of health and safety may waive the 30-day waiting period provision and publish the ROD simultaneously with the NOA for the final EIS. Other exceptions also exist. For example, some agencies have an established appeal process, allowing other agencies or the public to appeal a decision after the final EIS has been published. In such instances, if an opportunity exists to change the decision, it may be made and recorded at the same time that the EIS is published. The EIS must explain the public's right of appeal. Consequently, the period for the appeal of this decision and the 30-day waiting period may run concurrently (§ 1506.10[b]).

8.3 THE ROD

During public review of the CEQ's draft Regulations, one of the most widely supported provisions involved the requirement for agencies to issue an ROD on completion of the EIS process. By requiring an agency to make a concise record of its decision in the ROD, the decision-maker is procedurally forced to consider the analysis in the EIS.[43]

8.3.1 ISSUING THE ROD

Once the final EIS has been completed, the decision-maker is responsible for considering the analysis of alternatives before making a final decision on the proposed action. As mentioned earlier, this decision can neither be made nor recorded until the minimum 30-day waiting period following the publication of the NOA for the final EIS has expired (§ 1506.10[b]).

The agency involved is responsible for preparing the ROD, which it may then integrate into any other decision record it has prepared.[44]

The purpose of the ROD is to record the agency's final decision in the form of a concise statement that also discusses the agency's choice from among the various alternatives considered.

The ROD is a public document (§ 1505.2) and must be made public as prescribed under § 1506.6(b). Although there is no specific requirement within the Regulations that the ROD be published in the *Federal Register* (or elsewhere), it is considered good professional practice to do so.

8.3.1.1 Selecting a Course of Action

An agency may select any alternative provided it has already been adequately described and analyzed (§ 1505.5[e]). Care should be exercised to ensure that the agency's final decision is covered or bounded by the range of alternatives analyzed in the EIS.

An agency normally prepares the ROD at the time of making its final decision regarding a course of action, but it may actually be issued at any point after expiration of the minimum waiting period. If the agency has not decided on the course of action it will take by then, it may wait for a lengthy period of time before issuing the ROD.

An agency has discretion to change its initial decision regarding the preferred alternative cited in the final EIS. Furthermore, agencies have discretion to change a decision that was documented earlier in the ROD and to select a different alternative as long as it has been adequately analyzed within the document.[45]

8.4 REFERRALS

Because interagency disagreements sometimes arise, the Regulations provide a process known as referrals for resolving them. A federal agency that refers a matter to the CEQ, which it believes to be unsatisfactory from the standpoint of public health, welfare, or environmental quality is known as a referring agency (§ 1508.24).

8.4.1 PROCEDURE FOR REFERRALS

As described earlier, the EPA administrator is assigned responsibility under Section 309 of the CAA for reviewing and publicly commenting on EISs. The EPA administrator may refer to the CEQ any EIS deemed "unsatisfactory from the standpoint of public health or environmental quality" (§ 1504.1[b]).

Similarly, under the "predecision referral" provision, any federal agency has the authority to refer another agency's final EIS to the CEQ during the mandatory 30-day waiting period if it deems it to be unsatisfactory but it is expected to make a concerted effort to resolve differences with the lead agency before doing so (§ 1504.2). Such reviews must be made available to the CEQ, the president, and the public (§ 1504.1[c]). A referring agency has a period of 25 days from the date the final EIS is made publicly available in which to make a referral in writing to CEQ (§ 1504.3[b]). The CEQ will not accept referrals after this time except in cases where the lead agency has agreed to extend the referral period. During the same period, other agencies and the public may also submit written comments to the CEQ.

The referring agency must prepare a letter signed by the head of the agency, notifying the lead agency of the referral and requesting that no action be taken with respect to the proposal until the CEQ has acted upon it.

Once the referral has been made to the CEQ, the lead agency has a period of 25 days to deliver a written response to the CEQ and the referring agency. If the CEQ receives a referral, it may choose to (§ 1504.3[f])

1. conclude that the referral process has successfully resolved the problem,
2. initiate discussions to mediate the issue between the referring and lead agencies,
3. hold public meetings or hearings to obtain additional views and information,
4. determine that the issue is not one of national importance and request the referring and lead agencies to pursue their decision process,
5. determine that the issue should be further negotiated between the referring and lead agencies (until one or more agency heads report to the council that the agencies' disagreements are irreconcilable),
6. publish its findings and recommendations, and
7. submit the referral and the response together with the council's recommendation to the president for action.

Under Section 309, EPA has also been assigned more extensive review and referral authority than the other agencies. Section 309 grants EPA expanded authority to refer to CEQ a broader range of federal activities, not only actions for which EISs are prepared (§ 1504.1[b]).

8.4.2 THE REFERRAL PROCESS IN REALITY

In reality, since 1973 agencies have referred only a few dozen proposed federal actions to CEQ. EPA was responsible for somewhat more than half of these referrals. To date, in no case has CEQ made a formal referral to the Office of the President. Most often, CEQ has issued findings and recommendations. In few cases, the lead agency has withdrawn the proposal and, in three cases, the CEQ determined that the issue was not a matter of national importance.

The CEQ generally attempts to act as an arbitrator between agencies since in practice it has little actual power or control at its disposal to force an agency to alter its position. Instead, its true power lies in its ability to apply political pressure and create potential political embarrassment for an agency. In this way, rather than becoming directly involved in disagreements, the CEQ can encourage agencies to resolve their problems before they are elevated to the level at which the Council would have to become involved.

Experience has shown that agencies generally seek a resolution rather than face public embarrassment resulting from an environmentally questionable proposal. In reality, the referral process is rarely followed to completion. The CEQ has indicated that only a few dozen referrals have been filed and concludes that the referral process has been generally effective in resolving conflicts.[46]

8.5   POST EIS MONITORING AND ENFORCEMENT

Agencies are responsible for assuring that mitigation measures and other conditions committed to the ROD are implemented. Mitigation commitments are enforceable through litigation.[47]

The purpose of the post monitoring program is to ensure that the standards are met, mitigation measures are adequately incorporated, and no impacts that differ significantly from those originally predicted are encountered. Additional NEPA analysis may be required if the monitoring program reveals new information or unanticipated or significantly different impacts. Post monitoring efforts normally consist of measuring or sampling such metrics as air emissions, water discharges, noise levels, and vegetation disruptions.

Agencies are expected to implement a monitoring plan in important cases (§ 1505.2[c], § 1505.3). The Regulations, however, do not provide any definition of "importance." In practice, "important" cases normally involve circumstances where there is public interest and/or the potential for litigation.

Upon request, the lead agency must inform both the public and the cooperating or commenting agencies regarding the progress and status of the mitigation measures adopted in the ROD (§ 1505.3[c][d]). Agencies must put together a monitoring and enforcement program where a commitment has been made to adopt specific mitigation measures (§ 1505.2[c]).

8.6   LEGISLATIVE EISs

The NEPA process applies to proposals for congressional legislation and is expected to be integrated with the congressional legislative process (NEPA § 102[c], § 1506.8). Legislative EISs are subject to special requirements as described in § 1506.8 of the Regulations.

The term "legislation" includes bills and legislative proposals submitted to the Congress and "developed by or with the significant cooperation and support of a Federal agency," but does not include requests for appropriations. Legislation initiated by the Congress is not subject to NEPA (§ 1508.17).

The term "significant cooperation" means that it has been developed primarily by the federal agency rather than by another source. Drafting a proposal is not sufficient of itself to constitute significant cooperation. The agency having primary responsibility for the subject matter is responsible for preparing a legislative EIS (§ 1508.17).

8.6.1   PREPARING A LEGISLATIVE EIS

As discussed earlier, a typical EIS is prepared in two stages: draft and final. The legislative EIS forms part of a formal legislative proposal that is transmitted to the Congress and that may be transmitted up to 30 days after the formal proposal in order to allow time for completion of an accurate statement. Because the legislative EIS may be used in the congressional debate on the proposal, it must therefore be made available in time for the pertinent hearings and deliberations (§ 1508[a]).

The preparation of a legislative EIS must comply with the general requirements of the NEPA process, except for two provisions:

- The EIS scoping process is not required.
- A "draft" legislative EIS is prepared in essentially the same manner as that for a nonlegislative EIS. While the legislative EIS normally ends at the draft stage, such that a final EIS is not prepared, in a few extenuating circumstances, both the draft and the final documents may be required.

As mentioned above, usually only a draft legislative EIS is prepared and transmitted to the Congress and the public for review (§ 1506.8). This variation is based on the belief of the CEQ that legislative proposals are different from other proposed actions normally undertaken by an agency. Critical steps (e.g., hearings, votes) are not conducted by the Congress in the same controlled and predictable manner as those mandated within an agency. Instead, the Congress may vote and hold hearings as it deems appropriate and may hold hearings or request additional environmental information directly from an agency after it has received the EIS. Thus, to prevent delays and to support the congressional process, a final EIS for legislative proposals is generally not required.[43]

8.7   SUPPLEMENTAL EIS

In certain situations, an EIS must be revised or supplemented. Draft or final EISs must be supplemented (§ 1502.9[c]) when

- the agency makes substantial changes to the proposed action that are relevant to environmental concerns, or
- there is significant new information or circumstances relevant to environmental concerns that bear on the proposed action or its impacts.

A change, even a substantial one that is not relevant to environmental concerns, does not require the preparation of an S-EIS. An agency may also choose to prepare supplements either to a draft or to a final EIS if it believes these would further be the purpose of NEPA (§ 1502.9[c][2]).

An S-EIS follows the typical preparation, circulation, and filing process with one principal exception. The S-EIS does not need to repeat the formal scoping process (§ 1502.9[c][4]).

8.7.1   REQUIREMENTS GOVERNING S-EISs

Although the CEQ has not provided specific guidance for determining exactly what constitutes new information that is significant enough to necessitate the preparation of an S-EIS, the Ninth Circuit court has provided some insight into this issue by offering some factors that have been adopted by other courts in making their determinations. These factors include[48]

- significance of the new information in terms of environmental impacts,
- probable accuracy of new information,
- degree of care given in considering new information and determining and evaluating its impacts, and
- degree to which the agency supports its decision not to supplement an EIS.

The author has developed a systematic peer-reviewed tool, referred to as the Smithsonian Solution (described later), for assisting decision-makers in reaching such determinations.[49]

8.7.1.1   Additional Direction

In one prominent case, a plaintiff brought suit to enjoin (forbid) construction of a dam, partly because the Army Corps of Engineers did not prepare a second S-EIS to address concerns raised in two new reports regarding adverse downstream effects on fishing and increased turbidity.[50] The court held that an S-EIS was unnecessary for the following reasons:

- An agency has a duty to continue reviewing the environmental effects of a proposed action even after its initial approval. However, new information does not always compel an agency to prepare an S-EIS. "[A]n agency need not supplement an EIS every time new information comes to light after the EIS is finalized. To require otherwise would render agency decision-making intractable, always awaiting updated information only to find the new information outdated by the time a decision is made."

- An agency must take a "hard look" at possible new environmental effects and apply the rule of reason (described in Chapter 1) when it makes a decision regarding EIS supplementation even after a proposal has received initial approval. Its application thus depends on the value of the new information to the still pending decision-making process. In this case, the decision on whether to prepare an S-EIS is similar to the initial decision to prepare an EIS, that is, if major federal action remains and if the new information will affect the quality of the human environment in a significant manner or to a significant extent not already considered. In those cases where an agency decides not to prepare an S-EIS, it should carefully explain its reasoning.

A new statute or regulation does not necessarily constitute either a change in the proposed action or relevant new information,[51] nor does the mere passage of time compel an EIS to be supplemented.[52]

8.7.1.2   Reviewing the Agency's Decision

Although reviewing courts grant a wide degree of deference where an agency's decision is concerned, they must also carefully review the record. In reviewing a decision not to prepare an S-EIS, courts should take a careful look at the record to satisfy them that the agency has made a reasoned decision on the basis of its evaluation of the significance—or lack of significance—of the new information.

The arbitrary and capricious standard of the Administrative Procedure Act provides the standard for determining if an EIS or S-EIS is required.[53] In examining an agency's decision to decide if it has been arbitrary and capricious, courts consider whether or not it was based on a consideration of the relevant factors and if there has or has not been a clear error of judgment. Such an inquiry must be "searching and careful" but "the ultimate standard of review is a narrow one."[54]

8.7.2   Tiering and Supplementing EISs

As a streamlining measure, the CEQ suggests that EISs be prepared at the feasibility (i.e., go/go-no) stage with the option of supplementing them later if necessary (§ 1502.5[a]). For example, an EIS can be prepared during an agency's early planning. Later, when more information is available, an S-EIS can be prepared and tiered from it (§ 1508.28[b]).

In the CEQ's opinion, an S-EIS is required where an agency makes substantial changes to an action or where significant new information or circumstances having environmental implications exist. In such instances, tiering is not to be used as a means of avoiding preparation of an S-EIS.[6]

8.7.3   Smithsonian Solution: Determining When a Change Triggers Additional NEPA

Neither NEPA nor its subsequent Regulations provide definitive direction for determining:

> The degree to which a proposed action may change from the action described in the NEPA documentation before the action is no longer adequately covered, and a new analysis must be prepared.[55]

The author refers to this dilemma as the Potomac Paradox.* Experience indicates that no two decision-makers are likely to agree completely on the degree to which an action could vary before it is no longer covered or the existing NEPA documentation is invalidated. Because of this lack of a consistent methodology for reaching such decisions, the courts ultimately may be the avenue of last resort. Described below is a peer-reviewed general-purpose tool developed by the author for

---

* The dilemma derives its name from the fact that the author first began considering a solution to this problem while strolling along the Potomac River.

determining when a change to an action described in a NEPA document (EA or EIS) is sufficient to trigger preparation of additional NEPA documentation. This tool is consistent with the rule of reason and is referred to as the Smithsonian Solution.*

8.7.3.1 Procedural Mechanisms Have Not Eliminated the Potomac Paradox

The DOE has established a general mechanism for making such determinations in the form of a document referred to as a supplement analysis (SA).[56] Essentially, the SA compares a current project design with the original description presented in an EIS. Based on the results of this comparison, a decision-maker determines if the proposal has changed to such a degree that a new or an S-EIS is required.

Although the SA process provides a formal mechanism to be followed in determining when an EIS must be supplemented, it does not resolve the Potomac Paradox since an environmental planner or decision-maker is still faced with the dilemma of making determinations based largely on subjective opinions. Moreover, use of the SA is restricted to the domain of determining if an S-EIS must be prepared to address a change.

8.7.3.2 Rationale for a Defensible Tool

Under the Smithsonian Solution, the proposed change is reviewed against the existing NEPA document. If it is a new action (i.e., not previously analyzed), the requirement to prepare additional NEPA documentation is automatically triggered. If the issue in question is not new but instead represents a change in the description of the previously considered action/impact (including new information or circumstances), the decision-maker must comply with requirements prescribed in the Regulations (§ 1502.9[c][1]). Specifically, an EIS must be supplemented if

  (i) the agency makes substantial changes in the proposed action that are relevant to the environmental concerns, or

  (ii) there are significant new circumstances or information relevant to the environmental concerns and bearing on the proposed action or its impacts.

Because any action is potentially subject to the requirements of an EIS until proven otherwise, this provision can be considered equally applicable to actions that do not necessarily involve an EIS. While 40 CFR § 1502.9(c)(1) provides general information for determining if a change requires additional review, it fails to provide specific criteria for making such a determination. For example, what constitutes a "substantial change," and what do "significant new circumstances or information relevant to environmental concerns" mean? Criteria are proposed in the following section to assist the environmental planners and decision-makers in evaluating these two factors.

8.7.3.3 Criteria for Assessing Changes

Consistent with the rule of reason, the author has developed five discrete criteria for determining if a change to a proposed action requires additional NEPA documentation (Table 8.7). As part of the peer-review process, an environmental attorney performed a thorough legal verification review of over 200 NEPA cases involving S-EISs. As a result, the five criteria provided in Table 8.7 were found to be consistent with the existing case law.

A paper prepared by the author describes the Smithsonian Solution tool in more detail and provides a technical basis for the criteria used.[49] Criterion 2, based on the significant departure principle described in detail in Section 6.7, requires special explanation.

---

* The dilemma derives its name from the fact that the author developed the initial tool for resolving the Potomac Paradox at the Smithsonian Institute.

TABLE 8.7
Criteria for Determining When a Change to an Action Requires Additional NEPA Documentation

1. There is a change in a previously described action that might result in a significant new impact not previously investigated in the earlier NEPA document.
2. There is a change in a previously described action that might cause an analyzed impact to deviate significantly from projections described in the existing NEPA document (i.e., there is a reasonable possibility that the proposed change could significantly alter impact projections investigated earlier).
3. There is a reasonable expectation that new alternative(s) not already considered in the existing NEPA document could be identified for achieving the purpose and need of the proposed change that might affect the environment in a manner substantially different from the environmental effects that have already been determined.
4. Significant new circumstances or information relevant to environmental concerns have been obtained that could substantially change the agency's decision or allow the public to contribute comments that might substantially improve or affect the manner in which the proposed change to an action is implemented.
5. Significant new circumstances or information relevant to environmental concerns have been identified that could substantially change the public's understanding (or acceptance/rejection) of the proposed change in a manner substantially different from that existing when the NEPA document was prepared. The public would benefit from an additional NEPA process.

8.7.3.4   Smithsonian Solution Tool

The author's general-purpose Smithsonian Solution tool is depicted in Figure 8.5 for determining when a proposed change requires additional NEPA documentation (CATX, EA, EIS). This tool summarizes the criteria depicted in Table 8.7.

Applying the Smithsonian Solution Tool. The tool is initiated with the first rectangle at the top of Figure 8.5. A proposed change is reviewed against the existing NEPA document. If the proposed change is a new action not previously considered (see first diamond), the user automatically concludes that additional NEPA documentation must be prepared (CATX, EA, EIS). If the proposed change involves a previously considered action, the action is reviewed in terms of the five remaining tests. The user is encouraged to refer back to the corresponding and more detailed criteria in Table 8.7.

The next three tests are considered with respect to the question, "Could additional NEPA analysis of the proposed change reveal ...?" All three tests (second, third, and fourth diamonds, Figure 8.5) are considered in determining the outcome of this question. A "yes" response to any one of the three tests is sufficient to reach a determination that additional NEPA documentation must be prepared.

The remaining two tests are examined with respect to the question, "Could significant new circumstances or information, not previously considered and that is relevant to the potentially significant impacts, substantially change the ...?"

Answering "no" to all of six tests supports a decision that the proposed change could be implemented without preparing additional NEPA documentation. A "yes" answer to any one of the six tests provides a basis for concluding that additional NEPA documentation is necessary. Where the answer to any test is not clearly obvious, the user should err on the side of conservatism.

Restrictions, Advantages, and Limitations. The Smithsonian Solution tool must be implemented on a case-by-case basis. Each test is evaluated according to the decision-maker's best professional judgment. In responding to these tests, the decision-maker considers the degree to which the intensity and context could be affected by the proposed change. Decision-makers must also consider how the change would affect cumulative impacts in addition to direct and indirect effects.

Use of this tool is not restricted to EISs; it can also be used (possibly with some minor modifications) to assist decision-makers in reviewing proposed changes to the description/analysis of EAs. If these are considered substantial, preparation of additional NEPA documentation is warranted.

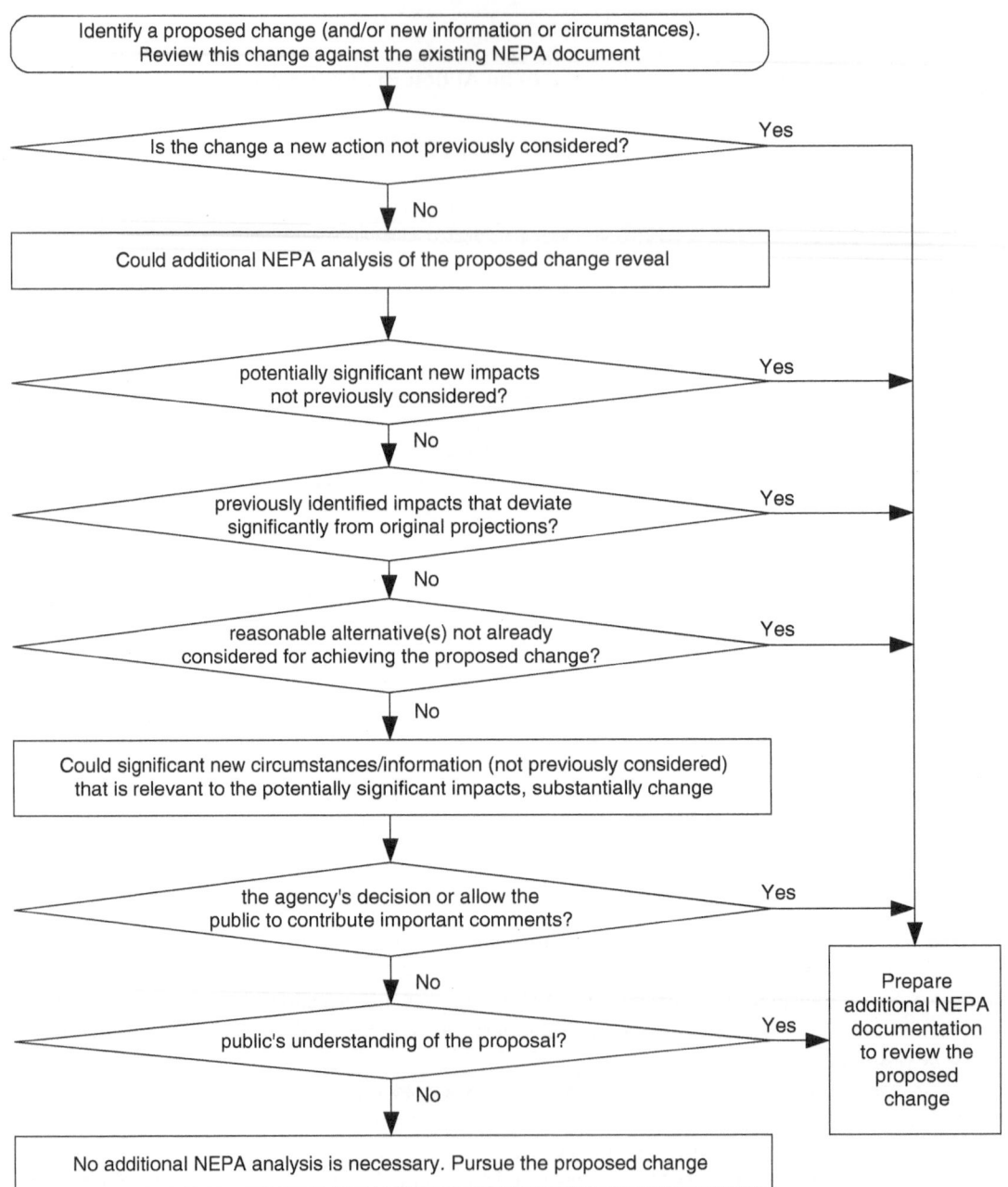

FIGURE 8.5 Smithsonian Solution tool. This tool provides a systematic methodology for determining when a proposed change requires additional NEPA documentation.

The tool does not completely eliminate subjectivity since decision-making by its very nature is subjective, but it does essentially reduce the vast array of potential considerations to six narrowly defined tests. Regardless of whether questions come from a decision-maker or from citizen groups opposed to the project, one must be prepared to justify why the proposed change does or does not meet one or more of the tests. Opinions based on other criteria or arguments are considered irrelevant unless compelling justification can be made as to why additional factors should be considered. Thus, views based only on vague or ambiguous arguments are not considered valid.

Rare exceptions might arise that this tool cannot completely address. In such instances, additional factors might need to be considered before reaching a final decision. However, decision-makers and critics alike would be expected to provide rational arguments and specific evidence justifying why additional factors beyond the six tests need to be considered.

8.8 PROGRAMMATIC EISs

A P-EIS may be prepared, and is sometimes required, for extensive federal actions such as the adoption of new agency programs or Regulations. P-EISs should be prepared so that they are relevant to policy decisions and are timed to coincide with meaningful points in agency planning (§ 1502.4[b]).

With regard to the preparation of a P-EIS, the Regulations state

> Environmental impact statements may be prepared, and are sometimes required, for broad Federal actions such as the adoption of new agency programs or regulations. Agencies shall prepare statements on broad actions so that they are relevant to policy and are timed to coincide with meaningful points in agency planning and decision-making (§ 1502.4 [b]).

Furthermore, the Regulations declare that

> Agencies are encouraged to tier their environmental impact statements to eliminate repetitive discussions of the same issues and to focus on the actual issues ripe for decision at each level of environmental review. Whenever a broad environmental impact statement has been prepared (such as a program or policy statement) and a subsequent statement or environmental assessment is then prepared on an action included within the entire program or policy (such as a site specific action) the subsequent statement or environmental assessment need only summarize the issues discussed in the broader statement and incorporate discussions from the broader statement by reference and shall concentrate on the issues specific to the subsequent action (§ 1502.20).

8.8.1 EXAMPLE OF A RECENT P-EIS

As an example of a P-EIS, the DOE is currently preparing a P-EIS (Global Nuclear Energy Partnership [GNEP] initiative) that will evaluate two international programmatic initiatives. First, the United States would cooperate with countries that have advanced nuclear programs to supply nuclear fuel services to those countries that refrain from pursuing technologies to enrich uranium or separate plutonium, both of which have application in the production of nuclear weapons. Second, the United States would promote proliferation-resistant nuclear power reactors suitable for use in developing economies.

During 2007, DOE scheduled a dozen scoping meetings across the nation. Participation averaged about 150 people at each meeting, and about a quarter of those attending provided oral comments. Most commentators expressed support for, or opposition to, the overall objectives of the GNEP proposal to recycle spent nuclear fuel and the proposed GNEP facilities. Commentators also addressed such issues as the origin of the spent nuclear fuel, disposal plans for wastes from the recycling processes, transportation, and various potentially affected resources such as water supplies.[57]

8.8.2 TIERING AND P-EISs

As a streamlining measure, the CEQ recommends that an agency consider preparing a high-level EIS for programs, policies, and plans so that tiering proceeds from statements of broad scope to those of narrower scope (§ 1500.4[i]). The P-EIS can establish high-level direction and provide high-level coverage for subsequent lower level NEPA documents that will evaluate the details and specific actions that are included within the scope of the P-EIS. Subsequent documents need only summarize the issues discussed in, and incorporate discussion from the P-EIS by reference, so that decision-makers may concentrate on issues specific to the subsequent action (§ 1502.20). As described in the next section, assessment of cumulative impacts is an important requirement that is receiving increasing attention.

8.8.3   Cumulative Impact Scoping Dilemma

Thousands of relatively innocuous decisions are made annually by federal agencies and private entities alike. Considered in isolation, the direct and indirect impacts of many, if not most, individual federal activities are typically reviewed and determined to be nonsignificant. Yet, when considered cumulatively, federal and private activities can collectively exact a sizable adverse impact on environmental quality.

In the author's professional assessment, much of, if not the preponderance of, environmental degradation results not from the direct and indirect impacts of large projects, but instead from the incessant compounding of diffuse and relatively innocuous impacts from a virtually immeasurable number of individual and small decisions.[58] Because many environmental resources have already been significantly perturbed by human activities the relatively small relentless actions that are an inherent part of modern society tend to exacerbate environmental degradation to a substantial degree.

8.8.3.1   High-Level versus Low-Level Scoping

The capability to consider seriously any realistic alternatives for mitigating the cumulative impacts of numerous and relatively diffuse actions tends to improve with a corresponding increase in the programmatic scope at which the planning analysis is performed.

Nevertheless, most NEPA analyses tend to be prepared from a relatively low decision-making standpoint (e.g., project specific) and from a narrowly focused planning perspective. While EAs facilitate efficiency in NEPA compliance, they are normally much narrower in scope and less technically demanding in their analyses than EISs. Consequently, their actual utility in terms of seriously investigating and mitigating cumulative impacts can often be considered suspect.

Accordingly, mismatches frequently result between the scope at which significant environmental degradation occurs versus that at which federal decisions are actually evaluated and made. This mismatch is evidenced by the fact that only about 30–50 P-EISs (broad scope) are prepared each year. Contrast this figure with the estimated 300–500 project-specific EISs and perhaps 30,000–50,000 EAs (narrow scope) prepared annually.[59]

As the impacts of relatively small or diffuse project actions tend to be considered in narrowly focused (low-level) NEPA analyses (such as narrowly scoped EAs or even CATXs), they correspondingly tend to be evaluated in relative isolation from one another; thus, from a pragmatic standpoint, the ability to evaluate rational alternatives and to address such impacts is impaired. This leads to the following dilemma.

8.8.3.2   Dilemma

It is unrealistic to expect most decision-makers to conclude that an EIS is required for relatively small actions that do not in themselves result in a significant direct or indirect environmental impact. This impractical situation results in a regulatory dilemma which the author has termed the Cumulative Impact Scoping Dilemma. The problem arises from the fact that the incessant number of small individual actions become cumulatively significant when taken together and viewed in their entirety. However, because these individually innocuous actions are generally evaluated from a relatively low-level planning perspective and in isolation from one another, it is unlikely that most environmental planners and decision-makers will even acknowledge their cumulative significance, let alone seriously consider alternatives for curtailing their cumulative effects.

8.8.3.3 Resolving the Dilemma

From a low-level planning perspective it can be very difficult if not impractical to combat problems such as air quality degradation and global greenhouse emissions in a narrowly defined proposal such as in an EA or narrowly scoped EIS. But viewed collectively, from a national or programmatic perspective, alternatives become increasingly more viable as the focus of analysis becomes more programmatic.

A more practical approach for seriously assessing and managing the Cumulative Impact Scoping Dilemma is to address such impacts from a relatively high-level or broadly focused planning perspective. A P-EIS provides perhaps the best mechanism for identifying and evaluating the impacts of all actions (e.g., connected, cumulative, and similar) associated with a broad program (§ 1508.25[a]). Once actions have been comprehensively identified at the program level they can then be adequately considered in evaluating the cumulative impacts for a program as whole. When analysis is undertaken from a more programmatic or perhaps even a strategic perspective, many alternatives become increasingly reasonable and cost-effective.

Thus, as planning is shifted from lower- to higher-level perspectives, many problems that are difficult to address initially become much more manageable. Many alternatives that appear impractical when viewed from a low-level planning perspective can be feasibly evaluated and mitigated at more programmatic levels.

This placing of a greater emphasis on performing analyses from a higher-level perspective provides a practical approach for resolving the Cumulative Impact Scoping Dilemma. Arguably, no other single approach to environmental planning holds the promise to enhance and safeguard environmental security more than implementing such a paradigm shift.

8.8.4 DETERMINING THE APPROPRIATE SCOPE OF A P-EIS

The Regulations provide no definitive guidance for determining the scope of issues or the amount of detail, discussion, and analysis that are appropriate to address within a P-EIS. Lack of definitive guidance may result in project delays, inconsistencies in the treatment of NEPA documents, and an increased risk that a project may be challenged as a result of inappropriate coverage or treatment of issues. Lacking systematic guidance, such determinations tend to be made on an *ad hoc* basis. To reduce the risk of a successful challenge, agencies frequently prepare P-EISs containing a level of detail much greater than is reasonably necessary to support programmatic decision-making. Such overkill can significantly increase cost, cause project delays, and unwisely reduce an agency's flexibility to implement detailed lower tier aspects of a policy or program.

8.8.4.1 The Basis for a Scoping Tool

Consistent with the rule of reason, the author has developed a set of criteria (or tests) for determining if a particular topic or issue should be addressed in a P-EIS (Table 8.8). A paper, also prepared by the author, describes the problem in detail and provides a technical basis for the criteria provided in the table.[60] These criteria are not ordered in terms of importance. Together, they provide the foundation for the peer-reviewed general-purpose tool shown in Figure 8.6.

For example, the third criterion in Table 8.8 can assist agency planners in ascertaining the appropriate scope of a P-EIS by determining if a particular topic should be included or excluded, based on whether it would substantially affect programmatic decision-making.

These criteria can also be used to address the question regarding the appropriate amount of detail necessary for a P-EIS. With respect to the third criterion in Table 8.8, this can be done by simply substituting the beginning phrase "Would discussion/analysis of the issue ...?" for the phrase "Would a more detailed discussion/analysis of the issue ...?"

TABLE 8.8
Criteria for Determining the Scope and Level of Detail Appropriate for a P-EIS

1. Can the issue be appropriately considered and analyzed at the programmatic level (i.e., ripe for decision) or should it be deferred to a later point in time when more information is available and events are better understood?

2. Would discussion/analysis of an alternative reduce flexibility (i.e., box the agency into a corner) to adapt or tailor specific lower tier actions to accommodate a dynamic and often evolving planning process where future events often change or new information becomes available?

3. Would discussion/analysis of the issue substantially improve or change high-level decisions regarding the program/policy or would it merely provide additional detail? Specifically, would the discussion allow the agency to "concentrate on (programmatic) issues that are truly significant to the action in question"?

4. Would the discussion/analysis provide a technical basis (e.g., provide descriptions, analysis, parameters, constraints, or bound the impacts) that would support or facilitate tiering of more detailed documents at a later date?

5. Would the analysis enhance the understanding of the potential environmental impacts from a programmatic perspective?

6. Would the discussion/analysis enhance the understanding or ability of an outside agency or the public to provide comments regarding the programmatic implications of the policy/program?

7. Would discussion/analysis identify/develop an alternative approach or course of action that could affect the outcome of the program/policy from a programmatic perspective, or would it merely "fine tune" programmatic decision-making?

8. Would the discussion/analysis result in programmatic mitigation measures that could be implemented to substantially mitigate or reduce programmatic environmental impacts?

8.8.4.2   A Tool for Determining Programmatic Scope

The following general-purpose tool provides a systematic and logical process for determining the scope of the issues, impacts, and alternatives that are most appropriate for a P-EIS and can also provide a basis for determining the level of detail that should be afforded to a particular topic.

While the tool is primarily intended to be used in the scoping process, it can also be employed at various other steps. During the time the P-EIS is being prepared it can also be used by analysts to determine if a particular topic or issue should be included in it. During the comment incorporation phase, the tool may also assist analysts and planners in determining the relevancy of a comment and the appropriate amount of discussion needed. In the event that the agency is challenged, the tool can provide a basis for demonstrating that the agency has addressed all the issues, impacts, and alternatives appropriate for consideration at the programmatic decision-making level.

Application. As shown in Table 8.8, application of the tool begins with a review of a topic, issue, impact, or alternative that is being considered for inclusion within the scope of the P-EIS. This examination is conducted by reviewing the particular topic in terms of each of the eight tests provided. Each test is evaluated with respect to the following question: "Would the discussion or analysis ...?" The first two tests provide an initial screening mechanism for identifying those discussions that are normally best deferred to lower-tier analyses.

The first test simply indicates that an issue should be deferred to lower-tier documentation if it cannot be appropriately considered and analyzed at the programmatic level (i.e., not ripe for decision). For example, if sufficient information does not exist to allow adequate consideration, the topic should be deferred for later consideration in lower-tier documentation.

As indicated in the second test, a discussion should be deferred if it would tend to reduce substantially the agency's flexibility to modify a policy/program or to pursue other options as events change and more information becomes available. Clearly, a P-EIS should provide programmatic high-level direction without overly constraining the detailed implementation of lower-tier decision-making. Ultimately, the goal should be to determine a programmatic direction, yet provide flexibility in determining the optimum implementation of the programmatic goals at the site-specific level.

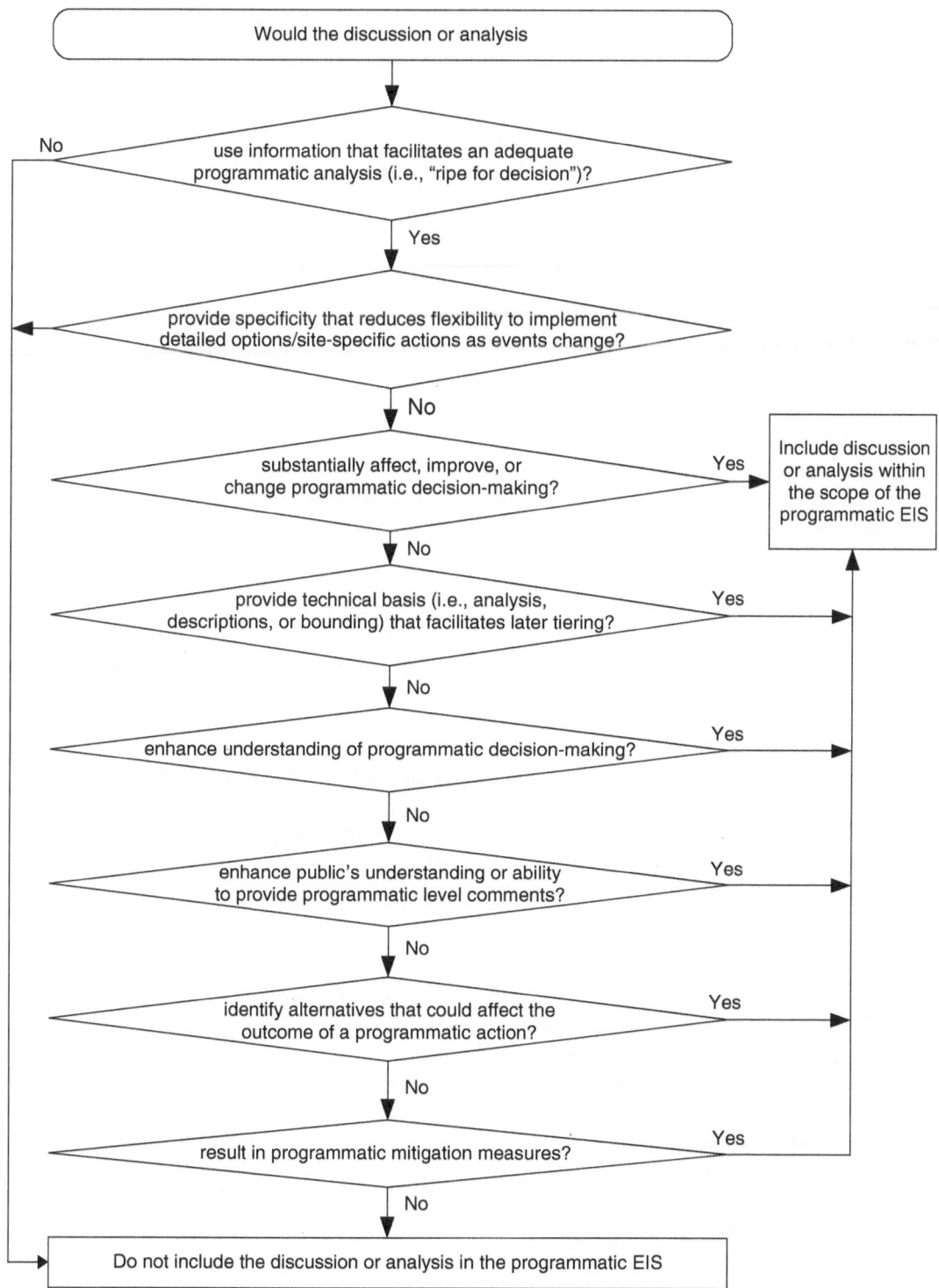

FIGURE 8.6 A tool for determining the appropriate scope of a P-EIS.

The remaining six tests provide criteria for determining if discussion of a particular issue, impact, or alternative is appropriate for consideration within a P-EIS. A "yes" answer to any of the remaining six tests supports a decision to include the discussion within the scope of the P-EIS. A "no" answer to all of the remaining six tests would indicate that a particular discussion should be deferred to lower-tier documents.

Each test must be evaluated on a case-by-case basis according to planners' best professional judgment. Planners and decision-makers must exercise professional judgment in cases where the answer to a particular test is not clearly obvious. When this tool is used, each individual issue, impact, and alternative that is considered for inclusion within the scope of the P-EIS can be reviewed to determine if it should in fact be included or else deferred to lower tier decision-making.

8.8.5   CASE LAW

Where an agency is challenged for failing to prepare subsequent site-specific analyses, a court is likely to scrutinize the P-EIS closely to determine if this has or has not been the case.[6] For example, in a P-EIS prepared by the Forest Service, the agency attempted to make decisions regarding wilderness and nonwilderness designations without a site-specific analysis. After reviewing the document the court concluded that its use had been improper in this case.[61] As detailed below, the Forest Service instead should have considered preparing and tiering lower-level site-specific analyses from the P-EIS.

8.8.5.1   Programs

Case law sheds some light on the types of agency programs and policies that require the preparation of a P-EIS. One case involved an animal productivity research program conducted by the U.S. Department of Agriculture.[62] After being challenged for not preparing a P-EIS for this program, the department countered that one was not required since the program was so diverse that it could not be considered either systematic or connected: in other words, it did not constitute a major federal action. The court agreed with the agency, basing its ruling on the fact that the project was indeed diverse, and that its components were separately operated and independent from one another.

8.8.5.2   Programmatic Alternatives

With respect to an EIS prepared for oil and gas drilling, the D.C. Circuit Court provided the following direction with respect to the evaluation of alternatives. An agency must look at reasonable alternatives but this requirement is not limited to measures the agency itself can adopt. When a proposed action forms an integral part of a coordinated plan to deal with a broader problem, the range of alternatives that must be evaluated is also broadened. While the Department of Interior did not have the authority to undertake certain alternatives such as the elimination of oil import quotas, such actions do fall within the purview of Congress and the president who receive the EIS. In other words, because an EIS is not prepared for use by the agency only, but is also for the guidance of others, it must provide all concerned parties with both the environmental effects of a proposal and its alternatives for their consideration.[63]

8.8.6   STRATEGIC EAs

Nations around the world are beginning to embrace the concept of developing strategic environmental assessments (SEAs) to establish national strategies for combating a host of environmental threats. While the definition and professional opinions vary, an SEA is frequently considered to be one level higher than a P-EIS. For example, an SEA might be prepared to establish a strategy for developing a national energy policy. Once the strategy has been formulated, a P-EIS can then be prepared to investigate various alternatives for developing an energy program to implement the national strategy.

Surprisingly, the concept of SEAs is barely acknowledged in the United States. Yet, NEPA contains all the requisite elements for assessing strategic issues of both national and international concern in an SEA.

For example, an SEA could be prepared for devising an energy plan that could be used to identify a scientifically based strategy for shifting the United States away from vulnerable and nonsustainable petroleum supplies. Besides enhancing national security, such a strategy might go a long way toward reducing environmental impacts such as global warming that, in one way or another, could affect the entire Earth. P-EISs could then be prepared and tiered from the SEA for developing specific national programs that would implement the high-level policy and strategy established in the SEA. Project-specific EAs and EISs could then be tiered from the P-EISs to implement project-specific decisions.

8.9 STREAMLINING THE REVIEW PROCESS

In many instances, delays and inefficiencies have been traced to the EIS process. For this reason, agencies should consider reviewing their procedures to prevent such problems from occurring.

The tools presented in this chapter provide environmental planners and decision-makers with rigorous and systematic methodologies that can not only streamline environmental planning but also provide a basis for preparing more systematic and defensible analyses.

In some cases, too many separate entities may be involved in an agency's internal review process. Emphasis should thus be given to simplifying the process and coordinating reviews so that they are performed in parallel rather than sequentially. One possibility is the use of specialized interdisciplinary review teams whose sole function would be to perform intensive integrated reviews of EISs, thus shortening the review time. The U.S. Air Force, for example, has completed EIS reviews in as little as 1 or 2 weeks by using an integrated review process in which all reviewers are required to meet together in a single conference room until the review has been completed.

PROBLEMS

1. Suppose an environmental impact statement (EIS) is prepared for a proposed action involving the construction and operation of an electrical power plant. The proposed action involves three boilers, each capable of producing 100 MW of electric energy. Together, the three boilers would produce a total of 300 MW. A water supply pipeline would also be constructed from a nearby lake to transmit cooling water to the plant

   After the ROD is issued, budgetary and other considerations require that certain design changes be made to the original concept. A new design consists of four 80 MW boilers using technology that is similar to but varies very slightly from the original design presented in the EIS. The new plant design also involves construction of another backup water pipeline that would run parallel but approximately 50 ft. adjacent to the pipeline described in the EIS resulting in marginally different environmental impacts. As a result of the new design, air quality impacts from sulfur and nitrogen oxide emissions would be approximately 3% lower than those presented in the EIS, but particulate impacts would be approximately 2% higher. The new plant design would also increase cooling water requirements by 12% from a major river. Can such variations be considered trivial enough that supplementing the EIS is not necessary? Justify your answer.

2. Suppose that the comment period for a draft EIS runs for a period of exactly 47 days starting on the day that the EPA publishes the NOA in the *Federal Register*. The agency promptly incorporates the comments into the draft and the EPA publishes the NOA for the final EIS, 9 days following completion of the mandatory 45-day comment period. How long must the agency wait before a final decision can be rendered?

3. Where is the NOI for an EIS published?
4. When should preparation of an EIS begin?
5. A security official has just written an opinion memo indicating that NEPA is a public process that requires open review of the proposal described in an EIS. However, because the subject proposal involves classified material, it cannot be openly disseminated and is therefore not subject to NEPA's requirements. Is this conclusion justified? If not, what can be done to comply with NEPA without compromising classified information?
6. Nine years ago, an EIS was prepared for the construction of a small air field that, due to budgetary limitations, was never built. Funding is now available but the NEPA compliance officer argues that the EIS is too old and must be redone. No significant changes have been made to the proposal and no new environmental issues have surfaced in the interim. Is the NEPA compliance officer correct?
7. Briefly explain EPA's alphabetic system for rating EIS.
8. If a final EIS is referred to the CEQ, can the lead agency legally be required to change its final decision?
9. What is the principal difference between a normal EIS and a legislative EIS?
10. What is the "Potomac Paradox"?

REFERENCES

1. Eccleston C. H., *Environmental Impact Statements: A Comprehensive Guide to Project and Strategic Planning*, John Wiley & Sons, New York, 2000.
2. CEQ, Forty most asked questions concerning CEQ's National Environmental Policy Act Regulations (40 CFR 1500–1508), question no. 14b, *Federal Register*, 46(55), 18026–18038, March 23, 1981.
3. *Environmental Law Handbook*, Government Institutes, Inc., 10th ed., Chapter 10.
4. CEQ, Forty most asked questions concerning CEQ's National Environmental Policy Act Regulations, question no. 14a, *Federal Register*, 46, 18026, March 23, 1981, as amended.
5. *Southern Utah Wilderness Alliance v. Norton*, not reported (D.D.C. 2002).
6. CEQ, Memorandum: Guidance regarding NEPA regulations, *Federal Register*, 48, 34263, July 28, 1983.
7. *Sohio Transportation Company v. United States*, 5 Cl.St. 620 (1984), aff'd, 766 F.2d 499 (Fed. Cir. 1985).
8. CEQ, Forty most asked questions concerning CEQ's National Environmental Policy Act Regulations, question no. 35, *Federal Register*, 46, 18026, March 23, 1981, as amended.
9. CEQ, Forty most asked questions concerning CEQ's National Environmental Policy Act Regulations, question no. 17a, *Federal Register*, 46, 18026, March 23, 1981, as amended.
10. CEQ, Forty most asked questions concerning CEQ's National Environmental Policy Act Regulations, question no. 17b, *Federal Register*, 46, 18026, March 23, 1981, as amended.
11. *Northern Crawfish Frog v. FHA*, 858 F. Supp. 1503, 1525 (D. Me. 1994).
12. *Sierra Club v. March*, 714 F. Supp. 539.
13. *STOP H-3 Association v. Lewis*, 538 F. Supp. 149 (D. Hawaii 1982).
14. CEQ, 1980, public memorandum titled, *Talking Points on CEQ's Oversight of Agency Compliance with the NEPA Regulations*.
15. CEQ, *Council on Environmental Quality Guidance on Pollution Prevention and the National Environmental Policy Act*, published at 58 FR 6478, January 29, 1993.
16. Pollution Prevention Act of 1990, Pub. Law 101-508, 6601 et seq.
17. Executive Order 11990, *Protection of Wetlands*, May 24, 1977.
18. Executive Order 11988, *Floodplain Management*, May 24, 1977.
19. Floodplain management guidelines of the U.S. Water Resources Council, *Federal Register*, 40, 6030, Feb. 10, 1978.
20. *Weinberger v. Catholic Action of Hawaii/Peace Education Project*, 454 U.S. 139 (1981).
21. CEQ, Forty most asked questions concerning CEQ's National Environmental Policy Act Regulations, question no. 32, *Federal Register*, 46, 18026, March 23, 1981, as amended.
22. *Coker v. Army* (CA5, 1991) 941 F.2d 1306.

23. *Northwest Coalition for Alternative to Pesticides v. Lying*, 844 F.2d 588 (9th Cir. 1988).
24. EPA 842-B-01-003, November 2002.
25. CEQ, Preamble to final CEQ NEPA regulations, *Federal Register*, 43, 55978, Section 1, November 29, 1978.
26. CEQ, Forty most asked questions concerning CEQ's National Environmental Policy Act Regulations, question no. 13, *Federal Register*, 46 18026, March 23, 1981, as amended.
27. *Citizens Against the Collider Here—Illinois, Inc. v. DOE*, 1988 WL 94142 (N.D. Ill. 1988).
28. Eccleston C. H., Decision-identification tree—a new NEPA scoping tool, *Journal of Environmental Management*, University of Massachusetts, 26(4), pp. 457–464, 2000.
29. *Edwardsen v. US Department of the Interior*, 268 F.3d 781 (9th Cir. 2001).
30. *Vermont Public Interest Research Group v. U.S. Fish and Wildlife Service*, 33 ELR 20062 (D. Vt. 2002), citing *County of Suffolk v. Secretary of the Interior*, 562 F.2d 1368, 1375 (2d Cir. 1977).
31. 5 U.S.C. 552.
32. *Del Norte County v. United States*, 732 F.2d 1462 (9th Cir. 1984), *cert. denied*, 469 U.S. 1189 (1985).
33. DOE, National Environmental Policy Act, *Learned Lessons*, Issue No. 32, September 3, 2002.
34. 116 Cong. Rec. S-20602, Dec. 18, 1970.
35. Sen. Rept. No. 91-1196, 91st Cong., 2d Sess. 43, 1970.
36. Fogleman V. M., *Guide to the National Environmental Policy Act Interpretations, Applications, and Compliance*, Quorum Books (1990).
37. EPA, Policy and Procedures for the Review of Federal Actions Impacting the Environment, October 3, 1984.
38. Clean Air Act, 42 U.S.C. § 7609, Section 309 (2000).
39. DOE, National Environmental Policy Act, *Learned Lessons*, Issue No. 24, September 1, 2000.
40. CEQ, *Forty most asked questions concerning CEQ's National Environmental Policy Act Regulations*, question no. 29a, *Federal Register*, 46, 18026, March 23, 1981, as amended.
41. CEQ, *Forty most asked questions concerning CEQ's National Environmental Policy Act Regulations*, question no. 29b, *Federal Register*, 46, 18026, March 23, 1981, as amended.
42. CEQ, *Forty most asked questions concerning CEQ's National Environmental Policy Act Regulations*, question no. 34b, *Federal Register*, 46, 18026, March 23, 1981, as amended.
43. CEQ, *Preamble to final CEQ NEPA regulations, Federal Register*, 43 55978, Section 4, November 29, 1978.
44. CEQ, *Forty most asked questions concerning CEQ's National Environmental Policy Act Regulations*, question no. 34a, *Federal Register*, 46, 18026, March 23, 1981, as amended.
45. CEQ, *Forty most asked questions concerning CEQ's National Environmental Policy Act Regulations*, question no. 5a, *Federal Register*, 46, 18026, March 23, 1981, as amended.
46. CEQ, *Special Report: Environmental Referrals and the Council on Environmental Quality*, CEQ 17th Annual Report, 1988.
47. Mandelker D. R., *NEPA Law and Litigation*, Clark Boardman Callaghan, New York, 4.12.
48. *Warm Springs Dam Task Force v. Gribble*, 621 F.2d 1017, 1024 (9th Cir. 1980).
49. Eccleston C. H., Determining if a change to a proposal requires additional NEPA documentation—the Smithsonian Solution, *Federal Facilities Environmental Journal*, pp. 119–132, Spring 1999.
50. *Marsh v. Oregon Natural Resources Council*, 490 U.S. 360, 109 S.Ct. 1851 (1989).
51. *National Indian Youth Council v. Watt*, 664 F.2d 220 (10th Cir. 1981), citing *Concerned Citizens v. Secretary of Transportation*, 641 F.2d 1, 6 (1st Cir. 1981).
52. *Coker v. Skidmore*, 941 F.2d 1306 (5th Cir. 1991); *Sierra Club v. United States Army Corps of Engineers*, 701 F.2d 1011, 1036 (2d Cir. 1983).
53. Administrative Procedure Act, §706(2)(A).
54. *Marsh v. Oregon Natural Resources Council*, 490 U.S. 360, 109 S.Ct. 1851 (1989); *Citizens to Preserve Overton Park, Inc. v. Volpe*, 401 U.S. 402, 416 (1971) Id. at 375, 376, 378.
55. 40 Code of Federal Regulations (CFR) 1500–1508, Regulations for implementing the procedural provisions of the National Environmental Policy Act, *Code of Federal Regulations*, November 29, 1978.
56. DOE, 10 CFR 1021.314(c), U.S. Department of Energy, National Environmental Policy Act, Implementing procedures and guidelines revocation; Final Rule and Notice, April 24, 1992.
57. DOE, *NEPA Lessons Learned*, March 1, Issue No. 50, 2007.
58. CEQ, *Considering Cumulative Effects Under the National Environmental Policy Act*, January, p. 1, 1997.

59. Eccleston C. H., *Effective Environmental Assessments: How to Manage and Prepare NEPA EAs*, Lewis Publishers Inc., Boco Raton, FL, 2001.
60. Eccleston C. H., Determining the scope and level of detail appropriate for a programmatic environmental impact statement, *Federal Facilities Environmental Journal*, Spring 1996.
61. *California v. Block*, 18 ERC 1149, 1982.
62. *Foundation on Economic Trends v. Lyng*, 817 F.2d 882 (D.C. Cir. 1987).
63. *Natural Resources Defense Council v. Morton*, 458 F.2d 827 (D.C. Cir. 1972).

9 The Scope of Analysis

The purpose of scoping is to obtain input from both the lead agency and outside entities regarding the range of actions, impacts, and alternatives to be considered. This enables the subsequent analysis to be sharply focused on issues of genuine concern. A well-orchestrated scoping process provides agencies with a particularly powerful tool for achieving this goal. Just as important, however, is its use to de-emphasize insignificant issues or irrelevant details so as to efficiently narrow the scope of the analysis. The author refers to this objective as "de-scoping." The sufficiency-test tool presented in Section 2.7 is a systematic and practical tool that can assist practitioners in determining the level of detail appropriate for a given analysis.

9.1 DE-EMPHASIZING THE SCOPE OF ANALYSIS

To the extent feasible, the analysis should be integrated with all other permits, agreements, plans, and policies relating to it so that processes run concurrently rather than consecutively. Scoping can be instrumental in identifying such requirements. Consistent with the rule of reason, the author has developed a set of criteria for assisting analysts in limiting the scope of an analysis (Table 9.1).

9.2 MANAGING PUBLIC SCOPING MEETINGS

Never assume a casual attitude toward public hearings and scoping meetings. This warning is given since the people who attend them, whether they are local homeowners, employees concerned with their jobs, or members of special interest groups, are usually the ones with the most vested interest in the outcome. They are also often well educated and technically sophisticated. This tends to be true especially of activists representing special interests. Attendance at the meetings also tends to increase with the level of controversy involved. In cases where this is expected to be high, managers may want to start off by holding a dry run that includes a question-and-answer session and perhaps even mock-media interviews.

To fully appreciate public sentiment, the lead agency's decision-maker should attend all scoping meetings held. As explained in the next section, the agency may find it advantageous to obtain the services of an independent and neutral moderator because the presence of such a person can often mitigate public hostility, particularly in circumstances where the agency or proposed project may lack credibility. The facilitator should, of course, be thoroughly briefed in advance on all aspects of the proposal.

Attendees often have the misconception that the scoping meeting is being held to make a decision about the proposal. An independent facilitator can dispel such misconceptions by explaining clearly at the outset that the purpose of the scoping meeting is only to elicit input and comments.

9.3 ACTIONS, ALTERNATIVES, AND EFFECTS

As detailed in the Council on Environmental Quality's (CEQ) implementing regulations (Regulations), an environmental impact statement (EIS) involves consideration of three basic concepts:

- Actions
- Alternatives
- Effects

TABLE 9.1

Criteria Useful in Limiting the Scope of the Analysis

- Is the issue potentially significant? If not, provide only enough analysis to demonstrate that the issue is not significant.
- Would additional analysis further the decision-making process? If not, analysts should consider whether the inclusion of any further analysis or information is warranted.
- Is the issue ripe for an analysis and a decision? If not, the agency should consider eliminating it from the analysis. Caution must be exercised in making this determination.
- Are the impacts of the action reasonably foreseeable? If not, the need to discuss them may be unjustified.

TABLE 9.2

Three Types of Actions, Alternatives, and Effects

Three Types of Actions
1. Connected actions
2. Cumulative actions
3. Similar actions

Three Types of Alternatives
1. No-action alternative
2. Other reasonable courses of action
3. Mitigation measures (not in the proposed action)

Three Types of Impacts
1. Direct impacts
2. Indirect impacts
3. Cumulative impacts

Together, these concepts delimit the scope of analysis (§ 1508.25). As indicated in Table 9.2, each of these concepts, in turn, is composed of three additional components. The scope of the analysis, therefore, refers to the range of actions, alternatives, or impacts to be reviewed. The following sections describe each of these in more detail.

9.4 THREE TYPES OF ACTIONS

The three types of actions defined under the National Environmental Policy Act (NEPA) are described in the following sections.

9.4.1 CONNECTED ACTIONS

Actions are said to be connected when they are related closely enough to be discussed in the same NEPA analysis. Actions are said to be connected if they (§ 1508.25[a][1])

i. automatically trigger other actions that may require the preparation of EISs,
ii. cannot or will not proceed unless other actions are taken either before them or simultaneously, and
iii. are interdependent parts of a larger action and thus dependent on that action for their justification.

A commonly cited example of a connected action involves a case in which an environmental assessment (EA) was prepared for a proposal involving the construction of a bridge to an island.

Opponents challenged it, arguing that the analysis did not adequately address the impacts that would result from the subsequent opening of the island to large-scale development (the connected actions). In response, the court sided with the opposition, concluding that an EIS was indeed necessary to evaluate the impacts.[1]

A second example involves a Hawaiian Geothermal Project (HGP) comprised of four phases: (1) exploring and testing geothermal reservoirs, (2) researching the feasibility of transmitting power via underwater cables, (3) drilling exploration wells, and (4) constructing geothermal power plants. The Department of Energy (DOE) provided funds for the first two phases. In 1988, when Congress appropriated money for Phase III, a congressional conference report stated that because this phase was "research," it was not a major federal action subject to NEPA, but the DOE should earmark some of the funds to prepare an EA or an EIS later for the project. Plaintiffs then sued the DOE for not preparing an EIS immediately. The court agreed rejecting Congress' characterization and holding that Phases III and IV were in fact connected actions that needed to be considered together in a single EIS. The court further held that the "research work" contemplated by Phase III "alone easily satisfies the statutory standards for 'major federal action' based simply on the extent of federal funding."[2]

9.4.2 Cumulative Actions

Cumulative actions involve activities that (§ 1508.25[a][2])

... when considered along with other proposed actions have cumulatively significant impacts and should, therefore be evaluated within the same analysis.

9.4.2.1 Cumulative Actions versus Cumulative Impacts

As illustrated by the following case, the reader should note that the concept of cumulative impacts can easily be confused with the related concepts of connected, cumulative, and similar actions. It is important to understand the difference because some courts have reached decisions based on the distinction between these two terms.

Confusing Connected Actions with Cumulative Actions. In one case, a court upheld a license issued by the Federal Energy Regulatory Commission (FERC) for the first phase of a hydroelectric plant. The EIS prepared for the project looked only at the environmental impacts of Phase I, although construction of Phase II, while not inevitable, was reasonably foreseeable. The plaintiffs challenged this, asserting that FERC had not assessed the potential impacts of Phase II before deciding to approve Phase I.[3]

In this case, the court rejected the plaintiffs' assertion, reasoning that Phase II had not yet been proposed and that "NEPA merely requires an agency to consider all other proposed actions that may, along with the proposed action at issue, have a cumulative [effect]."

This appears to be an example where the court confused the requirement to consider all connected or cumulative actions together in the same comprehensive EIS (§ 1508.25[a]) with the requirement to assess the cumulative impacts of the proposal and other reasonably foreseeable future actions (see §§ 1508.7, 1508.8, and 1508.25[c]). As the court noted, only actual proposals as defined in the Regulations (see § 1508.23) need to be considered together in a single EIS, but once the scope of the EIS has been determined, the agency is required to look at the cumulative impacts of other past, present, and reasonably foreseeable future actions (§ 1508.7).

Army Corps Case. In another case, the Army Corps was sued for its decision to prepare an EA on Section 404 permit to fill in a wetland area for a development project.[4] By all accounts, although there were plans for future development, these were not yet being actively evaluated by the Corps. Furthermore, it had already been acknowledged that the proposal would not result in a significant impact.

In ruling for the Corps, the court drew a distinction between the requirements to analyze cumulative actions and cumulative impacts. With respect to the former, the court noted that the

Regulations require connected, cumulative, and similar actions to be considered together in the same EIS. Moreover, proposals that are subject to consideration and that are functionally or economically related to each other must also be considered in a single EIS.

As defined in the Regulations, only actual proposals are considered sufficiently related to require evaluation in a single NEPA document (§ 1508.23); thus, only proposals that are ready for decision, for example, several Section 404 permits pending before the Army Corps in one geographic region, need to be evaluated in a single analysis. In contrast, the obligation to address cumulative impacts in a single EIS is not limited to actual proposals.

In other words, with respect to cumulative impacts the court noted that the impacts were not limited to those from actual proposals but also had to include impacts from actions merely being contemplated (i.e., not yet ripe for decision). However, in this case the court noted that contemplated actions must be "reasonably foreseeable" in the sense that they are not speculative and not in the distant future.

9.4.3 Similar Actions

Similar actions encompass activities that (1508.25[a][3])

> ... when viewed with other reasonably foreseeable or proposed agency actions, have similarities that provide a basis for evaluating their environmental consequences together, such as common timing or geography. An agency may *wish* to analyze these actions in the same impact statement. It should do so when the best way to assess adequately the combined impacts of similar actions or reasonable alternatives to such actions is to treat them in a single impact statement (emphasis added).

9.4.4 Categories of Federal Actions

With respect to NEPA, federal actions tend to fall into one of the four broad categories (§ 1508.18[b]):

* Policies
* Plans
* Programs
* Projects

These four types of actions are sometimes referred to as the "4 Ps."[5] Each of these categories is discussed below.

9.4.4.1 Policies

This first category involves the adoption of new policies or regulations that often have far-reaching environmental implications. Other actions falling under this category include the adoption of rules and regulations related to treaties and international conventions or agreements (§ 1508.18[b][2]). During informal rule making, a draft EIS is expected to accompany the proposed rule (§ 1502.5[d]).

9.4.4.2 Plans

Actions may involve adoption of official documents that guide or prescribe alternative uses of federal resources and on which future agency actions will be based. For example, the adoption of a plan for managing a national forest (§ 1508.18[b][2]).

9.4.4.3 Programs

An EIS may be prepared and is sometimes required for broad federal actions such as the adoption of new agency programs or regulations (§ 1508.18[b][3]). Agencies must prepare EISs that are relevant to policy and are timed to coincide with meaningful points in agency planning and decision-making (§ 1502.4[b]).

9.4.4.4 Projects

Federal actions falling under this category include approval of specific projects such as construction, operation, or management activities within a defined geographic area. Such projects include actions involving the approval of a permit or other regulatory decision as well as federal and federally assisted activities (§ 1508.18[b][4]).

9.4.5 SEGMENTATION

The term "segmentation" refers to the process of breaking an activity into smaller activities that are then analyzed individually under separate NEPA analyses. This practice can dramatically reduce the chance that the overall effect will be deemed significant, requiring the preparation of an EIS. The Regulations state (§ 1502.4[a])

Proposals or parts of proposals which are related to each other closely enough to be, in effect, a single course of action shall be evaluated in a single impact statement

Furthermore, the Regulations add (§ 1508.27[b][7])

... significance cannot be avoided by ... by breaking it down into small component parts.

9.4.5.1 Case Law

In one case, a district court held that the Bureau of Land Management (BLM) improperly violated NEPA by segmenting its analysis of an oil pipeline project from New Mexico to Utah.[6] The two pipelines were originally proposed as one project. The court concluded that the pipeline segment did not have independent utility from another proposed pipeline project between Texas and New Mexico. The court found that the BLM had acted arbitrarily in deciding that the two projects were not connected actions.

In another case, two EAs were prepared for a proposed hotel or casino on an Indian reservation and an interchange and access road connecting the reservation to a highway. The district court held that the decision to segment the review into two EAs did not violate NEPA. In this case, the two EAs were prepared because of jurisdictional considerations. Additionally, the interchange EA incorporated the casino EA by reference and considered the cumulative impacts of the project as a whole. The court also found that the EA adequately addressed potential environmental impacts and reasonably concluded that the impacts would not be significant (i.e., an EIS was not required).[7]

9.5 THREE TYPES OF ALTERNATIVES

An alternative that is actually analyzed, in addition to simply being identified as a reasonable alternative, is often referred to as an "analyzed alternative." The number of analyzed alternatives is generally smaller than the range of reasonable alternatives.

The Regulations state the following with respect to the ability to make informed decisions:

- Section on alternatives provides "a clear basis for choice among options by the decision-maker and the public" (§ 1502.14).
- EIS "shall inform decision-makers and the public of the reasonable alternatives which would avoid or minimize adverse impacts" (§ 1502.1).

The following sections describe the three types of actions defined in the Regulations.

9.5.1 No-Action Alternative

An EIS analysis must include the alternative of taking no action (§ 1502.14[d]). A discussion of the no-action alternative serves two purposes:

- First, this requirement forces decision-makers to seriously consider the possibility of not moving ahead with any action.
- Second, it provides the only reliable baseline against which the impacts of pursuing a particular action may be compared with the consequences of taking no action.

The no-action alternative must normally be evaluated in an EIS even if the agency has been commanded by Congress or is under a court order to pursue a particular course of action. This is because, as stated above, it provides a baseline for comparing the environmental impacts of pursuing an action.[8]

9.5.1.1 The No-Action Alternative versus the Affected Environment

The concepts of the no-action alternative and the affected environment are frequently confused. At times, an erroneous premise is made in which the impacts of the no-action alternative are equated to a description of the affected environment, but in fact these are different concepts that serve different purposes. The affected environment describes the setting within which a proposed action would take place or affect. It constitutes a snapshot of present conditions of resources and the geographic area that could potentially be affected by a proposed action and its alternatives. Thus, the affected environment defines an environmental baseline for assessing potential impacts of a proposal.

In contrast, the impacts of the no-action alternative provide a different environmental baseline that allows both decision-makers and the public to compare future impacts of taking no action against the corresponding effects of the analyzed alternatives. Although the no-action alternative is often described as representing the status quo, this does not mean that taking no action can be regarded as a static condition. The potential impacts of the no-action alternative are estimated from a projection of current conditions into the future, under the influence of activities that would continue and of those that are based on decisions made previously. To facilitate a meaningful comparison, the time period used to assess the impacts of the analyzed no-action alternative must be comparable to the time period used to analyze the impacts of the analyzed alternatives (including the proposed action).

For example, discussion of the air quality of an affected environment might describe its general climate, wind, temperature, rainfall, ambient concentrations of air pollutants, and current emissions and emission rates. This discussion would also, as appropriate, identify existing air quality permits and specify the attainment status for criteria pollutants.

In contrast, analysis of the no-action alternative might involve projecting the future site emissions and emission rates if a proposed action (or alternative) did not take place. The impact assessment would also identify the impacts of such future emissions *vis-á-vis* compliance with applicable air quality regulations and permits, the attainment status for criteria pollutants, and their effects on human health and the environment. Over time, the air quality conditions for the affected environment baseline could be very different from that of no action.

9.5.1.2 Describing the No-Action Alternative

A thorough examination of the no-action alternative may help an agency gain public acceptance for a proposal that might otherwise be unpopular, for example, a community landfill that is nearly filled to capacity. Such projects tend to be unpopular and controversial, but what would be the environmental and socioeconomic impacts if no action was taken to resolve the problem? A proper analysis of the no-action alternative could involve projecting the impacts of doing nothing into the future. For example, people would lack a sufficient garbage disposal site without a reasonable replacement for the landfill. Socioeconomic experts could be consulted to identify the likely responses of both individuals and the community at large if no viable replacement was provided. For instance, some people might dig holes in their backyards to dispose of their garbage. Others might dump garbage along roadsides or in nearby wilderness areas. The resulting impacts of taking no action could thus be very significant. Project proponents can often gain public acceptance simply by presenting a clear picture of what the potential consequences could be of taking no action.

9.5.2 REASONABLE ALTERNATIVES

As viewed by the CEQ, reasonable alternatives are those considered to be practical and feasible from both technical and economic standpoints. Thus, alternatives that a particular individual or group find desirable but which are neither technically or economically practical would not lie within the range of reasonable alternatives requiring consideration.[9] In one case, an EIS was found to be inadequate partly because it had improperly eliminated a project alternative based on an inadequate estimate of its cost.[10]

The range of alternatives must encompass those that will be or could be considered by the decision-maker. Moreover, agencies are expected to undertake a rigorous exploration and an objective evaluation of all reasonable alternatives, including those that avoid or minimize adverse impacts:

- Rigorously explore and objectively evaluate all reasonable alternatives, and for alternatives which were eliminated from detailed study, briefly discuss the reasons for their having been eliminated (§ 1502.14[a]).
- Use the NEPA process to identify and assess the reasonable alternatives to proposed actions that will avoid or minimize the adverse effects of these actions on the quality of the human environment (§ 1500.2[e]).
- The range of alternatives discussed in environmental impact statements shall encompass those to be considered by the ultimate agency decision-maker (§ 1505.1[e], § 1502.2[e]).

It is important to evaluate a broad range of alternatives in an EIS or an EA to provide responsible decision-makers with the flexibility to respond to changing circumstances. Through close coordination with project planners and engineers, document preparers can ensure that an EIS or an EA covers new ideas that may emerge on better, cheaper, and faster ways to accomplish the agency's purpose and need for action.

The option of alternative locations should not be overlooked. This is especially true in those instances where actions may take place in environmentally sensitive areas. Where feasible, good professional practice dictates and the courts may mandate that alternative locations be considered.

During the public review period for the draft Regulations, commenters voiced concern that the phrase "all reasonable alternatives" should be narrowed down because it was unduly broad. In rejecting these comments, the CEQ stated that the phrase was not meant to be interpreted to mean "that an infinite or unreasonable number of alternatives be analyzed."[11] This interpretation has been supported in case law.

In one example, a court concluded that an EIS cannot be deemed inadequate simply because the agency failed to consider every conceivable alternative. Rather, as a process evolves, agencies may consider "... more or fewer alternatives as they become better known and understood."[12]

Consistent with the philosophy of a sliding-scale (see Chapter 2), the number of alternatives that are eventually analyzed may vary with the complexity and potential of the particular project to cause environmental degradation.

9.5.2.1   Dismissing Unreasonable Alternatives

Consistent with the rule of reason, agencies are not expected to use a crystal-ball approach in performing their analyses. For example, some courts have ruled that agencies are not required to examine alternatives that are considered remote or speculative.[13]

Agencies are required, however, to consider alternatives outside their immediate control. While these alternatives are normally subject to NEPA's requirements, some courts have ruled that those outside an agency's "full control" are not. For example, an EIS prepared for an antimissile defense system would probably not have to look at the alternative of eliminating nuclear weapons under an international disarmament treaty because in this instance "full control" by the agency would not be possible. Nonetheless, caution should always be exercised before dismissing alternatives since all EISs must explain briefly why certain alternatives were determined to be unreasonable and therefore eliminated from detailed study (§ 1502.14[a]).[14] While this direction does not explicitly extend to EAs, professional practice dictates that the same procedure be followed in their case.

9.5.2.2   Alternatives beyond an Agency's Control

A potential conflict with a local or federal law does not necessarily render an alternative unreasonable, although such conflicts must be considered in its evaluation (§ 1506.2[d]).[9]

NEPA may be triggered by the act of granting a permit, approval, or authorization to a private applicant. In such cases, the CEQ has stated that an EIS must evaluate all reasonable alternatives, even if carrying out them is beyond the capability of the applicant.[9]

This direction applies equally to situations where reasonable alternatives may lie outside the legal jurisdiction of an agency (§ 1502.14[c]).[15] For instance, an agency may propose the construction of a new road. One of the reasonable routes suggested may pass through an area that is controlled by a local agency, thus requiring a land use permit enabling such development. Although implementing such an alternative may lie beyond the jurisdiction of the federal agency, it must nonetheless be considered in the analysis.

Another circumstance involves consideration of those alternatives that are outside the scope of what Congress may have authorized or funded. The rationale behind this requirement is that analyzing such alternatives may provide a sound basis that can inform Congress of the need to reconsider its authorization to support their implementation or to modify funding (§ 1500.1[a]).[9] In such cases, the agency should clearly explain why the alternative has not been selected or why the law should be changed so that it would be able to pursue that alternative.

9.5.2.3   Court Direction on Preparing Alternatives

In the case of an EIS prepared for oil and gas drilling, the Circuit Court of the District of Columbia provided the following direction with respect to the evaluation of alternatives:[16]

- Agency must look at reasonable alternatives, but this is not limited solely to the measures the agency itself can adopt. While an agency may not have the authority to undertake certain alternatives (such as the elimination of oil import quotas), such actions do fall within the purview of Congress and the president who are the recipients of the document. An EIS, therefore, is not prepared only for the agency's purposes but for the guidance of others. For this reason, it must provide those parties with information regarding the probable environmental effects of both the proposal and its alternatives for their consideration.
- Discussion of alternatives need not be exhaustive. What is required is information sufficient to permit a reasoned choice of alternatives so far as the environmental aspects are concerned. Nor is it appropriate to disregard alternatives merely because they do not offer a complete solution to the problem.
- Discussion of reasonable alternatives does not require a crystal-ball inquiry. The "rule of reason" must be applied.

With regard to the first of the bulleted paragraphs listed above, it should be noted that some courts have provided different directions.

9.5.2.4   Defining a Range of Reasonable Alternatives

The range of alternatives should include all those requiring objective evaluation and rigorous exploration as well as any that have been eliminated from detailed study.[14] Determining a range of reasonable alternatives may involve identifying options as diverse as the consideration of using alternative sites, modes of transportation, and technologies. The alternatives considered by decision-makers must be encompassed by the "... range of alternatives ..." described in the NEPA analysis (§ 1502.2[e], § 1505.1[e]).

There are occasions when the range of reasonable alternatives may indeed appear to be virtually unlimited. For example, a proposal for logging a forested area could involve a very large number of potential courses of action to ascertain the impacts of logging anywhere from 0% to 100% of the site. However, an EIS only needs to consider a reasonable number of scenarios that serve to cover the full spectrum of possibilities. An adequate analysis might therefore be limited to a range of alternatives that would analyze the different impacts resulting from developing or cutting 1%, 10%, 30%, 50%, 70%, 90%, or 100% of the forest.[17]

9.5.2.5   Describing Alternatives

All reasonable alternatives listed in an EIS must be given substantial attention. Descriptions must contain sufficient detail to provide analysts with the information they need to perform an effective identification and assessment of potential impacts. The alternatives should be clearly described so that decision-makers have a thorough understanding of what could take place once the project is under way. Where possible, alternatives should be described and evaluated throughout their life cycle (site preparation, construction, operation, and closure). Any private actions enabled by the federal agency concerned such as issuing grants and contracts for various activities should also be clearly described.

In the process of identifying potential alternatives, analysts should not forget options such as leasing a service or facility from the private sector since these are often not only reasonable, but also very economical courses of action. Moreover, if an agency is able to avoid having to construct a new facility, the environmental impacts may be reduced.

9.5.2.6   Evaluating Hypothetical Scenarios

In 2003, the BLM completed an EIS on its plan to lease up to 8.8 million acres of federal land in northern Alaska for oil and gas exploration. The EIS investigated five alternatives, including that of taking no action. The four action alternatives considered a spectrum ranging from making 47–100% of the BLM-administered lands available for leasing and assumed different types of management actions and mitigation measures.

With respect to potential environmental impacts associated with drilling, the BLM did not analyze specific parcels because, the agency contended, it had no way of knowing which areas, if any, subsequent exploration would find suitable for drilling. Instead, the analysis considered two hypothetical scenarios: one assuming development of the total available resources and the second assuming exploration of half the available parcels but no actual development.

Plaintiffs challenged the adequacy of the EIS for failing to include an analysis of site-specific environmental impacts.[18] On July 26, 2006, the U.S. Court of Appeals for the Ninth Circuit upheld a lower-court decision when it agreed with BLM that "... no such drilling site analysis is possible until it is known where the drilling is likely to take place, and that can be known only after leasing and exploration." Additionally, the court concluded that environmental consequences at specific sites can be assessed in connection with later applications for drilling permits at those sites.

It is recommended that the reader consult legal counsel in determining to what extent an analysis of hypothetical sites may be appropriately used in NEPA analyses.

9.5.3 MITIGATION MEASURES

Courts have long held that mitigation measures must be discussed in EISs. In 1987 the Ninth Circuit Court of Appeals went one step further, ruling that NEPA imposed a substantive requirement to mitigate adverse impacts.[19] In 1989, however, the Supreme Court reversed this decision.

The descriptions of alternatives should clearly indicate and describe any mitigation measures that are evaluated during their analysis and provide sufficient detail to enable analysts to rate their effectiveness. There are instances when some mitigation measures may be common to a proposed action and to all of its analyzed alternatives. In these cases, for brevity's sake, it is suggested that such common measures be placed together in a separate section.

9.6 THREE TYPES OF IMPACTS

As witnessed earlier, the Regulations recognize three categories of impacts. These are explained in detail in the following sections.

9.6.1 DIRECT IMPACTS

Direct impacts are those effects that occur at the same time and place as the action producing them (§ 1508.8[a]) and are directly attributable to that action. Direct impacts are generally easier to identify and evaluate than indirect or cumulative effects.

9.6.2 INDIRECT IMPACTS

Indirect impacts, or secondary effects as they are sometimes called, are not directly linked to the original action but are removed from it by distance or time (§ 1508.8[b]). Indirect effects often include socioeconomic impacts such as community growth and changes in population pattern as a result of implementing a proposed action (§ 1508.8).

A typical example involves the effects resulting from residential and commercial development that will be triggered by the construction of a new highway off ramp. The impacts of this future development may occur well after the ramp has been built and at some distance from the highway itself.

Because indirect impacts are frequently more difficult to identify and evaluate than direct impacts, it should come as no surprise that they are often given inadequate attention. This is not only unfortunate but potentially illegal, given the fact that the effects of indirect impacts ultimately often far exceed those of direct impacts.

9.6.3 CUMULATIVE IMPACTS

The science of cumulative impact assessment (CIA) is, at this point, still an emerging discipline. Practitioners frequently complain of being overwhelmed by its scoping and analytical details. The key to performing a successful CIA is to focus on important cumulative issues, recognizing that a better decision, not a perfect analysis, is the ultimate goal.

Cumulatively significant impacts may result from actions that are minor when analyzed individually but significant when viewed collectively. A significant cumulative impact may occur even if an individual action takes place over a long period of time. The following definition explains how cumulative impacts can include the effects of both federal and nonfederal actions (§ 1508.7):

> … the impact on the environment which results from the incremental impact of the action when added to other past, present, and reasonably foreseeable future actions regardless of what agency (Federal or non-Federal) or person undertakes such other actions. Cumulative impacts can result from individually minor but collectively significant actions taking place over a period of time.

For instance, an agency is preparing an EIS for the construction of extra access roads into a park area. In analyzing the potential new air, noise, traffic, and endangered species impacts, the agency would also need to evaluate the past impacts (e.g., roads and similar projects already constructed), present proposals (e.g., new projects within the immediate area), and reasonably foreseeable future projects (e.g., other known development projects being planned). Only through an analysis of the cumulative impacts can the agency obtain an accurate picture of what the total future effects of these impacts on the park may be.

While the effects of individual activities may appear negligible, the combined effects of numerous activities can be formidable. According to the Regulations

> Significance exists if it is reasonable to anticipate a cumulatively significant impact on the environment (§ 1508.27[b][7]).

Thus, the cumulative impacts associated with a proposal may be deemed to be significant even if the direct and indirect effects are nonsignificant. This concept is portrayed by the following example.

An EIS concluded that the direct and indirect radiation impacts on a population from a proposed action would result in 0.37 latent cancer fatalities (LCF). An LCF indicates the additional number of cancer fatalities that would be anticipated in a given population as a result of a radioactive impact. In other words, it was considered unlikely that even one additional cancer fatality would occur as a result of this action. However, numerous other nuclear activities were under way at that time and were expected to continue into the foreseeable future. The impact of the proposed action when added to those of the other ongoing and foreseeable future activities resulted in a value of 0.97 LCF. In other words, one person could be expected to die from cancer as a result of the proposed action when taken in combination with the other ongoing activities. This example clearly shows how a CIA has the potential to influence decision-makers and the public in a different manner from an analysis that focuses solely on the direct and indirect impacts.

9.6.3.1   Innocuous Activities Have Cumulatively Significant Impacts

The author has long argued that the greatest single adverse environmental impact actually tends to be the result of an incessant multitude of relatively small actions, which together extract a horrific toll on environmental resources. According to the CEQ, there is increasing evidence that the most destructive effects may actually result not from the direct effects of a given action but instead from the combination of individually minor effects of numerous actions over time.[20] William Odum has described environmental degradation from cumulative impacts as the "tyranny of small decisions."[21]

Some authorities maintain that most impacts can be viewed as cumulative because most systems have already been degraded by human actions. For example, according to the National Performance Review, the San Francisco Bay estuary has been severely affected by decisions made by a wide variety of government agencies; this report notes that a single mile of the San Francisco Bay delta may be affected by the decisions of more than 400 different local, state, and federal agencies.[22]

9.6.3.2   Defining Spatial and Temporal Bounds

One of the reasons that the analysis of cumulative impacts tends to be so much more challenging than the corresponding assessment of direct and indirect effects is simply the difficulty of defining the geographic (spatial) and time (temporal) boundaries. If these boundaries are defined too broadly, the analysis becomes unwieldy. Conversely, if they are defined too narrowly, the analysis may be insufficient to inform decision-makers of the potentially significant issues.

Spatial Domain. With respect to direct impacts, it is usually sufficient to limit the spatial bounds of the analysis to the immediate area in which the project would occur. The spatial domain used in the analysis of indirect effects frequently needs to be expanded beyond the bounds used for the analysis of direct effects. For a CIA, the geographic bounds typically have to be expanded beyond that which is deemed sufficient for evaluating either the direct or indirect effects.

Choosing the appropriate spatial domain for a CIA is critical and depends on the nature of the proposal and the potentially affected environmental resources. A cumulative boundary may involve considering an entire human community, groundwater system, air shed, water shed, ecosystem, or a basin.

Time Domain. The environmental impact of a specific project may end abruptly or diminish slowly with time. The timeframe for a project-specific analysis normally does not extend beyond the point where the project-specific effects diminish below the threshold level of significance. However, this same practice does not necessarily extend to the problem of assessing cumulative impacts.

The Regulations define a cumulative impact to be the "… incremental effect of the action when added to other past, present, and reasonably foreseeable future actions" (§ 1508.7). Defining the appropriate timeframe over which a CIA should be performed is often more difficult than establishing the corresponding spatial domain. The timeframe of a project-specific analysis may be helpful in determining how far to project the CIA into the future. For example, if the impacts of the project would extend 7 years into the future, this same timeframe might in some instances also be sufficient for performing the CIA. It is not uncommon, however, to find that the timeframe must be expanded substantially beyond that for the project itself.

Figure 9.1 shows a project's direct and indirect impacts diminishing until a point is reached, approximately 13 years into the future, where these effects drop below the point of significance.[23] The analysis of direct and indirect impacts would normally not extend beyond this point in time.

However, as shown in Figure 9.1, one or more future actions affecting this environmental resource would be triggered around the 16th year into the project.[23] The impacts of this future action(s) increase over time. The cumulative impact is therefore the summation of the dissipating project-specific impact and the increasing effect of this future action(s). As depicted in this figure, the cumulative impact slowly increases until it finally breaches the significance threshold

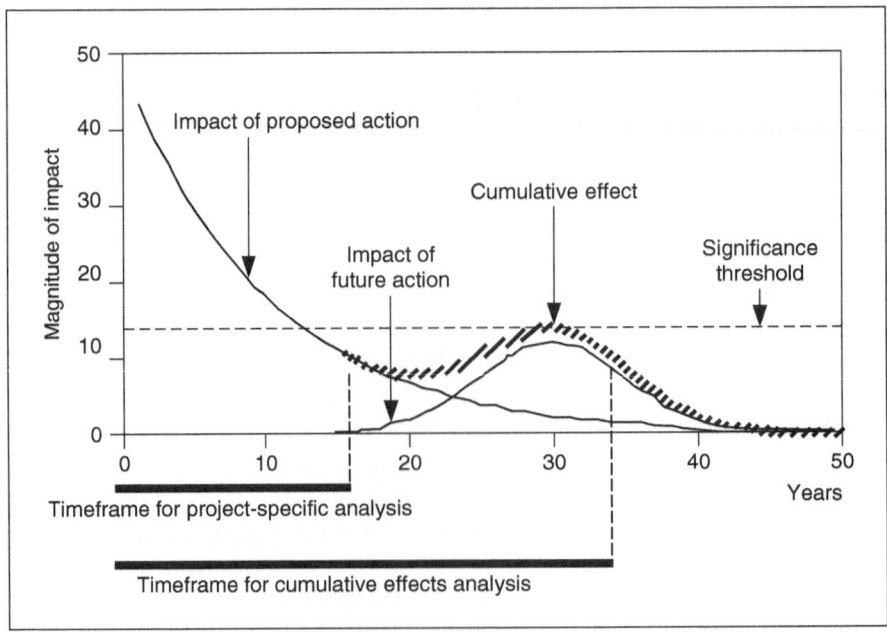

FIGURE 9.1   Timeframe for a cumulative impact assessment. (Courtesy of CEQ.)

approximately 30 years into the future. Thus, the timeframe over which the cumulative impact must be evaluated is substantially greater than that for the project-specific impacts by themselves.

One potential time constraint on the CIA timeframe is simply not to extend the analysis beyond the point in which the impacts of the reasonably foreseeable future actions can be identified or meaningfully evaluated; however, such a cut-off point must be defensible, that is, the analysts should be prepared to demonstrate that the impacts of these future impacts could not reasonably be identified or defined, or meaningfully evaluated.

CEQ Guidance on Assessing Impacts of Past Actions. The CEQ has recently issued guidance for addressing the effects of past actions in the CIA. The reader is referred to the CEQ's memorandum, *Guidance on the Consideration of Past Actions in Cumulative Effects Analysis.*[24]

This memorandum tackles a somewhat ambiguous provision in the CEQ's NEPA regulations that requires federal agencies to address cumulative impacts, including an assessment of "other past ... actions."[25] This memorandum is restricted to providing guidance and does not revise or change direction provided in the CEQ's NEPA regulations.

As detailed in the CEQ's memorandum, agencies have discretion (using the scoping process) to determine whether information regarding past actions should be incorporated into the agency's CIA. According to this memorandum, agencies can generally perform an adequate CIA by focusing on the "current aggregate effects of past actions without delving into the historical details of individual past actions." This memorandum goes on to advise that federal agencies may eliminate the specific listing of individual past actions, if it can otherwise describe the existing condition of a resource, even though the past actions may have caused the current condition.

The memorandum advises federal practitioners to focus on the present effects of past actions instead of the past actions themselves. The consideration of past actions may assist federal agencies in determining whether the effects of an agency's action could have a continuing significant additive effect in relation to those effects evaluated in the CIA. Alternatively, information about past direct and indirect effects of individual past actions may be useful in forecasting the direct and indirect effects of a proposal.

A word of caution is in order. The CEQ's guidance memorandum may be at odds with some recent court cases that have stressed the importance of addressing past actions in performing the EIS CIA.[26] The reader is encouraged to consult a legal counsel in determining the degree and extent to which past actions must be considered within a CIA.

9.6.3.3 Assessing Cumulative Effects

There are many approaches for investigating cumulative impact. Frequently, analysts determine the separate effects of past actions, present actions, other reasonably foreseeable future actions, and proposed actions (and reasonable alternatives). Cumulative effects can be calculated, once each group of effects has been determined.

For instance, with respect to air quality, one approach might involve evaluating all existing emission sources for which prevention of significant deterioration (PSD) permits are in the process of being reviewed or approved and those for which a PSD permit is planned but has not yet been submitted. The combined emissions create an effect on air quality, the significance of which can be determined by comparing the cumulative concentration of pollutants emitted to threshold concentrations specified in the National Ambient Air Quality Standards (NAAQS).

Once the impacts resulting from past, present, and reasonably foreseeable future air quality actions have been combined (i.e., cumulative impact baseline), the analysts may then add the impact of the proposed action to better understand how it affects the cumulative impact baseline.

Once all of the effects have been determined, a table can be used to organize and present itemized effects into categories of past, present, proposed, and future actions, and the resulting cumulative effect. Table 9.3 shows how such tables can be constructed. This table compares a narrative versus a quantitative description of the cumulative effects associated with an increase in SO_2 concentrations.

TABLE 9.3
Example of How a Table Can Be Used to Summarize a Cumulative Impact Analysis

	Past Action	Present Action	Proposed Action	Future Action	Cumulative Effect
Narrative description	No discernable effect on SO_2 levels	Notable deterioration in visibility during spring, but standards are met	Visibility further affected by the project, but standards are met	Increased vehicle emissions are expected	Standards are likely to be exceeded
Quantitative assessment		10% increase in SO_2 concentration, but standards are met	20% increase in SO_2 concentration, but standards are met	10% increase in SO_2 concentration	40% increase in SO_2 concentration will exceed regulatory standards

TABLE 9.4
A 14-Step Approach for Performing a Cumulative Impact Assessment

Key Impact Assessment Components	Cumulative Impact Analysis Steps
Scoping	1. Identify the significant cumulative impact issues associated with the proposed action and define the assessment goals
	2. Establish geographic scope for the CIA
	3. Establish the timeframe for the CIA
	4. Identify other past, present, and reasonably foreseeable future actions affecting the resources, ecosystems, and human communities of concern that must be considered in the CIA
Describing the affected environment	5. Characterize the resources, ecosystems, and human communities identified in scoping in terms of their response to change and capacity to withstand stresses
	6. Characterize the stresses affecting these resources, ecosystems, and human communities and their relationship to regulatory thresholds or other applicable threshold values
	7. Define and describe the baseline conditions (affected environment) for the resources, ecosystems, and human communities
Determining the environmental consequences	8. Identify and define environmental disturbances (e.g., emissions, effluents, noise, waste) produced by past, present, and reasonably foreseeable future actions that could affect the resources, ecosystems, and human communities
	9. Identify the important cause-and-effect relationships in which these environmental disturbances would affect human activities and resources, ecosystems, and human communities (e.g., how would the environmental disturbances affect humans and environmental resources)
	10. Combine or "add" the effects of the past, present, and reasonably foreseeable future actions to produce a cumulative impact baseline for the resources, ecosystems, and human communities that could be significantly affected
	11. Combine or "add" the impacts of the proposal to the cumulative impact baseline to determine how the project would affect the cumulative impact baseline
	12. Determine the magnitude and significance of the cumulative effect
	13. Modify or add alternatives or mitigation measures to avoid or reduce the cumulative effects
	14. Monitor the cumulative effects of the selected alternative and consider use an environmental management system or adaptive management approach

A rigorous assessment of cumulative effects can be a complex and iterative process. The following 14-step approach is recommended for assessing cumulative impacts (Table 9.4). The steps depicted in Table 9.4 are common to most CIAs.

For supplemental guidance on performing CIA, the reader is referred to the CEQ's publication, *Considering Cumulative Impacts*, which describes how to define the scope of a cumulative analysis, define the spatial or temporal baseline, and assess truly cumulatively significant issues.[27]

9.6.3.4 Litigation and Judicial Review

Historically, federal agencies have had a poor track record in defending their NEPA documents against legal challenges based on inadequate CIA. Over the past decade, litigation has increasingly focused on cumulative impacts. The number of cases where the courts have found the analysis of cumulative impacts to be inadequate has been proportionately higher than for those cases involving only the analysis of direct impacts.

Smith examined 25 recent judicial opinions from the Federal Ninth Circuit Court of Appeals involving such challenges.[28] With respect to challenges based on inadequate CIA, plaintiffs were successful in 60% of the cases over a 10-year period. In recent years, the success rate of plaintiffs has risen even further, to victories in 72% of the cases (8 out of 11 verdicts).

Based on Smith's study, the federal agency with the worst record was the U.S. BLM, which lost all three of its cases (100%); this was followed by the U.S. Forest Service, which lost 69% (9 of 13) of its cases and the U.S. Army Corps of Engineers, which lost 66% (2 of 3 cases) of the time.

The principal reasons federal agencies lost these court cases were because they

- failed to have any assessment of cumulative impacts in their NEPA document;
- left out obvious past, present, or reasonably foreseeable future actions; or
- presented unfounded assertions that their projects would not cause any significant cumulative impacts.

Examples of Court Direction. In considering cumulative impact, an agency must provide some quantified or detailed information; general statements about possible effects or some risks do not constitute a "hard look" (see Chapter 1) absent a justification regarding why more definitive information could not be provided.[29] More to the point, a cumulative analysis must be more than perfunctory; it must provide a useful analysis of the cumulative impacts of past, present, and future projects.[30]

In 1976 the U.S. Supreme Court ruled that whenever an EIS is prepared, it should include an analysis of activities that are related to or similar to the proposed action.[31] This concept paved the way for a later requirement to consider those cumulative impacts that could result from past, present, and reasonably foreseeable future actions. The CEQ, recognizing the importance of considering impacts in a cumulative sense, codified this requirement in the Regulations.

As a second example concerning cumulative impacts, the Sierra Club challenged an EA prepared by the DOE for the shipment of spent nuclear fuel through the Port of Hampton Roads, Virginia.[32] Under the proposed action, the fuel was to be transported in not one but several successive shipments, thereby increasing the risks to this population of a possible accident and its resulting impacts occurring. Yet, the analysis did not consider the cumulative potential risk of accidents and impacts for permitting several shipments. The court found the analysis to be flawed because of its failure to make this assessment.

Reasonably Foreseeable Actions. In addition to considering issues related to proposals, the Supreme Court has recognized the importance of considering reasonably foreseeable actions in terms of those needing identification and evaluation as a part of the CIA. Here, the Supreme Court provided a broad interpretation of the term reasonably foreseeable to include even the future impacts of projects that had not yet been formalized.[33]

One case involved a U.S. Forest Service proposal to construct a road through a roadless forest.[34] In response, a suit was brought stating that the agency needed to evaluate the cumulative impacts

TABLE 9.5

Supreme Court Direction on Performing a Cumulative Impact Analysis

- Identify the potential impacts expected to occur within the affected area
- Address other past, proposed, and reasonably foreseeable future actions that have affected or might affect the area
- Evaluate the identified impacts that have resulted or are expected to result from these actions
- Describe the cumulative impacts that can be expected if these individual impacts are allowed to accumulate
- Specify the area that would be affected by the proposed action

to the environment resulting from the logging and timber sales that would be triggered by the construction. The Forest Service argued that since no sales would be contemplated for many years into the future, such sales were therefore too uncertain to be evaluated as a part of the road building project. The court reasoned that if this were the case, their argument was tantamount to admitting that constructing the road was senseless. It went on to conclude that the Forest Service proposal not only precluded the analysis of other reasonable alternatives, but also swayed the final decision concerning future land use in favor of timber sales. The court held that if the cumulative impacts from both the road and the potential timber sales had been analyzed together, the agency might have reached a different conclusion. In the words of the court, "[If] sales are sufficiently certain to justify construction of the road, then they are sufficiently certain for their environmental impacts to be analyzed along with the road."

9.6.3.5   Court Direction on Performing CIA

The Supreme Court has stated that agencies may limit the scope of their cumulative effects analysis based on practical considerations.[35] As indicated in Table 9.5, the Supreme Court has also provided five criteria that a meaningful cumulative impacts analysis needs to meet.[33]

This direction by the Supreme Court does not necessarily imply that an agency must prepare a "full-blown" analysis of all the actions considered in a cumulative analysis, equivalent to the type of analysis performed in other reviews. Instead, the Supreme Court stated that other actions and their probable impacts had to be identified and considered in determining whether the proposal could result in a significant impact.[33]

In another case, the Federal Aviation Agency (FAA) was challenged for addressing only the incremental increase in noise and not the cumulative impacts that could result from replacing an airport near Zion National Park. There was no way to determine from the proposal if the FAA's estimated 2% increase would result in a significant impact if it was added to other existing noise impacts on the park.[36]

In yet another case, a plaintiff sued claiming that an EIS prepared for a postfire logging project in one section of a forest did not adequately disclose or analyze its potential cumulative impacts on the management of indicator species, fuel break maintenance, or fire-fighting tactics when added to the impacts of other post-fire logging projects in that forest. The court concluded that given the similarities of these projects on timing, geography, and purpose, such actions might result in a significant cumulative impact that needed to be addressed in a single EIS.[37]

9.6.4   Resolving Eccleston's Cumulative Impact Paradox

The importance of assessing cumulative impacts is underscored by one of the factors required to be considered in reaching a determination regarding potential significance:

> … whether the action is related to other actions with individually insignificant but cumulatively significant impacts. Significance exists if it is reasonable to anticipate a *cumulatively significant impact on the environment*"[38] (emphasis added).

An impasse, referred to in this book as Eccleston's Cumulative Impact Paradox (Eccleston paradox), can arise from the fact that the definition of cumulative impacts requires consideration of other impacts from past, present, and reasonably foreseeable future activities (to provide the cumulative impact baseline).[39]

9.6.4.1 The Paradox

According to the CEQ, approximately 30,000–50,000 EAs are prepared each year.[40] Many are prepared for locations or resources that have already sustained significant cumulative impacts.

Yet, a finding of no significant impact (FONSI), by definition, means an action

... will *not* have a *significant* effect on the human environment[41] (emphasis added).

A categorical exclusion (CATX) means

... a category of actions which do not individually or *cumulatively* have a *significant effect* on the human environment and which have been found to have no such effect ... and for which, therefore, neither an environmental assessment nor an environmental impact statement is required ...[42] (emphasis added).

The paradox arises from the fact that the definition of cumulative impacts requires consideration of effects from other past, present, and reasonably foreseeable future activities (cumulative impact baseline). This paradox is evidenced by the following example.

Consider a proposal to construct a relatively modest federal building in a crowded downtown business area of a large city. The area has already sustained a significant cumulative impact. For example, the downtown area is paved over with concrete, buildings, and skyscrapers. Streets run to-and-fro. The native vegetative state and animal habitat have been disrupted or destroyed. Visual resources have been significantly altered. The underground aquifer has been contaminated, and the water table has sustained a significant drawdown. The streets are crowded with commuter traffic, which has resulted in a high level of traffic noise. Ambient air quality has been significantly degraded. Fish and other aquatic species in a river bounding the city are on the decline. Destruction of wetlands and construction of an impermeable pavement have increased the risk of flooding within the city and downstream of the city. As a result of the impacts of past and present actions, a number of environmental resources have been significantly affected from a cumulative standpoint. Reasonably foreseeable future actions will only compound these problems. Would a full-blown EIS be required to construct something as mudane as a small building or parking lot?

As an additional example, consider a popular recreational campground located in a remote area. The campground serves approximately 7000 campers per season. The campers have extracted a heavy toll on the surrounding environment. The level of noise has increased to the point where it is affecting some species (and many campers are likewise upset). The population of an exotic flower has begun to decline, and the habitat of an animal listed as a threatened species is also on the decline. The campers have affected visual resources to the point where many campers are beginning to publicly complain. Finally, the water quality in a nearby stream has been significantly degraded by camping activities. The responsible federal agency would like to prepare an EA for revamping the access trail leading into the campground. Although the refurbished trail might increase the total number of campers, professional demographers have estimated (based on historical trends) that any such increase would be less than 10–20 recreationists per season. These additional recreationists would contribute a very small incremental impact when compared with the 7000 visitors already visiting the campground. Yet, as witnessed earlier, a strict interpretation of the Regulations appears to preclude issuance of a FONSI under these circumstances, theoretically resulting in a full-blown EIS. The cost of preparing an EIS could actually pay for a substantial mitigation effort to repair much of the environmental damage that has already taken place. Is it reasonable to expect an agency to prepare an EIS for such a small increase in campers simply because this resource has already sustained a significant cumulative impact, even though the direct and indirect impacts of the new proposal are virtually innocuous? And would there even be tangible alternatives to refurbishing the trail?

As witnessed in the aforementioned regulatory citations, a strict interpretation of the Regulations implies that a CATX or a FONSI cannot be applied to any proposal that adds any contribution to a cumulative impact that has already breached the threshold of significance.

If an environmental resource has already sustained a cumulatively significant impact, how can a decision-maker declare that a proposed action contributing any incremental impact (however, small) is eligible for a CATX or a FONSI? Yet, CATXs or FONSIs are routinely (i.e., technically unjustified) applied across virtually all federal agencies for proposed actions involving environmental resources that have already sustained cumulatively significant impacts.

Based on NEPA's regulatory definitions and requirements, many (if not most) activities for which EAs (and CATXs) are currently prepared should actually be ineligible for a FONSI, therefore requiring preparation of an EIS; yet, strict compliance with these regulatory provisions would result in an unreasonable and voluminous increase in the number of EISs required (even where the incremental impacts would be relatively small) and, in many areas, might render the concept of a CATX or a FONSI next to useless. For instance, in the example of the downtown area described above, a strict interpretation of cumulative significance could lead to the conclusion that a federal agency would have to prepare an EIS to construct something as mundane as a stoplight, walkway, or small parking lot. Clearly, a strict interpretation of NEPA's regulatory requirements can lead to absurd, unreasonable, and politically unacceptable results.

Consequently, Eccleston's paradox refers to

> The illogical, unreasonable, or absurd situation in which CATXs/FONSIs are routinely applied to relatively mundane actions in areas that have sustained cumulatively significant impacts, yet the application of these streamlining provisions (CATXs/FONSIs) violates the cumulative impact regulatory constraints placed on their very use.

9.6.4.2   Why It Is Important to Resolve This Paradox

This paradox is much more than one of mere passing or academic interest. From a practical standpoint, how can decision-makers be expected to make reasonable and consistent determinations regarding the significance of a cumulative impact when the very definition can, and frequently does, lead to absurd or unreasonable outcomes? The same paradox also applies to practitioners who must prepare CIA. The paradox must be resolved if the analysis of cumulative impacts is to be seriously and consistently implemented in a manner that truly safeguards environmental quality.

In considering this paradox, it is important to recognize that NEPA is governed by the "rule of reason." That is, "reason" should prevail when a regulatory requirement would result in an absurd outcome. A regulatory provision leading to the conclusion that an EIS is required, even in situations where it would contribute little or no substantive value to the decision-making process, contravenes the rule of reason.

The paradox also conflicts with NEPA's regulatory direction to reduce unnecessary paperwork and delay (i.e., unnecessary EISs) simply because a strict interpretation would require that EISs be prepared for a multitude of situations where the direct and indirect impacts as well as the incremental cumulative contribution of an action are relatively trivial and for which an EA or a CATX normally should suffice.[43] Perhaps, most importantly, it is simply unrealistic and unreasonable to expect federal agencies to prepare EISs for many trivial or even innocuous actions merely because the existing environment or cumulative environmental baseline may already be significantly affected from a cumulative standpoint.

9.6.4.3   Interpreting Significance

As described in Chapter 6, the Regulations require that the significance of an action be considered in terms of both the intensity and the context in which the impacts occur. Under the significant departure principle (SDP) described in the next section, both the intensity and the context must be considered in assessing the threshold of significance.

On context, the CEQ Regulations state that

...the significance of an action must be analyzed in several contexts such as society as a whole (human, national), the affected region, the affected interests, and the locality. Significance varies with the setting of the proposed action. For instance, in the case of a site-specific action, significance would usually depend upon the effects in the locale rather than in the world as a whole.[44]

With respect to intensity, the Regulations identify 10 intensity factors that need to be considered in making a determination of the significance or nonsignificance of an impact.[45] For example, one of these factors states that an agency should consider "the degree to which the proposed action affects public health or safety."[46]

While, 10 significance factors (in addition to context) have been developed to assist agencies in assessing significance, specific regulatory direction does not exist regarding how they are to be interpreted or applied in reaching such determinations. Federal agencies have, in fact, been given wide discretion in interpreting how such factors can be applied in reaching a determination of significance.

For example, six of the significance intensity factors state that decision-makers should consider the "degree" to which a given factor may affect some environmental attribute.[47] Agencies are given no direction, however, for interpreting or determining when an impact has affected an environmental resource to such a "degree" that it constitutes a significant environmental impact. Such judgment is left to the discretion of the decision-maker.

Thus, responsibility ultimately lies with the individual decision-maker for determining how significance factors are to be interpreted. Decision-makers must, therefore, exercise a considerable degree of professional judgment in making such determinations.

9.6.4.4 Significant Departure Principle

The SDP provides a cornerstone for developing an interpretation of significance that, in turn, provides a practical solution for resolving the paradox. The following solution to the paradox is based upon the fact that decision-makers have been granted a wide degree of discretion with respect to interpreting significance.

Would the Action Significantly Change the Cumulative Impact Baseline? Under the SDP, significance (with respect to the assessment of a CIA) can be viewed in terms of the degree to which a proposed action would change (i.e., depart from) the cumulative impact baseline. In other words, significance can be viewed as the degree to which a proposed action would affect or cause a cumulative impact to change or significantly depart from conditions that would otherwise exist if the proposed action was not pursued.

Based on this interpretation, an action could be considered nonsignificant (from the standpoint of a cumulative impact) as long as it does not cause a cumulative impact(s) to change or depart significantly from conditions that would exist if the action was not pursued. In other words, the incremental impact is of such nonsignificance that it would not appreciably change or contribute in an important manner to the cumulative impact if it were added to the effects of other past, present, and reasonably foreseeable future actions. Conversely, a proposed action resulting in a substantial or important cumulative change to the same environmental resource would be deemed significant.

The paradox is therefore resolved if the significance of a cumulative impact is considered in terms of how much the impact changes the cumulative environmental baseline, as opposed to a strict interpretation of assessing an impact from an *absolute* perspective. Significance is therefore interpreted from a *relative* perspective, as opposed to a strict, absolute assessment.

As used here, the term "absolute" denotes a strict or more standard interpretation in which the significance of an impact is assessed simply in terms of whether the threshold of significance would be or has been breached. In contrast, the term "relative" is used to denote the SDP interpretation in which significance is assessed in terms of the relative degree of increase or change (i.e., departure) in a cumulative impact.

Consider, for example, a situation where an agency needs to take an action that might result in the death of a certain species of fish. Assume that the fish and their habitat have already sustained a cumulatively significant decrease in their numbers and that the action would result in the loss of 10 additional fish out of the entire river. From a cumulative standpoint, this incremental loss may be considered nonsignificant if the fish species has a population numbering in the range of 100,000 (i.e., the action would decrease the fish population by 0.01%). Such a view is both justified and consistent with the rule of reason, because little or nothing may be gained by preparing an EIS for an action that would not substantially alter or change the cumulative impact baseline, even if the environmental resource has already been significantly affected. In contrast, the same action would very likely be deemed cumulatively significant (a significant departure) if the total fish population in the river was only 25 in number (40% decrease in the existing fish community).

One method for implementing the SDP simply involves "adding" the impacts of other past, present, and reasonably foreseeable future federal and nonfederal actions together to produce a cumulative impact baseline.[25] The incremental impact of the proposal can be added to this baseline. The SDP is then used to interpret whether the incremental change in the cumulative impact baseline is of such importance to this environmental resource that it can be deemed significant.

In applying the SDP, a review should be performed to ensure that the incremental impact would not trigger or breach a significance threshold (i.e., the straw that breaks the camel's back). If the incremental impact breaches or crosses over some threshold, clearly it should be considered significant. This threshold might involve the violation of an applicable regulatory limit or some other suitable threshold such as a scientific or an environmental constraint.

For supplemental guidance on performing CIA, the reader is referred to the CEQ's publication, *Considering Cumulative Effects under the National Environmental Policy Act*, which describes how to define the scope of a cumulative analysis, define the spatial or temporal baseline, and assess truly cumulatively significant issues.[48] The CEQ has also issued a memorandum, *Guidance on the Consideration of Past Actions in Cumulative Effects Analysis*.[24] With respect to using the SDP, this memorandum provides important guidance that may be helpful in establishing the cumulative impact baseline.

Criticisms. An argument can be raised that the SDP could allow a succession of many small projects to be implemented (without preparation of individual EISs) such that over a period of time, the cumulative incremental value of these small projects could amount to a large impact. It is conceivable that such situations might arise. As part of the CIA, however, the impacts of reasonably foreseeable future actions are expected to have already been included into the cumulative impact baseline. Notwithstanding, it is simply unrealistic to expect decision-makers to reach significance determinations (requiring preparation of lengthy and expensive EISs) for mundane projects with relatively innocuous, incremental impacts, and to which reasonable alternatives may not even exist.

Most practitioners and decision-makers understand that they must employ practical (not theoretically perfect) methods in assessing impacts. The SDP provides a more rational, objective, and practical approach for assessing significant cumulative impacts and resolving the Eccleston paradox. Thus, the SDP offers decision-makers a method for focusing on truly important cumulative incremental impacts while de-focusing attention on those small enough to be deemed unimportant to the decision-making process.

Applicability. Application of the SDP can provide a particularly valuable tool for interpreting significance in an EA or an EIS when an area or environmental resource has

1. already sustained a cumulatively significant environmental impact, but the impacts of the proposed action are so small as to contribute little or no appreciable change to the cumulative baseline; or
2. not been significantly affected and the incremental impact would not breach the threshold of significance, yet the impact might still be considered significant simply as a result of the sizeable increase in contribution to the cumulative impact baseline (this circumstance will be described shortly).

Examples of the Paradox. Application of the SDP is described in the following four cases involving cumulative noise quality impacts. These four examples assume that the significance threshold involves a scientifically established noise limitation.

Cases 1 and 2 represent situations where the threshold of significance has already been breached. Case 3 illustrates a situation where the cumulative impact baseline lies significantly below the threshold of significance, and the proposed action would substantially increase the cumulative baseline but would not actually breach the significance threshold. Case 4 depicts a situation where the cumulative baseline lies just below the threshold of significance, and the small contribution of the proposed action actually breaches the significance threshold.

Figures 9.2 through 9.5, correlating with each of the four cases, are included for conceptual purposes so as to clearly illustrate the concept of the paradox and the SDP; these figures are not drawn to scale. For simplification purposes, these four examples assume that the context is factored into the assessment of the significance threshold values and the assessment of the environmental impacts. Although the figures portray noise impacts, the SDP can be applied to virtually any cumulative impact.

The author fully acknowledges that there are many situations (perhaps the majority of cases) in which it may not be possible to assess a cumulative impact or its significance threshold in terms of an explicit number (or threshold value), as has been done in the following four cases. Nevertheless, even if an impact or threshold value cannot be described quantitatively (e.g., many visual, socioeconomic, or some biological impacts), the SDP can still be applied from a qualitative standpoint.

Case #1: Cumulative Significance Threshold Has Been Breached But Incremental Impact Is Nonsignificant. As denoted by Figure 9.2, assume that the average daytime noise level (as a result of past, present, and reasonably foreseeable future actions) within a city park is 65 dB. A noise quality significance threshold has already been established for this area, which calls for a maximum level of 35 dB. Because the existing baseline level (65 dB) already exceeds the established

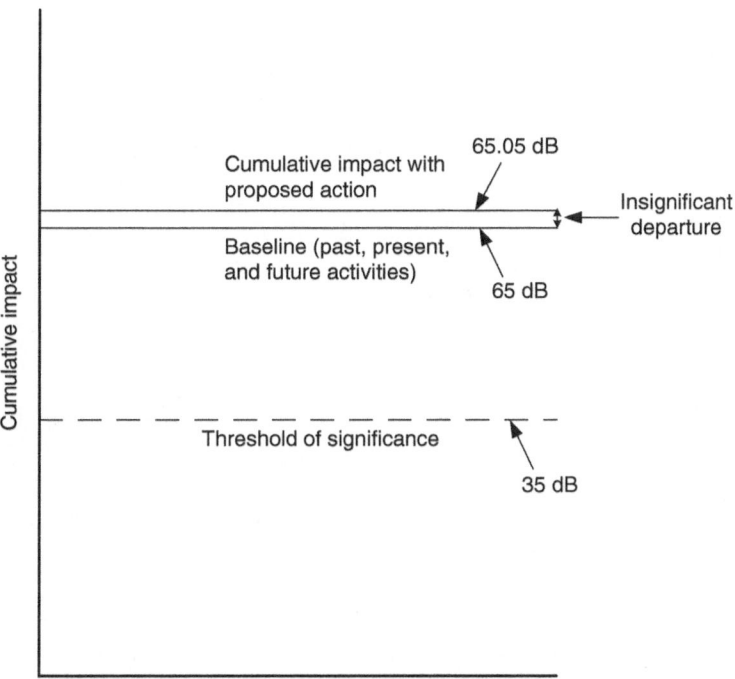

FIGURE 9.2   Assessing cumulative impacts using the significant departure principle, Case #1. The cumulative significance threshold has been breached, but the incremental impact is nonsignificant.

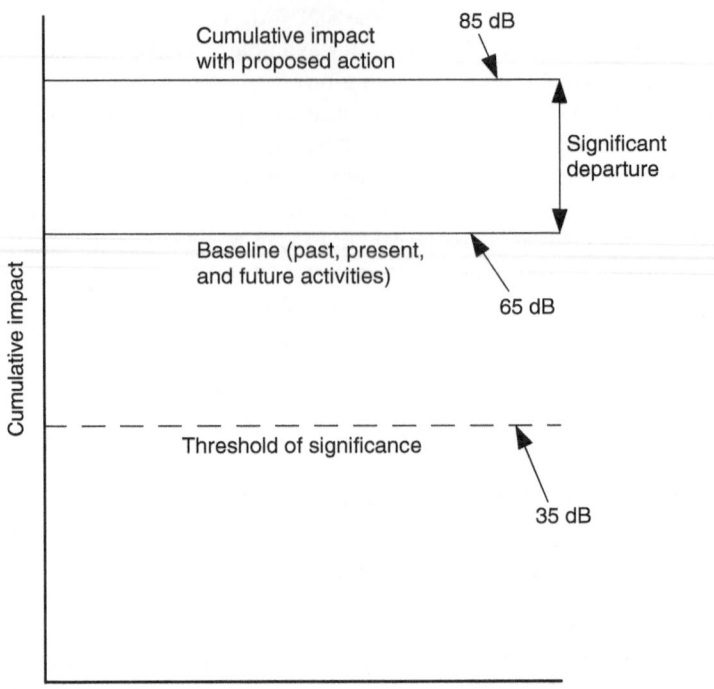

FIGURE 9.3   Assessing cumulative impacts using the significant departure principle, Case #2. The cumulative significance threshold has been breached, and the incremental impact is significant.

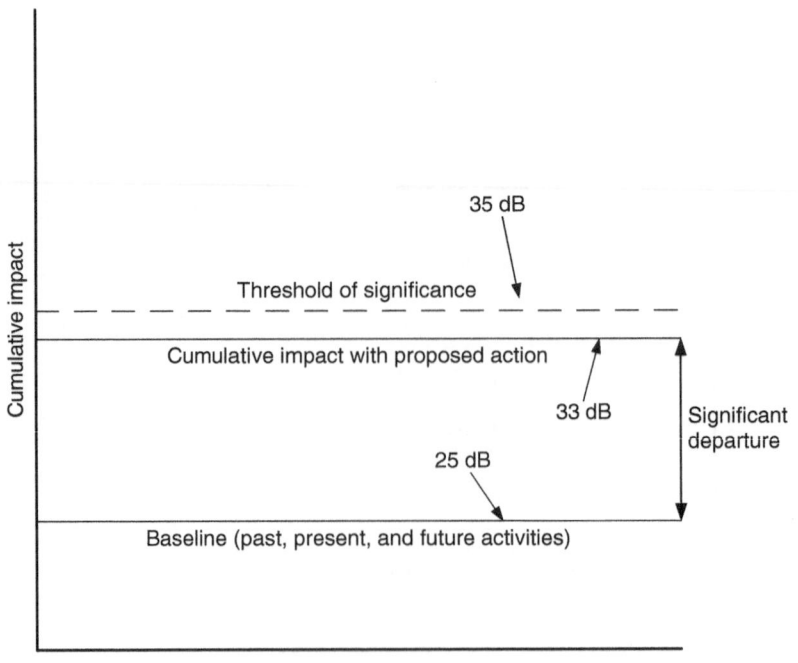

FIGURE 9.4   Assessing cumulative impacts using the significant departure principle, Case #3. The cumulative significance threshold has not been breached, and the incremental impact is significant.

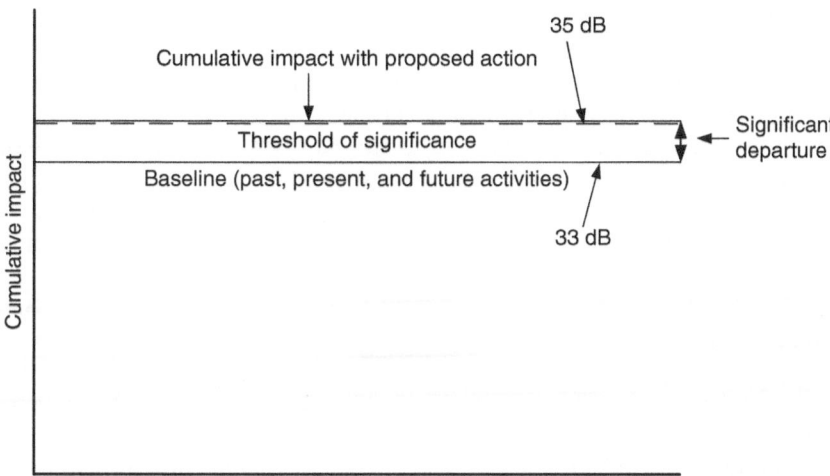

FIGURE 9.5 Assessing cumulative impacts using the significant departure principle, Case #4. A small incremental impact is sufficient to breach the significance threshold.

significance threshold (35 dB), this area can be viewed as having already sustained a cumulatively significant impact. (The reader should note that noise intensity is measured in decibel units on a base 10 logarithmic scale, and thus this is a logarithmic rather than a linear phenomenon; human perception of loudness also conforms to a logarithmic scale.)

Further, assume that the streets bordering the park are in severe need of repair. A proposed maintenance project is planned to repave the surrounding streets. Because a federal agency has jurisdiction over the property and adjacent roads, NEPA is triggered; also, assume that such a small action would normally be covered under a CATX. Further, assume that a traffic expert concludes that as a result of this maintenance improvement project, the traffic is expected to increase slightly, raising the average daytime noise level from 65 dB to 65.05 dB, for a 0.05 dB increase. Because the threshold of significance has already been breached (35 dB), a strict or standard interpretation of significance suggests that any additional contribution must also be significant (regardless of how trivial an increase); an EIS could therefore be deemed necessary for this maintenance project as well as for similar projects (e.g., installing a traffic light that results in a slight increase in noise as a result of stop-and-go or diverted traffic) that cause an increase in the cumulative impact baseline, regardless of how nominal the effect.

Under the SDP, however, the decision-maker considers the proposed project from its relative incremental effect rather than from an absolute perspective. That is, the decision-maker considers whether the departure or change in an impact(s) (rather than the absolute magnitude of the effect) is significant with respect to the cumulative impact baseline. This involves examining the 0.05 dB incremental impact to see if it would make an important difference (i.e., change) in the cumulative impact baseline. Such an analysis involves examining the incremental impact in terms of the CEQ's 10 intensity factors as well as the context in which the incremental impact would take place.[49]

Thus, if a 0.05 dB cumulative change is considered significant, the entire project is deemed significant, and an EIS must be prepared. Conversely, if a 0.05 dB cumulative increase is considered inconsequential, posing no significant change to the environmental resources, an EIS is not required (with respect to this cumulative impact). The SDP, therefore, focuses on the degree to which the impact changes the environmental baseline.

Case #2: Cumulative Significance Threshold Has Been Breached and the Incremental Impact Is Significant. Figure 9.2 illustrated a case where the change was small and relatively nonsignificant (0.05 dB increase), while Figure 9.3 illustrates a case where the change is large (an increase of 20 dB, resulting in a total cumulative impact of 85 dB). In the latter case (Case #2), the cumulative impact baseline already exceeds the threshold of significance and, given these circumstances, a decision-maker might properly conclude that the proposal's incremental impact is cumulatively significant, requiring preparation of an EIS.

Case #3: Cumulative Significance Threshold Has Not Been Breached and the Incremental Impact Is Significant. Figures 9.2 and 9.3 represent situations where the threshold of significance had already been breached prior to the proposed action. Figure 9.4 illustrates a third case, where the cumulative noise baseline is 25 dB, which lies well below the threshold of significance (i.e., noise level of 35 dB). In this case, however, the proposed project would increase the cumulative impact baseline from 25 dB to 33 dB, which is near but still below the threshold level of 35 dB. Because the proposed project would substantially increase but not actually breach the threshold of significance, a strict interpretation of significance might lead to the conclusion that the cumulative impact is still nonsignificant, and therefore an EIS is not required.

Using the SDP, however, the proposed action might more correctly be deemed to constitute a significant departure or change in the cumulative impact baseline, even if it would not actually breach the threshold of significance. Such a conclusion may be justified because the proposed action would significantly increase the baseline (even though it has not actually exceeded the threshold of significance); a decision-maker could (perhaps should) be justified in concluding that the action would constitute a cumulatively significant increase in noise impact, requiring preparation of an EIS. In this case, preparation of an EIS may be warranted because alternatives or mitigation measures that eliminate or substantially reduce the significant increase in noise level may be identified.

Thus, the SDP can in some cases result in a determination of significance (indicating preparation of an EIS), where a more standard approach would result in the opposite conclusion (i.e., nonsignificance). The reader should note that this interpretation is consistent with one of the existing significance criteria:

> … whether the action *threatens* a violation of Federal, State, or local law or requirements imposed for the protection of the environment[50] (emphasis added).

This criterion could have more simply been written as "… whether the action violates a federal …," but instead the word "threatens" was included. This wording appears to imply that the authors of the Regulations believed that if an impact approached a regulatory threshold, the impact could be regarded as being significant even if it did not actually breach or violate the threshold. In other words, the act of significantly increasing an environmental baseline—making it that much easier for an actual violation to occur sometime in the future—might best be viewed as a significant impact.

Case #4: A Small Incremental Impact Is Sufficient to Breach the Significance Threshold. Figure 9.5 illustrates a final case, where the cumulative impact baseline noise level is 33 dB, slightly below the threshold of significance (35 dB). In this case, however, the proposed project would increase the cumulative impact baseline from 33 dB to 35 dB.

Although, the increase in noise level is relatively small, it is nevertheless sufficient to breach the significance threshold of 35 dB. Because the impact of this proposal would breach the significance threshold, it is automatically considered to constitute a significant impact.

In this case, preparation of an EIS is warranted because alternatives or mitigation measures that can reduce the impact enough to prevent the significance threshold from being breached might be identified.

Factors Used in Assessing Significance. Consistent with the strict or more standard interpretation, both the context and intensity factors described in § 1508.27 must be considered in reaching a SDP determination of significance. As illustrated by the four cases above, one of the CEQ's 10 significance factors (i.e., § 1508.27[b][2]) could be assessed in terms of the degree to which the project would cause a change in existing public health and safety (i.e., noise). An action's effect on public health and safety could be considered cumulatively nonsignificant as long as the cumulative impact on public health and safety does not substantially change or depart from conditions that would exist if the action were not pursued.

Specific federal agency guidelines could even be issued to assist decision-makers in determining how the 10 significance factors should be interpreted in terms of the SDP. Specific criteria might also be developed to assist agencies in determining when an impact causes a significant departure in the environmental impact baseline. For instance, the severity or magnitude of a cumulative impact on a particular species might depend on the amount of habitat that has been disrupted or will be disrupted in the future. Similarly, the magnitude of a cumulative effect on archaeological resources might be quantified or measured by estimating the number of past, present, and future sites or artifacts that have been or will be disrupted. Likewise, the threshold of cumulative significance for a visual impact might be measured based on the perceptions (e.g., low, medium, or high) of visitors to a recreational area that is the site for a proposed action. In many cases, computer models can be applied to predict cumulative changes such as socioeconomic patterns, finical conditions, or hydrological conditions.

Fred March summarizes seven specific tests for significance.[51,52] These are based on the significance factors defined in the CEQ's NEPA regulations.[49] Based on March's scheme, the SDP does not suggest an eighth independent test; instead, it can be applied in assessing and interpreting some of his seven tests.

9.7 SPECIAL IMPACT ISSUES

Some special environmental impact issues are addressed in the following sections.

9.7.1 ADEQUACY OF THE ANALYSIS

The need to pursue a rigorous analysis, leaving no stone unturned, was underscored in a case where a court concluded "… ignoring the possible environmental consequences will not suffice."[53]

However, in another case, a court ruled that NEPA did not require that an analysis contain a complete assessment of all the potential impacts of a proposed project.[54] The court noted that it was doubtful that any projects would ever be launched if a requirement were imposed that an analysis needs to correctly evaluate all the environmental effects associated with the project. In deferring judgment regarding the adequacy of the analysis to the agency, the court concluded that

> … What appellants seek is for this court to substitute its judgment for that of the [Agency's] Secretary, who is charged by NEPA with preparing a thorough statement of the environmental consequences of a proposed project, [and] as to what particular information will be required to complete that statement. We decline to assume that role.

One of the more complicated factors that must be considered by decision-makers is the risk of uncertainty. They must balance this risk and the cost against the benefits of moving forward. One court ruled that when these have been considered and the decision-maker has decided that the benefits exceed the cost, then the courts should defer judgment to the agency.[55]

9.7.2 SOCIOECONOMIC IMPACTS

Social and economic impacts are not, of and by themselves, sufficient to require the preparation of an EIS. However, when an EIS is prepared, and socioeconomic effects are interrelated with other effects on the physical environmental, then all of these impacts should be addressed together in the EIS (§ 1508.14). This requirement to assess socioeconomic impacts does not strictly apply to the preparation of EAs.

In one case, a citizens' group alleged that converting a mental hospital campus into a federal prison hospital would induce bringing of weapons and drugs into the area, increasing crime and halting community development.[56] The court reasoned that these impacts would not result from the physical changes created by the proposed action but instead from social changes brought about by the change in the use of the facility and the types of people treated there. With respect to socioeconomic impacts, an EIS is normally required only when an action poses a threat to the actual physical aspects of an area (e.g., traffic, population patterns, water supply, and historical sites). In this case, because the actual impacts resulting from these changes would be small (i.e., security fences and a small road), the court concluded that an EIS was not required.

9.7.2.1   Analysis of Socioeconomic Impacts

The public is often more interested in "bread and butter" issues than they are in the "bugs and bunnies." Not surprisingly, socioeconomic impacts are often the most emotional and controversial issues of concern. For this reason, it is recommended that a concerted effort be taken to address and, if possible, mitigate these impacts.

With respect to base closures, for example, the Department of Defense has taken the position that socioeconomic impacts must translate into an actual impact on the natural environment before the effect would be evaluated in an EIS.[57] In such a case, an EIS for the proposed action would not be warranted if the socioeconomic impacts were substantial, but their actual effects on the physical environment were considered to be insignificant.

9.7.3   Short-Term versus Long-Term Impact

The Regulations require consideration of both short-term and long-term impacts (§ 1508.27). The distinction between these terms is subjective, and no definitive criteria exist for determining the differences in timeframes.

To simplify the problem, the author recommends that the term "short-term" be used to denote impacts that occur while the proposal is under way. Factors such as noise and fugitive dust generated at a construction site would thus be considered short-term impacts. In contrast, the term "long-term" should be interpreted to refer to impacts that continue for an extended period of time after the construction and operational activities have ceased. In reality, the difference is academic as an analysis normally should not separate short-term impacts from those that are of a longer term.

9.7.4   Bounding Analysis

A quantitative methodology referred to as "bounding" is sometimes used to simplify the evaluation of environmental impacts in the preparation of NEPA analyses. A bounding analysis typically uses conservative assumptions and analytical methods to estimate (i.e., bound) the maximum value of a potential environmental impact.

Saylor and McCold provide an overview of the situations in which a bounding approach is appropriately applied.[58] A bounding analysis is a legitimate and particularly efficient method for demonstrating, with respect to a given set of potential actions under consideration, that an impact would be nonsignificant regardless of which course of action was taken. This application is consistent with the regulatory mandate to "... eliminate from detailed study the issues which are not significant ..." and "... focus on significant environmental issues ... and reduce ... accumulation of extraneous background data" (§ 1501.7[a][3] and § 1502.1).

9.7.5   Are Accident Scenarios Subject to NEPA's Requirements?

A former general counsel of the CEQ, Nicholas Yost, has stated that "NEPA essentially requires analysis of both the lesser risks of greater harm and the greater risks of lesser harm before actions are taken to bring about the risks."[59] Some agencies routinely address impacts of high-consequence accidents in their NEPA analysis, while the issue is virtually ignored by others.

Questions have arisen regarding an agency's responsibility to evaluate the consequences of potential accidents, particularly with respect to low-probability, high-consequence events. Such analyses may be required for actions involving incomplete or unavailable information (§ 1502.22). Beyond the direction provided in § 1502.22, the Regulations contain no explicit requirement to perform analyses of potential accidents, especially with respect to high-consequence, low-probability events. Lack of explicit direction should not be construed, however, to imply that an analysis of accidents is unnecessary.

This section investigates the circumstances under which an accident scenario should be evaluated. A peer-reviewed systematic tool for reaching such determinations is described shortly. For more information, the reader is directed to a paper by the author on this issue.[60] Additional information can also be found in Chapter 10 and in the companion book, *Effective Environmental Assessments*.[61]

9.7.5.1   Accident Analyses and Case Law

Few would argue that a reasonable scenario involving the potential catastrophic failure of a dam or meltdown of a nuclear reactor can be reasonably excluded from the scope of a NEPA analysis. Accident analyses under NEPA are often necessary for making a reasoned choice among alternatives and giving appropriate consideration to mitigation measures. Document preparers are also required to exercise considerable judgment to determine the scope of accident analyses. But in this lies the dilemma. Where does one draw the line on the spectrum of events that can run the gamut from simple vehicular accidents to nuclear meltdowns? The following cases are notable as they appear to indicate that at least under some circumstances an analysis of accident scenarios is within the scope of a NEPA analysis.

A systematic peer-reviewed tool developed by the author is presented in Figure 10.2, which can assist decision-makers in determining if a given accident scenario may be subject to NEPA requirements. Two court cases are also reviewed below.

Galileo Satellite. The first case involves the National Aeronautical and Space Administration's (NASA) launch of the Galileo satellite in 1989.[62] In challenging the EIS prepared for it, the Florida Coalition alleged that the plutonium used as the satellite's power source posed an environmental risk in the event of an accident. The EIS prepared by NASA indicated that if an accident (with a probability in the range of one chance per million) did occur it could result in nine deaths from cancer. In siding with NASA, the court explained that the EIS had adequately considered the potential consequences of low-probability accidents.

Spent Nuclear Fuel. A second case involves the EA (discussed earlier in Section 9.4.2.1) prepared by the DOE in 1991 for a shipment of spent nuclear fuel from Taiwan through the Port of Hampton Roads, Virginia.[32] The assessment only evaluated risks that the DOE considered credible. In its defense, the DOE argued that accidents falling into a category referred to as extremely unlikely involved severe but very improbable events with a probability ratio similar to that of a meteorite striking the transportation truck. The Sierra Club countered that the DOE failed to consider more probable accidents having severe consequences such as a collision with an oil tanker. The court concluded that disregarding such severe low-probability scenarios had biased the analysis in favor of the agency's proposed action. The court went on to stress the importance of being completely open and forthright concerning possible risks, particularly in projects involving radioactive material.

9.7.5.2   Sliding-Scale Approach

Consistent with a sliding-scale approach, impacts should be analyzed commensurate with their potential consequences. Key factors to consider in applying the sliding-scale principle to accident analyses are presented in Section 10.6.

PROBLEMS

1. With respect to the concept of scope, what are the three types of actions?
2. What are the "4 Ps"?
3. Must socioeconomic impacts be addressed in an EA?
4. An EA is prepared for the construction of a short highway spur. Is the assessment of potential fatalities from auto accidents a direct or an indirect impact? Explain.
5. A project engineer wants to build a small airport. However, he is informed that he must wait for approximately 18 months until an EIS has been completed. He decides that the control tower, fueling station, and office building can be prepared under separate CATXs. The landing strip will be constructed using an EA. Aircraft flight operations will be evaluated under a separate EA. Is this strategy appropriate? Explain your answer.
6. NEPA applies to the assessment of federal actions. But what about the assessment of cumulative impacts, that is, do the impacts of related private projects need to be included in assessing the cumulative effects of the proposal? Explain.
7. Do the courts generally require that an EIS analysis be absolutely accurate and precise?
8. Must an agency thoroughly evaluate every reasonable alternative that has been identified? Explain.
9. What is the SDP?
10. What purpose(s) does the assessment of the no-action alternative serve?

REFERENCES

1. *Mullin v. Skinner*, 756 F. Supp. 904 (E.D. N.C. 1991).
2. *Blue Ocean Preservation Society v. Watkins*, 745 F. Supp. 1450 (D. HI. 1991).
3. *National Wildlife Federation v. Federal Energy Regulatory Commission*, 912 F.2d 1471 (D.C. Cir. 1990).
4. *Fritiofson v. Alexander*, 772 F.2d 1225 (5th Cir. 1985).
5. Yost N. C. and Rubin J. W., *The National Environmental Policy Act* (unpublished).
6. *Hammond v. Norton*, No. 01-2345 (PLF), 35 ELR 20100 (D. D.C. 2005).
7. *El Dorado County v. Norton*, No. S-02-1818 GEB DAD, 35 ELR 20014 (E.D. Cal. 2005).
8. CEQ, 40 Questions, Question No. 3.
9. CEQ, 40 Questions, Question No. 2a.
10. *Utahans for Better Transportation v. U.S. Department of Transportation*, 305 F.3d. 1152 (10th Cir. 2002).
11. CEQ, Preamble to final CEQ NEPA regulations, *Federal Register*, 43, 55978, Section 4, November 29, 1978.
12. *Vermont Yankee Nuclear Power Corporation v. NRDC*, 435 U.S. 519, 1978; *Kleepe v. Sierra Club*, 427 U.S. 390 (1976).
13. *Environmental Law Handbook*, Government Institutes, Inc., 10th ed., Chapter 10.
14. CEQ, 40 Questions, Question No. 1a.
15. CEQ, 40 Questions, Question No. 2.
16. *Natural Resources Defense Council v. Morton*, 458 F.2d 827 (D.C. Cir. 1972).
17. CEQ, 40 Questions, Question No. 1b.
18. *Northern Alaska Environmental Center et al. v. U.S. Bureau of Land Management et al.*
19. Mandelker D. R., NEPA alive and well: the Supreme Court takes two, *Environmental Law Reporter*, September 1989.
20. CEQ, *Considering Cumulative Effects Under the National Environmental Policy Act*, p. 1, January 1997.
21. Odum W. E., Environmental degradation and the tyranny of small decisions, *Bioscience*, 33, 728–729, 1982.
22. National Performance Review, Creating a Government That Works Better and Costs Less, Washington, D.C., September 1994.
23. CEQ, *Considering Cumulative Effects Under the National Environmental Policy Act*, p. 17, January 1997 (figure is after Bisset 1983).

24. CEQ, *Guidance on the Consideration of Past Actions in Cumulative Effects Analysis*, June 24, 2005, http://ceq.eh.doe.gov/nepa/regs/Guidance_on_CE.pdf

25. 40 CFR § 1508.7.

26. *Lands Council v. Powell*, 395 F.3d 1019, 1027 (9th Cir. 2005) and *Natural Resources Defense Council v. United States Forest Service* (9th Cir. 2005).

27. CEQ, *Considering Cumulative Impacts*, January 1997.

28. Smith M. D., An analysis of recent NEPA cumulative impact assessment case law, *Environmental Practice Journal*, 8(4), December 2006.

29. *Neighbors of Cuddy Mountain*, 137 F.3d 1379–80.

30. *Kern*, 284 F.3d 1075 (quoting *Muckleshoot Indian Tribe* [quoting *Muckleshoot Indian Tribe v. United States Forest Serv.*, 177 F.3d 800, 810 {9th Cir. 1999}]).

31. *Kleepe v. Sierra Club*, 427 U.S. 390, 1976.

32. *Sierra Club v. Watkins*, 808 F. Supp. 852 (1991).

33. *Fritiofson v. Alexander* 772 F.2d 1225 (CA5, 1985).

34. *Thomas v. Peterson*, 753 F.2d 754 (9th Cir. 1985).

35. *Kleppe*, 427 U.S. at 414.

36. *Grand Canyon Trust v. Federal Aviation Administration*, 290 F.3d 339 (D.C. Cir. 2002).

37. *Sierra Club v. Bosworth*, 199 F. Supp. 2d 971 (N.D. Cal. 2002).

38. 40 CFR § 1508.27(b)(7).

39. Eccleston C. H., Applying the significant departure principle in resolving the cumulative impact paradox, *Journal of Environmental Practice*, 8(4), December 2006.

40. CEQ, *The National Environmental Policy Act: A Study of Its Effectiveness After Twenty-Five Years*, p. 19, January 1997.

41. 40 CFR § 1508.13.

42. 40 CFR § 1508.4.

43. 40 CFR § 1500.4, Reducing Paperwork; 40 CFR § 1500.5, Reducing Delay.

44. 40 CFR § 1508.28(a).

45. 40 CFR § 1508.27, Significantly.

46. 40 CFR § 1508.27(b)(2).

47. 40 CFR § 1508.27(b), Significantly.

48. CEQ, *Considering Cumulative Effects under the National Environmental Policy Act*, January 1997.

49. 40 CFR § 1508.27.

50. 40 CFR § 1508.27(b)(10).

51. March F., *NEPA Effectiveness: Mastering the Process*, Rowman & Littlefield Pub Inc., Section 3.3.7.

52. Eccleston C. H., *Effective Environmental Assessments: How to Manage and Prepare NEPA EAs*, Chapter 8, Lewis Publishers, Boca Raton, FL, 2001.

53. *Foundation on Economic Trends v. Heckler*, 756 F.2d 143 (D.C. Cir. 1985).

54. *Jicarilla Apache Tribe of Indians v. Morton*, 471 F.2d 1275 (9th Cir. 1973).

55. *Alaska v. Andrus*, 580 F.2d 465 (D.C. Cir. 1978), vacated in part as moot, 439 U.S. 922 (1978).

56. *Olmsted Citizens for a Better Community v. United States*, 793 F.2d 201 (8th Cir. 1986).

57. Thompson J. G. and Williams G., Social assessment: roles for practitioners and the need for stronger mandates, *Impact Assessment Bulletin*, 10(3), 1992.

58. Saylor R. E. and McCold L. N., Bounding analyses in NEPA documents: when are they appropriate? *The Environmental Professional, National Association of Environmental Professionals*, 16, 285–291, 1994.

59. Yost N. C., Administrative implementation of and judicial review under the National Environmental Policy Act, *Law of Environmental Protection*, S. Novick ed., 1987.

60. Eccleston C. H., Determining if and when impacts of potential accidents need to be evaluated in a NEPA analysis, *Environmental Practice—The Journal of the National Association of Environmental Professionals*, August 1999.

61. Eccleston C. H., *Effective Environmental Assessments: How to Manage and Prepare NEPA EAs*, Lewis Publishers, Boca Raton, FL, 2001.

62. *Florida Coalition v. Bush* (U.S.D.C. 1989).

10 Planning and Mitigating Effects of Natural Disasters and Terrorist Attacks

The legendary Chinese general Sun Tzu was a strategic military genius. In his classical book, *The Art of War*, Sun Tzu wrote

> The highest form of generalship is to balk the enemy's plans; the next best is to prevent the junction of the enemy's forces; the next in order is to attack the enemy's army in the field; and the worst policy of all is to besiege walled cities....[1]

This book is still widely read by people from a diverse range of backgrounds including generals, politicians, businessmen, and even terrorists. A growing number of terrorist groups, some of whom are drawing at least in part from Sun Tzu's strategies, have a common goal not only to undermine the political and economic structure of the West, but also to throw Western society into chaos.

This being the case, what can Westerners do to prevent this nightmarish scenario from becoming a reality? The answer is obvious: to thwart their plans at every turn. This implies being at least one step ahead of them and perhaps a little more clever. As Sun Tzu might have counseled, this can be achieved by developing superior plans while disrupting their objectives. Plans can be developed for scattering and splitting the enemies of peace to "prevent the junction of their forces"; and perhaps most importantly, plans and security measures can be forged for constructing walled cities that cannot be besieged.

10.1 CAN PLANNING FOR A DISASTER PREVENT A DISASTER?

Nearly all environmental statutes regulate or place substantive constraints on what may be done and how it is to be done. The National Environmental Policy Act (NEPA) is unique in that it neither regulates nor mandates substantive contraints.[1] NEPA provides the only comprehensive federal planning process that is applicable to virtually all federal actions. Its purpose is not to place strict limitations on what can be done, but instead provides a rigorous planning process for ensuring that actions and alternatives are appropriately considered before a final decision is made, and before other highly prescriptive environmental laws and regulations are triggered that dictate precisely under what conditions actions may be carried out.

NEPA is a planning process that might provide a 21st century framework for implementing Sun Tzu's strategy. Performed correctly, NEPA and other similar planning processes can provide a cutting-edge tool for helping secure the Western homeland.

10.1.1 CATASTROPHIC EVENTS AND THE HUMAN ENVIRONMENT

Detractors may question the relevance of including potential terrorist acts or natural disaster scenarios into what has been more traditionally and strictly an environmental analysis. Some might even question whether it is legal, let alone wise, to analyze such scenarios within a NEPA or similar planning process; for that matter, do potentially catastrophic events even fall within the scope of NEPA?

As prescribed in Section 102 of NEPA, an environmental impact statement (EIS) must be prepared for all

"… major Federal actions significantly affecting the quality of the *human environment* …"[2] (emphasis added).

Consequently, federal actions that have the potential to significantly affect the human environment are potentially subject to the EIS requirement. According to the Council on Environmental Quality's (CEQ), NEPA regulations (Regulations), the human environment shall be interpreted (§ 1508.14)

… comprehensively to include the natural and physical environment and the relationship of people with that environment … When an environmental impact statement is prepared and *economic* or *social* and *natural* or *physical environmental effects* are interrelated, then the environmental impact statement will *discuss all of these effects on the human environment.* (Emphasis added.)

Based on this, the phrase "human environment" is interpreted comprehensively to include not only the physical environment but also its relationship with people; thus, significant socioeconomic impacts that are interrelated to environmental impacts must be evaluated within an EIS. Moreover, under the Regulations, the section of the EIS describing environmental consequences is to address (§ 1502.16[g])

… Urban quality, historic and cultural resources, and the *design of the built environment*, including the reuse and conservation potential of various alternatives and mitigation measures.

This provision implies that if urban quality, historic and cultural resources, or the design of the built environment (i.e., human-made structure) could be significantly impacted, such effects should be investigated in the EIS. Thus, it appears that potential terrorist acts, homeland security issues, and natural disasters might indeed fall within the scope of NEPA.

10.1.2 SIGNIFICANCE

As described in Chapter 6, the Regulations define 10 factors that are to be assessed in determining the significance of a potential impact (§ 1508.27[b]). Depending on the particular circumstance, virtually each one of the 10 significance factors could be triggered by various types of potential attacks or human-induced disasters. An EIS might be required if such events could trigger any one of them.

Table 10.1 compares the relationship between potential terrorist acts and natural disasters, and NEPA's 10 significance factors.

10.2   NRC RULES TERRORISM REVIEWS ARE NOT REQUIRED

What is an agency's responsibility under NEPA to consider intentionally malevolent acts, such as those directed at the United States on 9/11? The U.S. Nuclear Regulatory Commission (NRC) addressed this question in four orders issued in 2002, each holding that NEPA did not require the NRC to consider impacts of terrorism in rendering licensing decisions.

10.2.1 NUCLEAR FUEL STORAGE CASE

The NRC provided a detailed rationale for its conclusion in an order involving a proposal by Private Fuel Storage, LLC (PFS) to build an independent spent fuel storage installation on the Skull Valley Goshute Indian Reservation in Utah. The proposed facility was to store spent nuclear fuel from commercial nuclear power plants pending disposal in a repository.

TABLE 10.1

How Potential Terrorist Acts or Natural Disasters Might Trigger NEPA's 10 Significance Factors

NEPA's 10 Significance Factors	Effects of Potentially Catastrophic Events
1. Impacts that may be both beneficial and adverse. A significant effect may exist even if the federal agency believes that on balance the effect will be beneficial.	A catastrophic event is obviously an adverse impact, potentially subject to NEPA's requirements.
2. The degree to which the proposed action affects public health or safety.	A catastrophic event can affect public health and safety, resulting in untold fatalities and casualties. For example, an attack on a dam, nuclear reactor, or petroleum storage facility might result in disastrous impacts.
3. Unique characteristics of the geographic area such as proximity to historic or cultural resources, parklands, prime farmlands, wetlands, wild and scenic rivers, or ecologically critical areas.	A catastrophic event can affect historic and cultural sites, particularly landmarks and national emblems as well as ecologically sensitive resources. For example, a new agricultural policy or decision might leave a major food source vulnerable to acts of agricultural terrorism, resulting in catastrophic impacts.
4. The degree to which the action effects on the quality of the human environment are likely to be highly controversial.	The impacts of potentially catastrophic events are uncertain, which can raise significant scientific controversy concerning the severity of the actual effects.
5. The degree to which the possible effects on the human environment are highly uncertain or involve unique or unknown risks.	The impact of many potentially catastrophic scenarios is uncertain or involves unique or unknown risks.
6. The degree to which the action may establish a precedent for future actions with significant effects or represents a decision in principle about a future consideration.	Many federal policies, plans, or decisions may establish precedents that could have far-reaching implications in terms of the risks of potential terrorism. For example, a decision to employ a new nuclear technology might result in setting a precedent that could have grave homeland implications in terms of future terrorist acts or natural disasters.
7. Whether the action is related to other actions with individually insignificant but cumulatively significant impacts. Significance exists if it is reasonable to anticipate a cumulatively significant impact on the environment. Significance cannot be avoided by terming an action temporary or by breaking it down into smaller component parts.	Obviously the impacts of a catastrophic event can result in significant cumulative impacts. Moreover, segmented federal actions might result in significant cumulative impacts if a terrorist attack was targeted at one or more of these segmented actions. For example, making a decision regarding an isolated border crossing without considering the larger context of border security might trigger this factor.
8. The degree to which the action may adversely affect districts, sites, highways, structures, or objects listed in or eligible for listing in the National Register of Historic Places or may cause loss or destruction of significant scientific, cultural, or historical resources.	Evidence indicates that terrorists have actively considered attacking significant national scientific, cultural, and historic icons and institutions (e.g., Pentagon, World Trade Tower, Statue of Liberty, and Congress). Natural disasters can also significantly affect such resources.
9. The degree to which the action may adversely affect an endangered or threatened species or its habitat that has been determined to be critical under the Endangered Species Act of 1973.	A terrorist attack or natural disaster can obviously result in grave impacts to endangered species. Some federal proposals and plans should be considered in terms of potential impacts on species because of a terrorist attack. For example, a biological terrorist attack might eradicate an entire species.
10. Whether the action threatens a violation of federal, state, or local law or requirements imposed for the protection of the environment.	A significant terrorist attack or natural disaster could breach any number of safety and environmental laws.

In abbreviated orders issued for the three other companion cases, the NRC referred to its rationale expressed in the PFS order. One of the companion cases involved a proposed mixed oxide (MOX) fuel fabrication facility. The NRC order in the MOX case reversed a decision of the Licensing Board to admit for licensing, after hearing an intervenor's contention that NEPA required the NRC to evaluate terrorism impacts at the proposed MOX facility. The Licensing Board had stated

> Regardless of how foreseeable terrorist acts that could cause a beyond-design-basis accident were prior to the terrorist attacks of September 11, 2001, it can no longer be argued that terrorist attacks ... are not reasonably foreseeable ...

10.2.1.1 Basis for NRC's Conclusion

As explained in the PFS case, the NRC concluded that

> ... the possibility of a terrorist threat ... is speculative and simply too far removed from the natural or expected consequences of agency action to require a study under NEPA ... As a practical matter, attempts to evaluate that threat even in qualitative terms are likely to be meaningless and consequently no use in the agency's decision-making.

In reaching this conclusion, the NRC noted two federal court of appeals decisions that addressed the issue of terrorism and NEPA in the area of nuclear regulation. Both the decisions upheld an agency's refusal to consider terrorism under NEPA as reasonable.[3] It should be pointed out that both of these court cases were rendered years before the attack of 9/11.

Further, the NRC observed that the risk of a terrorist attack (generally thought of as the product of the probability of an occurrence and the consequences) cannot be adequately determined because "the likelihood of attack cannot be ascertained using any state-of-the-art methodology."

An intervenor in the PFS proceedings asked the Commission to assume an attack with a large jumbo jet and to analyze the consequences without the consideration of probability. The NRC, however, concluded that such an analysis

> ... amounts to a form of 'worst case' analysis, which the Supreme Court, in Robertson v. Methow Valley Citizens Council [490 U.S. 332 (1989)], determined is not required under NEPA.

The NRC went on to write

> ... presumably all other kinds of terrorism, if conceivable, would require NEPA review as well ... Such an open-ended approach to NEPA is unworkable ... As the Supreme Court noted in Robertson, it is always possible to 'conjure up' progressively more disastrous scenarios.

In further arguments that NEPA is not an appropriate forum for considering terrorism, the NRC noted

> The public aspect of NEPA processes conflicts with the need to protect certain sensitive information ... In our view, the public interest would not be served by inquiries ... into where and how nuclear facilities are vulnerable ...

The NRC did not entirely close the door to analyzing terrorism in NEPA documents, as mentioned in a footnote:

> This is not to suggest that an environmental review should never consider [the] threat of terrorism ... In fact, the NRC has briefly considered, as a matter of discretion, the issue of terrorism in generic environmental reviews [for nuclear power plant license renewal].

Yet, some local critics maintain that planning for events such as terrorist acts are precisely the types of scenarios and issues that Congress intended agencies to consider when it passed NEPA. Although potential terrorist events may not have been reasonably foreseeable prior to the attack of 9/11, such an argument is more difficult to defend in a post-9/11 world.

10.2.1.2 Court Rejects Reasoning

More recently, plaintiffs petitioned the court to review NRC's approval of the proposed dry cask storage facility of spent nuclear fuel. The plaintiffs' NEPA claims challenged the 2003 decision by NRC not to evaluate terrorism-related impacts in an environmental assessment (EA) completed for the proposed storage facility. On June 2, 2006, U.S. Court of Appeals for the Ninth Circuit concluded that NRC erred in its determination that NEPA does not require an analysis of impacts resulting from a potential terrorist attack.[2]

The court did not provide direction on how NRC was to evaluate terrorism-related impacts and instead left this matter to the agency's discretion. The court concluded that NRC's justification for not considering impacts of potential terrorist attacks "… either individually or collectively, do not support the NRC's categorical refusal to consider the environmental effects of a terrorist attack."

On January 16, 2007, the Supreme Court declined to review the decision by the Ninth Circuit Court of Appeals. As is common for this type of action, the Supreme Court provided no explanation for its denial of the appeal request. It is recommended that the reader consult legal counsel in determining to what degree the impacts of potential terrorist attacks be considered in NEPA analyses.

10.2.2 DOE Practice

The U.S. Department of Energy (DOE) sometimes finds it appropriate to consider potential environmental impacts of intentional destructive acts (e.g., sabotage or terrorism) in its NEPA documents, although the Department has not expressed a conclusion regarding whether or not such analyses are required under NEPA.

In its guidance document, Recommendations for Analyzing Accidents under NEPA (July 2002), the DOE stated

> In identifying the reasonably foreseeable impacts of a proposed action and alternatives, past DOE NEPA documents have addressed potential environmental impacts that could result from intentional destructive acts. Analysis of such acts poses a challenge because the potential number of scenarios is limitless and the likelihood of attack is unknowable.

The guidance further states that

> Intentional destructive acts are not accidents. Nevertheless … the consequences of an act of sabotage or terrorism could be discussed by a comparison to the consequences of a severe accident …. When intentional destructive acts are reasonably foreseeable, a qualitative or semi-quantitative discussion of the potential consequences of intentional destructive acts could be included in the accident analysis.

DOE's guidance provides two examples of qualitative discussions of intentional destructive acts that might be appropriate in an EIS.

Regarding security concerns, DOE conducts reviews of its environmental documents to ensure that security-sensitive information is protected. For example, some DOE EISs have contained a nonsensitive summary of the results of an analysis of intentional destructive acts; in such cases, details of the analysis, which may contain nonclassified security-sensitive information, have been segregated into a separate EIS appendix whose distribution was appropriately limited.

10.2.2.1   DOE Litigation

In a recent case involving a challenge to the DOE, EA for construction and operation of a biosafety level-3 facility in which the impacts of a potential terrorist attack were not addressed, the court concluded:[3]

> Concerning the DOE's conclusion that consideration of the effects of a terrorist attack is not required in its Environmental Assessment, we recently held to the contrary in San Luis Obispo Mothers for Peace v. Nuclear Regulatory Commission. In Mothers for Peace, we held that an Environmental Assessment that does not consider the possibility of a terrorist attack is inadequate. Similarly here, we remand for the DOE to consider whether the threat of terrorist activity necessitates the preparation of an Environmental Impact Statement. As in Mothers for Peace, we caution that there "remain open to the agency a wide variety of actions it may take on remand [and] … [w]e do not prejudge those alternatives" (citations omitted).

As a result of such court decisions, the DOE issued an interim guidance on December 1, 2006, indicating that all DOE EISs and EAs, whether for nuclear or nonnuclear proposals, should include explicit consideration of the potential environmental impacts of sabotage and terrorism. The DOE is currently developing additional guidance on considering sabotage and terrorism in NEPA documents.

10.3   WHY NEPA CAN PROVIDE AN IDEAL FRAMEWORK FOR EVALUATING TERRORIST AND NATURAL DISASTER SCENARIOS?

NEPA is the only federally mandated planning process that is applicable to virtually all major federal proposals. There are many advantages of using NEPA as a comprehensive planning process for screening actions in terms of potential terrorist attacks or other high-consequence events.

Planning processes such as NEPA, State Environmental Policy Acts (SEPAs), and other similar impact assessment processes provide an ideal framework for ensuring that a rigorous analysis of potential threats is performed. As depicted in Table 10.2, the NEPA planning process incorporates every essential element necessary for ensuring that a comprehensive, scientific, rigorous, and analytical process is used in evaluating threats such as terrorist acts or natural disasters.

10.4   USING NEPA TO PLAN FOR POTENTIAL TERRORIST ACTS AND NATURAL DISASTERS

NEPA is sometimes perceived as just another obstacle that federal officials must surmount before implementing a proposal. Yet, executed in a streamlined manner, NEPA presents an ideal framework for providing a decision-maker with information regarding the consequences of potential terrorist attacks. Table 10.3 describes how NEPA and NEPA-like impact assessment processes can be applied to safeguard communities, and government projects and installations.[4,5]

10.4.1   Strategic and Programmatic Reviews

A NEPA analysis can be prepared to actively identify the vulnerabilities and weaknesses of an entire program or mission. Here, a strategic, programmatic, or site-wide EIS can be prepared to assess the risks of an entire program.

Experts can be consulted and the public can be actively engaged in seeking comments that would help the agency identify potential risks and weaknesses, as well as terrorist and natural disaster scenarios. In some cases, these comments and the information obtained might need to be restricted from public distribution. This input would then be evaluated to determine threats and impacts.

TABLE 10.2

Why NEPA Provides All Essential Elements Necessary for Comprehensively Evaluating and Countering Potential Terrorist Threats

Essential Elements	Description of How NEPA's Elements Support a Successful Analysis
Rigorous planning process	NEPA requires that a rigorous process be used in assessing impacts of potential actions.
Scoping	NEPA requires a thorough scoping process to identify the range of actions, impacts, and alternatives that need to be considered. NEPA's procedures can help ensure that a reasonable range of potential terrorist threats, and all significant issues are identified and flushed out for investigation.
Scope of decision-making	NEPA is applicable to the adoption of official policy, formal plans, programs, and approval of specific projects.
Consultation	A successful analysis of potential terrorist acts typically necessitates consultation with other agencies and experts (e.g., cognizant agency, Department of Homeland Security, Federal Bureau of Investigation, and Department of Defense) possessing special expertise with respect to these issues.
Interdisciplinary	An interdisciplinary approach is typically required to successfully identify and assess potential terrorist scenarios.
Systematic	A logically ordered and systematic process is typically required to successfully identify and assess potential terrorist scenarios.
Scientifically-based process	NEPA provides a scientifically based process for rigorously ensuring that all impacts and threats are appropriately analyzed.
Impact analysis	Requires analysis of direct, indirect, and cumulative impacts that could result from a potential terrorist attack.
Alternatives	All reasonable alternatives and mitigation measures must be considered for avoiding or reducing significant impacts of potential terrorist acts.
Review and commenting	The NEPA regulations establish specific procedures (i.e., reviewing and commenting on the draft EIS) for ensuring that a scientifically competent analysis has been performed. A procedural process is also established for addressing and responding to internal and public comments concerning the adequacy of the analysis.
Handling classified information	The NEPA regulations establish procedures for ensuring that any sensitive information is appropriately classified to prevent harm to national security.
Decision-making	NEPA established a formal and procedural process for ensuring that the analysis is considered by the appropriate decision-maker in reaching a final decision.

Such information can be used to identify and evaluate programmatic alternatives and mitigation measures for addressing potential threats.

10.4.2   PROJECT-SPECIFIC REVIEWS

As witnessed earlier, virtually all federal proposals are subject to a NEPA review. At the minimum, a cursory review of potential terrorist acts should be an integral element of any proposal involving potentially high-value targets or high-consequence events. As described below, this can be efficiently implemented for the three principal levels of NEPA review.

10.4.2.1   Environmental Impact Statements

If an EIS is prepared, a task force should be selected to screen the proposal for potential terrorist and natural disaster threats. NEPA acts as an umbrella planning process for integrating relevant threats that need to be considered comprehensively during the early planning process. As appropriate, experts should be consulted to consider potential scenarios and their potential impacts. If this

TABLE 10.3

How NEPA, SEPA, and Other Similar Planning Processes Can Be Used in Assessing Potential Terrorist Acts and Natural Disasters

Properly integrated and executed planning processes such as NEPA provide an ideal tool not only for analyzing traditional impacts of proposed projects, but also for evaluating terrorist and natural-disaster scenarios associated with proposed projects. Alternatives and mitigation measures can be assessed for reducing or eliminating such threats.

Federal agencies can prepare strategic or programmatic EISs for developing master plans for identifying and securing high-value targets across new or existing broad programs. These analyses can be used in evaluating programmatic alternatives and mitigation measures for countering or reducing such threats.

Nearly one-half of the states in the US have a NEPA-like process (SEPA), a number of which contains a requirement to prepare an analysis of potentially significant proposals. These planning processes can be used by states in preparing programmatic counter-terrorist plans for fortifying potential terrorist targets or mitigating the impacts of natural disasters.

At the city and community level, a NEPA-like process can be applied in identifying targets and evaluating potential threats. For example, the analysis can be used to identify and prioritize high-risk terrorist scenarios (water reservoirs, chemical factories, national monuments, airports, etc.). Here again, the results of such studies can be used in developing alternatives and measures for mitigating potential impacts.

Most Western governments already have an Environmental Impact Assessment (EIA) process in place similar to that of NEPA. From the standpoint of the international community, EIAs can be prepared to identify potential threats, evaluate their impacts, and consider alternatives and measures for mitigating them.

---

screening process concludes that a reasonable threat exists that could pose a significant impact, a detailed analysis of the threats would be performed.

An EIS analysis of potential threats would typically be carried out in much the same way that accident analyses are performed on high-profile proposals such as nuclear reactors, hydroelectric dams, and chemical or radioactive processing facilities. If the analysis finds that the impacts could indeed be significant, an effort should be launched to identify and evaluate alternatives and mitigation measures that could reduce or eliminate potential threats. Section 9.2 provides specific guidance for assessing mitigation measures.

10.4.2.2   Environmental Assessments

As appropriate, the proposed action described in an EA should be briefly screened for potentially significant terrorist acts or threats from natural disasters. As appropriate, experts should be consulted to consider potential scenarios and impacts.

If there is a reasonable possibility that a terrorist threat related to the proposal could result in a significant impact, an EIS can be prepared.

Conversely, if the screening process concludes that there is no reasonable scenario for a significant terrorist or natural disaster threat and that there are no other significant environmental impacts, the proposal qualifies for a finding of no significant impact (FONSI).

10.4.2.3   Categorical Exclusions

Where a categorical exclusion appears to be appropriate, the action should be screened in terms of any significant impact that could result from a potential terrorist attack or natural disaster. In the vast majority of cases, this would not be the case. However, if the screening review concludes that such a threat could result in significant impacts, this would meet the criteria for extraordinary circumstances, and an EA or EIS can be prepared.

For instance, moving a tank of chlorine gas from a fortified building onto an outdoor pad might appear to be an innocuous activity until one considers that a terrorist act might easily breach the tank, resulting in potentially catastrophic results for the nearby workers or inhabitants.

10.5   DIFFERENCES BETWEEN THE USE OF AN EA AND AN EIS

As described below, there is a distinct difference in the way potentially catastrophic events should be analyzed in an EA versus an EIS.

10.5.1   MAXIMUM REASONABLY FORESEEABLE EVENTS

An analysis of a maximum reasonably foreseeable event or a maximum credible event represents potential acts of nature or accidents (or acts of terrorism) at the high-consequence end of the spectrum. A maximum reasonably foreseeable accident is therefore an event with the most severe consequences that can reasonably be expected to occur for a given action. Typically, this kind of event has a very low probability of occurrence.

Most facilities or operations have operational lifetimes measured in decades. Therefore, accident scenarios having frequencies less than 10^{-6} per year are so unlikely to occur during the lifespan of such facilities or operations that they are generally not considered important in making decisions. Nevertheless, analysis of scenarios in the range of 10^{-6} to 10^{-7} per year may need to be considered if the consequences could be catastrophic. As a practical matter, events with frequencies less than 10^{-7} per year rarely need to be examined.

In contrast to a standard accident analysis, an investigation of intentional destructive acts (terrorism or sabotage) poses a challenge because the number of potential scenarios is virtually limitless, and the likelihood of attack is correspondingly unknowable.

Fortunately, the situation is not hopeless. The physical effects of an intentionally destructive act—whether caused by a fire, explosion, missile, or something else—are frequently nearly the same as, or bounded by, the effects of accidents, particularly maximum reasonably foreseeable events. That is, the impacts (release of radioactivity, hazardous materials, explosions, fires) of an act of sabotage or terrorism on operations and facilities frequently do not exceed those of a severe accident.

10.5.1.1   Bounding

Under a bounding approach, the impacts of a potential event are generally bounded by the effects of a maximum credible event; likewise, since experience indicates that the consequences of an intentionally destructive act are generally "bounded" by those of a severe accident scenario. The same approach may also apply to the analysis of natural disasters.

Prudence must be exercised in performing a bounding analysis, as the decision-maker may be unable to make a reasoned choice among alternatives. This is because a bounding analysis tends to mask their differences.

A similar procedure to that described above can also be followed in preparing SEPA, environmental impact assessments (EIA), or other related planning analyses.

10.5.2   AN EA VERSUS AN EIS ANALYSIS

Since an EA can be used primarily to determine if an action (e.g., a terrorist attack, an accident, or a natural disaster scenario) could result in a significant impact, the maximum credible event may need to be evaluated. Once various scenarios have been screened in an effort to identify the event, it can then be evaluated in detail.

Where the maximum credible event can be shown to be nonsignificant, no further review of such events is warranted. However, if the maximum credible event is deemed to be potentially significant, two options exist:

1. Mitigate the potential impacts to the point of nonsignificance.
2. Prepare an EIS to evaluate potential alternatives that might reduce these significant impacts.

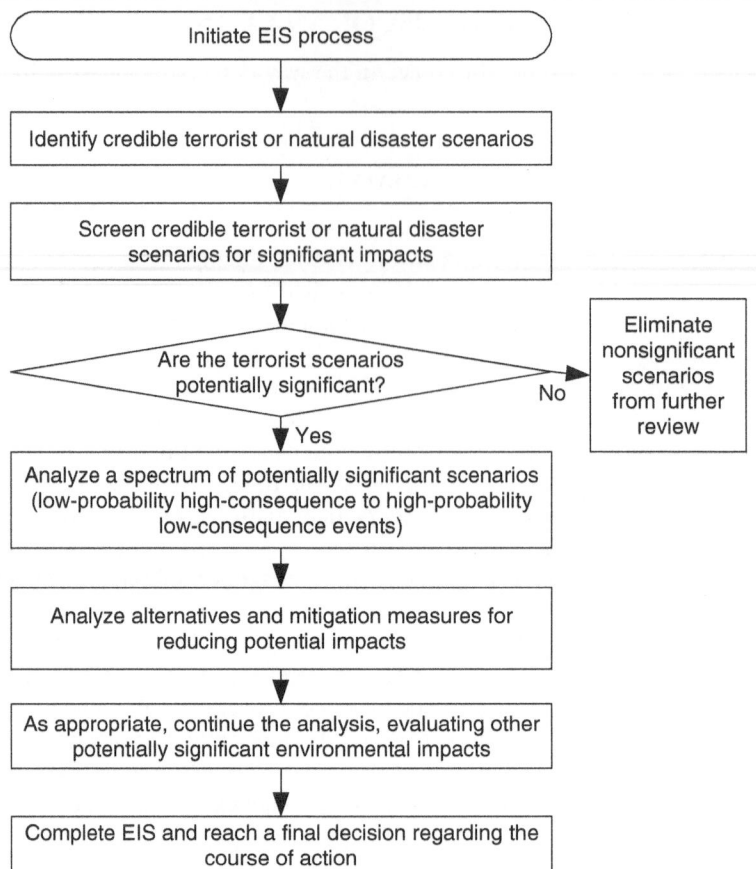

FIGURE 10.1 Using an EIS to evaluate potential terrorist and natural disaster threats.

In contrast, the purpose of an EIS is to provide the decision-maker with information on which to base informed decisions. It should, therefore, come as no surprise that planning for and evaluating potentially significant terrorist or natural disaster events in an EIS would be performed in a different manner from an EA. As shown in Figure 10.1, various scenarios are screened in the EIS process in an effort to identify a spectrum or reasonable range of potentially significant threats. The reasonable range of potentially significant threats normally includes low-probability high-consequence scenarios as well as high-probability low-consequence events. As appropriate, a reasonable range of these scenarios should be evaluated.

10.6 ANALYTICAL METHODOLOGY

With little or no modification, methodologies currently used for evaluating potential accidents (e.g., nuclear reactors, hazardous facilities, dams) can also be adopted to screen terrorist and natural disaster scenarios, evaluate their potential consequences, and, if appropriate, develop alternatives and measures for mitigating potential threats.

10.6.1 EVALUATING POTENTIAL THREATS

The purpose of scoping is to solicit input so that the analysis can be more clearly focused on issues of genuine concern. Experience has shown that a well-orchestrated scoping process provides a particularly effective tool not only for focusing on impacts and issues that are truly of concern, but also for dismissing the unimportant impacts and issues from further study.

To this end, an interdisciplinary team should be assembled to determine the appropriate scope of the study (including potential terrorist and natural disaster threats) and evaluate potentially significant impacts on the environment, human health, and safety.

10.6.1.1 Accident Analyses

The approach used for performing a standard accident analysis in NEPA and other EIA processes (e.g., in cases concerning nuclear reactors, hazardous chemical or waste facilities, and dams) is virtually identical to the same methodology that can be used in screening potential terrorist scenarios, evaluating their consequences, and, if appropriate, developing alternatives and mitigation measures for reducing potential impacts to effectively safeguard high-threat targets.

Some agencies such as the DOE have extensive experience in considering potential accidents and intentionally destructive actions (i.e., facility, operations, and transportation scenarios) in NEPA documents. In nearly every case, it has found that the consequences of intentionally destructive acts were "bounded" by those of severe accidents that had also been analyzed. Thus, the methodology now used for evaluating potential accidents can be adopted to the problem of evaluating the impacts of potential terrorist acts and natural disasters on proposals. The companion text, *Effective Environmental Assessments*, describes the methodology for analyzing potential accidents.[6] Section 9.3 provides a specific guidance for assessing potential impacts.

Sliding-Scale Approach. The Regulations state that impacts are to be evaluated in proportion to their significance (§ 1502.2[b]); that is, the degree of effort expended on the analysis is commensurate with the level of risk. Thus, as described in Section 2.2, analysts should implement a sliding-scale approach in performing such analyses.

Key factors to be considered in applying the sliding-scale principle to accident analyses are

- probability or frequency of an event to occur,
- severity of potential consequences,
- context of the proposed action and its alternatives,
- degree of uncertainty of the event, and
- level of technical controversy.

Table 10.4 illustrates how this level of effort varies, according to a sliding-scale approach, as the risk of impacts increases for workers due to a potential explosion (terrorist attack or otherwise) at a chemical processing facility. For relatively lower levels of risk, a qualitative analysis may be appropriate. Correspondingly, a quantitative analysis may be necessary for higher levels of risk.

In the context of analyzing potential accidents or terrorist scenarios, the environment includes biota and environmental media, such as the land and water, which may also be affected.

10.6.2 Determining a Reasonable Range of Scenarios

An inverse relationship tends to exist between the probability and consequences of a potential terrorist attack, natural disaster, or an accident. That is, the higher the probability of an event, the lower the consequences tend to be. Conversely, the lower the probability, the higher the probable consequences. For this reason, a set of potential scenarios should be considered, representing a range or spectrum of reasonably foreseeable events, including both

- low-probability/high-consequence events, and
- high-probability/low-consequence events.

TABLE 10.4

Applying a Sliding-Scale Approach in Determining the Level of Analysis Appropriate for Analyzing Impacts to Workers from a Chemical Explosion

←——————————————————————————————————————→

Level of Analysis (Sliding Scale)		
Low Risk		**Greater Risk**
Qualitative	**Semiqualitative**	**Quantitative**
Narrative discussion of potential consequences • Estimated number of workers at risk • Summary of potential acute and chronic consequences	Gross estimates of chemical concentrations at receptor location(s).	In addition to semiquantitative considerations
	Comparison of chemical concentrations to appropriate health-related references, such as • emergency response planning guidelines, • temporary emergency exposure limits, • reference concentrations.	• An explicit consideration of mitigation measures, such as engineering controls, inventory reduction, and design changes is performed.
A detailed discussion of engineering controls or other potential mitigation measures usually unwarranted.	Simplistic exposure assumptions (e.g., assume 15-min time-weighted average for events with long release periods unless there are counter indications, such as substances that cause immediate adverse effects).	• Detailed dispersion modeling is performed (e.g., time profile and frequency distribution of releases and concentrations at receptor locations). • Determine the potential for physical or chemical form changes (e.g., phosgene formation from carbon tetrachloride combustion; liquid-to-vapor changes). • Perform a detailed health effects analysis (e.g., potential impairment effects, dose-dependent effects, combined effects).

10.6.3 SIGNIFICANCE AND POTENTIALLY CATASTROPHIC SCENARIOS

The need to consider impacts of potential terrorist events, natural disasters, and accident scenarios is implied by at least 4 out of the 10 factors cited in the Regulations for determining significance:

- The degree to which the proposed action affects *public health or safety* (§ 1508.27[b][2], emphasis added)
- The degree to which the effects on the quality of the *human environment* are likely to be *highly controversial* (§ 1508.27[b][4], emphasis added)
- The degree to which the possible effects on the human environment are *highly uncertain* or involve *unique or unknown risks* (§ 1508.27[b][5], emphasis added)
- Whether the action *threatens a violation* of federal, state, or local law or requirements imposed for the protection of the environment (§ 1508.27[b][10], emphasis added)

10.6.3.1 Decision-Making Criteria

Consistent with the Rule of Reason and a sliding-scale approach, the author has identified seven tests or criteria described in the following sections that provide guidance useful in determining if a potential natural disaster or terrorist attack should be subject to the scope and requirements of a NEPA analysis. As in determining the applicability of other (albeit less complex) actions and impacts, NEPA's public scoping requirement also provides a useful procedure for determining if an analysis of a severe event needs to be performed.

Reasonably Foreseeable Adverse Impacts. Section 1502.22(b) specifies requirements that must be met when information relevant to the "reasonably foreseeable significant adverse impacts" is incomplete or unavailable. When such information cannot be obtained, the EIS must provide an "… evaluation of such impacts based upon theoretical approaches or research methods" that are

1. generally accepted in the scientific community.

The phrase "reasonably foreseeable significant adverse impacts" includes effects that have catastrophic consequences, even if their probability of occurrence is low, provided that analysis of these effects is

2. supported by credible scientific evidence,
3. not based on pure conjecture, and
4. within the rule of reason.

An analysis of potentially severe events typically involves incomplete or unavailable information concerning reasonably foreseeable significant adverse impacts. A scenario can also involve catastrophic consequences that may involve a low probability of occurrence. Accordingly, § 1502.22(b) provides important insight and direction useful in determining the applicability of potential scenarios involving reasonably foreseeable significant adverse impacts.

Remote and Speculative. An analysis of potentially severe events may require an unnecessary degree of speculation. However, a review of case law indicates that environmental impacts do not have to be evaluated if they are determined to be remote and speculative. Fogleman has identified factors used by the courts for determining when an impact should be deemed remote or speculative:[7]

5. Level or degree of confidence that the agency has in predicting the impact;
6. Amount of information available to the agency, which provides a basis for describing the impacts in a manner meaningful to the decision-maker.

According to Fogleman, an action is likely to be deemed reasonably foreseeable if it is a logical stepping-stone to potential local or regional development or accelerates such development. Conversely, the degree of speculation increases as a projected impact becomes removed or dissociated from the precipitating action. Adding an additional step in the causal chain of events tends to increase the degree of speculation, even if the incremental step (by itself) is considered reasonably foreseeable. A seventh test is presented in Section 10.6.4.

10.6.4 Risk–Uncertainty Significance Test

As we have seen, events involving impacts that are highly uncertain or involve unique or unknown risks are factors to be considered by a decision-maker in reaching a determination regarding potential significance (§ 1508.27[b][5]).

10.6.4.1 Uncertainty

Decision-makers need to understand the nature and extent of uncertainty in choosing among alternatives and considering potential mitigation measures. Where uncertainties preclude quantitative analysis (terrorism), the unavailability of relevant information should be explicitly acknowledged. The NEPA document should describe the analysis that is used and the effect that the incomplete or unavailable information has on the ability to estimate the frequency and probabilities or consequences of reasonably foreseeable events (§ 1502.22).

For events where the consequences are relatively low or for which numerical probability estimates are unavailable or difficult to be obtained, qualitative descriptions such as "very infrequent" or "highly unlikely" may be adequate if the basis for such usage is provided.

An approach proposed by March provides a systematic methodology for determining the significance of an impact that involves a degree of uncertainty.[8,9] Determining the significance of an event involving uncertainty commonly involves consideration of both the frequency and severity (consequences) of a potential event.

10.6.4.2 Risk

Even though there is no universally accepted definition, risk is often defined as

$$R = F \times C$$

where
  R = risk
  F = frequency (events expected/year)
  C = consequences

Similarly, the risk associated with a course of action can be more generally defined as

$$R = \sum_{i=1}^{n} F_i \times C_i$$

where i assumes values from 1 to n, and n is the number of potential events associated with a particular course of action.

With respect to assessing significance involving a degree of uncertainty (i.e., frequency of occurrence), this definition of risk appears to be the most useful.

10.6.4.3 Frequency

Table 10.5 displays a modified frequency scale originally developed by the DOE for assessing events involving uncertainty. Using a numerical range, Table 10.5 describes the number of times (frequency) a particular event is expected to occur over a given period of time.[10] A category, level, and description (e.g., "Frequent") are included for interpreting and describing the numerical value of the frequency. As appropriate, Table 10.5 may need to be modified to address special problems or circumstances unique to a particular mission.

TABLE 10.5
Frequency Scale for Assessing the Risk–Uncertainty Significance Criterion

Category	Level	Frequency (f)	Descriptions
Frequent	A	$f > 1$	Expected one or more times per year
Likely	B	$1 > f > 10^{-1}$	Once in 1–10 years
Occasional	C	$10^{-1} > f > 10^{-2}$	Once in 10–100 years
Unlikely	D	$10^{-2} > f > 10^{-3}$	Once in 100–1000 years
Remote	E	$10^{-3} > f > 10^{-6}$	Once in 1000–1,000,000 years
Very remote	F	$10^{-6} > f$	Less than once in 1,000,000 years

Where possible, the probability that adverse consequences will occur during the lifetime of a proposal should be presented, rather than only the annual frequency of initiating events (e.g., earthquakes, floods).

10.6.4.4 Severity

A modified severity (consequences) scale originally developed by the U.S. Department of Defense is presented in Table 10.6.[11] This table provides guidance for gauging the severity of potential consequences. Severity is designated using a severity descriptor (i.e., "Negligible" through "Catastrophic") and a numerical scale (i.e., I through IV). The column labeled "Description of Consequences" defines the severity in terms of both human and environmental consequences.

As no severity scale has been universally accepted, this table provides a reasonable starting point for most circumstances. As necessary, Table 10.6 may need to be modified to address special problems or circumstances unique to a particular mission.

10.6.4.5 Assessing Significance of a Potential Event

The frequency and severity scales (Tables 10.5 and 10.6) are combined to produce Table 10.7, which provides systematic guidance for assessing significance in terms of both variables (i.e., frequency and severity). The frequency designations are indicated in the first row of Table 10.7, while the severity scales are depicted in the first column. Originally advanced by March, Table 10.7 has been

TABLE 10.6
Severity Scale for Assessing the Risk–Uncertainty Significance Criterion

Severity	Scale	Description of Consequences
Catastrophic	IV	*Human*: Loss of 10 or more lives and/or large-scale and severe injury or illness
		Environmental: Large-scale damage involving destruction of species, ecosystems, infrastructure, or property with long-term effects, and/or major loss of human life
Critical	III	*Human*: Loss of less than 10 lives and/or small-scale severe human injury or illness
		Environmental: Moderate (medium-scale and short-term) damage to ecosystems, infrastructure, or property
Subcritical	II	*Human*: Minor human injury or illness
		Environmental: Minor (small-scale and short-term) damage to ecosystems, infrastructure, or property
Negligible	I	*Human*: No reportable human injury or illness
		Environmental: Negligible or no damage to ecosystems, infrastructure, or property

TABLE 10.7
Guidance for Determining Significance Based on the Severity and Frequency of an Event

	A: Frequent $(f > 1)$	B: Likely $(1 > f > 10^{-1})$	C: Occasional $(10^{-1} > f > 10^{-2})$	D: Unlikely $(10^{-2} > f > 10^{-3})$	E: Remote $(10^{-3} > f > 10^{-6})$	F: Very Remote $(10^{-6} > f)$
(IV) Catastrophic	Significant	Significant	Significant	Significant	Marginal	Marginal
(III) Critical	Significant	Significant	Significant	Marginal	Nonsignificant	Nonsignificant
(II) Subcritical	Significant	Significant	Marginal	Nonsignificant	Nonsignificant	Nonsignificant
(I) Negligible	Nonsignificant	Nonsignificant	Nonsignificant	Nonsignificant	Nonsignificant	Nonsignificant

modified slightly by the author. Table 10.7 leads to three possible outcomes with respect to the determination of the significance of an impact involving a frequency or probability of occurrence. The designation assigned to individual elements within the matrix (i.e., significant, nonsignificant, or marginal) has been made based on professional judgment and experience:

1. *Significant*: If an event falls within the category labeled "significant," the threshold of significance would normally be breached.
2. *Nonsignificant*: If an event falls within the category labeled "nonsignificant," the threshold of significance has normally not been breached.
3. *Marginal*: If an event falls within the category labeled "marginal," the threshold of significance is quantitatively indeterminate, and professional judgment must be exercised in determining if a potentially severe event scenario must be evaluated in an EIS.

For example, the significance of a fire breakout with a frequency between 0.1 and 0.01 $(10^{-1} > f > 10^{-2})$ and a severity level of II (subcritical) would be deemed marginal and therefore may need to be the subject of an analysis within a NEPA document.

Chapter 6 provides specific guidance for assessing the significance of events such as terrorist attacks.

10.6.5 MODIFICATIONS TO THE APPROACH

Although the proposed process would be performed in a manner quite similar to that of more typical NEPA analyses, a few methodologies and regulatory procedures for evaluating potential terrorist attack or natural disasters might need to be performed differently from the more standard approach. There are a multitude of innovative approaches that can be used to elicit input and inform interested American citizens while preventing sensitive information from falling easily into the hands of terrorists.

In particular, methods for enlisting public involvement, scoping terrorist scenarios, circulating the analysis for review, and classifying sensitive information may need to be modified from the way NEPA or related planning processes have been practiced in the past. However, it does not appear that (beyond developing supplemental orders or guidelines) any significant change would need to be made to either the NEPA statute or the Regulations.

10.7 A TOOL FOR DETERMINING IF AN ACCIDENT, NATURAL DISASTER, OR TERRORIST EVENT SHOULD BE EVALUATED

A systematic peer-reviewed tool developed by the author is presented in Figure 10.2, which may be useful in determining if a given accident, natural disaster, or terrorist event scenario is subject to NEPA's requirements.[12] This tool combines the risk–uncertainty significance test with the six criteria defined in Section 10.6.3.1.

Application of this tool is initiated with the first rectangle at the top of Figure 10.2. A preliminary review is performed to ascertain whether a proposal involves any potential scenarios that might need to be the subject of a NEPA analysis. The first test (first diamond) shown in Figure 10.2 considers the potential significance of adverse impacts associated with a severe event in terms of the risk–uncertainty significance test (Table 10.7). If the response to this test is "no," the potential impacts are considered nonsignificant and therefore not subject to an accident analysis; the six remaining tests need not be considered. If the response is "yes," the review continues down to the next test; the remaining tests are based on the six criteria cited in Section 10.6.3.1.

The second and third tests are considered with respect to the question: "Is the analysis of impacts reasonably foreseeable (i.e., not considered remote or speculative)?" A "no" to either test is sufficient to reach a determination that an analysis of the accident scenario is not warranted. If the response

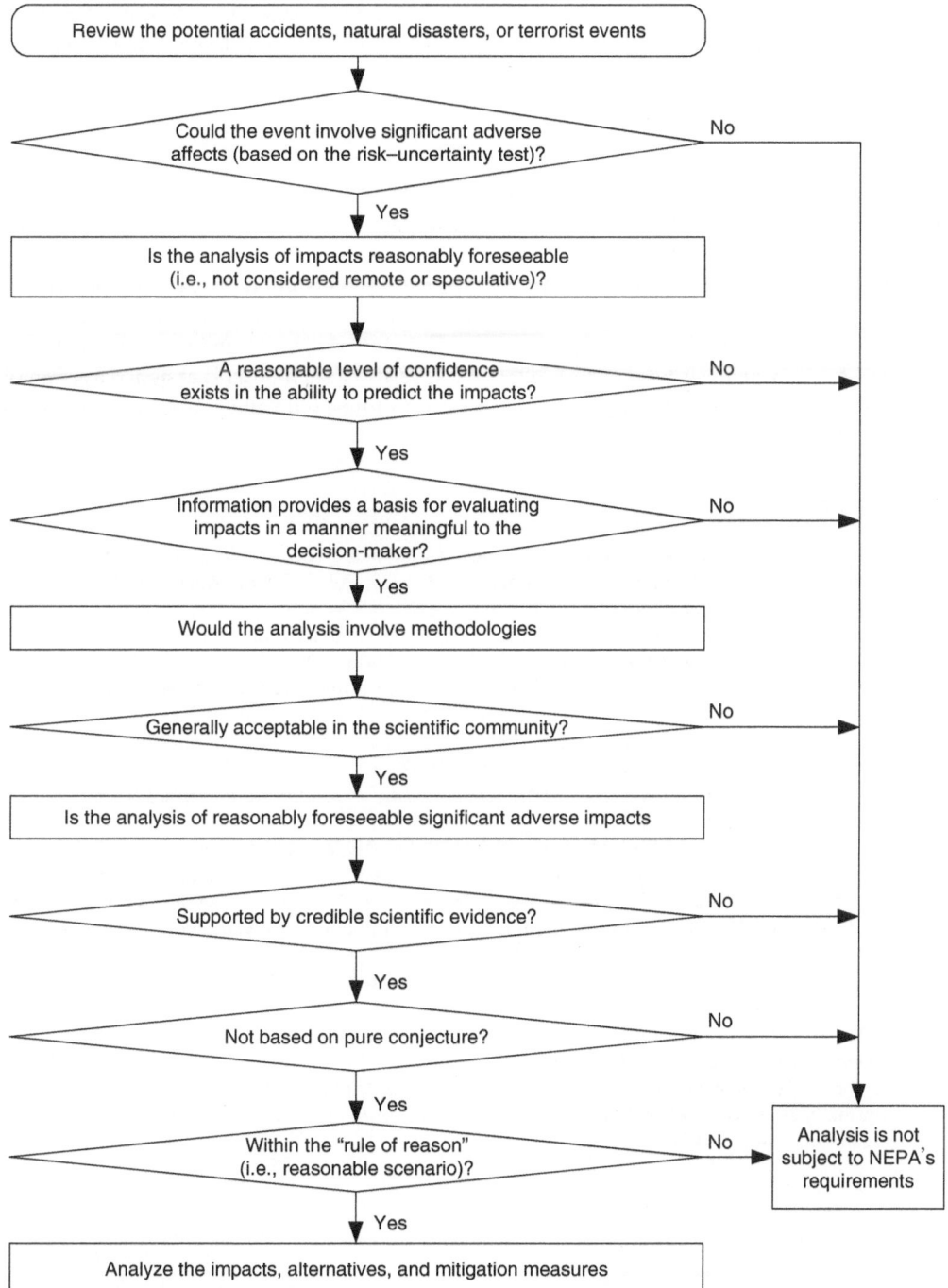

FIGURE 10.2 General-purpose tool for determining if an analysis of potential accidents, terrorist events, or natural disasters should be performed.

to both the tests is "yes," the user continues down to the fourth test which is considered with respect to the question: "Would the analysis involve methodologies?" If the response is "yes," the review continues down to the final question: "Is the analysis of reasonably foreseeable significant adverse impacts?" The final question is reviewed with respect to three distinct but related tests. A response of "yes" to all the three tests tends to support a conclusion that the scenario is subject to NEPA's impact assessment requirements. The reader should note that this tool is to be used in conjunction with other relevant information.

PROBLEMS

1. What is the most commonly accepted definition of risk?
2. Class project: Assume that there is a federal installation with 8000 workers and 100 facilities located on a 30 square mile site. Develop a strategy using NEPA for securing the installation against possible terrorist attacks. Define your own assumptions and parameters.
3. A proposed task would contain 2000 gallons of chlorine. An earthquake capable of rupturing the tank has a probability of between once in a hundred years and once in a thousand years. A safety analysis has concluded that the severity of such event is assigned a rating of "catastrophic." Would such an event trigger NEPA's threshold value requiring preparation of an EIS?

REFERENCES

1. The National Environmental Policy Act of 1969, as amended (Pub. L. 91-190, 42 USC 4321–4347, January 1, 1970, as amended by Pub L. 94-52, July 3, 1975, Pub L. 94-83, August 9, 1975, and pub L. 97-258, § 4(b), September 13, 1982).
2. *San Luis Obispo Mothers for Peace v. Nuclear Regulatory Commission.*
3. *Tri-Valley CAREs v. Department of Energy*, October 16, 2006.
4. Eccleston C. H., How NEPA can be used to counter terrorism, *Journal of Environmental Practice: Special Issue on the National Environmental Policy Act*, Vol. 5(4), December 2003.
5. Eccleston C. H., NEPA and terrorism: is it time for a paradigm shift? *Federal Facilities Environmental Journal* (Lead journal article), Vol. 13(2), Summer 2002.
6. Eccleston C. H., *Effective Environmental Assessments: How to Manage and Prepare NEPA's EA*, Lewis Publishers, Boca Raton, FL, 2001.
7. Fogleman V. M., *Guide to National Environmental Policy Act*, Section 3.5, 1990.
8. March F., Determining the significance of proposed actions, *National Association of Environmental Professionals 21st Annual Conference Proceedings*, NEPA Symposium, Session TC3, p. 421, 1996; March F., *NEPA Effectiveness: Mastering the Process*, Section 3.3.7, Government Institutes, Rockville, MD, 1998.
9. Eccleston C. H., *The NEPA Planning Process: A Comprehensive Guide with Emphasis on Efficiency*, Chapter 8, John Wiley & Sons Inc, New York, 1999.
10. U.S. Department of Energy, Order 5481.1B.
11. U.S. Department of Defense, MIL-STD-882B.
12. Eccleston C. H., Determining if and when impacts of potential accidents need to be evaluated in a NEPA analysis, *Environmental Practice—The Journal of the National Association of Environmental Professionals*, August 1999.

11 Emergency Planning for Continuity of Business Operations

As with the biblical story of Noah's Ark, ample physical and documentary evidence shows that throughout the course of human history, different forms of planning have been undertaken to protect communities from natural disasters. Today, in an increasingly crowded and conflicted world, all organizations, regardless of their size, face risks to their facilities and business operations. While a small organization may consider itself to be uncomfortably vulnerable to an incident or disaster, a very large organization might face truly unacceptable risks due to its sheer size and complexity. The process of assessing and managing the business of continuing operations, also known as business continuity planning (BCP), has thus become a crucial and necessary reality for ensuring the continuity and survival of government and business organizations. In an effort to demonstrate this need, the following discussion outlines the reasons why a business continuity program should be implemented, explains its benefits, and describes the processes that support its management throughout the lifetime of an organization.

In the past, conventional forms of business continuity might have included buying adequate fire insurance, having sufficient fire extinguishers on hand, and routinely conducting fire drills. However, in the current environment, several factors have combined to make insurance and passive deterrents inadequate for dealing with continuity issues. These factors include tougher legislation in certain areas, increased health and safety concerns, privacy and security issues, rise in insurance costs and insurance industry mandates for organizations to actively manage their business risks, a greater likelihood of liability exposure due to legal action, and a high reliance on technology and sometimes its disparate infrastructure.

Understandably, the importance of business continuity has changed over time. Operational risks are many and varied, ranging from terrorist incidents to natural disasters and pandemics. Because of the rapid changes taking place within today's operating environment, the process of assessing and managing business risk and operational continuity needs to be both continuous and ongoing.

Many organizations have not yet fully embraced the procedure of developing a business continuity program. Some organizations still rely solely on the "it won't happen to me" concept. Evidence that organizations do not invest sufficient time and resources to BCP preparations is demonstrated by disaster survival statistics. For instance, fires permanently close 44% of affected businesses.[1] Among the 350 businesses affected in the 1993 World Trade Center bombing, 150 failed to survive. In contrast, the firms affected by the September 11, 2001, attack were back in business within days of the attack because of their well-developed and tested BCP manuals.[2]

Risk should be identified, assessed as to its importance, and remedied by the development of preventive measures to mitigate its effect. For operations where mitigation efforts would be too costly or unfeasible, contingency, response, and resumption processes should be established to deal with any problems that may arise.

Ultimately, the responsibility for introducing a business continuity program lies with the executive management of the organization. A successful program should be driven down from the top. This support should include an agreement based on the need to implement such a program, a commitment for the necessary allocation of resources to operate it once approval is obtained, and the development of policies that include directing the top management to become actively involved in all its aspects. A completed BCP cycle results in a formal printed manual that is available for reference before, during, and after a disruption has occurred.

11.1 ESSENTIAL CONCEPTS AND BENEFITS

The three terms discussed throughout this chapter are:

- *BCP*: a methodology used to create business continuity processes and plans for how an organization will assess the risk, mitigate the risk, and resume partially or completely interrupted critical function(s) within a predetermined time after a disruption or disaster.
- *Risk*: the possibility of the occurrence of an undesirable event.
- *Essential functions*: those services or products that an organization offers.

A business continuity program is defined as

A program supported and funded by executive management to ensure business continuity requirements are assessed, resources are allocated, mitigation is implemented, and contingency planning, response, recovery, and continuity strategies and procedures are completed and tested. Continuity strategies are a process of developing advance arrangements and procedures that enable an organization to respond to an event in such a manner that essential functions continue with planned levels of interruption or essential change.

11.1.1 Developing the Business Continuity Plan

As with all plans, a business continuity plan includes five critical elements: its people, processes, technology, facilities, and infrastructure.

The sustainability aspect of planning is often ignored. Sustainability means simply that the organization has the resources, motivation, and focus of the management to follow a plan, make necessary updates to that plan, and practice its use.

Table 11.1 depicts an outline of the stages and processes of the program typically involved.

11.1.2 Business Risk

Probability and severity are two of the primary factors used to measure and quantify the risks that need to be managed. Many environmental planners and safety engineers can play an important part in developing a BCP as they have extensive experience of working with these two concepts. With respect to the discipline of environmental impact assessment, these concepts were described in detail in Chapter 10.

The use of severity and probability factors provides a practical way to initially assess and prioritize risk. For example, some risks are low in severity and happen frequently, such as minor workstation failures. Though these risks are very probable, they have a low severity impact. On the other hand, a more serious event such as the disruption of electrical power from a key supplier has a higher severity impact that could impede mission-critical business operations, but may also have a lower probability of occurrence. If some kind of event had occurred in the past, probability and severity ratings can be more accurately determined.

11.1.3 Benefits of Business Continuity

Although business continuity applies to all organizations, the benefits are not easily quantified. Some organizations are thus more likely to benefit from implementing business continuity programs, but they are particularly necessary to any organizations with the following characteristics:

- Multiple sites
- Size that precludes any single individual knowing the details of every risk

TABLE 11.1

Outline of the Typical Program Stages and Processes

Business Continuity Program Stages	Processes Organized by Phases	Staff Involvement
Project management	• Preproject phase • Start-up phase	• Program manager • Project manager • Business analyst
Risk management	• Inventory phase • Risk assessment phase • Business impact analysis phase	• Project manager • Business analyst • Inventory compiler • Database administrator
Mitigation	• Mitigation strategy phase • Mitigation planning phase	• Project manager • Business analyst • Auditor • Facilitator • Test process manager • Testing resources
Contingency	• Contingency identification phase • Contingency planning phase	• Project manager • Business analyst • Facilitator • Trainer
Response operations	• Detection phase • Response phase/crisis management	• Project manager • Business analyst • Facilitator • Test process manager • Testing resources
Business resumption	• Recovery phase • Resumption phase • Personnel training phase	• Project manager • Business analyst • Human resources • Facilitator • Test process manager • Testing resources

TABLE 11.2

Potential Benefits from Implementing a Business Continuity Program

Areas of Impact	Benefit of Business Continuity
Health and safety	Avoid worker litigation; reduce insurance premiums; ensure public safety
Business interruption	Avoid loss of service, business failure, and legal liability (where applicable) for not planning for such an event; gain operational reliability
Technical	Avoid failures of obsolete methods or technologies; avoid a service stoppage
Computer	Prevent inability to communicate; avoid lack of access to information
Theft and fraud	Prevent loss of money, assets, or intellectual property

- Widely diversified business processes
- Uses many contractors, suppliers, or business partners who are not under the direct control of the organization

Generally, the larger or more complex the organization or program, the more it benefits from a formal business continuity program management process. Table 11.2 provides some potential benefits from implementing a proactive program.

11.2   FOCUSING ON CONTINUITY OF OPERATIONS

This section focuses on those aspects of business continuity that are concerned with managing risks related to operations. It ensures that if a serious incident occurs, the organization will continue to function at a level acceptable to the executive management. With that focus in mind, the question becomes not one of "Do we need to have a business continuity program?" but rather one of "To what extent do we need a business continuity program?"

Fundamentally, business continuity is about avoiding loss of business operations. To accomplish it, two questions should be answered:

1. What can be done to eliminate a risk before it occurs?
2. If a risk cannot be eliminated, what can be done to minimize the impact after it occurs and to restore normal operations quickly after an interruption?

Executive management has a responsibility for ensuring that essential functions under its control are adequately protected. To that end, a cost-effective business continuity strategy should be developed that is consistent with the organization's current business strategies. It should focus on risks related to unplanned interruptions of mission-critical business operations. In the event of a serious incident, it should also enable essential functions to continue at a predetermined level acceptable to the management.

In many cases, acceptable protection can be achieved through the proactive formulation of preventive measures and the strengthening of system and equipment reliability. Mitigation planning is the process of developing a plan that can either prevent or reduce the likelihood of the occurrence of a performance failure or that is designed to reduce the impact of a performance failure.

In case a disastrous event occurs, the organization should be prepared to respond and recover from its impact. Contingency planning is the process of developing a plan to ensure the continued availability of essential functions, programs, and operations, including all the resources necessary to operate the organization at a predetermined level, in response to the loss of operational capability. This process contains procedures for emergency response, backup, postdisaster recovery, reconstitution, and resumption to ensure the continuity of mission-critical business operations.

11.2.1   GETTING STARTED

If the business continuity philosophy is being introduced to an organization for the first time, it needs support at the executive management level. Awareness of the need for business continuity can be raised by

- highlighting potential risks to the organization, possibly by drawing comparisons with other organizations that have suffered serious business disruption and have successfully weathered the crisis;
- illustrating potential impacts to the organization in terms of key performance indicators, such as customer (interorganizational and outside customers) service levels, costs, staff turnover, and revenues generated; and
- drawing attention to commitments to business continuity made by comparable organizations, federal, state, and local governments, and industry.

Table 11.3 outlines the essential steps necessary to establish and operate a successful business continuity program.

11.2.2   PLANNING AND COMMUNICATING THE PROGRAM

An organization-wide business continuity team must be formed to monitor and guide the program. This team will be responsible for ensuring that any potential problems likely to cause operational

TABLE 11.3
Essential Steps Necessary to Establish a Successful Business Continuity Program

1. Form an enterprise-wide business continuity team
2. Form subteams within each organization
3. Communicate the purpose of the business continuity program to employees
4. Create an enterprise-wide inventory of assets and business operations
5. Conduct a high-level risk assessment and report the results
6. Create an enterprise-wide inventory of essential elements that supports business operations
7. Conduct a legal assessment
8. Conduct interviews with key staff from each functional area
9. Collect, store, and analyze the risk data and report the results
10. Plan, develop, and budget for risk prevention measures with mitigation and event-detection processes
11. Test, train, and implement preventive measures and processes
12. Monitor results of preventive measures and revise new processes as necessary
13. Develop contingency plans for risks that cannot be provided with adequate protection
14. Implement event warning, detection, and response processes
15. Develop resumption plans to resume business as usual
16. Train, test, and audit the contingency plans

failures and revenue reduction are minimized. Organizations should form teams and subteams with personnel who possess business expertise and skills in such areas as business analysis, environmental management, communications, legal and contract administration, strategic and tactical planning, financial management, project management, information technology, and staff training.

All the employees of the organization should be made aware of the program and provided with a general introduction to the issues and risks that the organization intends to address. They should be educated regarding

- the business implications of these risks,
- who the contact person of the business continuity program is within each organization, and
- the development of a plan to deal with identified risks.

Initial employee communication should include a description of the resources employed to support the business continuity program and a general outline of how the program is expected to proceed. Ongoing awareness can be accomplished in many ways such as by including a business continuity program column in the organization's internal newsletter, developing a specialized business continuity program newsletter, sending out periodic electronic mail messages from the program's sponsor, establishing a collaborative business continuity web site, or publishing progress information on an intranet Web page.

11.2.3 GATHERING INFORMATION

The next step in the business continuity program is to develop a strategy for conducting an enterprise-wide inventory of

- business operations and
- essential elements that support the operations.

This strategy will establish general objectives concerning the risk exposures on which the organization intends to focus its efforts. Risks that are inherent to an organization typically originate from three sources:

- Mission, structure, and culture of the organization
- Assets and resources either owned by or under the control of the organization
- Business partners of the organization

11.2.3.1   Inventory of Essential Elements

For each of these business operations, an inventory of the essential elements that provide direct or indirect support should be conducted. Generally, an inventory of essential elements in the following categories is necessary to facilitate an effective risk assessment and business impact analysis (BIA) (Table 11.4).

An inventory strategy describes the level of inventory detail that should be collected prior to a risk assessment is being performed. The selected inventory approach should provide data essential for enabling the more specific identification of potential risks to mission-critical business operations. Results of this inventory process will help establish

- scope of the business continuity program,
- overall strategy of the organization's business continuity program, and
- impact on the organization.

Inventory Approaches. Multiple approaches for collecting these inventory data should be examined, such as

- performing only a high-level (macro) inventory;
- performing a complete and detail-level (micro) inventory; and
- performing a combination of both a high-level and, as needed, a detail-level inventory.

In a situation where limited time and financial resources are available for commitment to the business continuity program, an approach for developing an inventory of essential elements, to perform only a high-level inventory, might be followed by a risk assessment (that will be described later) based on the summarized inventory data. This approach has the advantage of enabling the expeditious collection of inventory, which can then be used to begin the risk assessment process. However, it has the disadvantage of introducing the possibility of overlooking critical operations or inventory elements, resulting in an incomplete baseline from which the assessment is conducted.

TABLE 11.4

An Inventory of Essential Elements Normally Involves Seven General Categories

- Business partners, including suppliers, vendors, customers, or other third-party organizations that regularly provide services or products
- Organizational structure
- Organization-based performance measurements
- Facilities and office equipment
- Telecommunication systems
- Computer software and equipment
- Contracts, agreements, insurance, and investments

TABLE 11.5

Steps Performed during a Typical Risk Assessment

1. Identify, define, and prioritize the organization's essential functions (services or products)
2. Identify mission-critical business operations and associated risks
3. Perform a high-level analysis that highlights the severity of impact on the organization, given the loss of a mission-critical business operation(s)
4. Identify immediately apparent areas of vulnerability, such as the use of single-source suppliers or an outdated technology infrastructure
5. Prioritize mission-critical business operations
6. Estimate the scope and cost of proceeding with recovery strategies, risk mitigation, and contingency planning

The organization should therefore be diligent in weighing the advantages and disadvantages of each inventory approach before making its decision.

11.2.4 RISK ASSESSMENT

Once the inventory is complete, a high-level risk assessment is performed. Its purpose is to assess and prioritize essential functions and their associated risks.

If necessary, the different organizations and each of their respective divisions should complete their own risk assessment report. These should then be amalgamated to form the enterprise-wide report. Thereafter, the risk assessment should be updated on an annual or alternate year basis.

A risk assessment typically takes about 1–2 months to complete depending on the size of the organization. As detailed in Table 11.5, a BCP risk assessment typically involves six discrete steps.

Throughout the mitigation process, risk management should include risk assessments and business impacts for each mitigation strategy. These assessments and impacts should be completed specifically for the particular body of work and should be limited to the scope of the project. Project risk management should also identify project risks and impacts to the organization.

11.3 BUSINESS IMPACT ANALYSIS

A full-scope BIA should be performed to ensure that both dependencies and interdependencies of mission-critical business operations are identified and where necessary to employ preventive measures for mitigating impacts and disruptions.

When performing a BIA, the mission-critical business operations are defined and evaluated together with their respective essential elements, including dependent and interdependent variables. The impact analysis can be performed by

1. identifying all business operations, processes, and elements;
2. developing a questionnaire that will help identify, define, and prioritize the mission-critical business operations and their respective essential elements;
3. meeting with management to approve the questionnaires;
4. collecting and tabulating questionnaire responses with business and technical personnel; and
5. producing a prioritized list of essential elements and processes, including their dependencies and interdependencies based upon tabulated questionnaire responses.

An impact analysis is a way to quickly pinpoint those areas that would suffer the greatest financial and operational impact in the event of a disruption. Using the severity of impact (of an operation's interruption) as the primary rating factor, the management should rate the impact that an interruption of an operation would have on the critical success factors that enable the success of the organization. These critical success factors include, but are not limited to, the following:

- *Safety and security*: Would the safety and security of the staff or the physical assets of the organization be in danger?
- *Service and/or product fulfillment*: Would the organization's ability to generate revenue and to service its customers be affected?
- *Legal*: Would the organization be in violation of regulatory requirements or contractual agreements?
- *External reporting*: Would this affect the organization's ability to generate external reports, such as financial statements, tax reports, and so on?
- *Communications*: Would the organization's ability to communicate by e-mail or telephone (e.g., electronic data interchange) with its partners be interrupted?
- *Internal controls*: Would the organization's internal controls, measurements, and reporting be jeopardized?

It can be seen from the above factors that the management of risk related to essential functions of the organization becomes the primary focus of the business continuity program. The establishment of the best and most practical priorities for mitigating risk associated with the essential functions is the ultimate goal of this process. Accomplishing it means the realization of the most effective and efficient use of the organization's resources (staff, time, and money).

11.3.1 REGULATORY, LEGAL, AND CONTRACTUAL REVIEW

In some cases, due to poor planning, the management of an organization can be held personally liable for its failure or poor performance in carrying out response and recovery operations. For this reason, a legal assessment of potential liability related to an interruption of mission-critical business operations is an important part of any business continuity program.

Mandated legal requirements that involve environmental, health, safety, security, and emergency management are all possible risk areas. These requirements include a detailed review of all contracts, agreements, and documented performance standards, as well as the management's liability to service level agreements, contracts, and customer services. The latter requirement encompasses a review of mandated requirements and of all contractual relationships with third parties, including vendors and suppliers. It also includes identifying obligations related to maintenance or other outsourced services that are being delivered to the organization.

11.3.1.1 Legal Risk Management Strategy

After a risk-reduction mitigation strategy has been prepared to respond to issues discovered during the legal review, it should be presented to the management for its approval. The aim of the legal risk management strategy is to provide executive management with sound advice and viable alternatives as they strive to make responsible business decisions relative to the goals of the business continuity program.

During the development of the legal risk strategy, special attention should be paid to the following conditions:

- Areas where the impact of an interruption to the organization far outweighs the remedies available
- Whether the odd occurrences of such a problem seem likely
- Whether recovery from the potential problem is difficult and costly to the organization
- Where specific legal mandates are required

11.3.1.2   Potential Recommendations

To validate the efforts of the organization and to ensure that current activities and plans achieve the goals of the program, an operational audit of the business continuity program could be one of the recommendations of the legal risk strategy. Other recommendations could include

1. an outline of the policies and procedures related to business partner management,
2. changes to insurance coverage,
3. operational and procedural changes required to avoid injury and improve safety risks,
4. business continuity program activities required for regulatory compliance,
5. financial practices required to comply with reporting and disclosure guidelines, and
6. ongoing legal activities required to support the business continuity program.

11.3.2   Assessing and Analyzing Risks

As soon as mission-critical business operations have been identified and prioritized, and an inventory of essential elements that support those operations has been collected, the team can proceed with the next step of the project. At this stage, practical alternatives and guidelines should be defined that will be used to

- gather risk assessment and business impact information;
- store the accumulated data in a manner (electronic database) that allows impact analysis and reporting to be performed; and
- assess, quantify, and evaluate risk.

11.3.2.1   Severity and Probability

Developing a model that can be used to assess risk involves the identification of risk measurement criteria. These criteria consist of factors used to assess the severity and probability of a business operation or essential element failure. The factors described in Table 11.6 should be considered in rating the impact severity of a performance failure.

Severity. A precise and easily understood rating scale is needed for assigning severity impact to the interruption of an operation or an essential element failure; for example,

- 1 = negligible impact (on the organization or supported operation),
- 2 = minor impact,
- 3 = moderate impact,

TABLE 11.6
Factors That Should Be Considered in Rating the Impact Severity of a Performance Failure

- Impairment level of the failure represents the maximum impact resulting from the failure if it is not quickly resolved.
- Time horizon from failure to full impairment, where there could be a time difference between the event of failure and the full realization of its effects. For example, failure of the general ledger system may ultimately cause severe impairment to an organization's ability to produce financial budgets, but the full effect of the loss of that system might take weeks to be fully realized.
- Failure tolerance is an indication of the maximum length of time that the loss of an essential element or operation can be reasonably tolerated.
- Mitigation implies reducing the impact (e.g., changing a process, failover, backup, or other strategies). Those that cannot be mitigated are passed on to contingency planning.
- Contingency planning serves to reduce the ultimate impact experienced by a performance failure involving mitigated and nonmitigated processes that require human intervention.

- 4 = considerable impact, and
- 5 = total impairment.

The application of severity ratings to business operations and essential elements provides the input data needed to conduct a performance failure impact analyses. In many cases, severity impact ratings may provide enough information for the management to make informed decisions regarding mitigation and contingency strategies.

The impact of a business operation or the failure of an essential element provides a clear indication of their importance to the organization. However, the likelihood of a failure actually occurring should not alter the level of their importance. Therefore, a rating model based upon severity of impact can provide a straightforward means to establish a prioritized list of business operations and supporting essential elements.

Probability. In addition, rating the probability of a performance failure helps to highlight potential failures that pose real or very likely threats to an organization. This separate and distinct rating measurement helps to focus on appropriate levels of resources on mitigation and contingency planning efforts. As previously stated, some risks are low in severity but occur quite frequently, while other risks may be severe but rarely occur. Gathering failure frequency data from staff, vendors, or suppliers responsible for an essential element can usually provide failure probability estimates for most items under their scope of responsibility.

11.3.2.2 Developing a BIA Process

After the BIA process has been established, it is used to guide the development and use of an effective assessment survey tool. A set of comprehensive and business-unit-specific questions is developed for use during a series of BIA interviews that are conducted with key staff from each functional area of the organization. These interviews help identify and quantify risks related to the potential for failure of an essential element, and also

- provide insight concerning dependencies that exist between mission-critical business operations and supporting essential elements, and
- provide information on which to base mitigation and contingency-planning activities.

Database. Ideally, a database application should be developed that would serve as the master data repository for the business continuity program, storing data from the

- inventory lists,
- risk assessment surveys,
- BIA surveys, and
- other project-related information.

Database Reports. The database should provide reporting and query capabilities to support risk assessment and BIA as well as mitigation and contingency-planning efforts. The assessment and analysis reporting requirements should specify a set of metrics for assessing, selecting, and developing mitigation and contingency plans.

For example, reports might be structured in the following manner:

1. Identify business operations and assess their severity impact on the critical success factors of the organization (e.g., safety and security, service or product fulfillment, revenue generation, legal issues, communications, and so on).
2. Identify essential elements (e.g., suppliers, vendors, customers, information technology systems, documents, data, staffing, equipment, and facilities) and assess their severity impact in the event of the failure of any of these elements.

The first report can be used to determine the overall scope of the project and its priorities. The second report can be applied to each business operation in the order of business operation severity, to the extent that time, budget, and resource constraints permit. It should be noted here that an issue raised concerning the above analysis approach is that it could generate a large number of items with the same severity rating values.

Additional metrics can be used to refine the precision of the BIA and provide executive management with more comparative information for decision-making. These may include

- the time horizon from the moment of performance failure to full impairment,
- the estimated maximum duration that an operation can reasonably tolerate the loss of one of its essential elements (failure tolerance),
- the time required to launch the contingency plans or to implement backup systems, and
- the number of operations dependent on or supported by the essential element or operation.

The risk assessment and impact analysis process should be fully documented and presented to the management for their approval.

11.4   MITIGATION STRATEGIES

Now that the essential functions, mission-critical business operations, and supporting essential elements of the organization have been identified and their importance and criticality to its overall success have been prioritized, this information can be used in the development of mitigation strategies and implementation plans.

As previously mentioned, mitigation strategies and implementation planning together is the process of developing a planned action designed either to

- prevent or
- reduce the likelihood of the occurrence of a performance failure or to reduce the impact of a performance failure.

In the context of business continuity and risk management, prevention can be done proactively to avoid the occurrence of negative impacts on the organization.

11.4.1   EXECUTIVE DECISION-MAKING

Executive management should now make decisions regarding

- the allocation of capital for the remainder of the business continuity program,
- the priority to be given to the program, and
- the impact such priorities will have on other efforts throughout the organization.

It is imperative that sound business rules be established regarding what risks

- are to be mitigated,
- are considered to be acceptable, and
- justify the retirement of a business operation and its supporting essential elements.

Establishing business rules at the outset of this process will help avoid decisions made in an arbitrary or prejudicial manner. This should also help the executive management to view elements of the business continuity program as a series of business decisions, allowing them to focus on appraising the value of managing each identified risk.

Cost and benefit guidelines and constraints must be clearly defined with stated procedures for justifying mitigation and contingency planning efforts (such as service levels, product delivery, or trust impact). If the scope of the project is too broad for the resources allocated, meaningful results will probably not be possible, resulting in a poor return on investment.

The prevention planning process should examine existing capabilities within the organization. Existing mitigation and contingency plans should be leveraged to the greatest extent possible in an effort to avoid the "reinventing the wheel" syndrome. For example, plans may already exist that contain procedures for dealing with system failures within information technology departments or divisions of an organization. Plans of this type are generally referred to as disaster recovery, business resumption, or business continuity plans (in this context a low-level department or division specific).

The need for new processes, awareness, and training should be minimized as much as reasonably possible. Such plans should be updated and rolled up to complement the organization's overall business continuity plan that in turn affect the enterprise-wide business continuity plan.

11.4.2 MITIGATION IMPLEMENTATION PLAN

As with other plans, a mitigation implementation plan and its subordinate and functionally specific action plans must

- reflect the organization's philosophy,
- be dynamic, and
- be sustainable.

11.4.2.1 Plan Outline

A plan outline used for the mitigation implementation plan must be established and based upon the strategy and project scope decisions established when the business rules and guidelines were adopted at the beginning of the mitigation planning process. An example of an outline for a mitigation implementation plan that addresses the selected mitigation strategies is depicted in Table 11.7.

11.4.2.2 Mitigation Budget

To avoid any delays in the implementation of the newly developed mitigation action plan, it is necessary at this point to estimate, justify, and formally allocate the budget needed to implement it. At the minimum, this budget should include the funds required to purchase equipment, compensate vendors for services, and pay for new facilities or infrastructure, or whatever other expenses that would be incurred during the effort of executing the plan.

11.4.3 POTENTIAL FIXES

Various methods, or "fixes," that address risk issues can be employed during mitigation planning. These methods should first be clearly defined and then assigned individually to each essential element being subjected to the planning process. Potential fixes include the following methods:

- *Quick fix*: Adjustment or correction to an essential element that requires significantly less time than other potential remedies.
- *Partial replacement*: Usually applies to a system and involves replacing a nonworking part or component within a system with a working part or function.
- *Full redundancy or replacement*: Actually two approaches. Full redundancy refers to prepositioning a working part or component to be used upon failure of the incumbent part or component. Replacement refers to the total replacement of any failed system or essential element with a functioning one.

TABLE 11.7

Example of a Mitigation Implementation Plan Outline

Section No.	Section Title Executive Summary	Section Description Summary of the Entire Plan
Plan Body		
1	Purpose statement	Why is the plan being developed?
2	Scope statement	What areas are affected by the plan?
3	Responsibilities identification	Who is responsible for each part of the plan?
4	Supporting organizations	What areas are charged with support development?
5	Coordinating organizations	What areas should review the completed plan?
6	Plan review and revision schedule	How often and by whom is the plan to be reviewed?
7	Legal review and comment	How does the plan affect the enterprise's legal relationships?
Plan Support		
8	Critical data	Data to be used as input to the plan
9	Supporting data	Data to be used as support to the plan
Action Plan Guidance **(created for each selected mitigation strategy)**		
10	Action plan's required contents	Detailed description of the content of individual action plan, including a description of resources, staff roles, procedures, and timetables needed for implementation
11	Administrative and reporting	Defines expected level of administrative and management reporting related to action plan implementation
12	Associated fiscal data	Contains estimated costs related to action plan implementation
13	List of action plans	A list of all mitigation action plans
14	Reference list, support documents	Identifies manuals and standard procedures needed to develop the plan

- *Outsourcing*: Refers to the utilization of a third-party organization to correct failures of a given essential element or to provide ... (provide what?—assistance, technical expertise, etc.?).
- *Hire and train additional staff*: A manual alternative to the above methods that can be used to replace all or part of a failed automated process.

11.5   IMPLEMENTING AND TESTING PREVENTIVE MEASURES

The objective of mitigation action plan testing is to evaluate whether the plans are

- capable of providing the desired level of support to the organization's essential functions and
- whether the plans can be implemented within the estimated period of time.

11.5.1   TESTING AND ACTION PLANS

Test planning and the testing of mitigation action plans either during or after implementation are a critical part of the business continuity program. Formal acceptance testing guarantees the functional

TABLE 11.8
Questions Representative of Quality Assurance Issues

- How or where is the test environment established?
- If a separate test environment does not exist, what are the risks associated with inadvertent damage to the production environment?
- What are the differences between the test and production environments?
- How are the baseline test standards established?
- What are test results and where will they be saved for future comparisons?
- What organization is responsible for conducting the tests?
- Who will create test documents and test scripts?
- Is there a standard database(s) for system-wide testing?
- What types of tests are required?
- What constitutes acceptable test results?

performance of each action plan. The formal test plan for each action plan is unique and specific to a mission-critical business operation and its related processes or specific projects.

This is not a onetime process. Instead, as organizational and operational changes occur, they should be documented with complete BCP processes implemented from inventory to resumption. A comprehensive testing and implementation strategy should be established for this purpose.

In some cases, a specific action plan might require a special testing and implementation process because of a unique situation. In the case of remedying software application(s) from (for?) a new project, this might involve unique data interaction requirements or the need to acquire additional hardware or software. The testing plan and strategy development utilized should be guided by quality assurance standards in use by the organization.

11.5.2 Quality Assurance

Testing and quality assurance issues should be addressed to determine if any changes are necessary to the organization's quality assurance practices. Table 11.8 provides questions representative of quality assurance issues that should first be answered.

11.5.3 Training

Prior to implementing mitigation action plans, training the staff regarding new processes and procedures will be required. The amount of training can vary widely depending upon the extent of operational changes needed to accommodate the action plan. For example, for some employees, changing old habits can be an extremely difficult task. In a case where a crucial legacy software application is being replaced after many years of use, a significant training effort will be needed for the transition to succeed.

A training "needs assessment" should be conducted to answer questions concerning who needs training and the specific training that will be required. This assessment should evaluate possible training alternatives, including

- mentoring within the project staff,
- using subject matter experts from outside the project to hold classroom training,
- individual distribution of a training document to be used by staff, or
- formal classroom training presented by the organization's training department staff.

TABLE 11.9

Contingency Planning Objectives

- Ensure that threats to the safety of the organization's employees and visitors are minimized or eliminated
- Provide a sense of security, knowing that mission-critical and other business operations can continue to function during various situations
- Minimize damage to, or loss of, organizational assets
- Minimize the risk of delay in setting up an alternative processing location for restoration of mission-critical business operations and their respective essential elements
- Minimize the need for unplanned decision-making during critical situations
- Provide a standard for testing and updating contingency plans
- Ensure the availability of necessary resources, based on the essential function(s), to help the organization to continue meeting its needs during an interruption

11.6   DEVELOPING CONTINGENCY PLANS

Apart from making the best risk-avoidance efforts, the organization should also be prepared to cope with various complex "what if" negative impact scenarios. For example, if multiple incidents occur across organizational and geographical boundaries, accompanied by communication and power disruptions, the organization needs to have an alternative method to collect, filter, prioritize, and escalate issues up the management chain, as appropriate.

11.6.1   CONTINGENCY PLANNING GOALS

The ultimate goal of the contingency planning effort is to develop cost-effective contingency plans. Operational stability and reliability, representing the primary objective of contingency planning, should be maintained to ensure the survival of essential functions. To that end, contingency planning objectives include the elements depicted in Table 11.9.

As with mitigation planning, the previously defined essential functions and supporting mission-critical business operations of the organization are used to drive the development of the complete span of contingency plans. As previously mentioned, contingency planning is the process to ensure the continued availability of essential functions, programs, and operations, including all the resources necessary to operate the organization at a level predefined by executive management, in response to the loss of operational capability due to any event.

11.6.2   CONTINGENCY PLANNING FACTORS

Data obtained from the risk assessment and BIA of the probable external or internal impacts that an organization is exposed to is fundamental to the contingency planning process. The nature of an event can vary based upon several factors including, but not limited to,

- the geographic location of the organization,
- the degree of physical accessibility to the organization,
- the track record of local utility companies in providing uninterrupted services, and
- the history of the area's susceptibility to technological or natural threats.

11.6.3   POTENTIAL SOLUTIONS

As depicted in Table 11.10, the team should consider a wide range of possible solutions to deal with the failure of a business operation, process, or an essential element.

TABLE 11.10

Possible Solutions for Dealing with a Business Disruption

- Stockpile extra supplies from a key supplier
- Make arrangements for space to store additional supplies or raw materials
- Make arrangement to have supplies delivered by an alternate mode of transportation
- Acquire cellular, radio, and satellite telephones for emergency communications
- Revert to the old manual procedures for a process that has been automated
- Consider using retired employees to provide additional staffing resources

11.6.4 TYPES OF CONTINGENCY PLANS

There are different basic types of contingency plans that must be developed by the organization. These include

- incident management plan that encompasses detection criteria and incident and emergency response processes,
- recovery operations that include information technology disaster recovery and infrastructure recovery processes, and
- identification of alternative processes necessary for a mission-critical business operation to continue functioning until the failure has been resolved.

A department that is responsible for cross-functional support, like facilities, information technology, or telecommunications, normally develops the failure response plan. The second type of plan is developed by a department responsible for a specific function, like finance, marketing, human resources, or engineering, and addresses the need for the mission-critical business operation to function despite a failure.

11.6.5 STAFFING RESOURCES

Critical staffing resources that are necessary and able to respond in the event of a disaster must be identified. This can be accomplished by developing an organizational chart showing the command and control structure of the incident management team (IMT) and the relationship of its members to the organizational structure of the enterprise. IMT members are identified, and their roles and responsibilities are defined by establishing standard operating guidelines for each of the team's assignments. This ensures that the enterprise has in place a command and control structure that will be able to respond successfully to an event, minimizing the impact on mission-critical business operations.

11.6.6 WRITING THE PLAN

When detailed function recovery procedures of the plan are set out in writing, they should be written at a level that is clear and detailed enough to allow the plan to be followed just by reading them. Minimizing the need to make rushed, *ad hoc* decision-making during a disastrous situation is one of the major goals of contingency planning.

The plan should encompass all the activities that have to be carried out from the time of the interruption through the return to normal operations. It is also important to focus on the impact of the business interruption, as opposed to its cause. Many contingency plans have been written to address only a specific type of interruption, and consequently fail when a disaster of a different nature occurs. In addition, the plan for each function to be recovered and the plan for the enterprise as a whole should both incorporate the costs of implementation in terms of personnel and financial resources.

11.6.7 AUDITING AND TESTING

Prior to finalizing the plan, it is necessary for auditors from the enterprise to become involved in it. They must conduct a thorough review and audit to determine the fitness of the plan for protecting mission-critical business operations and must attest to the plan's compliance with laws or regulations regarding contingency planning. Auditors may also be able to expose possible risks from competitors. For example, in the event of a regional disaster, a direct competitor that already has a contingency plan in place could conceivably win over additional market share because of its ability to maintain adequate service levels for its operations throughout the impact period.

It is very important that the plan is thoroughly tested before a disastrous event occurs. This point cannot be overemphasized. Only by testing the plan, it can be proved that each step has been well thought out and that nothing has been overlooked. A scenario for testing the plan in the most realistic manner possible should be developed and carried out on a regular basis. The results of each test should be measured and documented to determine the plan's effectiveness since subsequent evaluations may reveal areas that need to be updated. The installation of a master data repository that can relate the causes of incidents to their impacts on essential elements and mission-critical business operations can really boost the effectiveness of the testing process.

11.6.8 MANAGEMENT APPROVAL

Executive management must approve the contingency plan and issue a blanket authorization for its funding and execution if certain conditions exist. Necessary agreements, letters of intent, and memos of understanding should be signed and put in place so that the IMT's response efforts will not be impeded.

Developing the initial contingency plan is only the beginning of the process. As already recommended, ongoing changes in technology, staffing, and business goals and objectives require that the plan be regularly reviewed, tested, and updated to remain an effective risk management tool.

11.7 MONITORING RESULTS: A RECAP

Progress tracking of the business continuity program involves providing accurate reports to the organizational management. This is essential to facilitate decisions regarding any newly implemented mitigation measures. The primary objective of this reporting is to assist with identifying problem areas that could result from the implementation of new procedures as well as monitoring their effectiveness.

The business continuity program is an ongoing effort. After procedure tracking processes have been put in place, regular follow-up review and testing of contingency plans are required to ensure the readiness of the organization to deal with an unplanned interruption.

Processes for the monitoring and detection of potential threats to the organization should also be put in place. Criteria and associated metrics, parameters, and alert mechanisms that could be used as indicators of actual or impending impairments to mission-critical business operations should be identified. Element monitoring criteria allow oversight personnel to evaluate the operational quality of an essential element (for example, suppliers, customers, facilities, equipment, data, staff, communications, hardware, and software).

When an event or the threat of an event has been detected, the IMT is notified and it responds accordingly.

11.7.1 Response and Resumption Resources

A response and resumption strategy for each service and facility supporting an essential function can be developed at this time. These resources typically fall into one of the following categories:

1. Facilities—include development of a facility recovery plan, identification of alternate physical work environments, inventory items, and any other fixed assets required to resume essential functions.
2. Information systems—include duplication of all the necessary computing equipments, the required operating environment, and the data recovered from off-site storage.
3. Telecommunications—include notification and resumption of voice and data communications.
4. Operations—include staffing and supplemental staffing if necessary. Direct customer service functions (internal and external) are normally given a high priority within this category.
5. Key business partners—include suppliers, vendors, or other third-party organizations providing crucial products or services to the organization.

11.7.1.1 Critical Staffing Resources

Critical staffing resources necessary to respond to an event should be identified. This is achieved by developing an organizational chart showing the command and control structure of the IMT and the relationship of its members to the organizational structure. Members of the IMT are identified, and their roles and responsibilities are defined by establishing standard operating guidelines for each of the team's assignment. This ensures that the organization has a command and control structure in place that can successfully respond to an event.

11.7.2 Writing Procedures

When writing detailed functional processes and procedures, write at a level detailed enough to allow the process to be followed just by reading them. Minimizing the need for unplanned decision-making during a situation is one of the goals of the planning process. Processes and tasks should encompass all necessary activities from the initiation of an interruption through the return to normal operations.

Many plans have been written to address only a specific type of interruption and consequently fail when an event of a different nature occurs. Plans should focus on the impact of the business interruption as opposed to the cause of the interruption. In addition, the plans for each function or mission-critical business operation to be recovered and the plans for the organization as a whole should incorporate the costs of implementation in terms of personnel and financial resources.

11.7.3 Auditing the Documentation Process

Prior to finalizing the BCP processes and supporting documentation, auditors from the organization should become involved in conducting a thorough operational audit to determine the plan's fitness for protecting mission-critical business operations and to certify the plan's compliance with laws or regulations regarding business continuity.

The planning processes and tasks should be thoroughly tested before an event occurs. A scenario for testing the plans in as realistic a manner as possible should be developed and carried out on a regular basis. Results of the tests should be measured and documented to determine the effectiveness of the tasks implemented.

Finally, executive management should approve the plans and issue a blanket authorization for their funding and execution, provided certain conditions exist. Necessary agreements, letters of intent, and memos of understanding should be signed and put in place so as not to impede the business continuity program efforts.

Developing the initial plans is only the beginning of the process. Ongoing changes in technology, staffing, and business goals and objectives require that plans be regularly reviewed, tested, and updated in order to remain effective.

PROBLEM

1. Class project: Assume that there is a large corporate site with 15,000 workers and 60 major facilities. Develop a strategy for securing corporate business continuity operations against potential natural disasters. Define your own assumptions and parameters.

REFERENCES

1. http://www.iwar.org.uk/infocon/business-continuity-planning.htm
2. http://howe.stevens.edu/Research/ATT/ReportAllSep1004_v3.pdf

12 International Environmental Impact Assessment

Not long after the now historic National Environmental Policy Act (NEPA) was enacted into law in the United States, the international community began to realize that environmental degradation was not only an American problem, but also a global one. Many nations around the world began to appreciate how the environmental impact assessment (EIA) element inherent in the U.S. NEPA process could facilitate sound economic development while providing the methodology to establish plans and policies that would enhance public participation and help their own decision-makers avoid making costly and damaging environmental mistakes.

12.1 NEPA AND THE INTERNATIONAL COMMUNITY

Throughout the 1970s, and continuing into the following decades, many countries moved quickly to adopt their own versions of national environmental policies and assessment procedures.[1] So many, in fact, that today some have postulated that NEPA may have become the most emulated statute in the world. As described in the Introduction, the president's Council on Environmental Quality (CEQ) has written that, in one form or another, NEPA has been used as a pattern or copied by more than 80 countries worldwide.[2] However, the number of nations that have adopted some version of NEPA's policy or EIA process may actually be much higher than this estimate. For example, Canter reports that over 100 countries have instituted some form of EIA measures.[3] A book by John Cronin and Robert F. Kennedy, Jr. cites an even larger number—over 125 nations.[4]

Table 12.1 provides a chronological outline of some representative nations that have followed the American footprint.[5]

12.1.1 STATUS OF EIA LEGISLATION IN DEVELOPING COUNTRIES

It is remarkable how many foreign EIA processes have come to mirror NEPA's original model, which still remains virtually unchanged after more than 35 years. Some countries that did not initially incorporate American EIA principles, such as considering alternatives, encouraging public participation, or investigating cumulative impacts, eventually revised their EIA processes to include such elements. A former senior policy advisor for the president's CEQ, Ray Clark, has written that this is indeed a tribute to the vision that was forged in NEPA by the U.S. Congress.[6]

Credit should also be given to CEQ for its own successful effort, during the Carter administration, to convert its early nonbinding NEPA guidance into formal, legally binding regulations for implementing NEPA's procedural requirements. CEQ's NEPA regulations (40 CFR Parts 1500–1508), published in November 1978, clarified many ambiguities regarding how federal agencies should interpret the statutory language, set out a recommended format for environmental impact statement (EIS) documents, and defined key terms such as effects, mitigation, scope, and significance. These regulations remain in effect today virtually unchanged and have served as the model for EIA implementation procedures in many other countries around the world.

Whereas some developing countries, such as the Philippines, required EIAs to be prepared for major development projects as early as the 1970s, a few of the leading industrialized countries, such as Japan and the Federal Republic of Germany, adopted similar requirements only many years later.[5] By one account, developing Asian countries alone have already performed more than 15,000 EIA studies.[5]

TABLE 12.1
Some Representative Nations (in Chronological Order) That Have Adopted an EIA Process Similar to That of NEPA

Nation	Year EIA Process Was Adopted	Notes
U.S.	1969	National Environmental Policy Act
Canada	1973	Environmental Assessments Review Process (EARP)
Australia	1974	Environmental Protection (Impact of Proposals) Act, 1974
Malaysia	1974	EIA required under Section 34A, Environmental Quality Act, 1974
France	1976	National Environmental Assessment Legislation
Philippines	1978	As per Presidential Decree No. 1586
Japan	1984	Environmental Assessment implemented via a cabinet resolution
U.K.	1985	Town and Country Planning (Assessment of Environmental Effects) Regulations 1988 (S. No. 1199)
Indonesia	1986	AMDAL (EIA) process established by law through Government Regulation No. 29 of 1986
Netherlands	1986	
New Zealand	1986	
Sri Lanka	1988	National Environmental Act No. 47 of 1980 was amended to include an EIA provision
CEC	1988	EU Directive on Environmental Assessment for 12 Member States
Norway	1989	Under the Planning Act of 1989
Germany	1990	National Environmental Assessment Legislation
Thailand	1992	Sections 46 and 47 under National Environmental Quality Act, 1992
Nepal	1993	In the form of National EIA Guidelines issued by National Planning Commission Secretariat
India	1994	Before January 1994, obtaining Environmental Clearance from Central Ministry was only an administrative requirement intended for mega projects but from 1994 the EIA notification was issued

12.1.2 International Organizations

As early as 1974, the Organization of Economic Cooperation and Development (OECD) recommended that its member states adopt EIA processes. The OECD now uses an EIA process similar to that of NEPA in granting aid to developing nations.[7]

Throughout the 1980s, many developing countries continued to establish EIA processes as an essential element of environmental policy and project planning.

12.1.2.1 United Nations

The United Nations Environmental Program (UNEP) developed guidance for performing EIAs and since that time has strongly encouraged member states to establish EIA processes.[8] According to the UNEP, EIA provisions now exist in the environmental legislative framework of 55 developing countries.

12.1.2.2 World Bank

The World Bank eventually ruled in 1989 that an EIA process should normally be prepared for those projects it provides funding. In 1991 the bank published a three-volume EIA sourcebook that provided practical guidance for the preparation of EIA documents for various types of development projects.[9] The bank's most recent procedure clarifies the need for the nature and levels of environmental assessment to be applied to investment projects. However, this procedure does not apply to macroeconomic adjustment lending.[10]

12.1.2.3   European Union

The European Economic Community now requires its members to comply with an environmental process similar in nature to NEPA. Another example involves the North American Free Trade Agreement (NAFTA), which incorporates an EIA process modeled after NEPA for evaluating extraterritorial impacts that cross the borders of the United States, Canada, and Mexico.

As an example, a project referred to as "Nord Stream" involved construction of a 1300-mile natural gas proposal (570 miles in Russia and 750 miles under the Baltic Sea) consisting of two parallel natural gas pipelines with an estimated capacity of around 55 billion cubic meters (2 trillion cubic feet) per year from Russia to Germany.[11]

The pipeline project was subject to the European Union (EU) EIA Directive and Baltic Marine Environment Protection Commission (HELCOM or Helsinki Commission) recommendations.

One of the environmental concerns was that pipeline construction might disturb the seabed and dislodge toxic materials, including chemical munitions deposited in the Baltic Sea during and after World Wars I and II. Other environmental groups raised concerns about potential effects on pipeline construction activities on bird and marine life in the Baltic Sea.

12.1.2.4   NAFTA and Executive Order 13141

Under a 1994 North American agreement on environmental cooperation, the United States, Canada, and Mexico agreed to develop recommendations covering proposed projects "likely to cause significant adverse transboundary effects."

The Commission for Environmental Cooperation (CEC) is a trinational organization created at the time NAFTA was signed to address regional environmental concerns, prevent potential trade and environmental conflicts, and promote enforcement of environmental law. In 1997, the council of the CEC agreed to begin developing a transboundary EIA agreement.

Additionally, Presidential Executive Order 13141, issued during the Clinton administration, directs responsible agencies to assess and consider environmental impacts of trade agreements carefully through a process of ongoing assessment and evaluation.[12] A provision of the executive order designates the U.S. Trade Representative and the Chair of the CEQ to develop procedures for conducting environmental reviews in consultation with appropriate foreign policy, environmental, and economic agencies.

12.2   FOSTERING INTERNATIONAL DEMOCRACY

As a result of NEPA, in a way never possible before, American citizens are now able to participate in and influence proposed federal actions that may affect their lives during the early planning process for such actions. Arguably, no other single U.S. law has contributed so much toward opening up the federal planning and decision-making process to its citizens.

As described in the Introduction, the effect NEPA has had in fostering international democracies and promoting democratic principles is particularly noteworthy. The adoption of international EIA processes similar to NEPA by so many countries has opened up government decision-making processes to tens of millions of citizens around the world. A book by John Cronin and Robert F. Kennedy Jr. has pointed out this fact:[4]

NEPA, which has now been adopted in some form by over 125 countries, has become one of the great promoters of democracy around the world....

12.3   INTRODUCTION TO THE EIA PROCESS

The International Association for Impact Assessment (IAIA) was founded in 1980 and has become the premier international professional association for EIA and strategic environmental assessment (SEA) scholars and practitioners. It has over 2500 members from 125 countries, publishes a professional journal, holds annual meetings and training workshops (English is its primary working language), and has helped to spread and improve the practice of EIA around the world. Its Web site is www.iaia.org. The IAIA, in cooperation with the Institute of Environmental Assessment, U.K., defines EIA generally to mean[13]

> The process of identifying, predicting, evaluating and mitigating the biophysical, social, and other relevant effects of development proposals prior to major decisions being taken and commitments made.

EIA thus aids in promoting improved decision-making and superior project development. It is applied to proposals as diverse as land use development projects, power generation and transmission infrastructure facilities, waste management facilities, and transportation infrastructure projects.

The requirement to implement EIA studies for activities that are likely to significantly affect the environment has been reflected in

* Principle 17 of the Rio Declaration on Environment and Development,[14]
* Article 5 of the Legal Principle for Environmental Protection and Sustainable Development,[15] and
* The 1987 UNEP Goals and Principles of Environmental Impact Assessment.[16]

12.3.1   TYPICAL EIA PROCESS

Although the process involved in making decisions that affect the environment is not and has never been a smooth or uniform one, the lack of prominence devoted to environmental considerations is changing. As depicted in Table 12.2, this process has undergone a series of evolutionary changes to reach its current state of the art.[17]

Paoletto details the principal steps that an EIA process should typically include (Table 12.3).[18]

The IAIA has gone beyond this fundamental guidance by producing a set of 14 basic principles for all EIA processes (Table 12.4). These principles apply to all stages of an EIA or SEA process.

The IAIA has also established 10 operating principles for all EIA processes (Table 12.5).[13] These operating principles describe how the basic principles outlined in Table 12.4 should be applied to the main steps and specific activities of the EIA process (i.e., screening, scoping, identification of impacts, and assessment of alternatives).

For additional information, the reader is referred to the text by Lee and Clive, who provide a review from an economic and environmental context of the processes and practice of EIA in six different parts of the developing world (Chile, Indonesia, Russia, Nepal, Jordan, and Zimbabwe, and also three institutional studies of the World Bank, Asian Development Bank, and OECD). They conclude with a chapter on strengthening future EIA practice from an international perspective.[19]

TABLE 12.2
Evolution of the EIA Process

1970s	Beginning with NEPA, early EIA processes focused primarily on the natural environment.
1980s	Eventually social-economic assessments were accepted as elements of the process.
1990s	Integrated Environmental Management (IEM) promotes principles of transparency, accountability, and informed decision-making throughout the life cycle of a project.
1990s to present	SEA emerged as a proactive tool for addressing the environment in plans and policies.

TABLE 12.3
Principal Steps in a Typical (Project-Level) EIA Process

1. *Impact identification*: The EIA process typically involves a broad analysis of the impacts of project activities with a view to identifying those that are worthy of a detailed study.
2. *Baseline study*: Involves collection of detailed information and data on the condition of the project area prior to the project's implementation.
3. *Impact evaluation*: Is performed whenever possible in quantitative terms and should include the working-out of potential mitigation measures.
4. *Assessment*: Assessing the environmental losses and gains with economic costs and benefits for each analyzed alternative.
5. *Documentation*: A document is prepared detailing the EIA process and conclusions regarding the significance of potential impacts.
6. *Decision-making*: The document is transmitted to the decision-maker, who will either accept one of the project alternatives, request further study, or reject the proposed action altogether.
7. *Post audits*: These are made to determine how close to reality the EIA predictions were.

TABLE 12.4
Fourteen Basic Principles Underlying an EIA Process

Purpose	Support informed decision-making and result in appropriate levels of environmental protection and community well-being
Rigorous	Apply "best-practicable" science, employing methodologies and techniques appropriate to address the problems being investigated
Practical	Result in information and outputs that assist with problem solving and that are both acceptable to proponents and able to be implemented by them
Relevant	Provide sufficient, reliable, and useable information for development planning and decision-making
Cost effective	Achieve the objectives of EIA within the limits of available information, time, resources, and methodologies
Efficient	Impose minimum cost burdens in terms of time and finance on proponents and participants consistent with meeting accepted requirements and objectives of EIA
Focused	Concentrate on significant environmental effects and key issues; that is, the matters that need to be taken into account in making decisions
Adaptive	Adjust to the realities, issues, and circumstances of the proposals under review without compromising the integrity of the process, and be iterative, incorporating lessons learned throughout the proposal's life cycle
Participative	Provide appropriate opportunities to inform and involve the interested and affected public, and their inputs and concerns should be addressed explicitly in the documentation and decision-making
Interdisciplinary	Ensure that the appropriate technique and experts in the relevant bio-physical and socio-economic disciplines are employed, including use of traditional knowledge as relevant
Credible	Should be carried out with professionalism, rigor, fairness, objectivity, impartiality, and balance, and be subject to independent checks and verification
Integrated	Address the inter-relationships among social, economic, and biophysical aspects
Transparent	Should have clear, easily understood requirements for EIA content; ensure public access to information; identify the factors that are to be taken into account in decision-making; and acknowledge limitations and difficulties
Systematic	Result in full consideration of all relevant information on the affected environment, or proposed alternatives and their impacts, and of the measures necessary to monitor and investigate residual effects

TABLE 12.5
Ten Operating Principles Underlying an EIA Process

Screening	Determine whether or not a proposal should be subject to EIA and, if so, at what level of detail
Scoping	Identify the issues and impacts that are likely to be important
Examination of alternatives	Establish the preferred or most environmentally sound and benign option for achieving proposal objectives
Impact analysis	Identify and predict the likely environmental, social, and other related effects of the proposal
Mitigation and impact management	Establish measures necessary to avoid, minimize, or offset predicted adverse impacts and, where appropriate, incorporate these into an environmental management plan or system
Evaluation of significance	Determine relative importance and acceptability of residual impacts (i.e., impacts that cannot be mitigated)
Preparation of EIS or report	Document clearly and impartially the impacts of the proposal, the proposed measures for mitigation, the significance of effects, and the concerns of the interested public and the communities affected by the proposal
Review of the EIS	Determine whether the report meets its terms of reference, provides a satisfactory assessment of the proposal(s), and contains the information required for decision-making
Decision-making	Approve or reject the proposal and establish the terms and conditions for its implementation
Follow-up	Ensure that the terms and conditions of approval are met; monitor the impacts of development and the effectiveness of mitigation measures; strengthen future EIA applications and mitigation measures; and, where required, undertake environmental audit and process evaluation to optimize environmental management.[a]

[a] Whenever monitoring, evaluation, and management plan indicators are designed, it is desirable, whenever feasible, that they also contribute to local, national, and global monitoring of the state of the environment and to sustainable development.

Wood also presents a well-researched comparative analysis of the legal basis for, and the practice of, EIA in six developed countries (U.S., U.K., the Netherlands, Canada, Australia, and New Zealand) and one transitional country (the Republic of South Africa). He also includes chapters comparing report review, decision-making, consultation and participation, benefits and costs of EIA systems, and the evolution of SEA in each of these countries.[20]

12.3.2   EIA Document Content

Paoletto suggests a set of minimal elements that a typical EIA document should contain (Table 12.6).[18]

12.3.3   EIA Problems and Limitations

The EIA process normally involves reviewing both the existing state of the environment and the characteristics of the proposed action and its alternatives. In practice, application of the EIA process tends to be limited to projects (although it has also been infrequently applied to programs and strategic planning). A project in the context of EIA is "an individual development or other scheme as distinct from a suite of schemes or a strategy for development of a particular type or in a particular region."[21] When the EIA process is applied to broader programs or regional planning, it is often done through the related analytical process of SEA (see below).

TABLE 12.6

Basic Elements Addressed in a Typical EIA Document

1. Provide a brief nontechnical summary of the information provided in the following items
2. Indicate any uncertainties and gaps in information
3. Describe the proposal
4. Describe the affected environment
5. Describe practical alternatives (as appropriate)
6. Assess potential environmental impacts of the proposal (proposed action and alternatives), including short-term and long-term effects, and the direct, indirect, and cumulative impacts
7. Indicate whether the environment of any other state/province or areas beyond national jurisdiction are likely to be affected by the proposal
8. Identify and describe practical measures (including their effectiveness) for mitigating significant adverse environmental impacts of the proposed activity and alternatives.

Although the EIA process does not necessarily prevent a project from having an impact on the environment, it frequently manages to minimize the severity of its adverse impacts.[21] Nonetheless, there are some fundamental problems with most processes, even though EIA is widely established as the method by which environmental impacts are studied. For example, some EIA processes only address alternatives to the proposed project in a limited manner; that is, by the project assessment stage, a number of options having potentially different environmental consequences from the chosen one are likely to already have been eliminated.

12.3.4 DISADVANTAGES OF PROJECT-SPECIFIC EIAS

Sadler describes some disadvantages of project-specific EIAs:[22]

- Provide an analysis in a "stand-alone" process, which may be poorly related to the project cycle
- Restricted ability to address cumulative impacts, particularly for large development projects where secondary development could occur
- Restricted opportunities for effective public participation in planning or decision-making processes

Because project-level EIA often precludes consideration of alternative strategies, locations, and designs, at least one EIA practitioner argues that, in effect, "an EIA at the project level is essentially damage control."[23] Application of EIA at a more strategic level can promote a more effective assessment of alternatives and cumulative impacts at an earlier stage in the decision-making process. It can also facilitate consideration of a wider range of actions over a greater area.[21]

12.4 INTRODUCTION TO SEA

The formulation of policy and plans plays an important role in shaping the direction of frameworks and guidelines for development and resource management. A relatively recent innovation involves the concept of SEA. In essence, it extends the application of EIA to the level of policies, plans, and programs (PPPs). A key distinction between EIA and SEA is that SEA can be applied to PPPs at an earlier stage than individual projects. Thus SEA allows for environmental considerations and objectives to be viewed positively as inherent elements of the planning process, rather than just as problems to be mitigated after other development decisions have been made.

The term SEA can be defined as

> A process of anticipating and addressing the potential environmental consequences of proposed initiatives at higher levels of decision-making. It aims at integrating environmental considerations into the earliest phase of policy, plan or program development, on a par with economic and social considerations.[21]

SEAs are being used increasingly at the initial stages of decision-making to assess the consequences of PPPs. Countries such as the United States, Australia, Canada, Denmark, and New Zealand have already applied SEA in developing plans and policies.[24] Some aid agencies in Africa have also started to use them.[25]

Slowly, but ever more frequently, SEA is being recognized as a proactive tool for promoting sustainable development that may also serve to reduce the number of required project-specific EIAs. Planners may use it as a method to assess different ways for accomplishing sustainability policies. Chapter 13 provides an overview of the concept of sustainability.

12.4.1   Goals of SEA

The disadvantages and weaknesses of project-specific EIAs have led to the development and application of SEA. Sadler announces the rationale for SEA as the need to

- facilitate the application of sustainability principles and guidelines, for example, by focusing on the maintenance of a chosen level of environmental quality rather than by minimizing individual impacts;
- focus on project-specific EIA by ensuring that issues of need, proposal generation, and alternatives are addressed at the appropriate policy, plan, or program level; and
- improve the scope and assessment of cumulative impacts, particularly where large projects stimulate secondary development and where many small developments not requiring EIAs may occur.[23]

12.4.1.1   Performance Criteria

As depicted in Table 12.7, the IAIA has established performance criteria that SEA analyses should meet.[13]

12.4.2   Relationship between SEA and EIA

The concepts of EIA and SEA differ fundamentally in both the scope and nature of their approaches. The difference between the two processes is evident in the scale of their frameworks. The scope of an SEA tends to be much broader, both temporally and geographically, than for project-specific EIAs and allows consideration of alternatives and a better programmatic view of the "bigger picture."

In an ideal world, project-specific EIAs should be prepared once a policy has been established via an SEA. The EIA provides information about the likely environmental impacts of an individual project and is useful in implementing mitigation measures.

For example, if a government agency decides to develop a national wind power program, EIAs can be used to minimize the environmental damage from building specific power stations, but cannot practically address the more fundamental questions regarding design of the national wind power program. In contrast, an SEA could effectively lay out the overall policy and investigate the programmatic impacts associated with such a policy.

TABLE 12.7
SEA Performance Criteria

Integrated	• Ensure adequate environmental assessment of all strategic decisions relevant for the achievement of sustainable development
	• Address interrelationships of biophysical, social, and economic aspects
	• Tier to policies in relevant sectors and (transboundary) regions and, where appropriate, to project EIA and decision-making
Sustainability-led	• Facilitate identification of development options and alternative proposals that are more sustainable
Focused	• Provide sufficient, reliable, and usable information for development planning and decision-making
	• Concentrate on key issues of sustainable development
	• Customize analysis to the characteristics of the decision-making process
	• Cost- and time-effective
Accountable	• Is the responsibility of the leading agencies for the strategic decision to be taken?
	• Is carried out with professionalism, rigor, fairness, impartiality, and balance?
	• Is subject to independent checks and verification?
	• Documents and justifies how sustainability issues were taken into account in decision-making
Participative	• Informs and involves interested and affected public and government bodies throughout the decision-making process
	• Addresses their inputs and concerns in documentation and decision-making
	• Provides clear, easily understood information requirements and ensures sufficient access to all relevant information
Iterative	• Ensures availability of the assessment results early enough to influence the decision-making process and inspire future planning
	• Provides sufficient information on the actual impacts of implementing a strategic decision, to judge whether this decision should be amended and to provide a basis for future decisions

12.4.2.1 Comparison of SEA and EIA

McDonald and Brown have written that[26]

EIA tends to focus on the mitigation of impacts of proposed activities rather than determining their justification and siting.

Perhaps the most significant way in which SEA differs from EIA is that SEA is a proactive tool for environmental management, whereas EIA tends to be used reactively to assess specific development proposals. Some fundamental differences between SEA and EIAs are summarized in Table 12.8.

SEA and EIA also tend to be applied at different stages of plans and policies and to different levels of decision-making. Such a tiered approach is employed in New Zealand, the EU, and the United States.[17] Under a tiered approach, SEA is used to formulate strategies and policies in a proactive way. These policies and strategies create a framework against which specific development proposals and projects can then be assessed using EIA.

Swedish planners have used SEA to ensure that plans and environmental goals encourage sustainable development.[27]

TABLE 12.8
Comparison of SEA with EIA

SEA	EIA
Proactive and informs stakeholders about development proposals	Reactive; applied to the development of site-specific proposals
Evaluates impacts on development needs and opportunities	Evaluates impacts of proposed development on the environment
Considers areas, regions, or sectors of development	Considers specific projects
Tends to be a continuing process over a life cycle that is aimed at providing information at the right time	Has a defined beginning and end
Evaluates cumulative impacts and identifies issues for sustainable development	Focuses on direct impacts and benefits
Focuses on maintaining a chosen level of environmental quality	Focuses on mitigating impacts
Has a broader perspective and a correspondingly lower level of detail, providing an overall vision	Has a narrow perspective with a higher level of detail
Creates a basis against which impacts and benefits can be measured	Focuses on project-specific impacts

PROBLEMS

1. Does the World Bank require a process similar to NEPA for funding international development projects?
2. Why has it been said that NEPA promotes international democracy?
3. Does the typical EIA process described in Table 12.3 parallel that of NEPA? Explain.
4. Which characteristics outlined in Table 12.4 are inherent in NEPA?
5. What is the closest equivalent under NEPA to an SEA?

REFERENCES

1. Biswas A. K. and Agarwala S. B. C. (eds.) (1992). *Environmental Impact Assessment for Developing Countries*, Butterworth-Heinemann, Oxford, U.K.
2. House of Representatives, Committee on Resources, 105th Congress, Problems and Issues with the National Environmental Policy Act of 1969. *Oversight Hearing before House Committee on Resources*, One Hundred Fifth Congress, Second Session, March 18, 1998. Serial no. 105–102, U.S. Government Printing Office, Washington, D.C.
3. Canter L. W. (1996). *Environmental Impact Assessment*, 2nd ed., McGraw-Hill, New York.
4. Cronin J. and Kennedy R. F. Jr. (1997). *The Riverkeepers*, Scribner, New York, pp. 37 and 175.
5. Prasad M. and Biswas A. K. (1999). *Conducting Environmental Impact Assessment in Developing Countries*, United Nations University Press, Tokyo, p. 375.
6. Clark R. and Canter L. (eds.) (1997). *Environmental Policy and NEPA: Past, Present, and Future*, St. Lucie Press, Boca Raton, FL, chapter 7.
7. OECD (1992). Good Practices for Environmental Impact Assessment of Developing Projects. Development Assistance Committee, Paris.
8. UNEP (1988). Environmental Impact Assessment: Basic Procedures for Developing Countries, Regional Office for Asia and the Pacific, Bangkok.
9. World Bank (1991). *Environmental Impact Assessment Sourcebook* (vols. 1–3), World Bank, Washington, D.C.
10. Operational Policy (OP)/Bank Procedures (BP) 4.01, January 1999.
11. DOE, *NEPA Lessons Learned*, March 1, 2007, issue no. 50.
12. Executive Order 1314, Environmental Review of Trade Agreements, 64 FR 63167, November 18, 1999.

13. International Association for Impact Assessment and Institute of Environmental Assessment, UK, Principles of Environmental Impact Assessment Best Practice, January 1999, http://iaia.org/Members/ Publications/Guidelines_Principles/Principles%20of%20IA.pdf.

14. United Nations, A/CONF.151/26 (vol. I), 12 August 1992, Report of the United Nations Conference on Environment and Development (Rio de Janeiro, 3–14 June 1992).

15. Adopted by the Experts Group on Environmental Law of the World Commission on Environment and Development.

16. UNEP, Working Group on Environmental Law, adopted by the UNEP Governing Council at its 14th session.

17. CSIR, CSIR Report, Strategic Environmental Assessment (SEA): A Primer, ENV/S-RR 96001, September 2, 1996.

18. Paoletto G. Lecture Notes, Environmental Impact Assessments (EIA), http://www.gdrc.org/uem/eia/ lecture-notes.html.

19. Norman L. and George C. (eds.) (2000). *Environmental Assessment in Developing and Transitional Countries*, John Wiley & Sons, Chichester, U.K.

20. Christopher W. (2003). *Environmental Impact Assessment: A Comparative Review*, 2nd ed., Prentice Hall (Pearson Education), Harlow, U.K.

21. Therivel R., Wilson E., Thompson S., Heaney D., and Pritchard D. (1992). *Strategic Environmental Assessment*, Earthscan, London.

22. Sadler B. (1995). Towards the Improved Effectiveness of Environmental Assessment. Executive Summary of Interim Report Prepared for IAIA 95. Durban, South Africa.

23. Robert B. S. (former member of the President's Council on Environmental Quality) (2007), personal communication.

24. Sadler B. and Verheem R. (1996). Strategic Environmental Assessment: Status, Challenges and Future Directions, the Netherlands: Ministry of Housing, Spatial Planning and the Environment.

25. Goodland R., Mercier J. -R., and Muntemba S. (eds.) (1996). Environmental Assessment (EA) in Africa: A World Bank Commitment. Proceedings of the Durban (South Africa) Workshop, June 25, The World Bank, Washington, D.C.

26. McDonald G. T. and Brown L. (1995). Going beyond environmental impact assessment: Environmental input to planning and design. *Environmental Impact Assessment Review* 15:483–495.

27. Eggiman B. (2000). Fysisk planering med strategisk miljöbedömning (SMB) för hållbarhet. En teoretisk diskussion och förslag till SMB-process med Stockholms stad sommodell. Karlskrona and Stockholm, Swedish Board of Housing, Building and Planning and Swedish Environmental Protection Agency.

13 Environmental Policy, Decision-Making, and Economics

This chapter provides an introduction to some fundamental concepts of environmental policy. It presents some elemental, albeit stand-alone, principles typically encountered in the context of environmental policy. For this reason, these topics are presented as separate autonomous concepts, and there has been no attempt to unify them into a single integrated theme. It is presumed that this introduction will allow the reader to appreciate some of the rudimentary issues commonly encountered and debated in the study of environmental policy.

We begin this chapter with an investigation of environmental sustainability and whether there is a need to establish policies to slow the growth in human population.

13.1 EASTER ISLAND AND THE TRAGEDY OF THE COMMONS

Easter Island, which lies on a dusty speck of rock some 2000 miles off the west coast of South America, is one of the most isolated yet still inhabited places on Earth. The first European to discover it was Admiral Roggeveen, who landed on this island in 1722. When he and his crew began to explore the island, they discovered a primitive society of about 3000 destitute individuals living in caves and reed huts. Instead of a lush, tropical paradise, Roggeveen found a nearly treeless island virtually denuded of vegetation. But even more perplexing were the 600 mysterious stone statues, each averaging 20 ft. in height, which sprawled across the landscape. The statues were a testament to the island's once thriving and relatively advanced society where human ingenuity had enabled the inhabitants to prosper for hundreds of years. But when Roggeveen arrived and explored the island, it became clear that at some point in time the once harmonious relationship between the islanders and their natural environment had been seriously disrupted.

Today, Easter Island is one of the world's most famous archaeological sites. To many archaeologists, the available evidence suggests that a small group of Polynesians, lost at sea, settled on the island perhaps as early as the fourth or fifth century A.D. Although the distance they might have sailed is breathtaking, these original settlers probably arrived in simple canoes and may have numbered less than 50 individuals. When they first arrived, they would have found a pristine natural environment endowed with lush forests dominated by palm trees. Despite starting out with a limited natural resource base, the inhabitants increased their numbers and eventually began to flourish. Related families formed clans, each of which developed its own center of religious and cultural activities involving elaborate rituals that included the construction of huge stone statues. Although their real purpose remains a mystery, one thing has become clear: the statues would provide a chilling testament to the islanders' downfall.

Immense amounts of human labor and environmental resources must have been needed to construct the statues. Massive stones, often weighing as much as 10 tons, were transported long distances to selected sites across the island. The islanders' engineering solution for solving this transport problem provides an important clue into the reasons for the demise of their society. Since they lacked beasts of burden, the islanders performed the heavy work themselves by dragging the statues across the island using tree trunks as rollers. Competition among the opposing clans for the available timber intensified as, in their attempts to secure greater prestige and status, they erected an increasing number of statues.

At its peak in the sixteenth century, the island's population exceeded 7000 inhabitants. Unfortunately, by this time almost the entire inventory of trees on the island had been cut down for fuel, housing, and to provide rollers to transport the stone monuments. The fragile environment began to break down. But the islanders were unable to escape. Without wood to build new canoes they became prisoners, trapped on the land they had ruined, and completely isolated from the rest of the world.

Today, archaeological teams continue research and excavation work on the island attempting to determine the causes, links, and time line of the human and environmental collapse. Despite the fact that some questions remain unanswered, most (but not all) archaeologists agree that evidence indicates that when the island had been completely deforested, chaos ensued. When wood was no longer available to build their homes, many inhabitants were forced to live in caves. Fishing would have become increasingly difficult because the supply of nets, previously manufactured from tree bark and vines, rapidly dwindled. In turn, as deforestation led to soil erosion and the subsequent leaching of vital nutrients, crop yields plummeted. At this point, the society began sliding into a steep decline. The slavery and poverty that seem to have followed were apparently exacerbated by nearly continuous warfare caused by conflicts over diminishing resources. Many of the magnificent stone statues were toppled and desecrated. As food supplies dwindled, the human population even appears to have turned to cannibalism. By the eighteenth century, the population had dropped to between one-quarter and one-tenth of its peak size.

For as long as a thousand years, the islanders' way of life enabled them not only to survive but also to flourish. Their utopia eventually collapsed because they failed to realize that their very existence depended on the limited natural resources of a small island.

What does the story of Easter Island teach the modern world about mounting environmental problems such as dwindling petroleum and water supplies, or global warming? The history of Easter Island is a vivid reminder of the consequences that human populations may face when vital environmental resources are irreversibly damaged. The fate of Easter Island's inhabitants may have perilous implications for our present global society. Many experts have argued that, as with Easter Island, the human population of the Earth is confined to an island, having no practical means of escape in the event of a catastrophe. The real lesson of Easter Island may be that rational societies can commit environmental suicide.

13.1.1 TRAGEDY OF THE COMMONS

The limited availability of most environmental resources allows ecologists to estimate a population parameter called *carrying capacity*. This term is frequently defined as the maximum population of a given species that can be supported indefinitely by its environment in a constrained habitat without permanently impairing the productivity of that habitat.

The term "commons" evolved from an old English custom. Until the era of the Enclosure Acts when a long series of parliamentary acts enabled powerful landowners to fence off their properties, turning them into privately held estates, many English villages included a "commons" or public area of land that could be freely used by any community member to graze their domestic livestock.

In the mid-nineteenth century, William Lloyd was the first to document what is now referred to as the Tragedy of the Commons.[1] In the 1960s, Garrett Hardin applied this concept to global environmental policy.[2] As he explained in an essay of the same title in 1968, when a village commons is managed judiciously, all users can benefit from it. But, unchecked, this prosperity inevitably leads to a dilemma in which the desire to maximize individual wealth results in overgrazing, eventually leading to the demise of the entire commons. This principle can, of course, be applied to many limited environmental resources far beyond that of simply grazing on a village commons.

The following example illustrates Lloyd's original principle. Consider a village commons on which 10 villagers graze their cattle. This village commons is a source of increased prosperity to anyone who is able to utilize its resources. Assume that the maximum carrying capacity (sustainability) of this commons is sufficient to support 100 cattle. As long as the total number of cattle does not exceed 100, each additional cow added to the commons increases an individual farmer's wealth without causing harm to the wealth of others. Over time, prosperity increases until each farmer has 10 cows grazing on the commons, each producing 1 unit of utility, for a total of 100 units. The carrying capacity or sustainability of the commons has been reached. From a macro perspective, it is no longer in the interest of the village to increase the number of cattle. In fact, the addition of each additional cow will actually reduce the total number of units that can be produced from the commons (the grass yield will begin to decrease, and the underlying soil will become increasingly compacted or be eroded away).

Now, consider the following scenario from a micro or short-term perspective. It is still in each individual farmer's short-term interest to add additional cows. Farmer Jones, for example, sees a short-term gain from adding an additional cow but fails to appreciate the long-term adverse implications to the community as a whole. From his individual perspective, Jones reasons: "I stand to gain by adding one more cow beyond the carrying capacity, because I will gain one more unit of wealth, yet pay only a small fraction of the total negative consequences to the commons."

For instance, by adding one more cow, farmer Jones stands to gain approximately an entire unit while diminishing the total utility of the commons by 1 unit. The total land utility is now 99 units; yet he gained nearly a full unit of value while he shares the one negative unit with the other nine farmers. Essentially, he ends up with approximately 10.9 units, while his neighbors' shares have each been reduced from 10 to approximately 9.9 units.

Of course, this slight imbalance does not end there since this same logic applies to each and every member of the commons. Soon, all the other farmers add more cows. Because its use is uncontrolled and unmanaged, the total productivity begins to collapse. As Hardin writes, "Freedom in a commons brings ruin to all." It is worth pointing out that there are alternative, though less well-known, analyses. One such analysis assumes that the carrying capacity (for cattle) is determined by one resource (grass) that remains static, while, in reality, the resources are multiple and their collective carrying capacity will vary over time in response to other variables such as weather, competition from or predation by other uncontrolled species, etc.

Hardin's principle can be applied to our modern world as well. In less than 300 years, we have moved from creating environmental problems that once wrought disaster in isolated villages to problems that are now wreaking environmental havoc on a global scale. So we are increasingly facing a pressing dilemma: How can we effectively assess limits on growth, let alone find common ground for cooperation that safeguards the global commons? One approach is to adopt national and international policies that are sustainable and enforceable.

13.2   LIMITS TO GROWTH, GAIA, AND SUSTAINABILITY

This section begins with a brief description of the Gaia hypothesis, a concept that has been used by some critics of environmental stewardship in arguing that there are no definitive limits to future growth.

13.2.1   MALTHUS, SIMON, AND LIMITS TO GROWTH

Are there really natural and physical limits to the growth of the human population? In 1980 economist Julian Simon and ecologist Paul Ehrlich, who had written a popular but controversial nontechnical book titled *The Population Bomb*, made a wager over what the price of certain metals would be by the end of that decade. Ehrlich selected a group of five metals—copper, chrome, tin, nickel, and tungsten—whose price he believed would rise significantly as their use by the growing populations led to increasing scarcity and depletion. Simon, who was willing to wager over the fall in the price of these metals, won the bet when there was a drop in the price of all five metals. Nonetheless, Ehrlich's supporters charged that much of the price drop resulted from an oil spike that had driven prices up in 1980 which was then followed by a recession that helped drive prices down in 1990;

moreover, they argued that the prices of these metals were not really critical indicators of environmental limitations.

In 1995, Simon issued Ehrlich a challenge to make a second bet on the prices of metals. Ehrlich refused, proposing instead that they bet on a metric for human welfare. Like the long-running debate over the limits to growth, the two failed to reach a consensus on their wager before Simon's death in 1997.

More recently, the debate has gained renewed interest as the exploding industrial growth in the world's two most populous nations, China and India, is placing a new strain on natural resources, including growing shortages (and resulting higher prices) of certain metals and fossil fuels, including oil and natural gas. The demand for wood is also escalating. China, for example, is presently arranging with Indonesia to clear-cut vital rainforests for wood in return for planting vast palm oil plantations. At the same time, the tropical rainforests in central African countries and in the Amazon basin (the largest of such forests on Earth) continue to be razed to make way for cattle ranching, for cash crops such as soybeans, and to grow corn and wheat for ethanol fuel production. In another sobering comparison to Easter Island, the vast deforestation of Haiti has added to the inhabitants' mounting poverty as arable soils are washed away, and mud slides bury entire villages during storms.

13.2.1.1 The Malthusian Growth Model

Economists and scientists have argued over limitations on growth ever since Thomas Robert Malthus (1766–1834) first popularized his hypothesis in the eighteenth century. Malthus has been referred to as the world's first professor of political economics. Malthus popularized his thesis on the limits to growth when his work *An Essay on the Principle of Population* was published in 1798. He based his principle on a simple mathematical concept after concluding from his studies that when left unchecked, population increases at an exponential rate (i.e., 1,2,4,8,16, …) while the food supply grows at a linear rate (i.e., 1,2,3,4, …). He attempted to prove from these inferences that nothing can indefinitely sustain exponential growth and thus, if population growth was not limited, the exponential increase in population would eventually outstrip the ability of society to feed itself.[3]

Malthus noted that his theory was frequently misrepresented; he took pains to point out that his hypothesis did not necessarily predict future catastrophe if people were willing to take action to prevent it. He pointed out[4]

> … this constantly subsisting cause of periodical misery has existed ever since we have had any histories of mankind, does exist at present, and will for ever continue to exist, unless some decided change takes place in the physical constitution of our nature.

Malthus held that his principle of population could provide a sound basis for predicting our future. For this reason, he believed it was critical that steps be taken to control population growth.

Although highly controversial, the Malthusian growth model has profoundly influenced the fields of socioeconomics and environmentalism. Prior to Malthus, many economists considered a high-fertility rate to be an economic plus since it increased the number of workers available to contribute to the growth of the economy. Following Malthus, many economists began to view fertility from a different perspective, arguing that while a large number of people might increase a nation's gross output, sheer numbers also tended to reduce the per capita output.

Malthus's concept continues to be the subject of lively debates to this day. For example, based partly on Malthusian concepts, Paul Ehrlich predicted in the late 1960s in his previously cited book, *The Population Bomb*, that hundreds of millions of people would die of starvation and disease from an overpopulation crisis that he anticipated would occur in the 1970s and that life expectancy in the United States would dwindle to only 42 years by the 1980s. Consistent with Malthus' premise, in 1972 the Club of Rome published equally dire predictions in its best seller, *The Limits to Growth*.[5]

Criticisms. Today, the Malthusian growth model of population growth versus food supply is nearly universally rejected since it can be demonstrated that for the last two centuries, largely due to increasing technological and scientific expertise not known in his day, food supply has generally kept pace with population growth. For example, at least in developed nations, as population has increased, the price of resources and foods relative to wages has generally declined.

Malthus's model has been proved incorrect because the analysis was premised on two partially or completely flawed assumptions:

1. It has been widely demonstrated that population growth is almost never exponential over the long term but is instead influenced by many factors that are inconsistent with such a simple mathematical model. Modern demographic analyses suggest that population growth rates tend to flatten and then invert as a function of economic prosperity. Malthus lived at a time when England was undergoing a geometric growth and it was sometime later before birth rates eventually began to flatten out. Moreover, Malthus had not studied large populations in Asia that had existed over multiple millennia and experienced such flattening of birth rates.
2. Growth of food production has never been restricted to the simple processes Malthus described. Modern studies reveal that the intensity of agricultural production rises in response to population increases and market demands. Production has also expanded greatly because of technological advances. However, in many parts of the world, evidence is accumulating to suggest that this may no longer continue to be the case.

Malthus clearly underestimated the power of technology and human ingenuity to increase the means of human subsistence. Modern human population growth, however, has been based on finite resources such as petroleum, potable water, and agricultural land, and reliance on these scarce natural resources may yet prove to be unsustainable. Despite continued advances, crop production in some countries can no longer keep pace with population growth. Increasing drought, protracted heat waves, intensified soil erosion, and loss of the remaining good arable land are all contributing to the problem. Few of the farmers in the world's poorest countries can afford the fertilizers needed to rejuvenate their soils, and considerable debate surrounds the subject of whether genetically modified crops will be able to contribute in the longer term to continued agricultural growth.

13.2.1.2   Julian Simon

As a professor of economics, Julian L. Simon is remembered for two things. He was the first one to suggest that airlines should provide rewards for travelers to give up their seats on overbooked flights, also known as "bumping." But this is a mere footnote in history. As described earlier, his real contribution was as a leading economic optimist and one of the harshest critics of the predictions of environmental doom by Ehrlich and others. His book published in 1984, *The Resourceful Earth*, co-authored by Herman Kahn, is a criticism of the conventional and theoretical limitations on population and economic growth.

Simon correctly noted that few of Ehrlich's 1968 predictions about rising prices and famines had actually occurred. He expressed the belief that humans "are not just more mouths to feed, but possess productive and inventive minds that help find creative solutions to man's problems, thus leaving us better off over the long run." In other words, the more the population increases, the greater is the chance that another Einstein will be born who will develop new ways to improve and replenish the Earth's dwindling resources. In support of his thesis, Simon cited statistics showing that some countries with rapid population growth, such as Singapore and South Korea, foster more economic prosperity than other nations.

Environmentalists and social scientists are divided over the issue of environmental degradation and the limits that nature may place on development and population growth. Detractors have presented Simon as an arrogant optimist and argued that social scientists, in particular, have failed to place sufficient emphasis on the intrinsic limitations of technology and nature. At the same time, many scientists continue to warn that limits exist on the number of people the planet can support.

However, there is a possibility that this ongoing debate may soon become muted as birth rates have been plummeting in many developed and in some developing countries. This decline has led to the projections that global human population growth might level off at somewhere around 10 billion people by the middle of this century. Even if this proves to be the case, it should still be borne in mind that with a present world population of over 6.5 billion people, much of the world is probably already overpopulated, and it is frequently in the poorest countries with the weakest economic development and corrupt governments where much of this growth continues unabated.

13.2.2   THE GAIA HYPOTHESIS

The Gaia hypothesis has been invoked by some environmental critics who charge that the issue of environmental quality is either overstated or not threatened at all. In the 1970s British scientist James Lovelock first proposed the Gaia hypothesis. Lovelock named it after the Greek goddess Gaia that drew the living world forth from Chaos, and he hypothesized that the Earth's life system functioned as if it were a single self-regulating living system or organism.

Lovelock's hypothesis ranges across a spectrum of two widely opposing concepts: the virtually undeniable (weak Gaia) to the much more sweeping (strong Gaia) hypothesis. Under the weak hypothesis lies the undeniable statement that life has dramatically altered planetary conditions. In contrast, the strong hypothesis goes much further in arguing that the Earth's biosphere effectively acts as a self-organizing system that works in a way to keep its systems in an approximate state of equilibrium conducive to life (however, geological history shows that the exact characteristics of this equilibrium have intermittently undergone rapid changes, which are believed to have caused extinctions). On the extreme side of the spectrum, some proponents hypothesize that the entire Earth is a single unified organism; under this strong hypothesis, the Earth's biosphere is considered to be consciously manipulating global processes to create conditions conducive to life. Most mainstream scientists contend that there is no evidence at all to support such a far-reaching or extreme view of the hypothesis.

Many authorities maintain that numerous global processes appear to be maintained by homeostatic mechanisms consistent with Gaia. For instance, a rise in the levels of atmospheric carbon dioxide enhances plant growth because the increased carbon dioxide concentration increases the ability of organisms to extract this greenhouse gas from the atmosphere—restabilizing the atmosphere; however, this process might also be overwhelmed leading to a chaotic response. Other examples are as follows:

- The atmospheric composition has remained relatively constant (79% nitrogen, 20.7% oxygen, and 0.03% carbon dioxide) over hundreds of millions of years (although experts argue that these concentrations have actually varied considerably over that time). Lovelock maintains that this composition should be unstable, and its stability can only have been accounted for by the actions and effects of biological organisms.
- Lovelock has also observed that since the origin of life, the sun's energy output has increased by 25–30%, yet the Earth's surface temperature has remained relatively constant over time. He believes that life and geological processes have maintained a reasonably stable climate conducive to life.
- A final example involves the salinity of the world's oceans, which has been relatively constant over a long period of geological history. This has posed a long-standing mystery, as rivers (carrying salts) should long ago have raised the ocean salinity to a much higher level. Salinity stability is vital as most life forms cannot tolerate values much higher than 5%. Again geological and biological forces must be working in unison to stabilize critical conditions in such a way as to maintain life.

Lovelock's hypothesis sparked almost instant controversy, not least of which was from the famous evolutionary biologist Richard Dawkins. In his work, *The Blind Watchmaker*, Dawkins argued that organisms cannot act in concert, as this requires forward planning. He rejected the possibility that feedback loops could stabilize global systems. Another opponent, Ford Doolittle, argued in a scientific paper in 1981 that there was nothing in the genome of organisms that could explain the feedback mechanisms required by Gaia, and therefore the hypothesis was unscientific.[6]

Despite this criticism, many supporters maintain that there is much to be said in favor of Lovelock's hypothesis. Echoing Lovelock's observations, Lewis Thomas, author of *Lives of a Cell*, writes:[7]

> I have been trying to think of the earth as a kind of organism, but it is no go. I cannot think of it this way. It is too big, too complex, with too many working parts lacking visible connections. The other night, driving through a hilly, wooded part of southern New England, I wondered about this. If not like an organism, what is it like, what is it most like? Then, satisfactorily for that moment, it came to me: it is most like a single cell.

The noted astronomer Carl Sagan is said to have joined the debate by even suggesting that from an astronomical perspective, space travel and planetary probes appear to provide a perspective in which the Earth, as a living organism, may be on the verge of seeding other planetary systems.[8]

Many, perhaps most, Earth scientists view the factors that stabilize the biosphere as an undirected aspect of the system; the combined actions resulting from competition among species, for example, tend to counterbalance the environmental perturbations. However, the opponents of Gaia argue that there are many examples where the effects of life have dramatically changed or even destabilized the biosphere (i.e., conversion of the Earth's atmosphere from a reducing environment to an oxidizing one); but proponents counter that in the long run, such changes promote an environment even more suitable to life.

Such intense scientific debate resulted in an international Gaia conference in 1988. A second international conference was held in 2000.

Throughout his career, Lovelock has generally been an adamant environmentalist. Yet, in his recent book, *The Revenge of Gaia*, the potential effects of global warming have led to his strong support of nuclear power as the only practical technology that can both meet the world's increasing energy demands while reducing climatic damage. Lovelock now believes that the global organism is sick, and drastic action must be taken.

Lovelock's pessimism about how climate change will affect the global community stems from his assessment of how Earth and life systems will respond in reestablishing the ecological balance. Earth will adjust to human-induced stresses, but it will do so with revenge. As a control system, Lovelock believes that counterbalancing forces that have generally worked in our favor are now beginning to turn against us. The effects of human activity, such as the rise of global temperature, will be harmful, perhaps with disastrous consequences.

A number of noted scientists suspect the existence of a threshold set by temperature and carbon dioxide levels, past which the Earth's atmosphere will be irreparably harmed. Activities such as increasing atmospheric carbon dioxide levels, destroying wetlands and forests, and overfarming do not simply produce linear increases in temperature; they can produce nonlinear effects that amplify the increase in temperature.

Lovelock believes that we are now approaching one of these tipping points; our future is like that of the passengers on a small raft quietly drifting toward Niagara Falls. Like a raft going over the falls, the global climate may abruptly flip into an entirely new equilibrium state that might force us to migrate to the poles, leaving the tropics uninhabitable. As Lovelock views it, Gaia has no reason to favor the human species over any other life-form. If global warming results in massive economic disruption or jeopardizes humanity, it will also presumably result in a reduction in the principal cause for global warming (i.e., human population). Just how Gaia would then react and "reset the thermostat" to maintain a new global ecosystem is problematic.

Whether there is any validity to the strong hypothesis remains to be proven. Notwithstanding, the hypothesis arguably was one of the first serious attempts to show that the Earth is not merely a compilation of unrelated biological processes and chemical reactions that work independently of one another; instead, many processes appear to work in unison to maintain stable environmental conditions.

13.2.3 SUSTAINABILITY

While there is, as yet, no universally accepted definition or concept of sustainability, various definitions have been proposed.* Most involve adopting a collection of economic, social, and environmental goals that are consistent with each other and mutually attainable.

13.2.3.1 Definitions

Sustainable development has become an accepted goal of many environmental policies, especially since 1987 when the World Commission on Environment and Development (WCED) released the Brundtland Report, *Our Common Future*. The commission defined sustainable development as being

> … development which meets the needs of the present without compromising the ability of future generations to achieve their needs and aspirations.[3]

However, the scope of sustainable development can be viewed more comprehensively than by simply considering natural resources. As a comprehensive concept, sustainability can be defined as

> … development that delivers basic environmental, social and economic services to all without threatening the viability of the natural, built and social systems upon which these services depend.[4]

The concept of sustainable development or sustainability means that the consumptive use of renewable resources does not exceed the regenerative capacity of the environment.[9] Social progress, environmental protection and preservation, conservation of resources, and economic maintenance are all the elements of sustainable development. Quality of life concerns, biological and cultural diversity considerations, and conservation and remedial compensations, not to mention philosophical questions for humanity, are also within these constraints. The welfare of future generations also fits into the sustainable development equation.

Sustainable yield can be thought of as the optimum (see below) level of production (e.g., timber, fisheries, and water) of a renewable resource that can be maintained indefinitely. In economic terms, it represents the maximum long-term level of income that can be derived from the use of a resource without causing eventual degradation or depletion of that resource.

It should be noted, however, that many ecologists largely reject the concept of "maximum sustained yield" (MSY) promoted in the last century by commercial forestry and agricultural and fishing interests because this concept assumes a long-term stability in the underlying ecosystems that usually cannot be demonstrated to exist; that is, natural systems are usually more complex, more variable, and less stable in response to disturbance, than the requirement of such a production strategy.

13.2.3.2 Agenda 21

The concept of sustainability gathered momentum to become the dynamic baseline for Agenda 21, the 40-chapter document that details the goals and programs resulting from the United Nations Conference on Environment and Development (informally known as the Earth Summit), held in

---

* Mr. Don Sayre contributed much of the material presented in the following section on sustainability.

TABLE 13.1

The Twenty-Seven Principles Contained in Agenda 21

1. Human beings are entitled to a healthy and productive life in harmony with nature.
2. States have the right to exploit their own resources but without damage to others.
3. The right to development must meet the needs of present and future generations.
4. Environmental protection is an integral part of the development process.
5. People must eradicate poverty to decrease disparities in standards of living.
6. Needs of the least developed and most environmentally vulnerable state must be a priority.
7. States must cooperate to conserve, protect, and restore Earth's ecosystem.
8. States are to eliminate unsustainable patterns of production and consumption.
9. States are to improve scientific understanding to strengthen capacity building.
10. Environmental issues are best handled with participation by all concerned.
11. States must enact effective environmental legislation.
12. States are to promote supportive, open economics for growth and development.
13. States must have laws to protect victims of pollution and environmental damage.
14. States are to cooperate to discourage and prevent severe environmental degradation.
15. The precautionary approach must be applied to threats involving serious damage.
16. Authorities are to promote "polluter pays" with due regard to the public interest.
17. Impact assessment must be undertaken for likely significant adverse impacts.
18. States must notify others of disasters or emergencies likely to harm others.
19. States must notify others of transboundary environmental effects.
20. Full participation of women is essential to achieve sustainable development.
21. World youth partnership is essential to achieve sustainable development.
22. Indigenous people and communities have a vital role in sustainable development.
23. The environment and resources of people under oppression are to be protected.
24. Warfare is inherently destructive to sustainable development.
25. Peace, development, and environmental protection are interdependent and indivisible.
26. States must resolve environmental disputes peacefully.
27. States and people must partner for sustainable development.

Rio de Janeiro, Brazil, in June 1992. The Rio conference was the follow-up to the U.N. Conference on the Human Environment, the first global conference ever convened on the environment, held in Stockholm, Sweden, in 1972. Today, the United Nations remains committed to the global goal of sustainable development, a mutual challenge to societies, economies, and environments around the world.

Agenda 21 provides 27 principles for implementation of its strategy (see Table 13.1). Nearly half of these sustainable development principles focus on actions undertaken by national governments. The remainder focuses on actions undertaken by individuals and organizations.[10]

In 2000, with Agenda 21 in mind, the United Nations identified eight millennium development goals. In 2002 the U.N. World Summit on Sustainable Development held in Johannesburg, South Africa, benchmarked the world against the goals and agenda, generating an improved implementation plan.

ICC on Sustainable Development
... sustainable development means adopting business strategies and activities that meet the needs of the enterprise and its stakeholders today while protecting, sustaining, and enhancing the human and natural resources that will be needed in the future.

Sustainable development is also at the core of ISO 14001, an increasingly popular international standard for environmental management systems (see Chapter 2). The ISO 14001 standard embraces Agenda 21 from the Earth Summit along with strategies of the International Chamber of Commerce (ICC) business charter for sustainable development.

13.2.3.3 Sustainable Development, NEPA, and EPA

While National Environmental Policy Act (NEPA) predates the modern concept of sustainable development, the rudimentary concept is nevertheless embedded in the Act. Consider the following two excerpts from NEPA:

> ... *productive and enjoyable harmony* between man and his environment; to promote efforts which will *prevent or eliminate damage* to the environment and biosphere and *stimulate the health and welfare* of man ... .[11] (emphasis added).
>
> ... it is the continuing policy of the Federal Government to use all practicable means and measures, including financial and technical assistance, in a manner calculated to foster and promote the general welfare, to create and *maintain conditions* under which man and nature can exist in *productive harmony*, and fulfill the social, economic, and other requirements of present and *future generations* of Americans[12] (emphasis added).

It should be noted that from a policy perspective, the federal courts have ruled that neither of these sections of NEPA contains provisions that are enforceable by law.

The U.S. Environmental Protection Agency (EPA) has taken its own path toward describing sustainable development. The EPA's Center for Sustainability promotes linking environmental, economic, and social goals to enhance quality of life and encourage livable communities that can someday realize a "New American Dream." It advises protecting vital resource lands, conserving energy and nonrenewable resources, and reversing unsustainable transportation trends.

13.2.3.4 Business and Dow Jones Concepts of Sustainability

The World Business Council for Sustainable Development is a coalition of 180 companies from around 35 countries. Each member shares the commitment to sustainable development through three pillars—economic growth, ecological balance, and social progress.

The Dow Jones Corporation, known for its business and financial indices and other publications, measures sustainability in the three dimensions of economic, environmental, and social responsibility (Table 13.2). It publishes a family of indexes to track performance of companies in terms of corporate sustainability as defined by the Dow Jones concept.

The Dow Jones concept of corporate sustainability has a business-like approach:[13]

> to create long-term shareholder value by embracing opportunities and managing risks deriving from economic, environmental and social developments. Corporate sustainability leaders harness the market's potential for sustainability products and services while at the same time successfully reducing and avoiding sustainability costs and risks.

The global 100 list of most sustainable corporations in the world is announced every year in Davos, Switzerland. The list is a compilation of publicly traded companies based on those companies that have the best developed abilities to manage risk, shareholder value, environmental, social, and strategic governance issues.

13.2.3.5 Adoption of Sustainability Policies

To achieve its goal, sustainability requires a proactive approach be taken between development and environmental quality. The concept of sustainability has received significant international attention, particularly within Europe and among other industrialized nations. For example, in 1991 the Resource Management Act of New Zealand was enacted. This act blazed a new precedent by

TABLE 13.2

Dow Jones Measurements of Sustainability

Environmental

1. Policy/management
2. Performance (ecoefficiency)
3. Reporting (content and coverage)

Economic

1. Codes of conduct, compliance, corruption, and bribery
2. Corporate governance
3. Customer relationship management
4. Investor relations
5. Risk and crisis management
6. Brand/supply chain/marketing practices criteria
7. Innovation/R&D/renewable energy criteria

Social

1. Citizenship/philanthropy
2. Stakeholder engagement
3. Labor practice
4. Human capital development
5. Social reporting
6. Talent attraction and retention
7. Product quality/recall management
8. Global sourcing
9. Occupational health/safety
10. Healthy living
11. Bioethic

articulating what some experts have called the world's first legislative statement promoting the principle of sustainability.

If the goal of a sustainable society could be achieved, many global environmental issues could be partially or perhaps even completely abated. Yet, of late, the concept of sustainability has received limited serious attention in the United States.

13.2.3.6 Basic Requirements

Three major elements are required to develop and achieve an international sustainability strategy that can protect the global commons:

- Commitment
- International cooperation
- Ability to assess global impacts and develop comprehensive plans for mitigating their effects

13.3 METHODOLOGIES FOR DEALING WITH POLICY AND UNCERTAINTY

Funtowicz and Ravetz have stated

Procrastination is as real a policy option as any other, and indeed one that is traditionally favored in bureaucracies; and inadequate information is the best excuse for delay.[14]

Failure to deal appropriately with uncertainty can lead to misleading or false conclusions. It should not come as a surprise that critics have seized upon the element of uncertainty as a tool for delaying or even distorting public debate. Thus, the science of uncertainty is not simply a subject of academic interest, but a pressing problem in search of practical solutions.

One principal source of uncertainty is a wildcard, that is, technological innovation. Although the new technology might resolve some or perhaps even many major environmental problems, many critics argue that if the society remains on its present course, even a plethora of new innovations would be insufficient to counter problems involving unsustainable use of resources. They reason, for example, that by the time a calamitous environmental problem such as global warming becomes unequivocally proven, it may be too late to reverse the trend, that is, the world may have passed beyond the point of recovery.

An excellent case involving uncertainty concerns the decision made in 1922 by chemical and automobile corporations to introduce tetraethyl lead into gasoline. When the decision was announced, a number of health experts warned that it was a dangerous idea and urged delay in order to allow time for scientific study. The chemical and automobile corporations countered that there was no scientific agreement concerning the threat; in the absence of solid scientific evidence to the contrary, they had the right to proceed. Tetraethyl lead became a standard gasoline additive. As a result, many medical scientists today believe that millions of children were adversely affected in a range of ways that may have included brain damage or even permanent impairment of their IQs. This example illustrates that prudence should be exercised when a major decision involving scientific uncertainty exists.

Scientific certainty is often defined as being 95% certain that a particular cause and effect have been correctly ascertained and described. In truth, it is often rare for even a large group of scientists to be 95% certain about most things, particularly when they concern complex environmental issues. This raises serious questions regarding how issues involving uncertainty should be addressed. It may well be that problems involving uncertainty will have to be resolved before global environmental problems such as climate change can be practically addressed.

13.3.1 Precautionary Principle

Prudence needs to be exercised in instances where major decisions involve significant scientific uncertainty. Like a pilot flying through fog, if you are not sure whether the shape looming just ahead is a mountain or simply a thick fog bank—exercise caution! This cautionary concept has been embedded in the precautionary principle, which generally announces a strategy of taking preventive action to avoid potential harm when an action involves scientific uncertainty.

Under this principle, the burden of proof lies with proponents, not the opponents, to demonstrate that a policy or action they advocate will not result in unacceptable environmental risks. The precautionary principle has frequently been applied, particularly within Europe, in circumstances involving environmental uncertainty, especially where there are far-reaching implications and consequences. It has also been embedded in many international and national statutes and agreements, such as the Rio Declaration on Environment and Development (1992), the U.N. Framework Convention on Climate Change (1992), Agenda 21 (agreed at the 1992 Rio conference), and the U.N. Convention on Biological Diversity (1992).

Apparently, exercising extreme caution may appear to be the wisest course of action when dealing with an uncertain issue that may have grave consequences. However, the principle has its own set of problems and limitations. The precautionary principle is not risk-free, and the choices can even lead to contradictory risks.

For example, the harmful effects of greenhouse gas emissions and the cost of reducing them involve significant uncertainties and create their own sets of risks. Thus, on the one hand, to protect the human environment, we may have to reduce emissions immediately but, on the other hand, to prevent undue economic hardship, perhaps, we should postpone taking any unnecessary actions until more information is available.

13.3.2   COST–BENEFIT ANALYSIS

The precautionary principle has been at the center of much controversy. Detractors argue that it is overly conservative and frequently results in unnecessary economic damage.

In the United States, an important alternative approach involves performing cost–benefit analysis (CBA).[15] A CBA has been used in an attempt to introduce a more rational approach to environmental decision-making. A CBA attempts to determine a rational course of action that will provide the greatest benefit with respect to cost. The best outcome is the one with the highest net benefit.

However, as with the precautionary principle, this approach is still plagued by uncertainty concerning future outcomes and limitations. Many scenarios are beset with uncertain assumptions. Moreover, the CBA sometimes does not offer clear-cut choices. An investigation of even simple problems can often involve very time-consuming and complex analyses.

To counter this problem, investigators sometimes explore an array of future scenarios rather than gambling on a single scenario or outcome. However, such approaches have their own limitations. They can address only a certain number of reasonable scenarios, and then there is the problem of deciding which scenarios to select—the one determined to be the most threatening or the one that is deemed to be the most probable?

Neither the precautionary principle nor most other analytical methodologies succeed in completely resolving the problem of the most effective way to plan, assess, and reach cost-effective decisions that can successfully mitigate significant impacts.

A promising technique places emphasis not on determining optimal strategies, but on evaluating robust or flexible ones. As described in the next section, a robust strategy is the one that performs well when compared with others across a wide range of reasonable scenarios and yields satisfactory results regardless of the actual outcomes or uncertainties involved in the analysis. Its principal advantage is that it attempts to address problems associated with uncertainty.

13.3.3   ROBUST PLANNING AND ASSESSMENT

NEPA and other similar environmental impact assessment (EIA) processes have often been relatively successful in analyzing trade-offs and predicting the impacts of policy choices that involve only modest uncertainty. However, many of the problems we face today involve significant uncertainty. As just witnessed, some of these problems are of such uncertainty and complexity (e.g., global weather patterns) that environmental scientists cannot make reliable predictions. In such situations, it may be more fruitful to find ways of managing uncertainty than seeking to eliminate it entirely.

A new methodology, known as robust decision-making, has been pioneered by the RAND Corporation. This methodology explores a wide spectrum of "what-if" scenarios.[16] Rather than focusing on identifying an optimal strategy, this approach concentrates on investigating the robust ones. That is, given a high degree of uncertainty, what actions should be pursued that can most effectively protect the future?

A robust decision-making approach need not be the optimal alternative since this strategy performs well when compared with alternatives across a wide spectrum of potential scenarios. Because this approach is designed to produce satisfactory outcomes under uncertain assumptions, it is adaptive and provides a hedge against various potential outcomes. It also provides flexibility to revise future plans as circumstances change or new information becomes available.

An interactive approach is used when a computer is used to "stress test" alternative strategies, searching for reasonable assumptions or conditions that could defeat them. In other words, a computer is used to generate and evaluate multiple future scenarios. The computer provides a tool for determining which strategy performs best across a sufficiently diverse set of reasonable scenarios.[17]

For example, with respect to greenhouse emissions, the interactive approach can evaluate a flexible alternative that imposes stringent emission limits but relaxes them if they cost too much.

Thus, if technological optimists prove correct, the cost threshold is never threatened, and polluters are able to meet the aggressive environmental goals. Conversely, if technological pessimists are correct, strict pollution limitations will exceed agreed-upon cost limits, and the strategy will provide polluters with more time to meet the environmental goals.

This approach may reduce a complex and controversial problem to a relatively simple set of more straightforward choices. As an alternative to endless debates concerning scenarios and uncertainties, analysts can focus on trade-offs. Such approaches may provide a path forward that most practical individuals can agree upon regardless of whose view of the future proves correct.

13.4   ENVIRONMENTAL DECISION-MAKING

Good information not necessarily results in rational choices. The following section describes a paradox that has potentially important implications in terms of environmental decision-making such as rationally assessing and choosing an alternative.

13.4.1   Ellsberg Paradox

Despite the fact that even his own biographer has questioned Daniel Ellsberg's true intentions and ethical motives for releasing the Pentagon Papers, no one questions the fact that he made an important contribution to the theory of decision-making.[18] The Ellsberg Paradox (the Paradox) arose out of a series of experimental games.[19] When first announced, its results surprised the world of decision-making theorists.

Consider two urns, each containing 100 poker chips. The poker chips are colored either red or black. The first urn contains 50 red and 50 black chips. The second urn also contains 100 chips, but the proportion of red to black in this urn is unknown: it might contain 100 red chips, or 100 black chips, or any proportion thereof.

A facilitator places a blind over your eyes, and asks you to draw one chip from either of the urns, without looking at it. If you draw a red poker chip, you win $10,000. Which urn do you choose to draw the chip from? Stop here and make your own mental choice before reading on.

Based on the information just supplied, there is no rational reason to believe that your chance of picking up a red chip is any higher from one urn than the other. Yet, most people choose the urn containing the 50 red and 50 black chips over the urn in which the proportion is unknown.

If you chose the first urn containing an equal proportion of red and black poker chips, it is logical to assume you had a hunch that the other urn contained more black chips. Now, consider a variation on this experiment. The facilitator makes a second wager: "It's clear that you must believe that the first urn is more likely to let you draw a red chip. Now, I'm offering you $10,000 if this time you draw a black chip." Given the facilitator's rational, you should logically draw your black chip from the second urn.

But, in experiments that have been performed, this does not appear to be the case. Again, subjects overwhelmingly choose the urn containing equal proportion of red and black poker chips. This is true despite the fact that the chance of picking either color is identical in both gambles.

This raises profound questions with respect to how people reach decisions. In essence, the Paradox suggests that people strongly prefer definite information over ambiguity and will make their decisions accordingly (i.e., ambiguity aversion). The value of such experiments resides in the fact that they illuminate the preference of people for choices that seem to involve least risk or uncertainty. One explanation for this is that people instinctively tend to avoid circumstances that may result in the worst possible outcome, as opposed to making a rational assessment of the choices or optimum course of action to pursue.

Additional research has found that uncertainty about technological and other risks tends to make less ambiguous technologies more acceptable to the public. The author believes that this same finding may well apply to the choices made by decision-makers as part of the NEPA process.[20]

13.4.1.1 NEPA and Environmental Decision-Making

While the aforementioned Paradox is widely discussed in fields such as economics, politics, and defense strategy, virtually no serious attention has been focused (as of this writing) on its implications with respect to NEPA. This is interesting given that environmental decision-making often involves a great deal of risk, ambiguity, and uncertainty.

In the author's opinion, this Paradox suggests that many NEPA decisions are probably reached by a process in which one option is chosen over another simply because decision-makers have a natural desire to avoid risk or uncertainty. Some superior alternatives and courses of action have probably even been rejected simply because they involved a greater degree of ambiguous information or circumstances. This implies that many bad choices have probably been made simply because they involved decisions that lacked a significant degree of risk or uncertainty. Thus, the way in which many decisions are made in relation to this Paradox may actually be endangering society.

For example, consider what the Paradox may have to say about a decision-maker who has to make a choice between two alternative courses of action. Is the public or a decision-maker more likely to accept a national coal-fired energy option that will definitely result in 2000 early deaths from cancer and lung ailments per year, but is well understood, over a second technological alternative that results in far fewer direct ailments but where the risk of a catastrophic accident is more ambiguous?

13.4.1.2 Dealing with Uncertainty in NEPA Documents

NEPA has often been referred to as an environmental full disclosure law because of its requirement to disclose all potentially significant environmental effects in environmental impact statements (EISs). While a rigorous investigation must be performed, such studies often involve a considerable degree of scientific uncertainty. Analysts must look into the future and sometimes even make educated guesses about the eventual consequences of proposed actions. Some examples of such consequences may include a nuclear reactor meltdown, an oil spill in a marine sanctuary, the application of a new herbicide, transportation accidents involving highly toxic chemicals, or the introduction of nanotechnology products into the market place. Reasonable forecasting and speculation are thus implicit in NEPA.

While the lack of sufficient scientific information need not halt an EIS or even prevent the implementation of a project, the Council on Environmental Quality (CEQ) NEPA regulations (Regulations) provide specific steps that must be complied with when dealing with important unknown information or uncertainties. Against this backdrop, the Regulations require agencies to disclose the uncertainty and to evaluate its possible impacts based on theoretical approaches or research methods generally accepted in the scientific community.[21]

Clearly, the aforementioned Paradox is an area where significant environmental research is needed. In the meantime, integration of the NEPA process with development of an environmental management system (see Section 2.6) may provide a very powerful framework for managing project implementation, particularly in cases where there is significant risk, ambiguous information, or uncertainty.

13.4.2 Decision-Making and the Delphi Method

Pioneered by the RAND Corporation during the 1950s and 1960s, the Delphi method is a systematic, interactive, and highly structured technique in which a blind panel of independent experts provides their assessment of likely future outcomes by responding to several rounds of questions. When properly implemented, this technique has often been very effective in generating an accurate consensus, particularly with respect to forecasting. The name Delphi is taken from the fabled oracle of Delphi whose prophecies were sought in ancient Greece.

13.4.2.1   The Process

A facilitator manages a panel of experts who are carefully chosen for their particular knowledge or views on a specific issue.[22,23] The facilitator first sends out a questionnaire to each participant and then collects the responses. The responses are then circulated anonymously to the team members. The identities of the panel members are usually not revealed even after the project has been completed.

The process is iterative. Each round of questioning is accompanied by feedback from the preceding round of replies. The facilitator also identifies common or conflicting viewpoints and sends these out to each group member for review. The experts are encouraged to revise their earlier answers based on comments from the participant of the previous round. The goal is to manage an evolutionary process in which the group converges toward a correct consensus.

13.4.2.2   Advantages and Disadvantages

Frequently in group settings, participants tend to hold irrationally and steadfastly to previously stated opinions; in other instances, weaker personalities may be swayed by stronger or more vocal individuals. But because all panel members maintain anonymity, the Delphi method avoids the negative effects of face-to-face discussions and personality conflicts, resolving many problems associated with traditional group dynamics. Specifically, this anonymity

- allows open critique of views and opinions,
- prevents strong members from dominating others,
- allows participants to freely express their opinions by eliminating the risk of being professionally embarrassed as a result of an incorrect view,
- allows participants to freely admit errors by revising their earlier judgments, and
- minimizes the bandwagon effect.

One of the disadvantages of the Delphi method is that it can be a lengthy and relatively expensive process. It also requires a skilled and experienced facilitator. Critics have also countered that there are many cases where the method has resulted in poor results.

Proponents respond that cases in which poor results have been recorded may not be the result of an inherent weakness in the method itself. Instead, such results may be due to the fact that in areas such as science, the degree of uncertainty can be so great as to make accurate predictions all but impossible; errors are thus to be expected regardless of the method used.

13.4.2.3   NEPA and Environmental Policy

There are many complex proposals where the Delphi method might yield a high-return value in reaching rational decisions because NEPA issues often involve a high degree of uncertainty, making a scientific consensus difficult to reach. Selection of the agency's preferred decision or the environmentally preferable decision often requires consideration of many diverse, ambiguous, competing, or conflicting factors. For instance, how does one assess the risk of health impacts associated with degradation in air quality against the extinction of a species or loss of wetlands? This method might also provide a systematic approach under NEPA for making a more rational action recommendation (preferred alternative) available to the decision-maker.

13.5   ENVIRONMENTAL ECONOMICS

For any new environmental policy or regulation, there are winners and losers. Good economic efficiency mandates that gains to the winners exceed losses imposed upon the losers. But how can this win–loss value be measured?

13.5.1 Cost–Benefit Analysis

A CBA is frequently used to compute monetary gains and losses. For example, consider a potential air quality improvement policy. This policy will result in gains to citizens, as well as losses incurred by businesses and consumers. The losses often tend to be easier to compute since they are direct quantifiable revenues that will be lost as a result of decreased production or higher production costs. However, the indirect benefits or gains from cleaner air may be more difficult to quantify, since they include such unknown factors as the amount of decrease in lung illnesses and increase in worker productivity, as well as intangible benefits such as increased enjoyment of life.[24,25]

13.5.2 Economic Surplus

A methodology known as economic surplus provides a theoretical basis for assessing economic benefits. Consumer surplus is a monetary measure of the net benefit a consumer gains from a transaction. For example, consider the purchase of a sports-utility vehicle (SUV). An economic surplus is simply the difference between the market price of the vehicle (the amount the vehicle costs) and what the consumer is able and willing to pay. Assume for instance, that a consumer is willing and able to pay a maximum of $45,000 for a Pluto SUV. Now assume that the actual negotiated price is $37,000. The consumer surplus is thus $8000, which is the difference between the consumer's actual negotiated price and the actual amount he is prepared to pay.

Similarly, commodities or services that are not purchased in markets (e.g., air quality) also have a consumer surplus. Consider a tourist who is willing (and able) to pay $10 for each additional opportunity to visit a scenic site. If a proposed regulation leads to an air quality improvement, enabling the tourist to visit three additional sites beyond the one normally experienced under the previously unregulated viewing conditions, the consumer surplus (i.e., monetary measure of sightseeing) increases by $30. But, since the sightseer did not pay directly for the air quality improvement, the $30 increase in consumer surplus is difficult to assess. Although this approach is difficult and far from perfect, it at least provides a theoretical model for evaluation of environmental quality.

PROBLEMS

1. Briefly describe the Malthusian growth model.
2. What is meant by the term "precautionary principle"?
3. Explain the concept of "robust decision-making."
4. A critic claims that there is no need to be concerned with preserving biodiversity since the Gaia effect will eventually rebalance any detrimental effect that occurs. How would you respond? Justify your response.
5. An EIS has been prepared for a flood control project. The proposed action involves constructing a levy control system that has a 50/50 chance over the next 20 years of being breached, which would result in limited flooding with deaths and damage amounting to $100,000,000. The alternative involves the construction of a flood control dam that would entirely eliminate the risk of any future flooding, but which could catastrophically fail under certain extreme seismic conditions. The chance of a dam failure is uncertain, but experts believe it is very remote; the consequences of a dam failure are likewise uncertain but might be potentially catastrophic. What decision-making factors should the public and officials be aware of before reaching a final decision to pursue either the levy or dam?

REFERENCES

1. Hardin G., The tragedy of the commons, *Science*, 162, 1968, 1243–1248.
2. Hardin G., The tragedy of the commons, *Science* 162, 1968; Hardin G., The Immigration Dilemma: Avoiding the Tragedy of the Commons, Federation for American Immigration Reform, Washington, D.C., 1995, 13–30.

3. Malthus T., *An Essay on the Principle of Population*, June 7, 1798. Available from www.amazon.com.
4. Malthus T., *An Essay on the Principle of Population*, Chapter 8, 1798. Available from www.amazon.com.
5. Donella M., Dennis M., Randers J., et al., *Limits to Growth* (commissioned by the Club of Rome), 1972.
6. Doolittle F., Is Nature Really Motherly? *CoEvolution Quarterly*, Spring 1981.
7. Thomas L., *Lives of a Cell*, Bantam Books, New York, 1980.
8. Wikipedia.com, Gaia hypothesis, http://en.wikipedia.org/wiki/Gaia_hypothesis (accessed March 26, 2007).
9. Sadler B., Towards the improved effectiveness of environmental assessment. Executive Summary of Interim Report Prepared for IAIA 1995. Durban, South Africa, 1995.
10. United Nations, Report of the United Nations Conference on Environment and Development, Agenda 21, Rio Conference 27 Principles (Rio de Janeiro, 3–14 June 1992), Annex I, Rio Declaration on Environment and Development (Rio de Janeiro, 3–14 June 1992).
11. The National Environmental Policy Act of 1969, as amended (Pub. L. 91-190, 42 U.S.C. 43214347, January 1, 1970, Section 2, Purpose.
12. The National Environmental Policy Act of 1969, as amended (Pub. L. 91-190, 42 U.S.C. 43214347, January 1, 1970, Title 1, Section 101(a).
13. Jones D., *Jones Sustainability North America Index Guide Book*, Section 3.1, Version 2.0, August 2006.
14. Funtowicz S. O. and Ravetz J. R., *Uncertainty and Quality in Science for Policy*, Kluwer Academic Press, Dordrecht, 1990.
15. Toth F. L. and Mwandosya M., Chapter 10: *Decision-Making Frameworks*, IPCC WGIII Third Assessment Report.
16. Steven W. P., Robert J. L., and Steven C. B., Shaping the future, *Scientific American*, April 2005.
17. Robert J. L., Steven W. P. and Steven C. B., Shaping the Next One Hundred Years: New Methods for Quantitative, Long-Term Policy Analysis, RAND MR-1626, 2003.
18. Wells T., *Wild Man: the Life and Times of Daniel Ellsberg*, Palgrave, New York, 2001.
19. Ellsberg D., Risk, ambiguity, and the Savage axioms, *Quarterly Journal of Economics*, 75, 1961, 643–669.
20. Paul S., Perception of risk, *Science*, 236, 1987, 280.
21. 40 Code of Federal Regulations (CFR) 1502.22.
22. Sahakian C. E., *The Delphi Method*, Corporate Partnering Institute, Skokie, IL, 1997.
23. Linstone H. A. and Turoff M., *The Delphi Method: Techniques and Approaches*, Addison-Wesley, Reading, MA, 1975.
24. Johansson P. O., *Cost-Benefit Analysis of Environmental Change*, Cambridge University Press, Cambridge, U.K., 2003.
25. Boardman A., Greenberg D., Vining A., et al., *Cost Benefit Analysis: Concepts and Practice*, 3rd ed., Prentice Hall, Upper Saddle River, NJ, 2005.

Glossary

The terms referenced in this glossary are defined according to their meaning as they are generally used and understood within the context of NEPA.

Act A synonym used in the Council of Environmental Quality Regulations to refer to the National Environmental Policy Act, as amended (42 U.S.C. 4321, et seq.).

Actions The CEQ NEPA regulations define three types of actions, other than unconnected single actions, that must be taken into consideration during a NEPA analysis. These three actions are connected, cumulative and similar actions.

Administrative Procedures Act A law specifying the requirements and procedures that must be followed in issuing regulations.

Administrative record All significant papers and files, as well as draft and final documents, that document the process used in reaching a final decision.

Air quality standards The level of regulated air pollutants that may not exceed during a given time in a proscribed area.

Alternatives The term "alternatives" as used in the CEQ NEPA regulations refers to other reasonable options that would meet the need of a proposed action. There are three types of alternatives: (1) no action alternative, (2) other reasonable courses of actions, and (3) mitigation measures (not in the proposed action).

Applicant A non-Federal party who has filed an application with a federal agency that is subject to a NEPA review before the agency may approve the application. Such applications normally involve required federal approvals or permits that must be obtained before the applicant may proceed with a specified action.

Attainment area An area with air quality that meets or exceeds the national ambient air quality standards as defined in the Clean Air Act.

Categorical exclusions Categorical exclusions (CATX) are classes of actions under NEPA which do not have a significant effect, either individually or cumulatively, on the human environment, and therefore do not require preparation of an environmental assessment or environmental impact statement.

CEQ See Council on Environmental Quality.

Connected actions The term "connected actions" as defined by the CEQ NEPA regulations refers to actions that are closely related and therefore should be discussed in the same impact statement. Actions are connected if they (1) automatically trigger other actions that may require environmental impact statements, (2) cannot or will not proceed unless other actions are taken first or simultaneously, and (3) are interdependent parts of a larger action that depend on the larger action for their justification.

Contaminant A chemical, physical, biological or radioactive substance that has an adverse effect on the environment.

Context As used in the CEQ NEPA regulations, this refers to a factor that must be considered in making a determination regarding the significance of an impact. The impacts must be analyzed in several contexts such as society as a whole (human, national), the affected region, the affected interests, and the locality. The extent of significance varies with the setting of the proposed action. For instance, in the case of a site-specific action, significance would usually depend upon the effects to the locale rather than to the world as a whole. Both short- and long-term effects are relevant.

Cooperating agency A federal agency other than a lead agency possessing jurisdiction by law or special expertise with respect to an environmental impact that is involved in a proposal (or a reasonable alternative) for legislation or other major federal action significantly affecting the quality of the human environment.

Council A synonym used for the Council on Environmental Quality.

Council on Environmental Quality The council created by Title II of the NEPA Act to oversee the NEPA process.

Council on Environmental Quality regulations The regulations issued by the CEQ (40 CFR parts 1500–1508) for implementing the procedural aspects of NEPA.

Credible accident As used in this book an accident whose probability of occurrence lies within the range of one chance in a million. Although considered physically possible, it is almost certain that such an accident will not occur in the proposed project.

Cumulative actions The term "cumulative action" as defined by the CEQ NEPA regulations describes actions that, when viewed together with other proposed actions, are found to have cumulatively significant impacts and should therefore be discussed in the same impact statement.

Cumulative impact The impact on the environment that results from the incremental impact of an action when it is added to other past, present, and reasonably foreseeable future actions, regardless of the agency (federal or nonfederal) or person responsible for them. This is an important concept to understand because individual impacts that may seem minor can accrue gradually over a period of time, resulting collectively in significant impacts.

Direct impacts Effects that are caused by the action and occur at the same time and place.

EA See environmental assessment.

Effects The terms "effects" and "impacts" as used in the NEPA regulations are synonymous. Effects may include the ecological impacts of an action (such as effects on natural resources and on the components, structures, and functioning of affected ecosystems), as well as the effects on esthetic, historic, economic, social, health, and cultural resources. The concept of effects (impacts) includes direct, indirect, and cumulative effects (impacts), and both beneficial and detrimental effects (impacts).

EIS See environmental impact statement.

Emission A pollution discharge into the atmosphere from smokestacks, vents, and other sources.

Endangered species Organisms that are threatened with extinction as a result of either man-made or natural changes to the environment.

Environment See human environment.

Environmental assessment An environmental assessment (EA) is a public document that is used to provide in a concise manner sufficient evidence and analysis for determining whether or not to prepare an environmental impact statement or a finding of no significant impact for a proposed action. An EA may be used to assist an agency to comply with NEPA when no environmental impact statement is necessary. An EA may also be used to facilitate preparation of an EIS when one is necessary. An EA must include brief discussions of the need for the proposal, alternatives, environmental impacts of the proposed action and alternatives, and a listing of agencies and persons consulted.

Environmental document As defined in the CEQ NEPA regulations, this document includes environmental assessment, environmental impact statement, finding of no significant impact, and notice of intent.

Environmental impact statement A detailed document that in accordance with the CEQ NEPA regulations is required to be prepared for any proposed action that may result in a significant environmental impact.

Federal agency As defined by the CEQ NEPA regulations, a federal agency refers to all agencies of the federal government. However, the Congress, the Judiciary, and the president (including the performance of staff functions in his Executive Office) are not included.

Finding of no significant impact The term "finding of no significant impact" (FONSI) describes a document issued by a federal agency that briefly presents the reasons why an action will not have a significant effect on the human environment and therefore will not require an environmental impact statement. The document must include the environmental assessment or a summary of it and must also note any related environmental documents.

FONSI See finding of no significant impact.

Foregone resources Those resources that, as a result of an action, are rendered inaccessible or unusable, such as land that is submerged under the water of a man-made lake.

Geographical information system (GIS) A software computer system designed for managing, sorting, analyzing, and displaying data in a geographical context.

Human environment The term "human environment" as defined by the CEQ NEPA regulations is interpreted comprehensively to include both the natural and physical environments and the relationship of people with those environments. Economic or social effects are not intended by themselves to require preparation of an environmental impact statement. However, when an environmental impact statement is prepared and it is found that the economic or social and the natural or physical environmental effects are interrelated, then the environmental impact statement will discuss all of these effects on the human environment.

Impacts See effects.

Implementation plan A document used by many federal agencies to record the results of the EIS scoping process. The IP also provides a plan for preparing the EIS.

Incredible accident An "incredible accident," as used in this book, is an accident with a probability of occurrence less than one chance in a million.

Indirect impacts These are reasonably foreseeable impacts that are caused by an action but occur at a later time or are removed by distance from the action. Indirect impacts may include effects related to changes induced in the pattern of land use, population density or growth rate, and their related effects on air, water, and other natural systems, including ecosystems.

Intensity The term "intensity" as used in the CEQ NEPA regulations refers to a factor that must be considered and analyzed in making a determination regarding the significance of an impact. The intensity is the degree to which the impact would affect the environment.

Interim action An action within the scope of a proposal that is the subject of an ongoing EIS, that is permissible under 40 CFR § 1506.1, and that an agency proposes to pursue before the record of decision is issued.

IP See implementation plan.

Irretrievable resources A resource that cannot be replaced after it is consumed. For example, timber is normally considered to be retrievable because it can be replanted and replaced as it is consumed.

Irreversible resource A resource, such as minerals, that realistically can never be replaced once they are consumed.

Jurisdiction by law The term "jurisdiction by law" as used in NEPA means agency authority to approve, veto, or finance all or part of a proposal.

Land use plans With respect to NEPA, the term "land use plans" includes any documents that have been formally adopted for land use planning or zoning. These include plans that have been proposed formally by a government body and are under active consideration.

Land use policies The term "land use policies" as used in reference to NEPA includes statements concerning land use policy that have been formally adopted and embodied in laws or regulations. It also includes land use policies that have been proposed formally but have not yet been adopted.

Lead agency The term "lead agency" as used in the CEQ NEPA regulations means the agency or agencies preparing or having primary responsibility for preparing the environmental impact statement.

Legislation The term "legislation" includes a bill or legislative proposal to Congress developed by or with the significant cooperation and support of a federal agency, but does not include requests for appropriations. The test that determines significant cooperation is whether the proposal has been prepared predominantly by an agency rather than by another source. The drafting of a proposal does not by itself constitute significant cooperation. Proposals for legislation include requests for ratification of treaties. Only the agency that has primary responsibility for the subject matter involved will prepare a legislative environmental impact statement.

Major federal action The term "major federal action" as used in the CEQ NEPA regulations includes actions with effects that may be major and therefore potentially subject to federal control and responsibility. The word "major" reinforces but does not have a meaning independent of the term "significantly." Actions include circumstances where the responsible officials fail to act. These failures can be reviewed as agency actions by courts or administrative tribunals under the Administrative Procedure Act or other applicable law.

Mitigation Measures that may be taken to avoid, minimize, rectify, reduce, or compensate for the adverse impacts of an action on the environment.

Mitigation action plan Refers to a document that describes the action plan for implementing commitments made in an EIS/ROD or EA/FONSI to mitigate adverse environmental impacts.

Ministerial actions Congressionally mandated actions over which an agency has no discretion.

Modeling A theoretical process that involves using physical or mathematical techniques to represent and predict the behavior and properties of physical phenomena. For example, mathematical models are being used increasingly to predict the behavior (flow) of groundwater and of contaminated air plumes.

Monitoring The process of observing and measuring environmental impacts on environmental resources to verify compliance with the description of the proposed action and any mitigation factors cited in a NEPA document.

NAAQS See National Ambient Air Quality Standards.

National Ambient Air Quality Standards National standards established by the EPA to protect the nation's air quality.

National Environmental Policy Act The federal statute passed by Congress in 1969 that established basic environmental policy for protection of the environment (42 U.S.C. 4321 et seq.). NEPA provides a systematic and interdisciplinary process that agencies are required to follow with the purpose of reducing or preventing environment degradation. The Act contains "action forcing" procedures (provisions?) that must be followed by federal agencies to insure that their decision-makers take environmental factors into full consideration before making a final decision regarding a proposed action.

NEPA See National Environmental Policy Act of 1969.

NEPA process The term "NEPA process" refers to all measures that are necessary for compliance with the requirements mandated in Section 2 and Title I of NEPA.

NEPA review The process followed in complying with Section 102(2) of NEPA.

NOA See notice of availability.

NOI See notice of intent.

Notice of availability A formal notice as defined in 40 CFR § 1508.22, published in the *Federal Register*, that announces the issuance and public availability of a draft or final EIS.

Notice of intent A formal notice, published in the *Federal Register*, that announces an agency's intent to prepare an EIS.

Nuclear power plant A plant that uses nuclear fuel to convert atomic energy into electrical energy.

P-EIS See programmatic EIS.

Pollutant A substance identified within the definition of pollutant in Section 101(33) of CER-CLA (42 U.S.C. 9601.101(33)). More generally, this term can also be used in referring to any substance that, if released into the environment, would result in adverse effects to the human environment.

Prevention of significant deterioration An EPA program in which a permit must be obtained for regulating emissions from new or modified sources within an area that meets or exceeds ambient air quality standards.

Program For the purposes of NEPA, a program can be defined as a sequence of connected or related actions as discussed in 40 CFR § 1508.18(b)(3) and § 1508.25(a).

Programmatic EA/EIS A broadly based EA/EIS prepared to evaluate a sequence of connected or related agency actions or projects as discussed in 40 CFR § 1508.18(b)(3) and § 1508.25(a).

Programmatic NEPA document A broadly scoped EA or EIS prepared to evaluate the environmental impacts of an agency program. This term can also be applied to related NEPA documents, such as a ROD or FONSI.

Project For the purposes of NEPA, a "project" refers to a specific agency effort, including actions approved by a license, permit, or regulatory decision, federal and federally assisted activities, or similar activities, as described in 40 CFR § 1508.18(b)(4).

Proposal A "proposal" as used in the CEQ NEPA regulations exists at that stage in the development of an action when an agency subject to the Act has a goal, is actively preparing to make a decision on one or more alternative means of accomplishing that goal, and the effects can be meaningfully evaluated. Preparation of an environmental impact statement on a proposal should be scheduled so that the final statement may be completed in time for the statement to be included in any recommendation or report on the proposal. A proposal may exist in fact as well as by agency declaration that one exists.

PSD See prevention of significant deterioration.

Public scoping This refers to that portion of the scoping process in which the public is invited to participate, as described in 40 CFR § 1501.7 (a)(1) and (b)(4).

RCRA The Resource Conservation and Recovery Act.

Record of decision A public document prepared on completion of an EIS. The document records the agencies' final decision and rationale for making the decision and any commitments with regard to monitoring and mitigation.

Referring agency A "referring agency" as used in the CEQ NEPA regulations means the federal agency which has referred any matter to the council after a determination has been made that the matter is unsatisfactory from the standpoint of public health or welfare or environmental quality.

Regulations The term as used in this book refers to the NEPA regulations issued by the CEQ.

Resources With respect to NEPA, environmental resources include all physical (e.g., geological, biological, and atmospheric), socioeconomic, and other related aspects of the environment that potentially may be affected by agencies' actions.

Risk As used in this book, "risk" is defined as the probability that an accident could occur, multiplied by the degree of the consequences resulting from that accident.

ROD See record of decision.

S-EIS See supplemental EIS.

Scope The term "scope" as used in the CEQ NEPA regulations is defined to consist of the range of actions, alternatives, and impacts to be considered in an environmental impact statement. The scope of an individual statement may depend on its relationships to other statements. To determine the scope of environmental impact statements, agencies must consider three types of actions, three types of alternatives, and three types of impacts.

Significance The degree to which an impact may affect the human environment. The term as used in the CEQ NEPA regulations requires consideration of both the context and intensity of an impact.

Similar actions The term "similar action" as defined by the CEQ NEPA regulations means actions that, when viewed with other reasonably foreseeable or proposed agency actions, have similarities that provide a basis for evaluating their environmental consequences together, such as common timing or geography. In such cases, an agency may wish to analyze these actions in a single impact statement. It should do so when this is determined to be the best way to assess adequately the combined impacts of similar actions or reasonable alternatives to such actions.

Site-wide NEPA document A broad-scope EIS or EA that is programmatic in nature and that evaluates environmental impacts of ongoing and reasonably foreseeable future actions at a federal agency site.

Special expertise The term "special expertise" as defined by the CEQ NEPA regulations means statutory responsibility, agency mission, or related program experience.

Supplement analysis Refers to a DOE document used either to determine whether a supplemental EIS should be prepared or to support a decision to prepare a new EIS.

Supplemental EIS An EIS prepared to supplement an existing EIS as described in 40 CFR § 1502.9(c). A supplemental EIS is prepared when a substantial change is made to the proposed action or when important new information is acquired regarding the action.

Synergistic An interaction between two or more chemicals or phenomena resulting in a combined effect greater than the two individual effects.

Tiering The term "tiering" refers to the coverage of general environmental issues in broader NEPA documents such as EISs (e.g., national program or policy statements) which are subsequently followed by narrower statements or environmental analyses (such as regional or site-specific analyses); these narrower NEPA analyses are tiered from the broader documents, that is, the tiered documents incorporate by reference the earlier general discussions so that they can concentrate solely on the specific issues at hand.

Tribal lands The area of "Indian country," as defined in 18 U.S.C. 1151, that is under the tribe's jurisdiction.

Wetlands An area that is saturated or partially saturated. An area need only be saturated during a small potion of the year to be designated a wetlands. In order to be designated as wetlands, the area must exhibit certain soil, hydrological, and vegetative characteristics.

Appendix A

THE NATIONAL ENVIRONMENTAL POLICY ACT OF 1969

The National Environmental Policy Act of 1969, as amended
(Pub. L. 91-190, 42 U.S.C. 4321-4347, January 1, 1970, as amended by Pub. L. 94-52, July 3, 1975, Pub. L. 94-83, August 9, 1975, and Pub. L. 97-258, § 4[b], Sept. 13, 1982)

An Act to establish a national policy for the environment, to provide for the establishment of a Council on Environmental Quality, and for other purposes.

Be it enacted by the Senate and House of Representatives of the United States of America in Congress assembled. That this Act may be cited as the "National Environmental Policy Act of 1969."

Purpose

Sec. 2 [42 USC § 4321]

The purposes of this Act are: To declare a national policy which will encourage productive and enjoyable harmony between man and his environment; to promote efforts which will prevent or eliminate damage to the environment and biosphere and stimulate the health and welfare of man; to enrich the understanding of the ecological systems and natural resources important to the Nation; and to establish a Council on Environmental Quality.

TITLE I CONGRESSIONAL DECLARATION OF NATIONAL ENVIRONMENTAL POLICY

Sec. 101 [42 USC § 4331]

a. The Congress, recognizing the profound impact of man's activity on the interrelations of all components of the natural environment, particularly the profound influences of population growth, high-density urbanization, industrial expansion, resource exploitation, and new and expanding technological advances and recognizing further the critical importance of restoring and maintaining environmental quality to the overall welfare and development of man, declares that it is the continuing policy of the Federal Government, in cooperation with State and local governments, and other concerned public and private organizations, to use all practicable means and measures, including financial and technical assistance, in a manner calculated to foster and promote the general welfare, to create and maintain conditions under which man and nature can exist in productive harmony, and fulfill the social, economic, and other requirements of present and future generations of Americans.
b. In order to carry out the policy set forth in this Act, it is the continuing responsibility of the Federal Government to use all practicable means, consistent with other essential considerations of national policy, to improve and coordinate Federal plans, functions, programs, and resources to the end that the Nation may
   1. fulfill the responsibilities of each generation as trustee of the environment for succeeding generations;
   2. assure for all Americans safe, healthful, productive, and aesthetically and culturally pleasing surroundings;
   3. attain the widest range of beneficial uses of the environment without degradation, risk to health or safety, or other undesirable and unintended consequences;

4. preserve important historic, cultural, and natural aspects of our national heritage, and maintain, wherever possible, an environment which supports diversity, and variety of individual choice;
5. achieve a balance between population and resource use which will permit high standards of living and a wide sharing of life's amenities; and
6. enhance the quality of renewable resources and approach the maximum attainable recycling of depletable resources.

c. The Congress recognizes that each person should enjoy a healthful environment and that each person has a responsibility to contribute to the preservation and enhancement of the environment.

Sec. 102 [42 USC § 4332]

The Congress authorizes and directs that, to the fullest extent possible: (1) the policies, regulations, and public laws of the United States shall be interpreted and administered in accordance with the policies set forth in this Act and (2) all agencies of the Federal Government shall

A. utilize a systematic, interdisciplinary approach which will insure the integrated use of the natural and social sciences and the environmental design arts in planning and in decision-making which may have an impact on man's environment;
B. identify and develop methods and procedures, in consultation with the Council on Environmental Quality established by title II of this Act, which will insure that presently unquantified environmental amenities and values may be given appropriate consideration in decision-making along with economic and technical considerations;
C. include in every recommendation or report on proposals for legislation and other major Federal actions significantly affecting the quality of the human environment, a detailed statement by the responsible official on
   i. the environmental impact of the proposed action,
   ii. any adverse environmental effects which cannot be avoided should the proposal be implemented,
   iii. alternatives to the proposed action,
   iv. the relationship between local short-term uses of man's environment and the maintenance and enhancement of long-term productivity, and
   v. any irreversible and irretrievable commitments of resources which would be involved in the proposed action should it be implemented.
      Prior to making any detailed statement, the responsible Federal official shall consult with and obtain the comments of any Federal agency which has jurisdiction by law or special expertise with respect to any environmental impact involved. Copies of such statement and the comments and views of the appropriate Federal, State, and local agencies, which are authorized to develop and enforce environmental standards, shall be made available to the President, the Council on Environmental Quality, and to the public as provided by Section 552 of title 5, United States Code, and shall accompany the proposal through the existing agency review processes;
D. any detailed statement required under subparagraph (C) after January 1, 1970, for any major Federal action funded under a program of grants to States shall not be deemed to be legally insufficient solely by reason of having been prepared by a State agency or official, if
   i. the State agency or official has statewide jurisdiction and has the responsibility for such action,
   ii. the responsible Federal official furnishes guidance and participates in such preparation,
   iii. the responsible Federal official independently evaluates such statement prior to its approval and adoption, and

iv. after January 1, 1976, the responsible Federal official provides early notification to, and solicits the views of, any other State or any Federal land management entity of any action or any alternative thereto which may have significant impacts upon such State or affected Federal land management entity and, if there is any disagreement on such impacts, prepares a written assessment of such impacts and views for incorporation into such detailed statement.

The procedures in this subparagraph shall not relieve the Federal official of his responsibilities for the scope, objectivity, and content of the entire statement or of any other responsibility under this Act; and further, this subparagraph does not affect the legal sufficiency of statements prepared by State agencies with less than statewide jurisdiction.

E. study, develop, and describe appropriate alternatives to recommended courses of action in any proposal which involves unresolved conflicts concerning alternative uses of available resources;

F. recognize the worldwide and long-range character of environmental problems and, where consistent with the foreign policy of the United States, lend appropriate support to initiatives, resolutions, and programs designed to maximize international cooperation in anticipating and preventing a decline in the quality of mankind's world environment;

G. make available to States, counties, municipalities, institutions, and individuals, advice and information useful in restoring, maintaining, and enhancing the quality of the environment;

H. initiate and utilize ecological information in the planning and development of resource-oriented projects; and

I. assist the Council on Environmental Quality established by title II of this Act.

Sec. 103 [42 USC § 4333]

All agencies of the Federal Government shall review their present statutory authority, administrative regulations, and current policies and procedures for the purpose of determining whether there are any deficiencies or inconsistencies therein which prohibit full compliance with the purposes and provisions of this Act and shall propose to the President not later than July 1, 1971, such measures as may be necessary to bring their authority and policies into conformity with the intent, purposes, and procedures set forth in this Act.

Sec. 104 [42 USC § 4334]

Nothing in Section 102 [42 USC § 4332] or 103 [42 USC § 4333] shall in any way affect the specific statutory obligations of any Federal agency (1) to comply with criteria or standards of environmental quality, (2) to coordinate or consult with any other Federal or State agency, or (3) to act or refrain from acting contingent upon the recommendations or certification of any other Federal or State agency.

Sec. 105 [42 USC § 4335]

The policies and goals set forth in this Act are supplementary to those set forth in existing authorizations of Federal agencies.

TITLE II   COUNCIL ON ENVIRONMENTAL QUALITY

Sec. 201 [42 USC § 4341]

The President shall transmit to the Congress annually beginning July 1, 1970, an Environmental Quality Report (hereinafter referred to as the "report") which shall set forth (1) the status and condition of the major natural, manmade, or altered environmental classes of the Nation, including, but not limited to, the air, the aquatic, including marine, estuarine, and fresh water, and the terrestrial environment, including, but not limited to, the forest, dryland, wetland, range, urban, suburban, and rural environment; (2) current and foreseeable trends in the quality, management, and utilization of such environments and the effects of those trends on the social, economic, and other requirements of the Nation; (3) the adequacy of available natural resources for fulfilling human and economic requirements of the Nation in the light of expected population pressures; (4) a review of the programs and activities (including regulatory activities) of the Federal Government, the State and local governments, and nongovernmental entities or individuals with particular reference to their effect on the environment and on the conservation, development, and utilization of natural resources; and (5) a program for remedying the deficiencies of existing programs and activities, together with recommendations for legislation.

Sec. 202 [42 USC § 4342]

There is a Council on Environmental Quality created in the Executive Office of the President (hereinafter referred to as the "Council"). The Council shall be composed of three members who shall be appointed by the President to serve at his pleasure, by and with the advice and consent of the Senate. The President shall designate one of the members of the Council to serve as Chairman. Each member shall be a person who, as a result of his training, experience, and attainments, is exceptionally well qualified to analyze and interpret environmental trends and information of all kinds; to appraise programs and activities of the Federal Government in the light of the policy set forth in title I of this Act; to be conscious of and responsive to the scientific, economic, social, aesthetic, and cultural needs and interests of the Nation; and to formulate and recommend national policies to promote the improvement of the quality of the environment.

Sec. 203 [42 USC § 4343]

a. The Council may employ such officers and employees as may be necessary to carry out its functions under this Act. In addition, the Council may employ and fix the compensation of such experts and consultants as may be necessary for the carrying out of its functions under this Act, in accordance with Section 3109 of title 5, United States Code (but without regard to the last sentence thereof).
b. Notwithstanding Section 1342 of Title 31, the Council may accept and employ voluntary and uncompensated services in furtherance of the purposes of the Council.

Sec. 204 [42 USC § 4344]

It shall be the duty and function of the Council

1. to assist and advise the President in the preparation of the Environmental Quality Report required by Section 201 [42 USC § 4341] of this title;
2. to gather timely and authoritative information concerning the conditions and trends in the quality of the environment both current and prospective, to analyze and interpret such information for the purpose of determining whether such conditions and trends are interfering, or are likely to interfere, with the achievement of the policy set forth in title I of this Act, and to compile and submit to the President studies relating to such conditions and trends;
3. to review and appraise the various programs and activities of the Federal Government in the light of the policy set forth in title I of this Act for the purpose of determining the extent to which such programs and activities are contributing to the achievement of such policy, and to make recommendations to the President with respect thereto;
4. to develop and recommend to the President, national policies to foster and promote the improvement of environmental quality to meet the conservation, social, economic, health, and other requirements and goals of the Nation;

5. to conduct investigations, studies, surveys, research, and analyses relating to ecological systems and environmental quality;
6. to document and define changes in the natural environment, including the plant and animal systems, and to accumulate necessary data and other information for a continuing analysis of these changes or trends and an interpretation of their underlying causes;
7. to report at least once each year to the President on the state and condition of the environment; and
8. to make and furnish such studies, reports thereon, and recommendations with respect to matters of policy and legislation as the President may request.

Sec. 205 [42 USC § 4345]

In exercising its powers, functions, and duties under this Act, the Council shall

1. consult with the Citizens' Advisory Committee on Environmental Quality established by Executive Order No. 11472, dated May 29, 1969, and with such representatives of science, industry, agriculture, labor, conservation organizations, State and local governments and other groups, as it deems advisable; and
2. utilize, to the fullest extent possible, the services, facilities, and information (including statistical information) of public and private agencies and organizations, and individuals, in order that duplication of effort and expense may be avoided, thus assuring that the Council's activities will not unnecessarily overlap or conflict with similar activities authorized by law and performed by established agencies.

Sec. 206 [42 USC § 4346]

Members of the Council shall serve full time and the Chairman of the Council shall be compensated at the rate provided for Level II of the Executive Schedule Pay Rates [5 USC § 5313]. The other members of the Council shall be compensated at the rate provided for Level IV of the Executive Schedule Pay Rates [5 USC § 5315].

Sec. 207 [42 USC § 4346a]

The Council may accept reimbursements from any private nonprofit organization or from any department, agency, or instrumentality of the Federal Government, any State, or local government, for the reasonable travel expenses incurred by an officer or employee of the Council in connection with his attendance at any conference, seminar, or similar meeting conducted for the benefit of the Council.

Sec. 208 [42 USC § 4346b]

The Council may make expenditures in support of its international activities, including expenditures for: (1) international travel; (2) activities in implementation of international agreements; and (3) the support of international exchange programs in the United States and in foreign countries.

Sec. 209 [42 USC § 4347]

There are authorized to be appropriated to carry out the provisions of this chapter not to exceed $300,000 for fiscal year 1970, $700,000 for fiscal year 1971, and $1,000,000 for each fiscal year thereafter.

The Environmental Quality Improvement Act, as amended (Pub. L. No. 91-224, Title II, April 3, 1970; Pub. L. No. 97-258, September 13, 1982; and Pub. L. No. 98-581, October 30, 1984.

42 USC § 4372

a. There is established in the Executive Office of the President an office to be known as the Office of Environmental Quality (hereafter in this chapter referred to as the "Office"). The Chairman of the Council on Environmental Quality established by Public Law 91-190 shall be the Director of the Office. There shall be in the Office a Deputy Director who shall be appointed by the President, by and with the advice and consent of the Senate.

b. The compensation of the Deputy Director shall be fixed by the President at a rate not in excess of the annual rate of compensation payable to the Deputy Director of the Office of Management and Budget.

c. The Director is authorized to employ such officers and employees (including experts and consultants) as may be necessary to enable the Office to carry out its functions; under this chapter and Public Law 91-190, except that he may employ no more than ten specialists and other experts without regard to the provisions of Title 5, governing appointments in the competitive service, and pay such specialists and experts without regard to the provisions of chapter 51 and subchapter III of chapter 53 of such title relating to classification and General Schedule pay rates, but no such specialist or expert shall be paid at a rate in excess of the maximum rate for GS-18 of the General Schedule under Section 5332 of Title 5.

d. In carrying out his functions the Director shall assist and advise the President on policies and programs of the Federal Government affecting environmental quality by
    1. providing the professional and administrative staff and support for the Council on Environmental Quality established by Public Law 91-190;
    2. assisting the Federal agencies and departments in appraising the effectiveness of existing and proposed facilities, programs, policies, and activities of the Federal Government, and those specific major projects designated by the President which do not require individual project authorization by Congress, which affect environmental quality;
    3. reviewing the adequacy of existing systems for monitoring and predicting environmental changes in order to achieve effective coverage and efficient use of research facilities and other resources;
    4. promoting the advancement of scientific knowledge of the effects of actions and technology on the environment and encouraging the development of the means to prevent or reduce adverse effects that endanger the health and well-being of man;
    5. assisting in coordinating among the Federal departments and agencies those programs and activities which affect, protect, and improve environmental quality;
    6. assisting the Federal departments and agencies in the development and interrelationship of environmental quality criteria and standards established throughout the Federal Government; and
    7. collecting, collating, analyzing, and interpreting data and information on environmental quality, ecological research, and evaluation.

e. The Director is authorized to contract with public or private agencies, institutions, and organizations and with individuals without regard to Section 3324[a] and [b] of Title 31 and Section 5 of Title 41 in carrying out his functions.

42 USC § 4373

Each Environmental Quality Report required by Public Law 91-190 shall, upon transmittal to Congress, be referred to each standing committee having jurisdiction over any part of the subject matter of the Report.

42 USC § 4374

There are hereby authorized to be appropriated for the operations of the Office of Environmental Quality and the Council on Environmental Quality not to exceed the following sums for the following fiscal years which sums are in addition to those contained in Public Law 91-190:

a. $2,126,000 for the fiscal year ending September 30, 1979.
b. $3,000,000 for the fiscal years ending September 30, 1980, and September 30, 1981.
c. $44,000 for the fiscal years ending September 30, 1982, 1983, and 1984.
d. $480,000 for each of the fiscal years ending September 30, 1985 and 1986.

42 USC § 4375

a. There is established an Office of Environmental Quality Management Fund (hereinafter referred to as the "Fund") to receive advance payments from other agencies or accounts that may be used solely to finance
   1. study contracts that are jointly sponsored by the Office and one or more other Federal agencies; and
   2. Federal interagency environmental projects (including task forces) in which the Office participates.
b. Any study contract or project that is to be financed under subsection (a) of this section may be initiated only with the approval of the Director.
c. The Director shall promulgate regulations setting forth policies and procedures for operation of the Fund.

Appendix B

CEQ NEPA IMPLEMENTING REGULATIONS

PART 1500—PURPOSE, POLICY, AND MANDATE

Authority: NEPA, the Environmental Quality Improvement Act of 1970, as amended (42 U.S.C. 4371 et seq.), sec. 309 of the Clean Air Act, as amended (42 U.S.C. 7609), and E.O. 11514, Mar. 5, 1970, as amended by E.O. 11991, May 24, 1977.

Source: 43 FR 55990, Nov. 28, 1978, unless otherwise noted.

Sec. 1500.1   Purpose

a. The National Environmental Policy Act (NEPA) is our basic national charter for protection of the environment. It establishes policy, sets goals (section 101), and provides means (section 102) for carrying out the policy. Section 102(2) contains "action-forcing" provisions to make sure that federal agencies act according to the letter and spirit of the Act. The regulations that follow implement section 102(2). Their purpose is to tell federal agencies what they must do to comply with the procedures and achieve the goals of the Act. The President, the federal agencies, and the courts share responsibility for enforcing the Act so as to achieve the substantive requirements of section 101.

b. NEPA procedures must insure that environmental information is available to public officials and citizens before decisions are made and before actions are taken. The information must be of high quality. Accurate scientific analysis, expert agency comments, and public scrutiny are essential to implementing NEPA. Most important, NEPA documents must concentrate on the issues that are truly significant to the action in question, rather than amassing needless detail.

c. Ultimately, of course, it is not better documents but better decisions that count. NEPA's purpose is not to generate paperwork—even excellent paperwork—but to foster excellent action. The NEPA process is intended to help public officials make decisions that are based on understanding of environmental consequences, and take actions that protect, restore, and enhance the environment. These regulations provide the direction to achieve this purpose.

Sec. 1500.2   Policy

Federal agencies shall to the fullest extent possible:

a. Interpret and administer the policies, regulations, and public laws of the United States in accordance with the policies set forth in the Act and in these regulations.

b. Implement procedures to make the NEPA process more useful to decision-makers and the public; to reduce paperwork and the accumulation of extraneous background data; and to emphasize real environmental issues and alternatives. Environmental impact statements shall be concise, clear, and to the point, and shall be supported by evidence that agencies have made the necessary environmental analyses.

c. Integrate the requirements of NEPA with other planning and environmental review procedures required by law or by agency practice so that all such procedures run concurrently rather than consecutively.

d. Encourage and facilitate public involvement in decisions which affect the quality of the human environment.

e. Use the NEPA process to identify and assess the reasonable alternatives to proposed actions that will avoid or minimize adverse effects of these actions upon the quality of the human environment.

f. Use all practicable means, consistent with the requirements of the Act and other essential considerations of national policy, to restore and enhance the quality of the human environment and avoid or minimize any possible adverse effects of their actions upon the quality of the human environment.

Sec. 1500.3   Mandate

Parts 1500 through 1508 of this title provide regulations applicable to and binding on all Federal agencies for implementing the procedural provisions of the National Environmental Policy Act of 1969, as amended (Pub. L. 91-190, 42 U.S.C. 4321 et seq.) (NEPA or the Act) except where compliance would be inconsistent with other statutory requirements. These regulations are issued pursuant to NEPA, the Environmental Quality Improvement Act of 1970, as amended (42 U.S.C. 4371 et seq.) section 309 of the Clean Air Act, as amended (42 U.S.C. 7609) and Executive Order 11514, Protection and Enhancement of Environmental Quality (March 5, 1970, as amended by Executive Order 11991, May 24, 1977). These regulations, unlike the predecessor guidelines, are not confined to section 102(2)(C) (environmental impact statements). The regulations apply to the whole of section 102(2). The provisions of the Act and of these regulations must be read together as a whole in order to comply with the spirit and letter of the law. It is the Council's intention that judicial review of agency compliance with these regulations not occur before an agency has filed the final environmental impact statement, or has made a final finding of no significant impact (when such a finding will result in action affecting the environment), or takes action that will result in irreparable injury. Furthermore, it is the Council's intention that any trivial violation of these regulations not gives rise to any independent cause of action.

Sec. 1500.4   Reducing paperwork

Agencies shall reduce excessive paperwork by:

a. Reducing the length of environmental impact statements (Sec. 1502.2(c)), by means such as setting appropriate page limits (Secs. 1501.7(b)(1) and 1502.7).

b. Preparing analytic rather than encyclopedic environmental impact statements (Sec. 1502.2(a)).

c. Discussing only briefly issues other than significant ones (Sec. 1502.2(b)).

d. Writing environmental impact statements in plain language (Sec. 1502.8).

e. Following a clear format for environmental impact statements (Sec. 1502.10).

f. Emphasizing the portions of the environmental impact statement that are useful to decision-makers and the public (Secs. 1502.14 and 1502.15) and reducing emphasis on background material (Sec. 1502.16).

g. Using the scoping process, not only to identify significant environmental issues deserving of study, but also to deemphasize insignificant issues, narrowing the scope of the environmental impact statement process accordingly (Sec. 1501.7).

h. Summarizing the environmental impact statement (Sec. 1502.12) and circulating the summary instead of the entire environmental impact statement if the latter is unusually long (Sec. 1502.19).

i. Using program, policy, or plan environmental impact statements and tiering from statements of broad scope to those of narrower scope, to eliminate repetitive discussions of the same issues (Secs. 1502.4 and 1502.20).

j. Incorporating by reference (Sec. 1502.21).

k. Integrating NEPA requirements with other environmental review and consultation requirements (Sec. 1502.25).
l. Requiring comments to be as specific as possible (Sec. 1503.3).
m. Attaching and circulating only changes to the draft environmental impact statement, rather than rewriting and circulating the entire statement when changes are minor (Sec. 1503.4(c)).
n. Eliminating duplication with State and local procedures, by providing for joint preparation (Sec. 1506.2), and with other Federal procedures, by providing that an agency may adopt appropriate environmental documents prepared by another agency (Sec. 1506.3).
o. Combining environmental documents with other documents (Sec. 1506.4).
p. Using categorical exclusions to define categories of actions which do not individually or cumulatively have a significant effect on the human environment and which are therefore exempt from requirements to prepare an environmental impact statement (Sec. 1508.4).
q. Using a finding of no significant impact when an action not otherwise excluded will not have a significant effect on the human environment and is therefore exempt from requirements to prepare an environmental impact statement (Sec. 1508.13).

[43 FR 55990, Nov. 29, 1978; 44 FR 873, Jan. 3, 1979]

Sec. 1500.5 Reducing delay

Agencies shall reduce delay by:

a. Integrating the NEPA process into early planning (Sec. 1501.2).
b. Emphasizing interagency cooperation before the environmental impact statement is prepared, rather than submission of adversary comments on a completed document (Sec. 1501.6).
c. Insuring the swift and fair resolution of lead agency disputes (Sec. 1501.5).
d. Using the scoping process for an early identification of what are and what are not the real issues (Sec. 1501.7).
e. Establishing appropriate time limits for the environmental impact statement process (Secs. 1501.7(b)(2) and 1501.8).
f. Preparing environmental impact statements early in the process (Sec. 1502.5).
g. Integrating NEPA requirements with other environmental review and consultation requirements (Sec. 1502.25).
h. Eliminating duplication with State and local procedures by providing for joint preparation (Sec. 1506.2) and with other Federal procedures by providing that an agency may adopt appropriate environmental documents prepared by another agency (Sec. 1506.3).
i. Combining environmental documents with other documents (Sec. 1506.4).
j. Using accelerated procedures for proposals for legislation (Sec. 1506.8).
k. Using categorical exclusions to define categories of actions which do not individually or cumulatively have a significant effect on the human environment (Sec. 1508.4) and which are therefore exempt from requirements to prepare an environmental impact statement.
l. Using a finding of no significant impact when an action not otherwise excluded will not have a significant effect on the human environment (Sec. 1508.13) and is therefore exempt from requirements to prepare an environmental impact statement.

Sec. 1500.6 Agency authority

Each agency shall interpret the provisions of the Act as a supplement to its existing authority and as a mandate to view traditional policies and missions in the light of the Act's national environmental objectives. Agencies shall review their policies, procedures, and regulations accordingly and revise them as necessary to insure full compliance with the purposes and provisions of the Act. The phrase "to the fullest extent possible" in section 102 means that each agency of the Federal Government shall comply with that section unless existing law applicable to the agency's operations expressly prohibits or makes compliance impossible.

PART 1501—NEPA AND AGENCY PLANNING

Authority: NEPA, the Environmental Quality Improvement Act of 1970, as amended (42 U.S.C. 4371 et seq.), sec. 309 of the Clean Air Act, as amended (42 U.S.C. 7609), and E.O. 11514 (Mar. 5, 1970, as amended by E.O. 11991, May 24, 1977).

Source: 43 FR 55992, Nov. 29, 1978, unless otherwise noted.

Sec. 1501.1   Purpose

The purposes of this part include:

a. Integrating the NEPA process into early planning to insure appropriate consideration of NEPA's policies and to eliminate delay.
b. Emphasizing cooperative consultation among agencies before the environmental impact statement is prepared rather than submission of adversary comments on a completed document.
c. Providing for the swift and fair resolution of lead agency disputes.
d. Identifying at an early stage the significant environmental issues deserving of study and deemphasizing insignificant issues, narrowing the scope of the environmental impact statement accordingly.
e. Providing a mechanism for putting appropriate time limits on the environmental impact statement process.

Sec. 1501.2   Apply NEPA early in the process

Agencies shall integrate the NEPA process with other planning at the earliest possible time to insure that planning and decisions reflect environmental values, to avoid delays later in the process, and to head off potential conflicts. Each agency shall:

a. Comply with the mandate of section 102(2)(A) to "utilize a systematic, interdisciplinary approach which will insure the integrated use of the natural and social sciences and the environmental design arts in planning and in decision-making which may have an impact on man's environment," as specified by Sec. 1507.2.
b. Identify environmental effects and values in adequate detail so they can be compared to economic and technical analyses. Environmental documents and appropriate analyses shall be circulated and reviewed at the same time as other planning documents.
c. Study, develop, and describe appropriate alternatives to recommended courses of action in any proposal which involves unresolved conflicts concerning alternative uses of available resources as provided by section 102(2)(E) of the Act.
d. Provide for cases where actions are planned by private applicants or other non-Federal entities before Federal involvement so that:
   1. Policies or designated staff are available to advise potential applicants of studies or other information foreseeably required for later Federal action.
   2. The Federal agency consults early with appropriate State and local agencies and Indian tribes and with interested private persons and organizations when its own involvement is reasonably foreseeable.
   3. The Federal agency commences its NEPA process at the earliest possible time.

Sec. 1501.3   When to prepare an environmental assessment

a. Agencies shall prepare an environmental assessment (Sec. 1508.9) when necessary under the procedures adopted by individual agencies to supplement these regulations as described in Sec. 1507.3. An assessment is not necessary if the agency has decided to prepare an environmental impact statement.
b. Agencies may prepare an environmental assessment on any action at any time in order to assist agency planning and decision-making.

Sec. 1501.4   Whether to prepare an environmental impact statement

In determining whether to prepare an environmental impact statement the Federal agency shall:

a. Determine under its procedures supplementing these regulations (described in Sec. 1507.3) whether the proposal is one which:
   1. Normally requires an environmental impact statement, or
   2. Normally does not require either an environmental impact statement or an environmental assessment (categorical exclusion).
b. If the proposed action is not covered by paragraph (a) of this section, prepare an environmental assessment (Sec. 1508.9). The agency shall involve environmental agencies, applicants, and the public, to the extent practicable, in preparing assessments required by Sec. 1508.9(a)(1).
c. Based on the environmental assessment make its determination whether to prepare an environmental impact statement.
d. Commence the scoping process (Sec. 1501.7), if the agency will prepare an environmental impact statement.
e. Prepare a finding of no significant impact (Sec. 1508.13), if the agency determines on the basis of the environmental assessment not to prepare a statement.
   1. The agency shall make the finding of no significant impact available to the affected public as specified in Sec. 1506.6.
   2. Certain limited circumstances, which the agency may cover in its procedures under Sec. 1507.3, the agency shall make the finding of no significant impact available for public review (including State and areawide clearinghouses) for 30 days before the agency makes its final determination whether to prepare an environmental impact statement and before the action may begin. The circumstances are
      i. The proposed action is, or is closely similar to, one which normally requires the preparation of an environmental impact statement under the procedures adopted by the agency pursuant to Sec. 1507.3, or
      ii. The nature of the proposed action is one without precedent.

Sec. 1501.5   Lead agencies

a. A lead agency shall supervise the preparation of an environmental impact statement if more than one Federal agency either:
   1. Proposes or is involved in the same action; or
   2. Is involved in a group of actions directly related to each other because of their functional interdependence or geographical proximity.
b. Federal, State, or local agencies, including at least one Federal agency, may act as joint lead agencies to prepare an environmental impact statement (Sec. 1506.2).
c. If an action falls within the provisions of paragraph (a) of this section the potential lead agencies shall determine by letter or memorandum which agency shall be the lead agency and which shall be cooperating agencies. The agencies shall resolve the lead agency

question so as not to cause delay. If there is disagreement among the agencies, the following factors (which are listed in order of descending importance) shall determine lead agency designation:

1. Magnitude of agency's involvement.
2. Project approval/disapproval authority.
3. Expertise concerning the action's environmental effects.
4. Duration of agency's involvement.
5. Sequence of agency's involvement.

d. Any Federal agency, or any State or local agency or private person substantially affected by the absence of lead agency designation, may make a written request to the potential lead agencies that a lead agency be designated.

e. If Federal agencies are unable to agree on which agency will be the lead agency or if the procedure described in paragraph (c) of this section has not resulted within 45 days in a lead agency designation, any of the agencies or persons concerned may file a request with the Council asking it to determine which Federal agency shall be the lead agency. A copy of the request shall be transmitted to each potential lead agency. The request shall consist of:

1. A precise description of the nature and extent of the proposed action.
2. A detailed statement of why each potential lead agency should or should not be the lead agency under the criteria specified in paragraph (c) of this section.

f. A response may be filed by any potential lead agency concerned within 20 days after a request is filed with the Council. The Council shall determine as soon as possible but not later than 20 days after receiving the request and all responses to it which Federal agency shall be the lead agency and which other Federal agencies shall be cooperating agencies.

[43 FR 55992, Nov. 29, 1978; 44 FR 873, Jan. 3, 1979]

Sec. 1501.6 Cooperating agencies

The purpose of this section is to emphasize agency cooperation early in the NEPA process. Upon request of the lead agency, any other Federal agency which has jurisdiction by law shall be a cooperating agency. In addition any other Federal agency which has special expertise with respect to any environmental issue, which should be addressed in the statement, may be a cooperating agency upon request of the lead agency. An agency may request the lead agency to designate it a cooperating agency.

a. The lead agency shall:
1. Request the participation of each cooperating agency in the NEPA process at the earliest possible time.
2. Use the environmental analysis and proposals of cooperating agencies with jurisdiction by law or special expertise, to the maximum extent possible consistent with its responsibility as lead agency.
3. Meet with a cooperating agency at the latter's request.

b. Each cooperating agency shall:
1. Participate in the NEPA process at the earliest possible time.
2. Participate in the scoping process (described below in Sec. 1501.7).
3. Assume on request of the lead agency responsibility for developing information and preparing environmental analyses including portions of the environmental impact statement concerning which the cooperating agency has special expertise.
4. Make available staff support at the lead agency's request to enhance the latter's interdisciplinary capability.
5. Normally use its own funds. The lead agency shall, to the extent available funds permit, fund those major activities or analyses it requests from cooperating agencies. Potential lead agencies shall include such funding requirements in their budget requests.

   c. A cooperating agency may in response to a lead agency's request for assistance in preparing the environmental impact statement (described in paragraph (b)(3), (4), or (5) of this section) reply that other program commitments preclude any involvement or the degree of involvement requested in the action that is the subject of the environmental impact statement. A copy of this reply shall be submitted to the Council.

Sec. 1501.7 Scoping

There shall be an early and open process for determining the scope of issues to be addressed and for identifying the significant issues related to a proposed action. This process shall be termed scoping. As soon as practicable after its decision to prepare an environmental impact statement and before the scoping process the lead agency shall publish a notice of intent (Sec. 1508.22) in the *Federal Register* except as provided in Sec. 1507.3(e).

   a. As part of the scoping process the lead agency shall:
1. Invite the participation of affected Federal, State, and local agencies, any affected Indian tribe, the proponent of the action, and other interested persons (including those who might not be in accord with the action on environmental grounds), unless there is a limited exception under Sec. 1507.3(c). An agency may give notice in accordance with Sec. 1506.6.
2. Determine the scope (Sec. 1508.25) and the significant issues to be analyzed in depth in the environmental impact statement.
3. Identify and eliminate from detailed study the issues which are not significant or which have been covered by prior environmental review (Sec. 1506.3), narrowing the discussion of these issues in the statement to a brief presentation of why they will not have a significant effect on the human environment or providing a reference to their coverage elsewhere.
4. Allocate assignments for preparation of the environmental impact statement among the lead and cooperating agencies, with the lead agency retaining responsibility for the statement.
5. Indicate any public environmental assessments and other environmental impact statements which are being or will be prepared that are related to but are not part of the scope of the impact statement under consideration.
6. Identify other environmental review and consultation requirements so the lead and cooperating agencies may prepare other required analyses and studies concurrently with, and integrated with, the environmental impact statement as provided in Sec. 1502.25.
7. Indicate the relationship between the timing of the preparation of environmental analyses and the agency's tentative planning and decision-making schedule.

   b. As part of the scoping process the lead agency may:
1. Set page limits on environmental documents (Sec. 1502.7).
2. Set time limits (Sec. 1501.8).
3. Adopt procedures under Sec. 1507.3 to combine its environmental assessment process with its scoping process.
4. Hold an early scoping meeting or meetings which may be integrated with any other early planning meeting the agency has. Such a scoping meeting will often be appropriate when the impacts of a particular action are confined to specific sites.

   c. An agency shall revise the determinations made under paragraphs (a) and (b) of this section if substantial changes are made later in the proposed action, or if significant new circumstances or information arise which bear on the proposal or its impacts.

Sec. 1501.8 Time limits

Although the Council has decided that prescribed universal time limits for the entire NEPA process are too inflexible, Federal agencies are encouraged to set time limits appropriate to individual actions (consistent with the time intervals required by Sec. 1506.10). When multiple agencies are involved the reference to agency below means lead agency.

    a. The agency shall set time limits if an applicant for the proposed action requests them: provided that the limits are consistent with the purposes of NEPA and other essential considerations of national policy.

    b. The agency may:

      1. Consider the following factors in determining time limits:

        i. Potential for environmental harm.

        ii. Size of the proposed action.

        iii. State of the art of analytic techniques.

        iv. Degree of public need for the proposed action, including the consequences of delay.

        v. Number of persons and agencies affected.

        vi. Degree to which relevant information is known and if not known the time required for obtaining it.

        vii. Degree to which the action is controversial.

        viii. Other time limits imposed on the agency by law, regulations, or executive order.

      2. Set overall time limits or limits for each constituent part of the NEPA process, which may include:

        i. Decision on whether to prepare an environmental impact statement (if not already decided).

        ii. Determination of the scope of the environmental impact statement.

        iii. Preparation of the draft environmental impact statement.

        iv. Review of any comments on the draft environmental impact statement from the public and agencies.

        v. Preparation of the final environmental impact statement.

        vi. Review of any comments on the final environmental impact statement.

        vii. Decision on the action based in part on the environmental impact statement.

      3. Designate a person (such as the project manager or a person in the agency's office with NEPA responsibilities) to expedite the NEPA process.

    c. State or local agencies or members of the public may request a Federal Agency to set time limits.

PART 1502—ENVIRONMENTAL IMPACT STATEMENT

Authority: NEPA, the Environmental Quality Improvement Act of 1970, as amended (42 U.S.C. 4371 et seq.), sec. 309 of the Clean Air Act, as amended (42 U.S.C. 7609), and E.O. 11514 (Mar. 5, 1970, as amended by E.O. 11991, May 24, 1977).

    Source: 43 FR 55994, Nov. 29, 1978, unless otherwise noted.

Sec. 1502.1 Purpose

The primary purpose of an environmental impact statement is to serve as an action-forcing device to insure that the policies and goals defined in the Act are infused into the ongoing programs and actions of the Federal Government. It shall provide full and fair discussion of significant environmental impacts and shall inform decision-makers and the public of the reasonable alternatives which would avoid or minimize adverse impacts or enhance the quality of the human environment.

Agencies shall focus on significant environmental issues and alternatives and shall reduce paper-work and the accumulation of extraneous background data. Statements shall be concise, clear, and to the point, and shall be supported by evidence that the agency has made the necessary environmental analyses. An environmental impact statement is more than a disclosure document. It shall be used by Federal officials in conjunction with other relevant material to plan actions and make decisions.

Sec. 1502.2   Implementation

To achieve the purposes set forth in Sec. 1502.1 agencies shall prepare environmental impact statements in the following manner:

a. Environmental impact statements shall be analytic rather than encyclopedic.
b. Impacts shall be discussed in proportion to their significance. There shall be only brief discussion of other than significant issues. As in a finding of no significant impact, there should be only enough discussion to show why more study is not warranted.
c. Environmental impact statements shall be kept concise and shall be no longer than absolutely necessary to comply with NEPA and with these regulations. Length should vary first with potential environmental problems and then with project size.
d. Environmental impact statements shall state how alternatives considered in it and decisions based on it will or will not achieve the requirements of sections 101 and 102(1) of the Act and other environmental laws and policies.
e. The range of alternatives discussed in environmental impact statements shall encompass those to be considered by the ultimate agency decision-maker.
f. Agencies shall not commit resources prejudicing selection of alternatives before making a final decision (Sec. 1506.1).
g. Environmental impact statements shall serve as the means of assessing the environmental impact of proposed agency actions, rather than justifying decisions already made.

Sec. 1502.3   Statutory requirements for statements

As required by sec. 102(2)(C) of NEPA environmental impact statements (Sec. 1508.11) are to be included in every recommendation or report.

On proposals (Sec. 1508.23).
For legislation and (Sec. 1508.17).
Other major Federal actions (Sec. 1508.18).
Significantly (Sec. 1508.27).
Affecting (Secs. 1508.3, 1508.8).
The quality of the human environment (Sec. 1508.14).

Sec. 1502.4   Major Federal actions requiring the preparation of environmental impact statements

a. Agencies shall make sure the proposal which is the subject of an environmental impact statement is properly defined. Agencies shall use the criteria for scope (Sec. 1508.25) to determine which proposal(s) shall be the subject of a particular statement. Proposals or parts of proposals which are related to each other closely enough to be, in effect, a single course of action shall be evaluated in a single impact statement.
b. Environmental impact statements may be prepared, and are sometimes required, for broad Federal actions such as the adoption of new agency programs or regulations (Sec. 1508.18).

Agencies shall prepare statements on broad actions so that they are relevant to policy and are timed to coincide with meaningful points in agency planning and decision-making.

   c. When preparing statements on broad actions (including proposals by more than one agency), agencies may find it useful to evaluate the proposal(s) in one of the following ways:

      1. Geographically, including actions occurring in the same general location, such as body of water, region, or metropolitan area.

      2. Generically, including actions which have relevant similarities, such as common timing, impacts, alternatives, methods of implementation, media, or subject matter.

      3. By stage of technological development including federal or federally assisted research, development or demonstration programs for new technologies which, if applied, could significantly affect the quality of the human environment. Statements shall be prepared on such programs and shall be available before the program has reached a stage of investment or commitment to implementation likely to determine subsequent development or restrict later alternatives.

   d. Agencies shall as appropriate employ scoping (Sec. 1501.7), tiering (Sec. 1502.20), and other methods listed in Secs. 1500.4 and 1500.5 to relate broad and narrow actions and to avoid duplication and delay.

Sec. 1502.5   Timing

An agency shall commence preparation of an environmental impact statement as close as possible to the time the agency is developing or is presented with a proposal (Sec. 1508.23) so that preparation can be completed in time for the final statement to be included in any recommendation or report on the proposal. The statement shall be prepared early enough so that it can serve practically as an important contribution to the decision-making process and will not be used to rationalize or justify decisions already made (Secs. 1500.2(c), 1501.2, and 1502.2). For instance:

   a. For projects directly undertaken by Federal agencies the environmental impact statement shall be prepared at the feasibility analysis (go-no go) stage and may be supplemented at a later stage if necessary.

   b. For applications to the agency appropriate environmental assessments or statements shall be commenced no later than immediately after the application is received. Federal agencies are encouraged to begin preparation of such assessments or statements earlier, preferably jointly with applicable State or local agencies.

   c. For adjudication, the final environmental impact statement shall normally precede the final staff recommendation and that portion of the public hearing related to the impact study. In appropriate circumstances the statement may follow preliminary hearings designed to gather information for use in the statements.

   d. For informal rulemaking the draft environmental impact statement shall normally accompany the proposed rule.

Sec. 1502.6   Interdisciplinary preparation

Environmental impact statements shall be prepared using an inter-disciplinary approach which will insure the integrated use of the natural and social sciences and the environmental design arts (section 102(2)(A) of the Act). The disciplines of the preparers shall be appropriate to the scope and issues identified in the scoping process (Sec. 1501.7).

Sec. 1502.7   Page limits

The text of final environmental impact statements (e.g., paragraphs (d) through (g) of Sec. 1502.10) shall normally be less than 150 pages and for proposals of unusual scope or complexity shall normally be less than 300 pages.

Sec. 1502.8   Writing

Environmental impact statements shall be written in plain language and may use appropriate graphics so that decision-makers and the public can readily understand them. Agencies should employ writers of clear prose or editors to write, review, or edit statements, which will be based upon the analysis and supporting data from the natural and social sciences and the environmental design arts.

Sec. 1502.9   Draft, final, and supplemental statements

Except for proposals for legislation as provided in Sec. 1506.8 environmental impact statements shall be prepared in two stages and may be supplemented.

a. Draft environmental impact statements shall be prepared in accordance with the scope decided upon in the scoping process. The lead agency shall work with the cooperating agencies and shall obtain comments as required in Part 1503 of this chapter. The draft statement must fulfill and satisfy to the fullest extent possible the requirements established for final statements in section 102(2)(C) of the Act. If a draft statement is so inadequate as to preclude meaningful analysis, the agency shall prepare and circulate a revised draft of the appropriate portion. The agency shall make every effort to disclose and discuss at appropriate points in the draft statement all major points of view on the environmental impacts of the alternatives including the proposed action.

b. Final environmental impact statements shall respond to comments as required in Part 1503 of this chapter. The agency shall discuss at appropriate points in the final statement any responsible opposing view which was not adequately discussed in the draft statement and shall indicate the agency's response to the issues raised.

c. Agencies:
   1. Shall prepare supplements to either draft or final environmental impact statements if:
      i. The agency makes substantial changes in the proposed action that are relevant to environmental concerns; or
      ii. There are significant new circumstances or information relevant to environmental concerns and bearing on the proposed action or its impacts.
   2. May also prepare supplements when the agency determines that the purposes of the Act will be furthered by doing so.
   3. Shall adopt procedures for introducing a supplement into its formal administrative record, if such a record exists.
   4. Shall prepare, circulate, and file a supplement to a statement in the same fashion (exclusive of scoping) as a draft and final statement unless alternative procedures are approved by the Council.

Sec. 1502.10   Recommended format

Agencies shall use a format for environmental impact statements which will encourage good analysis and clear presentation of the alternatives including the proposed action. The following standard format for environmental impact statements should be followed unless the agency determines that there is a compelling reason to do otherwise:

a. Cover sheet.
b. Summary.
c. Table of contents.
d. Purpose of and need for action.
e. Alternatives including proposed action (sections 102(2)(C)(iii) and 102(2)(E) of the Act).
f. Affected environment.
g. Environmental consequences (especially sections 102(2)(C)(i), (ii), (iv), and (v) of the Act).

h. List of preparers.
i. List of Agencies, Organizations, and persons to whom copies of the statement are sent.
j. Index.
k. Appendices (if any).

If a different format is used, it shall include paragraphs (a), (b), (c), (h), (i), and (j) of this section and shall include the substance of paragraphs (d), (e), (f), (g), and (k) of this section, as further described in Secs. 1502.11 through 1502.18, in any appropriate format.

Sec. 1502.11 Cover sheet

The cover sheet shall not exceed one page. It shall include:

a. A list of the responsible agencies including the lead agency and any cooperating agencies.
b. The title of the proposed action that is the subject of the statement (and if appropriate the titles of related cooperating agency actions), together with the State(s) and county(ies) (or other jurisdiction if applicable) where the action is located.
c. The name, address, and telephone number of the person at the agency who can supply further information.
d. A designation of the statement as a draft, final, or draft or final supplement.
e. A one paragraph abstract of the statement.
f. The date by which comments must be received (computed in cooperation with EPA under Sec. 1506.10).

The information required by this section may be entered on Standard Form 424 (in items 4, 6, 7, 10, and 18).

Sec. 1502.12 Summary

Each environmental impact statement shall contain a summary which adequately and accurately summarizes the statement. The summary shall stress the major conclusions, areas of controversy (including issues raised by agencies and the public), and the issues to be resolved (including the choice among alternatives). The summary will normally not exceed 15 pages.

Sec. 1502.13 Purpose and need

The statement shall briefly specify the underlying purpose and need to which the agency is responding in proposing the alternatives including the proposed action.

Sec. 1502.14 Alternatives including the proposed action

This section is the heart of the environmental impact statement. Based on the information and analysis presented in the sections on the Affected Environment (Sec. 1502.15) and the Environmental Consequences (Sec. 1502.16), it should present the environmental impacts of the proposal and the alternatives in comparative form, thus sharply defining the issues and providing a clear basis for choice among options by the decision-maker and the public. In this section agencies shall:

a. Rigorously explore and objectively evaluate all reasonable alternatives, and for alternatives which were eliminated from detailed study, briefly discuss the reasons for their having been eliminated.
b. Devote substantial treatment to each alternative considered in detail including the proposed action so that reviewers may evaluate their comparative merits.
c. Include reasonable alternatives not within the jurisdiction of the lead agency.

d. Include the alternative of no action.
e. Identify the agency's preferred alternative or alternatives, if one or more exists, in the draft statement and identify such alternative in the final statement unless another law prohibits the expression of such a preference.
f. Include appropriate mitigation measures not already included in the proposed action or alternatives.

Sec. 1502.15   Affected environment

The environmental impact statement shall succinctly describe the environment of the area(s) to be affected or created by the alternatives under consideration. The descriptions shall be no longer than is necessary to understand the effects of the alternatives. Data and analyses in a statement shall be commensurate with the importance of the impact, with less important material summarized, consolidated, or simply referenced. Agencies shall avoid useless bulk in statements and shall concentrate effort and attention on important issues. Verbose descriptions of the affected environment are themselves no measure of the adequacy of an environmental impact statement.

Sec. 1502.16   Environmental consequences

This section forms the scientific and analytic basis for the comparisons under Sec. 1502.14. It shall consolidate the discussions of those elements required by sections 102(2)(C)(i), (ii), (iv), and (v) of NEPA which are within the scope of the statement and as much of section 102(2)(C)(iii) as is necessary to support the comparisons. The discussion will include the environmental impacts of the alternatives including the proposed action, any adverse environmental effects which cannot be avoided should the proposal be implemented, the relationship between short-term uses of man's environment and the maintenance and enhancement of long-term productivity, and any irreversible or irretrievable commitments of resources which would be involved in the proposal should it be implemented. This section should not duplicate discussions in Sec. 1502.14. It shall include discussions of:

a. Direct effects and their significance (Sec. 1508.8).
b. Indirect effects and their significance (Sec. 1508.8).
c. Possible conflicts between the proposed action and the objectives of Federal, regional, State, and local (and in the case of a reservation, Indian tribe) land use plans, policies and controls for the area concerned. (See Sec. 1506.2(d).)
d. The environmental effects of alternatives including the proposed action. The comparisons under Sec. 1502.14 will be based on this discussion.
e. Energy requirements and conservation potential of various alternatives and mitigation measures.
f. Natural or depletable resource requirements and conservation potential of various alternatives and mitigation measures.
g. Urban quality, historic and cultural resources, and the design of the built environment, including the reuse and conservation potential of various alternatives and mitigation measures.
h. Means to mitigate adverse environmental impacts (if not fully covered under Sec. 1502.14(f)).

[43 FR 55994, Nov. 29, 1978; 44 FR 873, Jan. 3, 1979]

Sec. 1502.17   List of preparers

The environmental impact statement shall list the names, together with their qualifications (expertise, experience, professional disciplines), of the persons who were primarily responsible for preparing the environmental impact statement or significant background papers, including basic components of the statement (Secs. 1502.6 and 1502.8). Where possible the persons who are responsible for a particular analysis, including analyses in background papers, shall be identified. Normally the list will not exceed two pages.

Sec. 1502.18   Appendix

If an agency prepares an appendix to an environmental impact statement the appendix shall

a. Consist of material prepared in connection with an environmental impact statement (as distinct from material which is not so prepared and which is incorporated by reference (Sec. 1502.21)).
b. Normally consist of material which substantiates any analysis fundamental to the impact statement.
c. Normally be analytic and relevant to the decision to be made.
d. Be circulated with the environmental impact statement or be readily available on request.

Sec. 1502.19   Circulation of the environmental impact statement

Agencies shall circulate the entire draft and final environmental impact statements except for certain appendices as provided in Sec. 1502.18(d) and unchanged statements as provided in Sec. 1503.4(c). However, if the statement is unusually long, the agency may circulate the summary instead, except that the entire statement shall be furnished to

a. Any Federal agency which has jurisdiction by law or special expertise with respect to any environmental impact involved and any appropriate Federal, State, or local agency authorized to develop and enforce environmental standards.
b. The applicant, if any.
c. Any person, organization, or agency requesting the entire environmental impact statement.
d. In the case of a final environmental impact statement any person, organization, or agency which submitted substantive comments on the draft.

If the agency circulates the summary and thereafter receives a timely request for the entire statement and for additional time to comment, the time for that requestor only shall be extended by at least 15 days beyond the minimum period.

Sec. 1502.20   Tiering

Agencies are encouraged to tier their environmental impact statements to eliminate repetitive discussions of the same issues and to focus on the actual issues ripe for decision at each level of environmental review (Sec. 1508.28). Whenever a broad environmental impact statement has been prepared (such as a program or policy statement) and a subsequent statement or environmental assessment is then prepared on an action included within the entire program or policy (such as a site-specific action) the subsequent statement or environmental assessment need only summarize the issues discussed in the broader statement and incorporate discussions from the broader statement by reference and shall concentrate on the issues specific to the subsequent action. The subsequent document shall state where the earlier document is available. Tiering may also be appropriate for different stages of actions. (Sec. 1508.28).

Sec. 1502.21   Incorporation by reference

Agencies shall incorporate material into an environmental impact statement by reference when the effect will be to cut down on bulk without impeding agency and public review of the action. The incorporated material shall be cited in the statement and its content briefly described. No material may be incorporated by reference unless it is reasonably available for inspection by potentially interested persons within the time allowed for comment. Material based on proprietary data which is itself not available for review and comment shall not be incorporated by reference.

Sec. 1502.22   Incomplete or unavailable information

When an agency is evaluating reasonably foreseeable significant adverse effects on the human environment in an environmental impact statement and there is incomplete or unavailable information, the agency shall always make clear that such information is lacking.

a. If the incomplete information relevant to reasonably foreseeable significant adverse impacts is essential to a reasoned choice among alternatives and the overall costs of obtaining it are not exorbitant, the agency shall include the information in the environmental impact statement.
b. If the information relevant to reasonably foreseeable significant adverse impacts cannot be obtained because the overall costs of obtaining it are exorbitant or the means to obtain it are not known, the agency shall include within the environmental impact statement:
   1. a statement that such information is incomplete or unavailable;
   2. a statement of the relevance of the incomplete or unavailable information to evaluating reasonably foreseeable significant adverse impacts on the human environment;
   3. a summary of existing credible scientific evidence which is relevant to evaluating the reasonably foreseeable significant adverse impacts on the human environment; and
   4. the agency's evaluation of such impacts based upon theoretical approaches or research methods generally accepted in the scientific community. For the purposes of this section, "reasonably foreseeable" includes impacts which have catastrophic consequences, even if their probability of occurrence is low, provided that the analysis of the impacts is supported by credible scientific evidence, is not based on pure conjecture, and is within the rule of reason.
c. The amended regulation will be applicable to all environmental impact statements for which a Notice of Intent (40 CFR 1508.22) is published in the *Federal Register* on or after May 27, 1986. For environmental impact statements in progress, agencies may choose to comply with the requirements of either the original or amended regulation.

[51 FR 15625, Apr. 25, 1986]

Sec. 1502.23   Cost–benefit analysis

If a cost–benefit analysis relevant to the choice among environmentally different alternatives is being considered for the proposed action, it shall be incorporated by reference or appended to the statement as an aid in evaluating the environmental consequences. To assess the adequacy of compliance with section 102(2)(B) of the Act the statement shall, when a cost–benefit analysis is prepared, discuss the relationship between that analysis and any analyses of unquantified environmental impacts, values, and amenities. For purposes of complying with the Act, the weighing of the merits and drawbacks of the various alternatives need not be displayed in a monetary cost–benefit analysis and should not be when there are important qualitative considerations. In any event, an environmental impact statement should at least indicate those considerations, including factors not related to environmental quality, which are likely to be relevant and important to a decision.

Sec. 1502.24   Methodology and scientific accuracy

Agencies shall insure the professional integrity, including scientific integrity, of the discussions and analyses in environmental impact statements. They shall identify any methodologies used and shall make explicit reference by footnote to the scientific and other sources relied upon for conclusions in the statement. An agency may place discussion of methodology in an appendix.

Sec. 1502.25   Environmental review and consultation requirements

a. To the fullest extent possible, agencies shall prepare draft environmental impact statements concurrently with and integrated with environmental impact analyses and related surveys and studies required by the Fish and Wildlife Coordination Act (16 U.S.C. 661 et seq.), the National Historic Preservation Act of 1966 (16 U.S.C. 470 et seq.), the Endangered Species Act of 1973 (16 U.S.C. 1531 et seq.), and other environmental review laws and executive orders.

b. The draft environmental impact statement shall list all Federal permits, licenses, and other entitlements which must be obtained in implementing the proposal. If it is uncertain whether a Federal permit, license, or other entitlement is necessary, the draft environmental impact statement shall so indicate.

PART 1503—COMMENTING

Authority: NEPA, the Environmental Quality Improvement Act of 1970, as amended (42 U.S.C. 4371 et seq.), sec. 309 of the Clean Air Act, as amended (42 U.S.C. 7609), and E.O. 11514 (Mar. 5, 1970, as amended by E.O. 11991, May 24, 1977).

Source: 43 FR 55997, Nov. 29, 1978, unless otherwise noted.

Sec. 1503.1   Inviting comments

a. After preparing a draft environmental impact statement and before preparing a final environmental impact statement the agency shall

1. Obtain the comments of any Federal agency which has jurisdiction by law or special expertise with respect to any environmental impact involved or which is authorized to develop and enforce environmental standards.

2. Request the comments of

    i. Appropriate State and local agencies which are authorized to develop and enforce environmental standards;

    ii. Indian tribes, when the effects may be on a reservation; and

    iii. Any agency which has requested that it receive statements on actions of the kind proposed.

    Office of Management and Budget Circular A-95 (Revised), through its system of clearinghouses, provides a means of securing the views of State and local environmental agencies. The clearinghouses may be used, by mutual agreement of the lead agency and the clearinghouse, for securing State and local reviews of the draft environmental impact statements.

3. Request comments from the applicant, if any.

4. Request comments from the public, affirmatively soliciting comments from those persons or organizations who may be interested or affected.

b. An agency may request comments on a final environmental impact statement before the decision is finally made. In any case other agencies or persons may make comments before the final decision unless a different time is provided under Sec. 1506.10.

Sec. 1503.2   Duty to comment

Federal agencies with jurisdiction by law or special expertise with respect to any environmental impact involved and agencies which are authorized to develop and enforce environmental standards shall comment on statements within their jurisdiction, expertise, or authority. Agencies shall comment within the time period specified for comment in Sec. 1506.10. A Federal agency may reply that it has no comment. If a cooperating agency is satisfied that its views are adequately reflected in the environmental impact statement, it should reply that it has no comment.

Sec. 1503.3   Specificity of comments

a. Comments on an environmental impact statement or on a proposed action shall be as specific as possible and may address either the adequacy of the statement or the merits of the alternatives discussed or both.

b. When a commenting agency criticizes a lead agency's predictive methodology, the commenting agency should describe the alternative methodology which it prefers and why.

c. A cooperating agency shall specify in its comments whether it needs additional information to fulfill other applicable environmental reviews or consultation requirements and what information it needs. In particular, it shall specify any additional information it needs to comment adequately on the draft statement's analysis of significant site-specific effects associated with the granting or approving by that cooperating agency of necessary Federal permits, licenses, or entitlements.

d. When a cooperating agency with jurisdiction by law objects to or expresses reservations about the proposal on grounds of environmental impacts, the agency expressing the objection or reservation shall specify the mitigation measures it considers necessary to allow the agency to grant or approve applicable permit, license, or related requirements or concurrences.

Sec. 1503.4   Response to comments

a. An agency preparing a final environmental impact statement shall assess and consider comments both individually and collectively, and shall respond by one or more of the means listed below, stating its response in the final statement. Possible responses are to
   1. Modify alternatives including the proposed action.
   2. Develop and evaluate alternatives not previously given serious consideration by the agency.
   3. Supplement, improve, or modify its analyses.
   4. Make factual corrections.
   5. Explain why the comments do not warrant further agency response, citing the sources, authorities, or reasons which support the agency's position and, if appropriate, indicate those circumstances which would trigger agency reappraisal or further response.

b. All substantive comments received on the draft statement (or summaries thereof where the response has been exceptionally voluminous) should be attached to the final statement whether or not the comment is thought to merit individual discussion by the agency in the text of the statement.

c. If changes in response to comments are minor and are confined to the responses described in paragraphs (a)(4) and (5) of this section, agencies may write them on errata sheets and attach them to the statement instead of rewriting the draft statement. In such cases only the comments, the responses, and the changes and not the final statement need be circulated (Sec. 1502.19). The entire document with a new cover sheet shall be filed as the final statement (Sec. 1506.9).

PART 1504—PREDECISION REFERRALS TO THE COUNCIL OF PROPOSED FEDERAL ACTIONS DETERMINED TO BE ENVIRONMENTALLY UNSATISFACTORY

Authority: NEPA, the Environmental Quality Improvement Act of 1970, as amended (42 U.S.C. 4371 et seq.), sec. 309 of the Clean Air Act, as amended (42 U.S.C. 7609), and E.O. 11514 (Mar. 5, 1970, as amended by E.O. 11991, May 24, 1977).

Source: 43 FR 55998, Nov. 29, 1978, unless otherwise noted.

Sec. 1504.1   Purpose

a. This part establishes procedures for referring to the Council Federal interagency disagreements concerning proposed major Federal actions that might cause unsatisfactory environmental effects. It provides means for early resolution of such disagreements.

b. Under section 309 of the Clean Air Act (42 U.S.C. 7609), the Administrator of the Environmental Protection Agency is directed to review and comment publicly on the environmental impacts of Federal activities, including actions for which environmental impact statements are prepared. If after this review the Administrator determines that the matter is "unsatisfactory from the standpoint of public health or welfare or environmental quality," section 309 directs that the matter be referred to the Council (hereafter "environmental referrals").

c. Under section 102(2)(C) of the Act other Federal agencies may make similar reviews of environmental impact statements, including judgments on the acceptability of anticipated environmental impacts. These reviews must be made available to the President, the Council, and the public.

Sec. 1504.2   Criteria for referral

Environmental referrals should be made to the Council only after concerted, timely (as early as possible in the process), but unsuccessful attempts to resolve differences with the lead agency. In determining what environmental objections to the matter are appropriate to refer to the Council, an agency should weigh potential adverse environmental impacts, considering

a. Possible violation of national environmental standards or policies.
b. Severity.
c. Geographical scope.
d. Duration.
e. Importance as precedents.
f. Availability of environmentally preferable alternatives.

Sec. 1504.3   Procedure for referrals and response

a. A Federal agency making the referral to the Council shall
   1. Advise the lead agency at the earliest possible time that it intends to refer a matter to the Council unless a satisfactory agreement is reached.
   2. Include such advice in the referring agency's comments on the draft environmental impact statement, except when the statement does not contain adequate information to permit an assessment of the matter's environmental acceptability.
   3. Identify any essential information that is lacking and request that it be made available at the earliest possible time.
   4. Send copies of such advice to the Council.

b. The referring agency shall deliver its referral to the Council not later than twenty-five (25) days after the final environmental impact statement has been made available to the Environmental Protection Agency, commenting agencies, and the public. Except when an extension of this period has been granted by the lead agency, the Council will not accept a referral after that date.

c. The referral shall consist of
   1. A copy of the letter signed by the head of the referring agency and delivered to the lead agency informing the lead agency of the referral and the reasons for it, and requesting that no action be taken to implement the matter until the Council acts upon the referral. The letter shall include a copy of the statement referred to in (c)(2) of this section.

2. A statement supported by factual evidence leading to the conclusion that the matter is unsatisfactory from the standpoint of public health or welfare or environmental quality. The statement shall
    i. Identify any material facts in controversy and incorporate (by reference if appropriate) agreed upon facts,
    ii. Identify any existing environmental requirements or policies which would be violated by the matter,
    iii. Present the reasons why the referring agency believes the matter is environmentally unsatisfactory,
    iv. Contain a finding by the agency whether the issue raised is of national importance because of the threat to national environmental resources or policies or for some other reason,
    v. Review the steps taken by the referring agency to bring its concerns to the attention of the lead agency at the earliest possible time, and
    vi. Give the referring agency's recommendations as to what mitigation alternative, further study, or other course of action (including abandonment of the matter) are necessary to remedy the situation.
d. Not later than twenty-five (25) days after the referral to the Council the lead agency may deliver a response to the Council and the referring agency. If the lead agency requests more time and gives assurance that the matter will not go forward in the interim, the Council may grant an extension. The response shall
1. Address fully the issues raised in the referral.
2. Be supported by evidence.
3. Give the lead agency's response to the referring agency's recommendations.
e. Interested persons (including the applicant) may deliver their views in writing to the Council. Views in support of the referral should be delivered not later than the referral. Views in support of the response shall be delivered not later than the response.
f. Not later than twenty-five (25) days after receipt of both the referral and any response or upon being informed that there will be no response (unless the lead agency agrees to a longer time), the Council may take one or more of the following actions:
1. Conclude that the process of referral and response has successfully resolved the problem.
2. Initiate discussions with the agencies with the objective of mediation with referring and lead agencies.
3. Hold public meetings or hearings to obtain additional views and information.
4. Determine that the issue is not one of national importance and request the referring and lead agencies to pursue their decision process.
5. Determine that the issue should be further negotiated by the referring and lead agencies and is not appropriate for Council consideration until one or more heads of agencies report to the Council that the agencies' disagreements are irreconcilable.
6. Publish its findings and recommendations (including where appropriate a finding that the submitted evidence does not support the position of an agency).
7. When appropriate, submit the referral and the response together with the Council's recommendation to the President for action.
g. The Council shall take no longer than 60 days to complete the actions specified in paragraph (f)(2), (3), or (5) of this section.
h. When the referral involves an action required by statute to be determined on the record after opportunity for agency hearing, the referral shall be conducted in a manner consistent with 5 U.S.C. 557(d) (Administrative Procedure Act).

[43 FR 55998, Nov. 29, 1978; 44 FR 873, Jan. 3, 1979]

PART 1505—NEPA AND AGENCY DECISION-MAKING

Authority: NEPA, the Environmental Quality Improvement Act of 1970, as amended (42 U.S.C. 4371 et seq.), sec. 309 of the Clean Air Act, as amended (42 U.S.C. 7609), and E.O. 11514 (Mar. 5, 1970, as amended by E.O. 11991, May 24, 1977).

Source: 43 FR 55999, Nov. 29, 1978, unless otherwise noted.

Sec. 1505.1   Agency decision-making procedures

Agencies shall adopt procedures (Sec. 1507.3) to ensure that decisions are made in accordance with the policies and purposes of the Act. Such procedures shall include but not be limited to

a. Implementing procedures under section 102(2) to achieve the requirements of sections 101 and 102(1).
b. Designating the major decision points for the agency's principal programs likely to have a significant effect on the human environment and assuring that the NEPA process corresponds with them.
c. Requiring that relevant environmental documents, comments, and responses be part of the record in formal rulemaking or adjudicatory proceedings.
d. Requiring that relevant environmental documents, comments, and responses accompany the proposal through existing agency review processes so that agency officials use the statement in making decisions.
e. Requiring that the alternatives considered by the decision-maker are encompassed by the range of alternatives discussed in the relevant environmental documents and that the decision-maker considers the alternatives described in the environmental impact statement. If another decision document accompanies the relevant environmental documents to the decision-maker, agencies are encouraged to make available to the public before the decision is made any part of that document that relates to the comparison of alternatives.

Sec. 1505.2   Record of decision in cases requiring environmental impact statements

At the time of its decision (Sec. 1506.10) or, if appropriate, its recommendation to Congress, each agency shall prepare a concise public record of decision. The record, which may be integrated into any other record prepared by the agency, including that required by OMB Circular A-95 (Revised), part I, sections 6(c) and (d), and Part II, section 5(b)(4), shall:

a. State what the decision was.
b. Identify all alternatives considered by the agency in reaching its decision, specifying the alternative or alternatives which were considered to be environmentally preferable. An agency may discuss preferences among alternatives based on relevant factors including economic and technical considerations and agency statutory missions. An agency shall identify and discuss all such factors including any essential considerations of national policy which were balanced by the agency in making its decision and state how those considerations entered into its decision.
c. State whether all practicable means to avoid or minimize environmental harm from the alternative selected have been adopted, and if not, why they were not. A monitoring and enforcement program shall be adopted and summarized where applicable for any mitigation.

Sec. 1505.3   Implementing the decision

Agencies may provide for monitoring to assure that their decisions are carried out and should do so in important cases. Mitigation (Sec. 1505.2(c)) and other conditions established in the environmental impact statement or during its review and committed as part of the decision shall be implemented by the lead agency or other appropriate consenting agency. The lead agency shall

  a. Include appropriate conditions in grants, permits, or other approvals.
  b. Condition funding of actions on mitigation.
  c. Upon request, inform cooperating or commenting agencies on progress in carrying out mitigation measures which they have proposed and which were adopted by the agency making the decision.
  d. Upon request, make available to the public the results of relevant monitoring.

PART 1506—OTHER REQUIREMENTS OF NEPA

Authority: NEPA, the Environmental Quality Improvement Act of 1970, as amended (42 U.S.C. 4371 et seq.), sec. 309 of the Clean Air Act, as amended (42 U.S.C. 7609), and E.O. 11514 (Mar. 5, 1970, as amended by E.O. 11991, May 24, 1977).
  Source: 43 FR 56000, Nov. 29, 1978, unless otherwise noted.

Sec. 1506.1   Limitations on actions during NEPA process

  a. Until an agency issues a record of decision as provided in Sec. 1505.2 (except as provided in paragraph (c) of this section), no action concerning the proposal shall be taken which would:
     1. Have an adverse environmental impact; or
     2. Limit the choice of reasonable alternatives.
  b. If any agency is considering an application from a non-Federal entity, and is aware that the applicant is about to take an action within the agency's jurisdiction that would meet either of the criteria in paragraph (a) of this section, then the agency shall promptly notify the applicant that the agency will take appropriate action to insure that the objectives and procedures of NEPA are achieved.
  c. While work on a required program environmental impact statement is in progress and the action is not covered by an existing program statement, agencies shall not undertake in the interim any major Federal action covered by the program which may significantly affect the quality of the human environment unless such action:
     1. Is justified independently of the program;
     2. Is itself accompanied by an adequate environmental impact statement; and
     3. Will not prejudice the ultimate decision on the program. Interim action prejudices the ultimate decision on the program when it tends to determine subsequent development or limit alternatives.
  d. This section does not preclude development by applicants of plans or designs or performance of other work necessary to support an application for Federal, State, or local permits or assistance. Nothing in this section shall preclude Rural Electrification Administration approval of minimal expenditures not affecting the environment (e.g. long leadtime equipment and purchase options) made by non-governmental entities seeking loan guarantees from the Administration.

Sec. 1506.2   Elimination of duplication with State and local procedures

  a. Agencies authorized by law to cooperate with State agencies of statewide jurisdiction pursuant to section 102(2)(D) of the Act may do so.
  b. Agencies shall cooperate with State and local agencies to the fullest extent possible to reduce duplication between NEPA and State and local requirements, unless the agencies

are specifically barred from doing so by some other law. Except for cases covered by paragraph (a) of this section, such cooperation shall to the fullest extent possible include:
1. Joint planning processes.
2. Joint environmental research and studies.
3. Joint public hearings (except where otherwise provided by statute).
4. Joint environmental assessments.

c. Agencies shall cooperate with State and local agencies to the fullest extent possible to reduce duplication between NEPA and comparable State and local requirements, unless the agencies are specifically barred from doing so by some other law. Except for cases covered by paragraph (a) of this section, such cooperation shall to the fullest extent possible include joint environmental impact statements. In such cases one or more Federal agencies and one or more State or local agencies shall be joint lead agencies. Where State laws or local ordinances have environmental impact statement requirements in addition to but not in conflict with those in NEPA, Federal agencies shall cooperate in fulfilling these requirements as well as those of Federal laws so that one document will comply with all applicable laws.

d. To better integrate environmental impact statements into State or local planning processes, statements shall discuss any inconsistency of a proposed action with any approved State or local plan and laws (whether or not federally sanctioned). Where an inconsistency exists, the statement should describe the extent to which the agency would reconcile its proposed action with the plan or law.

Sec. 1506.3   Adoption

a. An agency may adopt a Federal draft or final environmental impact statement or portion thereof provided that the statement or portion thereof meets the standards for an adequate statement under these regulations.

b. If the actions covered by the original environmental impact statement and the proposed action are substantially the same, the agency adopting another agency's statement is not required to recirculate it except as a final statement. Otherwise the adopting agency shall treat the statement as a draft and recirculate it (except as provided in paragraph (c) of this section).

c. A cooperating agency may adopt without recirculating the environmental impact statement of a lead agency when, after an independent review of the statement, the cooperating agency concludes that its comments and suggestions have been satisfied.

d. When an agency adopts a statement which is not final within the agency that prepared it, or when the action it assesses is the subject of a referral under Part 1504, or when the statement's adequacy is the subject of a judicial action which is not final, the agency shall so specify.

Sec. 1506.4   Combining documents

Any environmental document in compliance with NEPA may be combined with any other agency document to reduce duplication and paperwork.

Sec. 1506.5   Agency responsibility

a. Information. If an agency requires an applicant to submit environmental information for possible use by the agency in preparing an environmental impact statement, then the agency should assist the applicant by outlining the types of information required. The agency shall independently evaluate the information submitted and shall be responsible for its accuracy. If the agency chooses to use the information submitted by the applicant in the environmental impact statement, either directly or by reference, then the names of the persons responsible for the independent evaluation shall be included in the list of preparers (Sec. 1502.17). It is the intent of this paragraph that acceptable work not be redone, but that it be verified by the agency.

b. Environmental assessments. If an agency permits an applicant to prepare an environmental assessment, the agency, besides fulfilling the requirements of paragraph (a) of this section, shall make its own evaluation of the environmental issues and take responsibility for the scope and content of the environmental assessment.

c. Environmental impact statements. Except as provided in Secs. 1506.2 and 1506.3 any environmental impact statement prepared pursuant to the requirements of NEPA shall be prepared directly by or by a contractor selected by the lead agency or where appropriate under Sec. 1501.6(b), a cooperating agency. It is the intent of these regulations that the contractor be chosen solely by the lead agency, or by the lead agency in cooperation with cooperating agencies, or where appropriate by a cooperating agency to avoid any conflict of interest. Contractors shall execute a disclosure statement prepared by the lead agency, or where appropriate the cooperating agency, specifying that they have no financial or other interest in the outcome of the project. If the document is prepared by contract, the responsible Federal official shall furnish guidance and participate in the preparation and shall independently evaluate the statement prior to its approval and take responsibility for its scope and contents. Nothing in this section is intended to prohibit any agency from requesting any person to submit information to it or to prohibit any person from submitting information to any agency.

Sec. 1506.6   Public involvement

Agencies shall

a. Make diligent efforts to involve the public in preparing and implementing their NEPA procedures.

b. Provide public notice of NEPA-related hearings, public meetings, and the availability of environmental documents so as to inform those persons and agencies who may be interested or affected.

  1. In all cases the agency shall mail notice to those who have requested it on an individual action.

  2. In the case of an action with effects of national concern notice shall include publication in the *Federal Register* and notice by mail to national organizations reasonably expected to be interested in the matter and may include listing in the 102 Monitor. An agency engaged in rulemaking may provide notice by mail to national organizations who have requested that notice regularly be provided. Agencies shall maintain a list of such organizations.

  3. In the case of an action with effects primarily of local concern the notice may include:

      i. Notice to State and areawide clearinghouses pursuant to OMB Circular A-95 (Revised).

      ii. Notice to Indian tribes when effects may occur on reservations.

      iii. Following the affected State's public notice procedures for comparable actions.

      iv. Publication in local newspapers (in papers of general circulation rather than legal papers).

      v. Notice through other local media.

      vi. Notice to potentially interested community organizations including small business associations.

      vii. Publication in newsletters that may be expected to reach potentially interested persons.

viii. Direct mailing to owners and occupants of nearby or affected property.

ix. Posting of notice on and off site in the area where the action is to be located.

c. Hold or sponsor public hearings or public meetings whenever appropriate or in accordance with statutory requirements applicable to the agency. Criteria shall include whether there is

1. Substantial environmental controversy concerning the proposed action or substantial interest in holding the hearing.

2. A request for a hearing by another agency with jurisdiction over the action supported by reasons why a hearing will be helpful. If a draft environmental impact statement is to be considered at a public hearing, the agency should make the statement available to the public at least 15 days in advance (unless the purpose of the hearing is to provide information for the draft environmental impact statement).

d. Solicit appropriate information from the public.

e. Explain in its procedures where interested persons can get information or status reports on environmental impact statements and other elements of the NEPA process.

f. Make environmental impact statements, the comments received, and any underlying documents available to the public pursuant to the provisions of the Freedom of Information Act (5 U.S.C. 552), without regard to the exclusion for interagency memoranda where such memoranda transmit comments of Federal agencies on the environmental impact of the proposed action. Materials to be made available to the public shall be provided to the public without charge to the extent practicable, or at a fee which is not more than the actual costs of reproducing copies required to be sent to other Federal agencies, including the Council.

Sec. 1506.7   Further guidance

The Council may provide further guidance concerning NEPA and its procedures including:

a. A handbook which the Council may supplement from time to time, which shall in plain language provide guidance and instructions concerning the application of NEPA and these regulations.

b. Publication of the Council's Memoranda to Heads of Agencies.

c. In conjunction with the Environmental Protection Agency and the publication of the 102 Monitor, notice of

1. Research activities;

2. Meetings and conferences related to NEPA; and

3. Successful and innovative procedures used by agencies to implement NEPA.

Sec. 1506.8   Proposals for legislation

a. The NEPA process for proposals for legislation (Sec. 1508.17) significantly affecting the quality of the human environment shall be integrated with the legislative process of the Congress. A legislative environmental impact statement is the detailed statement required by law to be included in a recommendation or report on a legislative proposal to Congress. A legislative environmental impact statement shall be considered part of the formal transmittal of a legislative proposal to Congress; however, it may be transmitted to Congress up to 30 days later in order to allow time for completion of an accurate statement which can serve as the basis for public and Congressional debate. The statement must be available in time for Congressional hearings and deliberations.

b. Preparation of a legislative environmental impact statement shall conform to the requirements of these regulations except as follows:

1. There need not be a scoping process.

2. The legislative statement shall be prepared in the same manner as a draft statement, but shall be considered the "detailed statement" required by statute; provided that when any

of the following conditions exist both the draft and final environmental impact statement on the legislative proposal shall be prepared and circulated as provided by Secs. 1503.1 and 1506.10.

   i. A Congressional Committee with jurisdiction over the proposal has a rule requiring both draft and final environmental impact statements.
   ii. The proposal results from a study process required by statute (such as those required by the Wild and Scenic Rivers Act (16 U.S.C. 1271 et seq.) and the Wilderness Act (16 U.S.C. 1131 et seq.)).
   iii. Legislative approval is sought for Federal or federally assisted construction or other projects which the agency recommends be located at specific geographic locations. For proposals requiring an environmental impact statement for the acquisition of space by the General Services Administration, a draft statement shall accompany the Prospectus or the 11(b) Report of Building Project Surveys to the Congress, and a final statement shall be completed before site acquisition.
   iv. The agency decides to prepare draft and final statements.
 c. Comments on the legislative statement shall be given to the lead agency which shall forward them along with its own responses to the Congressional committees with jurisdiction.

Sec. 1506.9   Filing requirements

Environmental impact statements together with comments and responses shall be filed with the Environmental Protection Agency, attention Office of Federal Activities (A-104), 401 M Street SW, Washington, DC 20460. Statements shall be filed with EPA no earlier than they are also transmitted to commenting agencies and made available to the public. EPA shall deliver one copy of each statement to the Council, which shall satisfy the requirement of availability to the President. EPA may issue guidelines to agencies to implement its responsibilities under this section and Sec. 1506.10.

Sec. 1506.10   Timing of agency action

 a. The Environmental Protection Agency shall publish a notice in the *Federal Register* each week of the environmental impact statements filed during the preceding week. The minimum time periods set forth in this section shall be calculated from the date of publication of this notice.
 b. No decision on the proposed action shall be made or recorded under Sec. 1505.2 by a Federal agency until the later of the following dates:
   1. Ninety (90) days after publication of the notice described above in paragraph (a) of this section for a draft environmental impact statement.
   2. Thirty (30) days after publication of the notice described above in paragraph (a) of this section for a final environmental impact statement. An exception to the rules on timing may be made in the case of an agency decision which is subject to a formal internal appeal. Some agencies have a formally established appeal process which allows other agencies or the public to take appeals on a decision and make their views known, after publication of the final environmental impact statement. In such cases, where a real opportunity exists to alter the decision, the decision may be made and recorded at the same time the environmental impact statement is published.

   This means that the period for appeal of the decision and the 30-day period prescribed in paragraph (b)(2) of this section may run concurrently. In such cases the environmental impact statement shall explain the timing and the public's right of appeal. An agency engaged in rulemaking under the Administrative Procedure Act or other statute for the purpose of protecting the public health or safety may waive the time period in paragraph (b)(2) of this section and publish a decision on the final rule simultaneously with publication of the notice of the availability of the final environmental impact statement as described in paragraph (a) of this section.

c. If the final environmental impact statement is filed within ninety (90) days after a draft environmental impact statement is filed with the Environmental Protection Agency, the minimum thirty (30) day period and the minimum ninety (90) day period may run concurrently. However, subject to paragraph (d) of this section agencies shall allow not less than 45 days for comments on draft statements.

d. The lead agency may extend prescribed periods. The Environmental Protection Agency may upon a showing by the lead agency of compelling reasons of national policy reduce the prescribed periods and may upon a showing by any other Federal agency of compelling reasons of national policy also extend prescribed periods, but only after consultation with the lead agency. (Also see Sec. 1507.3(d).) Failure to file timely comments shall not be a sufficient reason for extending a period. If the lead agency does not concur with the extension of time, EPA may not extend it for more than 30 days. When the Environmental Protection Agency reduces or extends any period of time it shall notify the Council.

[43 FR 56000, Nov. 29, 1978; 44 FR 874, Jan. 3, 1979]

Sec. 1506.11   Emergencies

Where emergency circumstances make it necessary to take an action with significant environmental impact without observing the provisions of these regulations, the Federal agency taking the action should consult with the Council about alternative arrangements. Agencies and the Council will limit such arrangements to actions necessary to control the immediate impacts of the emergency. Other actions remain subject to NEPA review.

Sec. 1506.12   Effective date

The effective date of these regulations is July 30, 1979, except that for agencies that administer programs that qualify under section 102(2)(D) of the Act or under section 104(h) of the Housing and Community Development Act of 1974 an additional four months shall be allowed for the State or local agencies to adopt their implementing procedures.

a. These regulations shall apply to the fullest extent practicable to ongoing activities and environmental documents begun before the effective date. These regulations do not apply to an environmental impact statement or supplement if the draft statement was filed before the effective date of these regulations. No completed environmental documents need be redone by reasons of these regulations. Until these regulations are applicable, the Council's guidelines published in the *Federal Register* of August 1, 1973, shall continue to be applicable. In cases where these regulations are applicable the guidelines are superseded. However, nothing shall prevent an agency from proceeding under these regulations at an earlier time.

b. NEPA shall continue to be applicable to actions begun before January 1, 1970, to the fullest extent possible.

PART 1507—AGENCY COMPLIANCE

Authority: NEPA, the Environmental Quality Improvement Act of 1970, as amended (42 U.S.C. 4371 et seq.), sec. 309 of the Clean Air Act, as amended (42 U.S.C. 7609), and E.O. 11514 (Mar. 5, 1970, as amended by E.O. 11991, May 24, 1977).
    Source: 43 FR 56002, Nov. 29, 1978, unless otherwise noted.

Sec. 1507.1   Compliance

All agencies of the Federal Government shall comply with these regulations. It is the intent of these regulations to allow each agency flexibility in adapting its implementing procedures authorized by Sec. 1507.3 to the requirements of other applicable laws.

Sec. 1507.2   Agency capability to comply

Each agency shall be capable (in terms of personnel and other resources) of complying with the requirements enumerated below. Such compliance may include use of other's resources, but the using agency shall itself have sufficient capability to evaluate what others do for it. Agencies shall:

a. Fulfill the requirements of section 102(2)(A) of the Act to utilize a systematic, interdisciplinary approach which will insure the integrated use of the natural and social sciences and the environmental design arts in planning and in decision-making which may have an impact on the human environment. Agencies shall designate a person to be responsible for overall review of agency NEPA compliance.
b. Identify methods and procedures required by section 102(2)(B) to insure that presently unquantified environmental amenities and values may be given appropriate consideration.
c. Prepare adequate environmental impact statements pursuant to section 102(2)(C) and comment on statements in the areas where the agency has jurisdiction by law or special expertise or is authorized to develop and enforce environmental standards.
d. Study, develop, and describe alternatives to recommended courses of action in any proposal which involves unresolved conflicts concerning alternative uses of available resources. This requirement of section 102(2)(E) extends to all such proposals, not just the more limited scope of section 102(2)(C)(iii) where the discussion of alternatives is confined to impact statements.
e. Comply with the requirements of section 102(2)(H) that the agency initiate and utilize ecological information in the planning and development of resource-oriented projects.
f. Fulfill the requirements of sections 102(2)(F), 102(2)(G), and 102(2)(I) of the Act and of Executive Order 11514, Protection and Enhancement of Environmental Quality, Sec. 2.

Sec. 1507.3   Agency procedures

a. Not later than eight months after publication of these regulations as finally adopted in the *Federal Register,* or five months after the establishment of an agency, whichever shall come later, each agency shall as necessary adopt procedures to supplement these regulations. When the agency is a department, major subunits are encouraged (with the consent of the department) to adopt their own procedures. Such procedures shall not paraphrase these regulations. They shall confine themselves to implementing procedures. Each agency shall consult with the Council while developing its procedures and before publishing them in the *Federal Register* for comment. Agencies with similar programs should consult with each other and the Council to coordinate their procedures, especially for programs requesting similar information from applicants. The procedures shall be adopted only after an opportunity for public review and after review by the Council for conformity with the Act and these regulations. The Council shall complete its review within 30 days. Once in effect they shall be filed with the Council and made readily available to the public. Agencies are encouraged to publish explanatory guidance for these regulations and their own procedures. Agencies shall continue to review their policies and procedures and in consultation with the Council to revise them as necessary to ensure full compliance with the purposes and provisions of the Act.

   b. Agency procedures shall comply with these regulations except where compliance would be inconsistent with statutory requirements and shall include:

      1. Those procedures required by Secs. 1501.2(d), 1502.9(c)(3), 1505.1, 1506.6(e), and 1508.4.

      2. Specific criteria for and identification of those typical classes of action:

         i. Which normally do require environmental impact statements.

         ii. Which normally do not require either an environmental impact statement or an environmental assessment (categorical exclusions (Sec. 1508.4)).

         iii. Which normally require environmental assessments but not necessarily environmental impact statements.

   c. Agency procedures may include specific criteria for providing limited exceptions to the provisions of these regulations for classified proposals. They are proposed actions which are specifically authorized under criteria established by an Executive Order or statute to be kept secret in the interest of national defense or foreign policy and are in fact properly classified pursuant to such Executive Order or statute. Environmental assessments and environmental impact statements which address classified proposals may be safeguarded and restricted from public dissemination in accordance with agencies' own regulations applicable to classified information. These documents may be organized so that classified portions can be included as annexes, in order that the unclassified portions can be made available to the public.

   d. Agency procedures may provide for periods of time other than those presented in Sec. 1506.10 when necessary to comply with other specific statutory requirements.

   e. Agency procedures may provide that where there is a lengthy period between the agency's decision to prepare an environmental impact statement and the time of actual preparation, the notice of intent required by Sec. 1501.7 may be published at a reasonable time in advance of preparation of the draft statement.

PART 1508—TERMINOLOGY AND INDEX

Authority: NEPA, the Environmental Quality Improvement Act of 1970, as amended (42 U.S.C. 4371 et seq.), sec. 309 of the Clean Air Act, as amended (42 U.S.C. 7609), and E.O. 11514 (Mar. 5, 1970, as amended by E.O. 11991, May 24, 1977).

   Source: 43 FR 56003, Nov. 29, 1978, unless otherwise noted.

Sec. 1508.1   Terminology

The terminology of this part shall be uniform throughout the Federal Government.

Sec. 1508.2   Act

"Act" means the National Environmental Policy Act, as amended (42 U.S.C. 4321, et seq.) which is also referred to as "NEPA."

Sec. 1508.3   Affecting

"Affecting" means will or may have an effect on.

Sec. 1508.4   Categorical exclusion

"Categorical exclusion" means a category of actions which do not individually or cumulatively have a significant effect on the human environment and which have been found to have no such effect in procedures adopted by a Federal agency in implementation of these regulations (Sec. 1507.3) and for which, therefore, neither an environmental assessment nor an environmental impact statement is required. An agency may decide in its procedures or otherwise, to prepare environmental assessments for the reasons stated in Sec. 1508.9 even though it is not required to do so. Any procedures under this section shall provide for extraordinary circumstances in which a normally excluded action may have a significant environmental effect.

Sec. 1508.5   Cooperating agency

"Cooperating agency" means any Federal agency other than a lead agency which has jurisdiction by law or special expertise with respect to any environmental impact involved in a proposal (or a reasonable alternative) for legislation or other major Federal action significantly affecting the quality of the human environment. The selection and responsibilities of a cooperating agency are described in Sec. 1501.6. A State or local agency of similar qualifications or, when the effects are on a reservation, an Indian Tribe, may by agreement with the lead agency become a cooperating agency.

Sec. 1508.6   Council

"Council" means the Council on Environmental Quality established by Title II of the Act.

Sec. 1508.7   Cumulative impact

"Cumulative impact" is the impact on the environment which results from the incremental impact of the action when added to other past, present, and reasonably foreseeable future actions regardless of what agency (Federal or non-Federal) or person undertakes such other actions. Cumulative impacts can result from individually minor but collectively significant actions taking place over a period of time.

Sec. 1508.8   Effects

"Effects" include:

a. Direct effects, which are caused by the action and occur at the same time and place.
b. Indirect effects, which are caused by the action and are later in time or farther removed in distance, but are still reasonably foreseeable. Indirect effects may include growth inducing effects and other effects related to induced changes in the pattern of land use, population density or growth rate, and related effects on air and water and other natural systems, including ecosystems.

Effects and impacts as used in these regulations are synonymous. Effects includes ecological (such as the effects on natural resources and on the components, structures, and functioning of affected ecosystems), aesthetic, historic, cultural, economic, social, or health, whether direct, indirect, or cumulative. Effects may also include those resulting from actions which may have both beneficial and detrimental effects, even if on balance the agency believes that the effect will be beneficial.

Sec. 1508.9   Environmental assessment

"Environmental assessment":

a. Means a concise public document for which a Federal agency is responsible that serves to:
    1. Briefly provide sufficient evidence and analysis for determining whether to prepare an environmental impact statement or a finding of no significant impact.
    2. Aid an agency's compliance with the Act when no environmental impact statement is necessary.
    3. Facilitate preparation of a statement when one is necessary.

b. Shall include brief discussions of the need for the proposal, of alternatives as required by section 102(2)(E), of the environmental impacts of the proposed action and alternatives, and a listing of agencies and persons consulted.

Sec. 1508.10   Environmental document

"Environmental document" includes the documents specified in Sec. 1508.9 (environmental assessment), Sec. 1508.11 (environmental impact statement), Sec. 1508.13 (finding of no significant impact), and Sec. 1508.22 (notice of intent).

Sec. 1508.11   Environmental impact statement

"Environmental impact statement" means a detailed written statement as required by section 102(2)(C) of the Act.

Sec. 1508.12   Federal agency

"Federal agency" means all agencies of the Federal Government. It does not mean the Congress, the Judiciary, or the President, including the performance of staff functions for the President in his Executive Office. It also includes for purposes of these regulations States and units of general local government and Indian tribes assuming NEPA responsibilities under section 104(h) of the Housing and Community Development Act of 1974.

Sec. 1508.13   Finding of no significant impact

"Finding of no significant impact" means a document by a Federal agency briefly presenting the reasons why an action, not otherwise excluded (Sec. 1508.4), will not have a significant effect on the human environment and for which an environmental impact statement therefore will not be prepared. It shall include the environmental assessment or a summary of it and shall note any other environmental documents related to it (Sec. 1501.7(a)(5)). If the assessment is included, the finding need not repeat any of the discussion in the assessment but may incorporate it by reference.

Sec. 1508.14   Human environment

"Human environment" shall be interpreted comprehensively to include the natural and physical environment and the relationship of people with that environment (see the definition of "effects" (Sec. 1508.8)). This means that economic or social effects are not intended by themselves to require preparation of an environmental impact statement. When an environmental impact statement is prepared and economic or social and natural or physical environmental effects are interrelated, then the environmental impact statement will discuss all of these effects on the human environment.

Sec. 1508.15   Jurisdiction by law

"Jurisdiction by law" means agency authority to approve, veto, or finance all or part of the proposal.

Sec. 1508.16   Lead agency

"Lead agency" means the agency or agencies preparing or having taken primary responsibility for preparing the environmental impact statement.

Sec. 1508.17   Legislation

"Legislation" includes a bill or legislative proposal to Congress developed by or with the significant cooperation and support of a Federal agency, but does not include requests for appropriations. The test for significant cooperation is whether the proposal is in fact predominantly that of the agency rather than another source. Drafting does not by itself constitute significant cooperation. Proposals for legislation include requests for ratification of treaties. Only the agency which has primary responsibility for the subject matter involved will prepare a legislative environmental impact statement.

Sec. 1508.18   Major Federal action

"Major Federal action" includes actions with effects that may be major and which are potentially subject to Federal control and responsibility. Major reinforces but does not have a meaning independent of significantly (Sec. 1508.27). Actions include the circumstance where the responsible officials fail to act and that failure to act is reviewable by courts or administrative tribunals under the Administrative Procedure Act or other applicable law as agency action.

a. Actions include new and continuing activities, including projects and programs entirely or partly financed, assisted, conducted, regulated, or approved by federal agencies; new or revised agency rules, regulations, plans, policies, or procedures; and legislative proposals (Secs. 1506.8, 1508.17). Actions do not include funding assistance solely in the form of general revenue sharing funds, distributed under the State and Local Fiscal Assistance Act of 1972, 31 U.S.C. 1221 et seq., with no Federal agency control over the subsequent use of such funds. Actions do not include bringing judicial or administrative civil or criminal enforcement actions.
b. Federal actions tend to fall within one of the following categories:
    1. Adoption of official policy, such as rules, regulations, and interpretations adopted pursuant to the Administrative Procedure Act, 5 U.S.C. 551 et seq.; treaties and international conventions or agreements; formal documents establishing an agency's policies which will result in or substantially alter agency programs.
    2. Adoption of formal plans, such as official documents prepared or approved by federal agencies which guide or prescribe alternative uses of Federal resources, upon which future agency actions will be based.
    3. Adoption of programs, such as a group of concerted actions to implement a specific policy or plan; systematic and connected agency decisions allocating agency resources to implement a specific statutory program or executive directive.
    4. Approval of specific projects, such as construction or management activities located in a defined geographic area. Projects include actions approved by permit or other regulatory decision as well as federal and federally assisted activities.

Sec. 1508.19   Matter

"Matter" includes for purposes of Part 1504: (a) With respect to the Environmental Protection Agency, any proposed legislation, project, action, or regulation as those terms are used in section 309(a) of the Clean Air Act (42 U.S.C. 7609). (b) With respect to all other agencies, any proposed major federal action to which section 102(2)(C) of NEPA applies.

Sec. 1508.20   Mitigation

"Mitigation" includes:

a. Avoiding the impact altogether by not taking a certain action or parts of an action.
b. Minimizing impacts by limiting the degree or magnitude of the action and its implementation.
c. Rectifying the impact by repairing, rehabilitating, or restoring the affected environment.

d. Reducing or eliminating the impact over time by preservation and maintenance operations during the life of the action.
e. Compensating for the impact by replacing or providing substitute resources or environments.

Sec. 1508.21   NEPA process

"NEPA process" means all measures necessary for compliance with the requirements of section 2 and Title I of NEPA.

Sec. 1508.22   Notice of intent

"Notice of intent" means a notice that an environmental impact statement will be prepared and considered. The notice shall briefly:

a. Describe the proposed action and possible alternatives.
b. Describe the agency's proposed scoping process including whether, when, and where any scoping meeting will be held.
c. State the name and address of a person within the agency who can answer questions about the proposed action and the environmental impact statement.

Sec. 1508.23   Proposal

"Proposal" exists at that stage in the development of an action when an agency subject to the Act has a goal and is actively preparing to make a decision on one or more alternative means of accomplishing that goal and the effects can be meaningfully evaluated. Preparation of an environmental impact statement on a proposal should be timed (Sec. 1502.5) so that the final statement may be completed in time for the statement to be included in any recommendation or report on the proposal. A proposal may exist in fact as well as by agency declaration that one exists.

Sec. 1508.24   Referring agency

"Referring agency" means the Federal agency which has referred any matter to the Council after a determination that the matter is unsatisfactory from the standpoint of public health or welfare or environmental quality.

Sec. 1508.25   Scope

Scope consists of the range of actions, alternatives, and impacts to be considered in an environmental impact statement. The scope of an individual statement may depend on its relationships to other statements (Secs.1502.20 and 1508.28). To determine the scope of environmental impact statements, agencies shall consider 3 types of actions, 3 types of alternatives, and 3 types of impacts. They include:

a. Actions (other than unconnected single actions) which may be:
  1. Connected actions, which means that they are closely related and therefore should be discussed in the same impact statement. Actions are connected if they:
    i. Automatically trigger other actions which may require environmental impact statements.
    ii. Cannot or will not proceed unless other actions are taken previously or simultaneously.
    iii. Are interdependent parts of a larger action and depend on the larger action for their justification.

2. Cumulative actions, which when viewed with other proposed actions have cumulatively significant impacts and should therefore be discussed in the same impact statement.

3. Similar actions, which when viewed with other reasonably foreseeable or proposed agency actions, have similarities that provide a basis for evaluating their environmental consequences together, such as common timing or geography. An agency may wish to analyze these actions in the same impact statement. It should do so when the best way to assess adequately the combined impacts of similar actions or reasonable alternatives to such actions is to treat them in a single impact statement.

b. Alternatives, which include:

1. No action alternative.

2. Other reasonable courses of actions.

3. Mitigation measures (not in the proposed action).

c. Impacts, which may be: (1) Direct; (2) indirect; (3) cumulative.

Sec. 1508.26   Special expertise

"Special expertise" means statutory responsibility, agency mission, or related program experience.

Sec. 1508.27   Significantly

"Significantly" as used in NEPA requires considerations of both context and intensity:

a. Context. This means that the significance of an action must be analyzed in several contexts such as society as a whole (human, national), the affected region, the affected interests, and the locality. Significance varies with the setting of the proposed action. For instance, in the case of a site-specific action, significance would usually depend upon the effects in the locale rather than in the world as a whole. Both short- and long-term effects are relevant.

b. Intensity. This refers to the severity of impact. Responsible officials must bear in mind that more than one agency may make decisions about partial aspects of a major action. The following should be considered in evaluating intensity:

1. Impacts that may be both beneficial and adverse. A significant effect may exist even if the Federal agency believes that on balance the effect will be beneficial.

2. The degree to which the proposed action affects public health or safety.

3. Unique characteristics of the geographic area such as proximity to historic or cultural resources, park lands, prime farmlands, wetlands, wild and scenic rivers, or ecologically critical areas.

4. The degree to which the effects on the quality of the human environment are likely to be highly controversial.

5. The degree to which the possible effects on the human environment are highly uncertain or involve unique or unknown risks.

6. The degree to which the action may establish a precedent for future actions with significant effects or represents a decision in principle about a future consideration.

7. Whether the action is related to other actions with individually insignificant but cumulatively significant impacts. Significance exists if it is reasonable to anticipate a cumulatively significant impact on the environment. Significance cannot be avoided by terming an action temporary or by breaking it down into small component parts.

8. The degree to which the action may adversely affect districts, sites, highways, structures, or objects listed in or eligible for listing in the National Register of Historic Places or may cause loss or destruction of significant scientific, cultural, or historical resources.

9. The degree to which the action may adversely affect an endangered or threatened species or its habitat that has been determined to be critical under the Endangered Species Act of 1973.
10. Whether the action threatens a violation of Federal, State, or local law or requirements imposed for the protection of the environment.

[43 FR 56003, Nov. 29, 1978; 44 FR 874, Jan. 3, 1979]

Sec. 1508.28   Tiering

"Tiering" refers to the coverage of general matters in broader environmental impact statements (such as national program or policy statements) with subsequent narrower statements or environmental analyses (such as regional or basinwide program statements or ultimately site-specific statements) incorporating by reference the general discussions and concentrating solely on the issues specific to the statement subsequently prepared. Tiering is appropriate when the sequence of statements or analyses is

a. From a program, plan, or policy environmental impact statement to a program, plan, or policy statement or analysis of lesser scope or to a site-specific statement or analysis.
b. From an environmental impact statement on a specific action at an early stage (such as need and site selection) to a supplement (which is preferred) or a subsequent statement or analysis at a later stage (such as environmental mitigation). Tiering in such cases is appropriate when it helps the lead agency to focus on the issues which are ripe for decision and exclude from consideration issues already decided or not yet ripe.

Index

A

accident analyses
    analyzing impacts of accidents, 256–257
    disaster and terrorist scenario evaluation, 271–272
    NEPA requirements and, 256–257
    potential, 159
accuracy, in cost-benefit analysis, 119
action proposals
    business continuity planning, preventive and
        corrective action, 291–292
    in environmental assessments, 172
actions of NEPA
    4 Ps, 234
    case law concerning, 153–156
    categories of, 232–235
        connected actions, 232–233
        cumulative actions, 233–234
        federal actions, 234–235
        segmentation, 235
        similar actions, 234
    classified information, 136, 196–197, 267
    environmental impact statement ROD provisions, 213
    executive orders and, 152
    extra-territorial, 152–156
    federal, 14, 75, 146–147
    federalization of, 147–152
    four categories of, 234–235
    inaction and, 152
    international actions, 152
    justification for, 71
    limitations on, 69–72
    segmenting, 157
    statistical analysis of streamlining, 73–74
    three types of, 232–235
    transboundary effects, 152–153
"acts of God" exemptions, 140
adaptive management (AM)
    advantages of, 48–49
    environmental management standards (EMS),
        ISO14001/NEPA, 47–52
    environmental management standards (EMS),
        ISO14001/NEPA and implementation of, 50–52
    five-step process, 48
    implementation requirements, 47–48
    NEPA documentation, 49–50
adequacy criteria, cultural impact assessment, 255
Administrative Procedure Act, supplemental
    environmental impact statement, 5, 15, 217
administrative record, judicial review standard and, 15
adoption of EISs, streamlining of NEPA and, 68–69
adoption of plans, programs, and projects, 151
Advisory Council on Historic Preservation (ACH),
    establishment of, 103
affected environment standard, no-action alternative
    vs., 236

agency programs and policies. *See also* federal agency
    alternative actions outside of, 238
    preferred alternatives, flexible planning process
        and, 83
    programmatic environmental impact statements, 226
Agenda 21, sustainability principles and, 318–321
air pollutant comparisons, cost-benefit analysis and, 120
air quality models, general conformity process (Clean Air
    Act), 93–94
alternatives
    business continuity planning, 294–295
    case law concerning, 238–239
    crystal ball, 235–239
    NEPA categories of, 235–240
    no-action alternative, 236–237
    reasonable alternatives, 237–240
    remote and speculative, 238
    three types, 231–232, 235–239
analytical methodology
    accident scenarios, 256–257, 271
    actions, alternatives, and effects overview, 231–232
    actions categories, 232–235
    adequacy of, 255
    alternatives, 235–240
    bounding analysis, 256
    carrying capacity analysis, 116–120
    catastrophic events and human environment, 261–278
    connected actions, 232–233
    cost benefit analysis, 118–122, 327
    cumulative actions, 233–234
    cumulative impacts, 240–246
    de-emphasis of, 231
    detail guidelines for, 52–56
    direct impacts, 240
    Eccleston's cumulative impact paradox, 246–255
    economic and technical analysis, 83
    environmental assessments, 171–172
    environmental checklists, 112–113
    environmental impact assessment methodologies,
        109–122
    environmentally preferable, 58
    evaluating hypothetical scenarios, 239
    federal actions categories, 234–235
    impacts, 240–255
    incomplete/unavailable information, 124–125
    indirect impacts, 240
    mitigation, 183–184, 215
    mitigation measures, 108, 240
    networks, 115–116
    no-action alternative, 236–237
    objectively evaluate, 122–123, 237
    public scoping meetings, 231
    reasonable alternatives, 237–240
    requirements for, 121–125
    segmentation, 235
    short-term *vs.* long-term impact, 256